Weather Analysis

Weather Analysis

DUŠAN DJURIĆ
Texas A&M University

Prentice Hall, Upper Saddle River, New Jersey 07458

Library of Congress Cataloging-in-Publication Data

DJURIC, DUSAN.
Weather analysis/Dušan Djurić.
p. cm.
Includes bibliographical references and index.
ISBN 0-13-501149-3
1. Dynamic meteorology. 2. Troposphere. 3. Weather forecasting.
I. Title.
QC880.D555 1994
551.6—dc20 93-44935
CIP

Acquisitions editor: *Ray Henderson*
Editorial/production supervision
and interior design: *Kathleen M. Lafferty*
Proofreader: *Bruce D. Colegrove*
Cover design: *DeLuca Design*
Manufacturing buyer: *Trudy Pisciotti*

Printed in the United States of America

10 9 8 7 6 5 4 3 2

ISBN 0-13-501149-3

Prentice-Hall International (UK) Limited, London
Prentice-Hall of Australia Pty. Limited, Sydney
Prentice-Hall Canada Inc., Toronto
Prentice-Hall Hispanoamericana, S.A., Mexico
Prentice-Hall of India Private Limited, New Delhi
Prentice-Hall of Japan, Inc., Tokyo
Pearson Education Asia Pte. Ltd., Singapore
Editoria Prentice-Hall do Brasil, Ltda., Rio De Janeiro

Contents

8 Fronts 122

9 Air Masses and Weather 154

10 Cyclones and Anticyclones 165

11 The Upper Troposphere and Jet Streams 184

12 Mesoscale Storms 201

Appendices 232

Preface

For the last several decades, meteorology students and teachers have been without an introductory textbook for weather analysis. Most teaching of this subject has been done "from instructor's notes" (actually, students' notes from classes were used in many cases). Because I was an instructor who had to work without a suitable text, my motivation to write this book can be understood. It is easier for students and teachers to follow one textbook than to consult different books on general and dynamic meteorology and to compare notes with other students. *Weather Analysis* will make the studies more systematic.

Weather Analysis is written for students in a first course on weather analysis. It describes tropospheric structures, mainly air masses and patterns in wind and other variables, large enough to play a role in weather analysis. As a prerequisite, readers are expected to have learned the basic ideas about weather systems as they are presented in many books of general meteorology. Knowledge of elementary mathematical descriptions of atmospheric processes, as expressed by the equations of motion, hydrostatics, continuity, the first law of thermodynamics, and geostrophic approximation, is necessary for a successful study of this book. Readers are also expected to be familiar with methods of observation, transmission of data, and basic construction (plotting) of synoptic weather charts.

Teachers using this book may let students analyze historical weather charts concurrently with attending the lectures. Another useful supplementary action is the discussion of current weather charts. It is very instructive and rewarding to recognize the similarity of current weather patterns with the models in this book. Similarly, students benefit greatly when they calculate various quantities of dynamic meteorology on weather charts.

I find it practical to work on upper-level charts first because they are easier to analyze due to fewer data. On these charts it is convenient to evaluate gradients and other computed quantities. The analysis of the surface chart should be done after the large-scale structure on the upper levels has already been established.

The preparation of weather charts is done increasingly by computers. However, this does not negate the usefulness of studying graphical manual construction of weather charts. Weather analysis does not consist of line drawing but of recognizing physical processes in the atmosphere. Charts with isopleths are only the tools for weather analysis, not the goal. Manual construction of weather charts brings the atmospheric processes closer to the attention of students. The purpose of this book is to show all three facets of weather analysis: techniques, the physics of weather processes, and the structure of circulation patterns.

Physical processes in the atmosphere are recognized in two important ways: by a descriptive similarity with previous cases and by knowing the basic dynamics of atmospheric processes. This second way allows the possibility of perceiving new, previously unknown mechanisms. The traditional approach of teaching weather analysis as the practical part of dynamic meteorology is thereby justified.

Chapter 1 gives an overview of atmospheric structures that are the topics of this science. Other chapters contain the details of these structures, the techniques of analysis, and the physics of observed patterns.

Acknowledgments

I received numerous comments from colleagues, which helped in writing this book. I would like to single out Marion Alcorn, Steve Lyons, and Dušan Zrnić for their help. I also appreciate the comments I received from the following reviewers: Tony Brazel, Arizona State University; Fred Lutgens, Illinois Central College; James T. Moore, Saint Louis University; and Grandikota V. Rao, Saint Louis University. I appreciate the help of Mara D. Hill for her assistance with the technical illustrations. Ray Henderson provided valuable advice on many questions concerning the publication of this book. Finally, I particularly appreciate the copyediting and production work done by Kathleen Lafferty.

Dušan Djurić

Weather Analysis

1

Conceptual Models of the Lower Atmosphere

This chapter contains an overview of the large-scale atmospheric features studied in the science of weather analysis. The main features in the atmosphere are large bodies of air, characteristic circulation patterns, and various weather phenomena. Meteorologists have developed conceptual models to describe, to explain, and ultimately to predict the weather. When meteorologists analyze weather observations, they recognize the models of phenomena and processes. A finished analysis is a chart with models adapted to the available data.

A word of caution: No model is perfect and no list of models can be exhaustive or final. As more evidence is collected, old models are revised and new models are introduced. Therefore, the analyst must be alert to observed phenomena that do not fit the models. Such phenomena, together with familiar models, should be marked on weather charts as well. When some unusual event or phenomenon is noticed repeatedly, that is a sign that a new model should be developed.

1-1 PRINCIPAL LAYERS OF THE ATMOSPHERE

Thermal Stratification

Temperature generally decreases with height in the lowest 7 to 17 km of the atmosphere. This is the *troposphere*. The top of the troposphere is the *tropopause*. The atmosphere is usually very stable above the tropopause, up to the elevation of about 50 km. This layer is the *stratosphere*.

Main layers of the atmosphere are formed by radiative balance and the convective adjustment. The temperature generally decreases with height in the troposphere because the atmosphere is heated primarily from below. Calculations of radiative balance show that the radiative decrease of temperature with height forms a very unstable lower atmosphere. As potential temperature θ (or wet-bulb potential temperature θ_w in the case of saturated water vapor) decreases with height, convection develops until a more or less moist neutral stratification is achieved, where θ or θ_w are approximately constant (adiabatic stratification). Potential temperature θ (or θ_w) equalizes during mixing, much as the temperature equalizes by mixing in water. Therefore, due to convective mixing, the tropospheric variation of temperature with height usually is between constant potential temperature, $\theta(z) \approx$ const., and constant wet-bulb potential temperature, $\theta_w(z) \approx$ const. Such variation of temperature with height shows that the troposphere is conditionally unstable.

Figure 1-1 illustrates the main processes that form the troposphere and the stratosphere. The dashed curve represents the distribution of temperature in a stagnant atmosphere that transmits and absorbs radiation. The solid line is the distribution of temperature that results after adiabatic vertical adjustment. This vertical adjustment is between the moist and dry adiabatic states since condensation of water vapor often accompanies large vertical displacements in the atmosphere. The transition between the lower, convectively mixed layer and the upper stable layer is rather abrupt at the tropopause. This is the level where the buoyancy of rising convective elements is exhausted.

The highest troposphere (about 17 km) is in the tropics. The main reason for a high tropical troposphere is strong heating of the atmosphere from the earth's surface. The convection penetrates highest under such conditions. The lowest troposphere (about 7 km) is in the polar regions.

The top of the troposphere is the *tropopause*. The tropopause often consists of an inversion of 0.5 to 1 km thickness. If the inversion is much thicker than 1 km, it is considered a part of the stratosphere; then the bottom of that inversion is the tropopause.

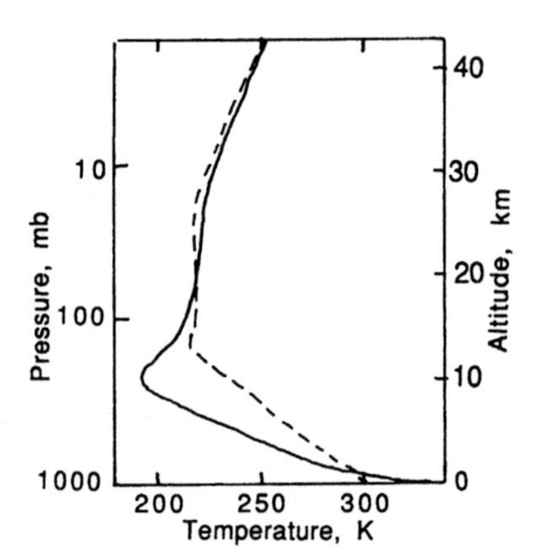

FIGURE 1-1. Vertical temperature distribution in the atmosphere under pure radiative equilibrium (solid) and thermal equilibrium (dashed), with a lapse rate of 6.5 K km^{-1}. Adapted from Manabe and Strickler (1964), by permission of the American Meteorological Society.

All "rules" about the atmosphere and models have exceptions. The vertical distribution of temperature frequently deviates from the described model. For example, there are inversions, multiple tropopauses and other deviations from a smooth "model" atmosphere.

Temperature in the High Atmosphere

Temperature increases with height between 20 and 50 km of the atmosphere. This layer is the *stratosphere*. Its lower boundary is the tropopause, as described above. The level of maximum temperature near 50 km is the *stratopause*. The heat in this layer is obtained by direct absorption of ultraviolet sun rays in the ozone. Depletion of ultraviolet rays in the ozonosphere protects life on the earth from harmful radiation.

The layer with decreasing temperature between 50 and 80 km is the *mesosphere*. The minimum temperature near 82 km signifies the *mesopause*, and the further indefinite increase of temperature toward outer space is in the *thermosphere*.

Weather processes occur almost exclusively in the troposphere. Only a few processes in the stratosphere can possibly be designated as "weather." One of them, the *stratospheric sudden warming* that appears at 15–25 km elevation in winter, is a temperature increase of 20–40 K over about half of the Northern Hemisphere, in the course of about one week. Warming may last one month and has been correlated with several weather processes. For example, the cyclones in North America form farther north than usual during the episodes of stratospheric sudden warming (McGuirk and Douglas, 1988).

Chemical Composition

Layers of the atmosphere may be classified by the chemical composition of the air. In the *homosphere*, up to about 100 km elevation, the atmospheric gases are very well mixed. Above this level is the *heterosphere* where, because of rare collisions between molecules, separation of gases occurs.

There are small chemical variations within the stratosphere, as well as between the stratosphere and the troposphere. The layer where ozone is relatively abundant, in the upper stratosphere, is the *ozonosphere*. Artificially or naturally induced trace materials (dust, smoke, radioactive substances, ozone) stay in the stratosphere much longer than in the troposphere because there is no rain in the stratosphere to remove the impurities. The rain in the troposphere washes these materials down.

The chemical composition of the atmosphere is of interest in weather analysis since the chemicals can serve as tracers of air masses. Movement of industrial or volcanic pollutants sometimes gives additional information of air flow. Emergence of stratospheric air in the troposphere can be recognized if detailed observations of chemical properties are available.

Atmospheric chemistry is also of interest in weather analysis because meteorologists are frequently called to evaluate the place of origin or the destination of various substances in the atmosphere.

High Atmospheric Phenomena

Several layers of ionized air constitute the layers of the *ionosphere*. The layers of the ionosphere appear at elevations between 80 and several hundred kilometers. The ionosphere is of great significance for the propagation of radio waves.

Mother-of-pearl clouds appear at an elevation of 25–30 km poleward from about 50° latitude. Still higher, at 80–85 km elevation, are the *noctilucent clouds*, also in higher latitudes. Both these cloud forms are not classified in the International Cloud Atlas. A connection of these clouds and tropospheric weather phenomena has not been found.

The spectacular *aurora* that appears at elevations between 200 and 500 km is not a tropospheric weather phenomenon of interest to weather analysts. It occurs only in the thermosphere and, as far as is known, is not related to weather processes. The aurora forms due to corpuscular solar radiation reaching the rarefied gases of the outer atmosphere and is studied by geophysicists. The aurora in the Northern Hemisphere is described by the Latin name *aurora borealis* for "northern aurora." Similarly, the aurora in the Southern Hemisphere is *aurora australis*.

1-2 LARGE-SCALE STRUCTURE OF THE TROPOSPHERE

Whereas the main layers of the atmosphere are classified by the thermal classification, the phenomena in the troposphere are classified by circulation patterns. Classification on the basis of atmospheric motions shows the dynamic structure of the troposphere and leads toward a physical interpretation of weather phenomena and processes.

Figure 1-2 represents the major dynamic structures found in the atmosphere. This figure covers the latitudes between about 70°N and 10°S. The arrows show the typical air flow. Polar and subtropical jets exist in the upper troposphere. The subtropical highs extend through the whole troposphere. The conveyor belt starts near the earth's surface and climbs to about 300 mb. All other models in Fig. 1-2 are shown roughly as they may appear at the earth's surface.

Westerlies, Easterlies, and General Circulation

The largest-scale motion in the troposphere is characterized by *westerlies* in middle latitudes and *easterlies* in the tropics. This situation is apparent to people living in the respective regions: It is common knowledge that weather changes come primarily from the west in middle latitudes and from the east in the tropics. Prevalent zonal winds were known to Christopher Columbus at the time of his first voyage in 1492. He sailed downwind to America at about 20°N and returned to Spain sailing at about 40°N. Other voyagers and traders also followed the easterlies on the travel to America; thus the easterlies came to be known as the *trade winds* or *trades*.

Easterlies and westerlies are separated by a belt of subtropical anticyclones at latitudes of about 25°N and 25°S. These large, warm, dynamic anticyclones extend through the whole troposphere and high into the stratosphere. Deserts are predominantly situated in the belts of these anticyclones. These anticyclones are "dynamic" since they are maintained through the geostrophic adjustment of westerlies and easterlies. The belt of subtropical highs is indicated by a serrated line (for a ridge) in Fig. 1-2, connecting the high centers.

Westerlies, easterlies, and the subtropical belt of anticyclones, and the basic mechanisms that drive them, are usually described together as the *general atmospheric circulation*.

Conservation of Angular Momentum

Formation of easterlies and westerlies is attributed to the law of conservation of angular momentum in rotating bodies. This law is most often illustrated by a spinning ice skater. The same law applies to the atmosphere that rotates together with the earth. The bodies that move away from the equator approach the earth's axis of rotation. The bodies approaching the equator increase their distance from the axis of rotation. The result is that the bodies moving north acquire an eastward motion component and the bodies moving south acquire a westward motion component.

Because the atmosphere has the property to mix, fueled by unequal heating over the globe, there are many air parcels that originate at other latitudes. In the

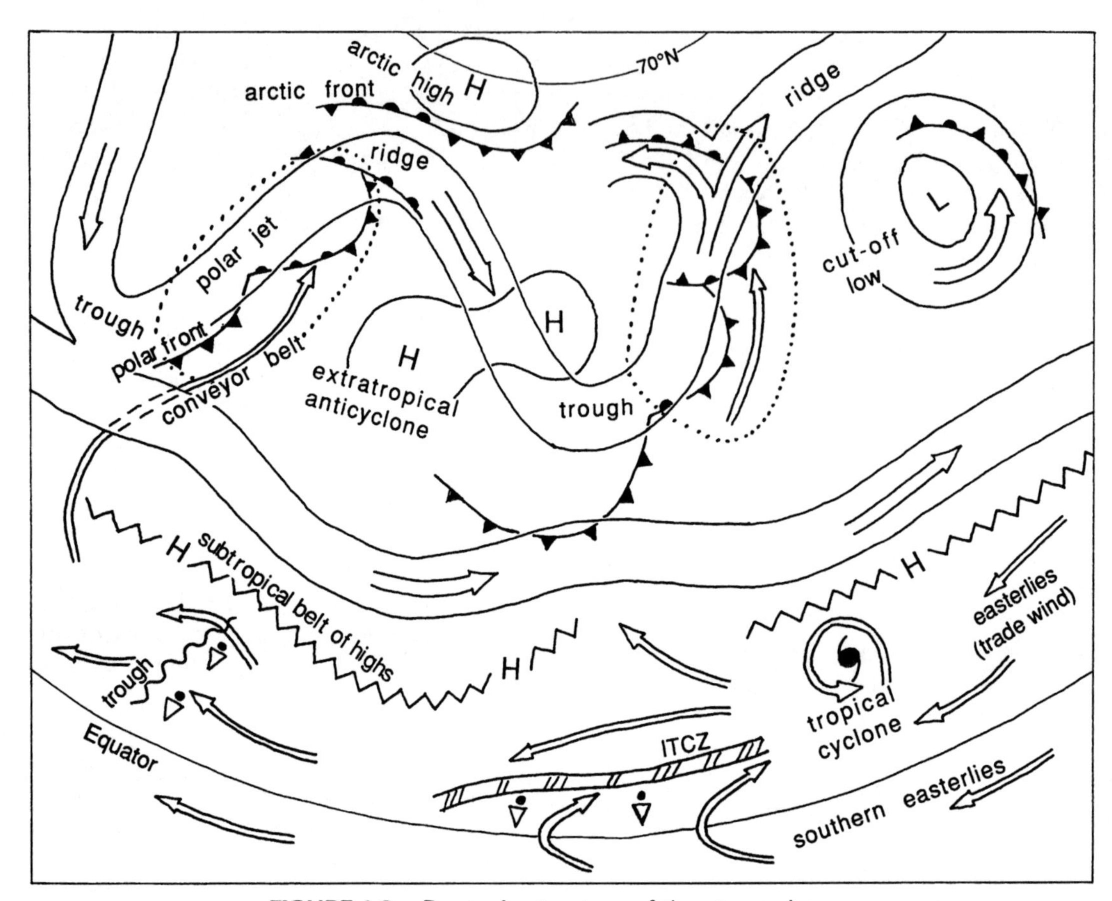

FIGURE 1-2. Dynamic structure of the atmosphere.

middle latitudes, air parcels (also air masses) that came from low latitudes are found. Similarly, in the low latitudes we find some middle-latitude air. (Arctic atmosphere is at a disadvantage in this exchange of mass and momentum because its volume is smaller than the volume of other regions.) Those recently arrived parcels of air tend to preserve their angular momentum (in agreement with the law) even after they mix with indigenous parcels at the new latitude. The result is westerlies in middle latitudes and easterlies in low latitudes.

A quantitative description of the conservation of angular momentum is

$$ur = \text{const.}$$

where u is the zonal motion component (wind plus the motion of the earth) and r is the distance to the earth's rotation axis. Therefore, when the air approaches the rotation axis by moving away from the equator, r becomes smaller and u increases. The motion of the earth is not influenced by this effect, but the wind is. This is the main mechanism of forming westerlies. Explanation of easterlies is analogous: Air parcels coming from middle latitudes lag in keeping up with the rotation of the earth. These parcels appear as an east wind.

Conservation of angular momentum does not appear in pure form in the atmosphere. Many effects modify this mechanism, such as friction, inertia in large eddies (cyclone and anticyclones), and instability of flow. Also, the air parcels push other parcels along the way, so their motion does not proceed with a "mathematical" uniformity. Thus there is not an immense tornado at the pole, where $r = 0$.

Processes contributing to transport of air across geographic parallels include instability of flow, Hadley circulation, and cyclogenesis. Most of these dynamic processes are explained only in a course on dynamics. Here only a short description is given.

Fronts and Jets

A basic feature of atmospheric motions is the occurrence of intense and comparatively narrow baroclinic zones. These are the *frontal zones*. The most baroclinic parts of frontal zones are slanted stable layers, the *fronts*. Above the frontal zones in the upper troposphere are *jet streams* or *jets*. There are two major westerly jet streams in the upper troposphere in each hemisphere (Northern and Southern). These are the *polar* and *subtropical* jet streams.

The polar front often extends through the whole troposphere. Figure 1-2 shows only a typical surface position, depicted by the conventional line with pointed or rounded tips. Frequently the polar front is developed only in the upper troposphere or only in the lower troposphere. At about 300 mb the polar front is located on the polar flank of the polar jet stream.

Polar and subtropical jets sometimes merge over a distance of 1000 to 5000 km. In the winter, such a situation does not last longer than a few days; the combined jet stream splits back to two jets soon after formation. A combined jet is more common in summer; then it may last several weeks. It occurs most often over eastern parts of Asia and North America, in the troughs of the climatological average upper-level polar jet.

The jet streams in the westerlies are shown by wavy bands in Fig. 1-2, with arrows along the cores. The waves in the polar jet are more prominent than the waves in the subtropical jet. Amplified waves often separate into large vortices. One such cyclonic vortex (*cut-off low*) is represented in the eastern part of Fig. 1-2.

Wind in the easterlies seldom attains jet speed. Only over southern Asia does an almost straight jet, the *tropical easterly jet*, form at about 150 mb. This is the *t*. The wind speed in it may reach 45 m s^{-1}. This jet forms only in summer, during Indian monsoon.

Waves in Jet Streams

Most extratropical flow patterns are dynamically related to the curvature in jet streams: *cyclonic* or *anticyclonic*. The region around maximum cyclonic curvature is the *trough*, and the region with maximum anticyclonic curvature is the *ridge* (Fig. 1-2). Dynamic meteorology shows that vertical motion and weather development are greatly different in the following two regions between troughs and ridges:

(a) The *active region* is downstream from the trough, where we normally find ascending air motion. The active region is favorable for cyclogenesis, frontogenesis, cloud formation, and precipitation. Two active regions are circled by dotted ovals in Fig. 1-2.

(b) The *inactive region* is upstream of the trough, where we find the anticyclone of middle latitudes, frontolysis near the earth's surface, and dissipation of clouds.

Synoptic-scale cyclones and mesoscale thunderstorms develop in the active region. This is the region where the polar front generates or intensifies near the earth's surface due to convergence and favorable wind shear.

The active region is the favored location for the *conveyor belt*, a prominent southerly air current that transports warm and humid tropical air from the subtropics into the westerlies of the middle latitudes. This current rises near extratropical cyclones and eventually joins the polar jet stream. Most precipitation of middle latitudes occurs in the regions where the conveyor belt lifts.

The upstream part of the conveyor belt is a broad air stream that often appears as the *low-level jet*. The maximum wind in the low-level jet is about 1 km above the earth's surface. The wind speed in the low-level jet normally attains 25–30 m s^{-1}.

When the atmosphere is statically stable, rain is continuous in the active region. Then the rain-producing cloud is nimbostratus. This stable lifting is called *frontal upgliding*, since fronts usually exist in such regions. There is a physical connection between fronts and rain because the processes that are favorable for frontogenesis are often favorable for formation of rain.

Convective rain and thunderstorms appear in the active region in the case of thermal instability. Severe thunderstorms form in such active regions where the polar and subtropical jets flow away from each other. This is the region with the most intense vorticity advection at the jet-stream level.

In the subtropical jet stream, active regions also develop downstream of the troughs. The subtropical active regions occur less frequently since the subtropical jet meanders less than the polar jet. The weather phenomena and the position of the front at the surface are often similar in the active regions of the polar and subtropical jets.

The Cyclone of Middle Latitudes

The westerly flow in the frontal zone is often unstable, especially with high wind speed in the jet stream. This is the *baroclinic instability* that causes the waves in the zonal flow to amplify and break up into vortices. The vortices that rotate cyclonically are the *cyclones* of the middle latitudes that bring rainy weather and storms. They are also called *extratropical cyclones* to distinguish them from the tropical cyclones. A great part of meteorological literature deals with cyclones.

Other names are also used for the extratropical cyclone. Since the pressure shows relatively lower values in the cyclone than in the surrounding area, the names *depression* and *low* are used. There is often a wave on the polar front in the cyclone. Such a cyclone is a *wave cyclone* or *frontal cyclone*.

Sometimes a wave of the polar jet stream contains several cyclones on the same frontal segment. That sequence of cyclones is the *family of cyclones*, and the associated wave on the jet stream is the *long wave*. When a wave on the jet stream is so short to contain only one cyclone, then it is a *short wave*.

Anticyclones

Anticyclonically rotating vortices in the westerlies are *anticyclones*, which come in three distinct kinds:

1. A *warm anticyclone* is over 10 km tall. Its usual position is in the subtropics between the westerlies and easterlies. It appears in middle latitudes (40–60°) with ridges and anticyclonic vortices of jet streams.
2. A *cold anticyclone* is confined in the lowest 1–3 km of the atmosphere. It coincides with the cold polar or arctic air, often moving toward the equator.
3. A *transitional anticyclone* is associated with the polar front and participates in rapid weather changes in middle latitudes. Such anticyclones in the westerlies are of a combined (warm and cold) type. A transitional anticyclone typically has two centers, as illustrated in the middle of Fig. 1-2. The eastern center is a cold anticyclone in the polar air, under a trough at upper levels. The western center is a warm anticyclone. A ridge at the jet-stream level is usually located above this center. The polar front usually dissipates near the earth's surface in such an anticyclone; however, it may be prominent in the middle and upper troposphere.

Passive Front

The southern extent of the polar air reaches low latitudes in the region of the trough in the westerlies. The polar air in these latitudes progresses as a gravity current, under its own weight. The polar air is heated from the underlying surface, and the contrast in temperature across the front diminishes. When the front reaches the latitude of 30° or 20°, it is often more similar to a shear line than to a boundary between cold and warm air. The polar front in these latitudes usually extends far from its jet stream. Sometimes a narrow line of convective clouds (a *rope cloud* on satellite photographs) appears along that front. Occasionally such shear lines or fronts reach the equator.

Another type of passive front, the *arctic front*, occurs at the edge of the thermal (cold) anticyclone, primarily in winter. The arctic air behind the arctic front is usually 1 to 3 km high. It is shown near 60°N in Fig. 1-2. The arctic front usually appears on the polar side of the polar jet stream. This front is not structurally associated with the main baroclinic frontal zone of the polar jet stream.

1-3 TROPICAL WEATHER

Easterlies

The most prominent phenomena that appear in the belt of easterlies are *troughs, tropical cyclones,* and the *intertropical convergence zone*. Ridges exist between the troughs in the easterlies; however, since they were never described as causes for significant weather phenomena, they are omitted from the above list.

In the easterlies we find waves, with troughs and ridges, similar to the waves in the westerlies. Wind speed in the easterlies, typically near 5 m s^{-1}, is much lower than in the westerlies. The troughs in the easterlies are often seats of widespread convective rain. Large clusters of convective clouds form on the eastern side (upstream!) of the troughs.

A zone of convergence between the easterlies in the Northern Hemisphere and easterlies in the Southern Hemisphere is called the *intertropical convergence zone* (ITCZ).

Hadley Circulation

Time-averaged easterlies do not show easterly direction exactly. In some places they are southeasterlies (Gulf of Mexico, China) and in other places they are northeasterlies (northwest Africa). When the wind in the easterlies is also zonally averaged, a relatively small (but important) average meridional wind component can be detected. In both hemispheres (Northern and Southern), the air approaches the equator in the lower troposphere and departs from the equator in the upper troposphere. These meridional components combine with lifting near the equator and sinking in the subtropics to form a direct thermal circulation: Warm air lifts near the equator and cold air sinks in the subtropical anticyclones. This is *Hadley circulation*.

Hadley circulation is directly related to the rain regimes: The ascending part of Hadley circulation stimulates formation of rain in the belt of equatorial rain forests. The descending branch of Hadley circulation is in the belt of subtropical anticyclones, where dry weather is dominant.

Hadley circulations from the Northern and Southern Hemispheres meet in the intertropical convergence zone at or near the equator. Convergence of Hadley circulations on both hemispheres enhances convection in the lower troposphere.

Monsoon Circulation

In some places the easterlies cross the equator because a thermal low develops over the low-latitude continents in the summer hemisphere. The air that crosses the equator gradually assumes the balance between Coriolis and pressure forces in the hemisphere where the air arrived. This balance requires a westerly wind. This is the origination of *equatorial westerlies*.

Equatorial westerlies occupy a part of the region that is occupied by trade-wind easterlies in other seasons. The boundary, or transition region, between the equatorial westerlies and remaining trade-wind easterlies on the same hemisphere is the intertropical convergence zone that is displaced away from the equator. The air mass between the intertropical convergence zone and the equator is the *equatorial air*, which is usually colder than trade wind air of the easterlies. A frontal interface may develop between the equatorial air of the equatorial westerlies and the tropical air of the easterlies. This interface is the *intertropical front*. Seasonal appearance of equatorial westerlies signifies the summer rain season that is the *monsoon* of Asia, Africa, or Australia.

The most prominent air current that crosses the equator is the *East African jet*. This is a strong southerly low-level jet that blows over the western Indian Ocean during the northern summer. The air in this current crosses the Arabian sea and enters India in the monsoon circulation.

The equatorial westerlies can only appear on one side of the equator at any one time. This is usually the summer side of the equator. Only over South America and adjacent tropical ocean surfaces do the equatorial westerlies almost always appear in the Northern Hemisphere.

Tropical Plume

The most prominent stream of tropical air into middle latitudes occurs in the *tropical plume*. Satellite images show the tropical plume as a mass of clouds that separates from the intertropical convergence zone and stretches into middle latitudes. This occurs after a trough in the subtropical jet stream approaches the equator. A sizable amount of tropical air bursts poleward in the jet branch that turns away from the equator. The plume extends several thousand kilometers from the tropics into the middle latitudes.

In the shape of the cloud mass and by advection of water vapor, the tropical plume is similar to the conveyor belt. The tropical plume and the conveyor belt are different from each other mostly by the origin of clouds in them. The clouds in the tropical plume originate as tops of tall tropical cumulonimbus. The clouds in the conveyor belt form in the air that originates from low-level easterlies and penetrate to the upper troposphere only in the storms of middle latitudes.

Tropical Cyclones

Some of the troughs in the easterlies develop into *tropical cyclones*. These occur over oceans in which the water temperature is above 27°C. Tropical cyclones are classified as *depressions*, *tropical storms*, and *hurricanes*, depending on their maximum wind speed. In various regions of the world, hurricanes have several other names. Those that appear over the Pacific Ocean near Asia and Australia are *typhoons*. In the Indian Ocean, they are *cyclones*. Tropical cyclones are described in Chapter 12.

1-4 MAIN MODELS ON THE VERTICAL SECTION

The vertical structure of the troposphere is summarized in Fig. 1-3. The following comments serve only as hints for application in weather analysis and do not constitute complete descriptions of the models.

Westerlies: Most of the atmosphere poleward of about 20° latitude moves predominantly with a westerly wind. The existence of the westerlies is apparent in climatological average wind distribution. This property of the atmosphere is so prominent that we talk about it as being "normal." Consequently, we often use the term *disturbances* or even *anomalies* for all deviations from the westerly flow.

Easterlies: The average and most frequent air flow is easterly near the equator, up to about 20° latitude. At the earth's surface this easterly flow is the trade wind. The deviations from easterly flow are called *tropical disturbances*, analogous to the disturbances in the westerly flow of higher latitudes.

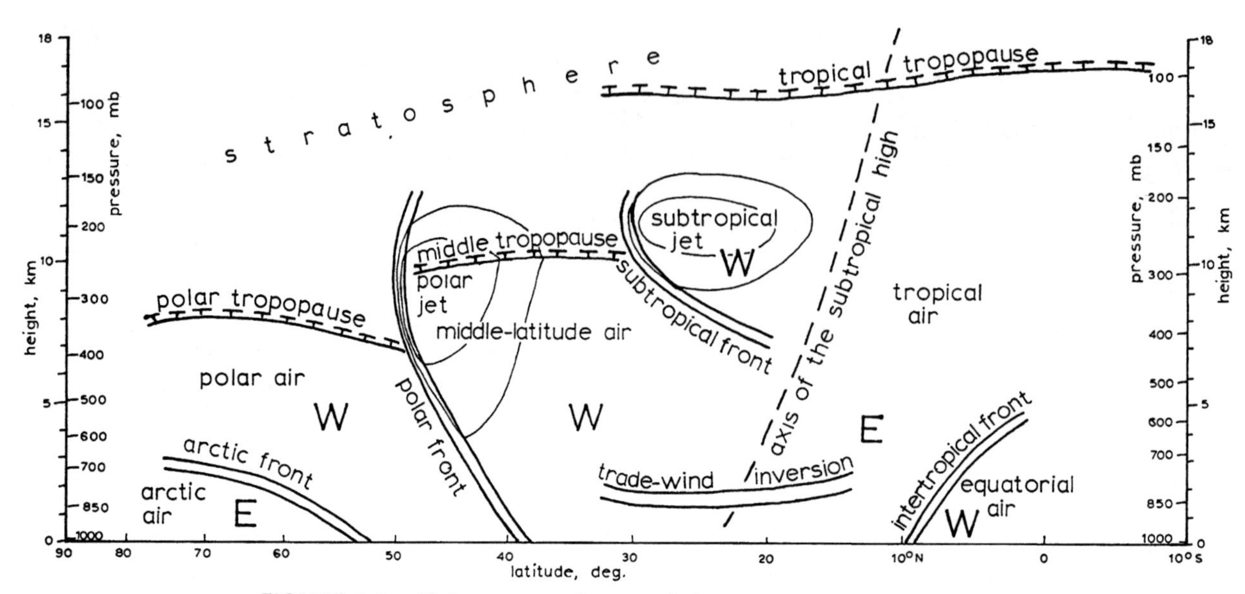

FIGURE 1-3. Main structural parts of the atmosphere in a vertical meridional section. Prevalent wind directions are indicated by E and W.

Subtropical highs: The transient highs and lows are not included in Fig. 1-3. However, the quasi-permanent nature of the belts of subtropical highs near 20° latitude justifies their inclusion among the principal constituents of the atmosphere. The subtropical high is slanted, as shown by the dashed line in Fig. 1-3.

Subtropical trade-wind inversion: A permanent feature of subtropical highs, the subtropical trade-wind inversion is a subsidence inversion, covering wide areas in the subtropics. It expands into the region of the trade wind; hence its name.

Troposphere: The region of the atmosphere under the tropopause is called the troposphere. The stability of the air in the troposphere is usually much smaller than in the stratosphere. The small stability facilitates vertical motion and the associated weather phenomena in the troposphere.

Tropopause: The top level of the troposphere, characterized by a transition from the less stable troposphere to the much more stable stratosphere, is the tropopause. The height of the tropopause is typical for the main air masses, and it changes abruptly near jet streams. Therefore, usually three tropopauses can be distinguished, corresponding to three main air masses: *polar*, *middle*, and *tropical*. The tropopause in Fig. 1-3 is shown by heavy lines with small T's attached.

Polar tropopause: Predominant in the area poleward of the polar jet stream, above the polar air, is the polar tropopause. The level of this tropopause is typically between 6 and 8.5 km, or between 450 and 300-mb. This tropopause sinks to lower levels (400–450 mb) in the troughs and cyclones of middle latitudes.

Middle tropopause: Typically appearing at an elevation between 9.5 and 11 km (210–270 mb) in the region between the two upper-tropospheric jet streams is the middle tropopause. It often rises to about 13–15 km in a region of severe thunderstorms.

Tropical tropopause: Covering the region between the subtropical jet streams on both hemispheres, the tropical tropopause's elevation is about 17 km (120–80 mb). A tropopause at about 20 km elevation is frequent over strong thunderstorms in Indonesia. The tropical tropopause sometimes extends over the subtropical jet stream into the region of the middle-latitude tropopause. In such cases there is a double tropopause poleward of the subtropical jet stream. A double tropopause also is frequent in the tropics where a *secondary tropical tropopause* often appears around 200–250 mb. The secondary tropical tropopause is less prominent than the main tropopause.

Air masses: Large volumes of air with somewhat uniform physical properties, air masses can be distinguished from one another by the presence of fronts. When the front weakens (*frontolysis*), the distinction between air masses diminishes. The three major air masses are *polar air*, the air mass on the poleward side of the polar front; *middle-latitude air*, the air mass between the polar and subtropical fronts; and *tropical air*, the air mass on the equatorial side of the subtropical front. Besides these, two other air masses, *arctic air* and *equatorial air*, can be found in appropriate seasons.

Arctic air: As mentioned earlier, arctic air is the air mass of the polar night. Over wide areas covered with snow in winter, a radiation inversion develops that does not disappear in the short daytime. The layer of air under this inversion becomes arctic air when it reaches a large size (1–3 km vertically) and covers areas wider than about 1000 km horizontally. Arctic air can be found regularly in winter at the surface and at 850 mb. It seldom appears at higher standard levels. When a large winter storm develops, the arctic air often enters the circulation of the cyclone and travels with it far south. The arctic front often reaches the polar front; in such cases the two fronts coincide on the surface chart. The two fronts can still be identified separately on vertical sections. We find *polar easterlies* within arctic air that spreads south with an easterly wind component. Polar easterlies are included in some climatological averages.

Equatorial air: The air mass of the equatorial westerlies, the equatorial air cannot be distinguished from the tropical air in the other hemisphere because the intertropical front appears only on one side of the equator. Equatorial air appears with the equatorial westerlies in the lowest few kilometers of the troposphere. This is particularly prominent during the monsoon in regions where monsoon occurs.

Fronts: Layers or surfaces of discontinuity that develop in the atmosphere, fronts form in convergent air currents. Many weather phenomena are associated with fronts and therefore weather analysis is very much concerned with locating fronts. Fronts are normally slanted toward the colder (denser) air. An extreme case is the vertical front at the jet-stream level where the differences in temperature and density disappear across the front. Distinction between fronts can be made by their structural association with air masses, jets, and tropopause.

Intertropical front: Between the tropical air in one hemisphere and the tropical air that crossed the equator from the other hemisphere, the intertropical front develops. Often, especially within 10° latitude away from the equator, this front is not easily discernible. In many cases this front expands to a larger transitional region between northern and southern trades. Then it is described as the *intertropical convergence zone*.

Polar front: The layer between the polar and middle-latitude air masses is the polar front. The baroclinicity of the westerlies is mostly concentrated in the frontal layer of the polar front. Because the polar jet stream is structurally coupled to the baroclinic zone of the westerlies, the polar front is also associated with the polar jet stream.

Subtropical front: Separating the tropical and middle-latitude air is the subtropical front. There is a horizontal convergence that enhances thermal contrasts on the upper poleward side of the Hadley circulation. Consequently, baroclinicity, frontogenesis, and thermal wind are enhanced in the region of convergence. The subtropical front is seldom observed at or below 500 mb; this makes it difficult to distinguish the tropical and middle-latitude air masses at lower levels. The subtropical front may consist of several frontal layers. It is difficult to find the subtropical front on isobaric charts due to small thermal contrast; it is usually easier to find it on vertical sections.

Arctic front: The boundary between the arctic air and the rest of the polar air is the arctic front. Normally this front is not structurally associated with a jet stream. Such a front in the southern hemisphere is the *antarctic front*.

Trade-wind inversion: In the domain of subtropical highs there is practically always a well-developed inversion that caps the convection in the mixed layer near the earth's surface. The trade-wind inversion spreads far north in the warm sectors of extratropical cyclones. It is usually present when the atmosphere is potentially unstable.

Jet streams: Zones of fast wind with approximately straight or wavelike streamlines are the jet streams. The meanders in the high-tropospheric jets may cut off into powerful vortices (highs and lows). Vortices of the jet streams can be distinguished easily from tropical cyclones, since the maximum wind in the jet stream is near the tropopause and the maximum wind in the tropical cyclone is near the earth's surface.

Polar jet, or *polar-front jet:* The jet stream of middle latitudes that is structurally associated with the polar front and with the break between the polar and middle tropopauses is the polar jet. Maximum wind in this jet is near the 300 mb level. The width of this jet is of the order of 1000 km. The polar jet occupies a thicker layer vertically than the subtropical jet and can be identified easily at levels between 500 and 200 mb, inclusive.

Subtropical jet: The jet associated with the subtropical front between the tropical and middle-latitude air masses is the subtropical jet. Maximum wind in this jet is normally near 11 km elevation or at about 250 mb. This jet is usually shallower than the polar jet, so it may be hidden between the standard levels of 300 and 200 mb. In such cases the subtropical jet can be detected by wind soundings passing through this jet.

The subtropical jet can usually be found between the latitudes of 25° and 30°. It meanders less than the polar jet, but in rare cases it has been observed as far north as 60°N.

There is abundant evidence that the above-listed phenomena and structures also exist in the Southern Hemisphere. An exception is the tropical easterly jet of the summer season in Asia. There are no large continents in the Southern Hemisphere to stimulate the formation of such a subtropical easterly jet.

All these structures may appear in altered form, greatly different from the models. These structures sometimes disappear from sizable parts of the atmosphere where they are expected to exist. Even the troposphere sometimes disappears. During the polar night, with very low temperature of about −60°C at the surface, the temperature may steadily increase with height into the stratosphere. In such cases the troposphere cannot be identified.

Another common difficulty in weather analysis is that air masses and atmospheric phenomena transform into each other. Then it is difficult or impossible to determine the exact limits as to when or where one stops and the other begins. Such circumstances make the meteorological profession difficult, but also more challenging.

Models cannot be applied like mathematical theorems that have no exceptions. Like all rules in natural sciences, the conceptual models have exceptions. Those few processes for which no exceptions have been found are described as laws of nature, or, more specifically, as laws of physics.

1-5 MESOSCALE STRUCTURES

Scale of Atmospheric Processes

The overview in preceding sections describes the phenomena that appear globally, at all longitudes and in both hemispheres (Northern and Southern). These structures are usually more than 10^6 m long in at least one dimension. However, weather is not governed only by processes of large and global dimensions. There are many small- and mesoscale processes that also must be scrutinized in weather analysis.

It is a matter of accepted terminology to refer to the scale of various phenomena and processes by their typical horizontal dimensions. The approximate classification is as follows:

Large scale: larger than 1000 km,
Small scale: less than 10 km,
Mesoscale: between the above two scales.

Large-scale processes are the main objects of weather chart analysis. This book is mostly concerned with these processes. However, larger mesoscale phenomena are also included since they are routinely shown on standard synoptic charts (fronts, larger thunderstorms, tropical cyclones, smaller extratropical cyclones). The scale of larger mesoscale phenomena is also called *synoptic scale*, since such phenomena are commonly represented on usual synoptic charts.

When the typical horizontal scale is about 10,000 km (or more), then we may use the term *planetary scale*. General atmospheric circulation and longitudinal extent of jet streams are of this scale.

There is a physical distinction between phenomena of various scales, besides the distinction in size. Most large-scale and planetary-scale phenomena are formed and modified by differences in heating on the planetary scale (between the poles and the equator). Larger mesoscale phenomena (some thunderstorms, low-level jet) are consequences of the configuration of large-scale flow or of distribution of static instability. Mesoscale phenomena, however, are often physically caused by geographical objects of appropriate scale, such as mountains or ridges, lakes or seacoasts.

Examples of small-scale processes and phenomena are smaller clouds, turbulence, transports through the boundary layer of the atmosphere, turbulence in free air, mountain waves, local accumulation of air in valleys, processes around vegetation, and many others. All these processes are important for weather development, but many appear on such a restricted scale that we cannot detect and follow them on weather charts. Therefore, we do not mark small-scale processes on synoptic charts. Generally, these phenomena will not be considered in this book, except for an occasional reference.

Mesoscale Phenomena

A number of larger mesoscale phenomena are of sufficient size to detect them on weather charts. The following processes among them stand out:

Thunderstorms: Thunderstorms are sufficiently frequent to be described in a separate chapter (Chapter 12).

Tropical cyclones: Comparatively rare phenomena, but often with extremely violent weather, tropical cyclones are also described in Chapter 12.

Land and sea breezes: Along the shores of seas and large lakes the diurnal variation of wind is a prominent weather factor. Land and sea breezes are also associated with coastal fronts.

Coastal front: The sea breeze enters the land behind the coastal front. Convection and thunderstorms may occur along this front in summer. In winter, marked differences in precipitation and fog may occur at this front.

Dryline: A trough that appears in the warm sector of the extratropical cyclone is the dryline. Its western side is markedly less humid than its eastern side. Thunderstorms sometimes develop on this line.

Steering of air currents around or along mountains: A variety of typical weather phenomena appear when stable air masses stream along the mountains. The trough in the lee and strong downslope wind are such phenomena.

Historical note. The discovery of the polar front and the introduction of the concept of air masses is credited to Jakob Bjerknes (1897–1975) and Halvor Solberg (1895–1974); they worked between 1917 and 1922 in the Institute of Meteorology of the University of Bergen, Norway. Other members of this group were their professor Vilhelm Bjerknes (1862–1951) and another prolific collaborator, Tor Bergeron (1891–1977). Their achievements are frequently referred to as the Norwegian School or *Bergen School.* The physical relation between the polar front and polar jet stream was worked out by a cooperative scientific research project under the leadership of Carl-Gustav Rossby (1898–1957) and Erik Palmén (1896–1985) in Chicago (1946–1948). The achievements of this group are referred to as the *Chicago School.*

2

Characteristics of Meteorological Observations

Natural sciences are based on observations, including measurements and experience. The observations originate from two important sources: from nature and from controlled experiments in the laboratory. Meteorology (like astronomy) uses practically no laboratory measurements, which gives more importance to the field observations than in most other sciences. This chapter gives an overview of instruments and ways of observation, and is primarily concerned with those properties of observations that can influence the work in weather analysis. A complete description of instruments and platforms that support them is given in other books and manuals.

2-1 TYPES OF OBSERVATIONS AND BASIC TERMINOLOGY

The main categories of observations with respect to the type of instruments are visual observations, direct measurements and remote sensing.

No instruments are used for visual observations. The elements observed are clouds, cloudiness, types of precipitation, various phenomena, damage from storms, and so forth.

For direct observations, the sensor of the instrument is located on the spot where the state of the atmosphere is observed.

In remote sensing, the sensor receives radiation (and sometimes sound) at a distance from the point of interest in the atmosphere. Remote sensing consists of two main categories: *active*, where radar or sound waves are emitted and echo is observed, and *passive*, where only naturally occurring signals are observed (light, infrared radiation).

Because it records visual observations, photography may be classified in the category of visual observations. Photographs taken from the surface are never used in routine weather analysis and seldom in research work, but are used more often in the study of cloud- and thunderstorm-scale weather processes.

Observations can be categorized as synoptic or asynoptic, depending on the time when made.

Synoptic means at the same time with observations at other stations. Such observations have been used since the telegraph became available for weather reports in the 1850s.

Asynoptic means at various times, as dictated by the property of the observing platform (aircraft, satellite). Analysis of asynoptic data is sometimes called *four-dimensional analysis* since, in addition to three space coordinates, the time coordinate enters into consideration.

Synoptic observations are not always made exactly at the same time. At surface stations, it takes about 10 min to collect all readings. A radiosonde takes about an hour for an ascent. Satellite and aircraft observations within 3 h of the observing hour are often taken as synoptic. This 3-h tolerance is acceptable for processes that last much longer than that time.

A traditional categorization of direct observations is in the two classes of *surface observations* and *upper-air observations*. The distinction comes from the times when the upper-air observations (by aircraft, radiosonde, and so forth) required disproportionally more effort, expertise, and expense than the surface observations. The surface observations are performed by the observer and instruments at the earth's surface, even if the clouds in the upper troposphere are observed. So, for example, much knowledge about the atmosphere has been derived from the nephoscope, a surface instrument for measurement of the direction and apparent speed of cloud motion; and most clouds are in the "upper air." One might also argue that the nephoscope is a remote-sensing instrument, since the reflected light from distant clouds is observed.

There is a proliferation of observing instruments and observing platforms. Improvement in radar and satellite technology and automation of many other instruments promise great new possibilities in weather analysis. These improvements pose a formidable practical task for weather analysts. All the new instruments and means of communication yield enormously more data than at the time when most data came from radiosonde and manual observations at the surface stations. Fortunately, the development of computers parallels the increase of observing capabilities. In this way most processing of weather data will necessarily be done by computers.

Transmission of standard weather data and plotting of the data on the weather charts are reviewed in Appendix E.

2-2 REPRESENTATIVENESS

A major problem encountered with all meteorological observations is the adequacy of the observation to represent the atmospheric phenomena or properties of interest. This is the *representativeness*. We ideally expect the instrument (or human observer) to give correct information about the atmosphere in the vicinity of the observing point. The "vicinity" often means up to halfway to the next observing point. Numerous problems, however, cause incorrect or incomplete information. Some of these problems are the following:

1. The instrument does not sense the properties of air adequately. For example, the thermometer may be heated by sun rays and thus not show the air temperature.
2. The instrument may be in an unusual spot where the processes of a smaller (or sometimes larger) scale obscure the desired process. For example, the wind vane or radiosonde may happen to be within a small-scale eddy at the time of observation. Under such conditions the instrument does not show the wind that is typical of a large-scale cyclone correctly.
3. The instrument senses too small a sample of the area around the station. For example, the rain gauge may not show the rain in the region around the station. Much of the rain naturally occurs in showers of limited horizontal extent.
4. The instrument covers too much area. Satellite sensors may have too wide a field of view and therefore not show desired detail.
5. The instrument is not properly adjusted. This seems to be the least problem of all. Present-day technology is very reliable. Still, some observations suffer from inaccuracies, such as using radar to measure precipitation.

Educated weather analysts are often familiar with deficiencies in representativeness. Some atypical observations can be useful in many cases. For example, one unusual observation of wind may indicate a thunderstorm between observing stations. In another example, one station may show warm wind in winter surrounded by several stations with calm, cold weather. This warm station may indicate that there is a warm air mass already above the region, while the majority of stations are still covered with a thin, stable layer of cold air. The cold surface layer is broken at that warm station.

While progress is being made by the construction of better instruments and by standardizing the methods of observation, we cannot expect that all problems with observations will ever be solved. Also, from a scientific point, we do not want to have observations of "everything, everywhere." That would give us too much information that we cannot process. In the routine work of weather analysis, a selection of observations is needed that gives information about the relevant atmospheric processes, without unnecessary detail. In scientific research work, progress is made with detailed new observations that will yield information about the suspected unknown phenomena and processes. Even the most diligent scientists do not analyze everything; instead, they select a sufficient number of observations that prove the hypothesis. This selection must be fair: We must not eliminate data that do not suit us. All evidence to the contrary of the hypothesis must also be taken into account. Selection may dismiss only irrelevant or redundant observations.

2-3 SURFACE OBSERVATIONS

The instruments for observation of temperature and humidity (psychrometer, hygrometer) are screened from direct insolation and are well ventilated. The old practice is to house the instruments in a wooden *instrument shelter*, also known as the *Stevenson screen*. This shelter is painted white, has a double roof, and has louvers on the sides for circulation of air. More recent shelters are small and metallic. Standardization of weather observations is coordinated through the World Meteorological Organization.

It is a common practice to keep the thermometer sensor at an elevation of 1.5 or 2 m above the ground. The ground under the instruments is covered with short grass. The shelter with instruments is located far from obstacles to ensure good air flow. Large buildings in the vicinity must be avoided, because they may radiate heat on a shelter with thermometers. The wind vane is preferably at an elevation of 10 m, again away from obstacles.

These conditions cannot be always satisfied, and this problem has global significance. For instance, it is difficult to determine if and how much the temperature has changed on our planet during the past century, in spite of careful observations. Weather stations that were in the fields in the nineteenth century are today in big cities and do not have the same representativeness as before. On the other hand, perhaps the climate has warmed because so many big cities have been built.

It is not clear whether the choice of instrument shelters is best for observing the atmosphere. The major advantage of standard shelters is that this is a worldwide practice and the observations from different countries are comparable to each other. Also, the instrument shelter is a rather inexpensive device that can be easily installed even with limited budgets.

Routine visual observations are a part of surface observations. They cover a wide range of phenomena (fog, lightning, thunder, rain showers or snow showers in the vicinity of the station, and others). It is unfortunate that some useful observations are not provided in the international code for weather reports. Among these are frost, rainbow, halo, and others. Furthermore, for economic reasons, there is an effort to eliminate many stations that have human observers. The automated stations will miss observing many phenomena. On the other hand, there are numerous other advantages of automatic stations. Besides being more economical, the automated stations will provide data more frequently and from locations that are very inconvenient for human observers.

The system being introduced in the United States in the early 1990s is the *Automated Surface Observing System* (ASOS), which will provide data on wind (direction and speed), temperature, dew point, pressure, cloud amount and height, visibility, precipitation amount, and some weather phenomena such as rain, freezing rain, snow, drizzle, fog, and haze. The observations will be timely reported by radio.

2-4 UPPER-AIR OBSERVATIONS

Direct upper-air (or upper-level) observations are made with instruments on balloons and aircraft. Other instruments are being developed that will yield upper-level observations by remote sensing. Upper-level observations are also called *aerological observations*.

Radiosonde

The simple, but marvelous, invention of the radiosonde in 1928 (by Väisälä in Finland and Molchanov in Soviet Union) opened the way for three-dimensional systematic exploration of the atmosphere. Many new instruments have been developed since that time, but the radiosonde is still firmly entrenched as the main instrument for the study of upper-atmospheric weather processes.

Prior to the radiosonde, vertical soundings were rare and expensive. They were performed by aircraft, balloons, and kites. The pilot-balloon for wind observations has been used since the 1870s, before the radiosonde, since it does not need much equipment. The pilot-balloon is tracked by a theodolite. It suffers from some inaccuracy due to the crude determination of its ascent speed. The pilot-balloon cannot be tracked in cloud or fog. However, it can be used at night with an attached lantern with a candle. It happens that the lantern is confused with a star.

The radiosonde gives readings of pressure, temperature and relative humidity up to the elevation of over 30 km, where the pressure is as low as 10 mb. Sometimes, the radiosonde reaches the level of 3 mb. The readings from the radiosonde are accurate to about 1°C, 2 mb, and 5 percent relative humidity. This is adequate for routine weather analysis.

The *rawinsonde* is a radiosonde that can be tracked by radar or by a movable directional antenna that senses the radio signal from the radiosonde. This system provides wind observations from subsequent positions of the sonde. The wind from the rawinsonde is more accurate than the wind from pilot balloons, since the elevation of the balloon is more accurately determined by the pressure sensor than by the theoretical speed of the rising balloon. The wind accuracy from a rawinsonde is within a 3 m s^{-1} vector error.

Dropsonde

A *dropsonde* is similar to a radiosonde. It has a parachute and is released from the aircraft to provide a sounding as it descends. The dropsonde is more rugged and heavier than the radiosonde since it must sustain a high wind speed at release. Because of its weight, the use of the dropsonde is limited to oceans where it is unlikely that it will do damage when it falls to the surface.

A dropsonde equipped with a wind measuring device is a *dropwindsonde*. Wind is measured by the retransmission of omega or loran (long range navigation) signals from a dropwindsonde to a base station. Omega signals are emitted from eight transmitters on the globe. These signals are used for precise positioning of ships and aircraft. Loran signals are more accurate, but they are available only on coasts with much sea traffic. Omega signals are precisely timed, so the location of a transmitter can be determined by the difference in time of signal arrival. Omega and loran signals can also be used to find the speed of the receiving station (dropwindsonde). This is achieved by measuring the frequency shift of the base signals, much as Doppler radar provides the speed of various objects.

A base station, which monitors the dropwindsonde, may be on board the aircraft that launched the sonde. A moving platform can be used since the timing of signals at the sonde does not depend on the motion of the base station.

Aircraft Observations

ASDAR (aircraft to satellite data relay) is an automatic observing station similar to the radiosonde but mounted on an airliner. It relays the observations via satellite to the ground, without assistance from the aircraft crew. ASDAR reports temperature, wind, and the position of the aircraft at desired intervals.

An instrument system similar to ASDAR is ACARS (ARINC communications and reporting system, ARINC being the company of Aeronautical Radio, Inc.). The ACARS observations are transmitted to the ground receivers, without satellite link.

With either ASDAR or ACARS, a choice of eight observations per hour yields a good spatial resolution ($\approx$ 75 km) along the flight path. These observations are available at elevations between 8.5 and 13 km, where most airliners travel. During airliner's ascent and descent, the observations may be made more frequently than eight per hour to provide good vertical soundings. The wind information comes from precision positioning of the aircraft (omega) and its velocity with respect to the air. ASDAR is particularly useful over long transoceanic flights and over other regions where no other direct observations are available. Accuracy of 0.5 K for temperature and 1 to 2 m s^{-1} for wind has been demonstrated. It is expected that the instruments will be modified to also report data on turbulence. Once an appreciable initial expense ($\approx$ \$5000) for the instrument and transmitter is met, ASDAR observations have a relatively low operating cost. The cost of communication is also involved: Satellite frequencies are leased and the user fee must be covered. While each report costs about \$2 to \$3, a great number of reports may escalate the total cost.

There are some disadvantages to ASDAR. Worldwide and time distribution of these observations is uneven. ASDAR observations are clustered around popular routes and major airports. Also, aircraft avoid cyclones and storms that are of primary interest to meteorologists.

Other problems with ASDAR arise in relation to the weather service and commercial airline companies. Sometimes airline companies do not want to release ASDAR data from their aircraft; the companies wish to keep the useful data for their own economic benefit and not make them available to the competition. It is hoped that such problems will diminish in the future.

2-5 RADAR

Physics of Radar

Radar was introduced in meteorology in the 1940s, when it was discovered that rain clouds give echoes of radio waves in the spectral range of 0.4 to 60 GHz. These echoes can be used to locate rain clouds very accurately. The principle of radar is to send rays of electromagnetic waves about 1°–2° wide, horizontally and above horizon, in pulses of about 1000 per second. The waves scatter mainly from rain drops. A part of the scattered energy is received by the radar between pulses. The time delay of the echo gives the indication of the distance to the target. At a distance of 100 km, the beam width is almost 2 km wide and tall. Each elementary sampling volume is about 150 m "deep" (long in radial direction, from the radar). The pulses are emitted from a rotating antenna. The echo gives information from scatterers within a circle around the radar, up to a distance of about 400 km. Farther distances are not observed, because straight rays exit from the troposphere, due to the curvature of the earth's surface. There are no hydrometeors and other interesting scatterers in the stratosphere. (The aurora and meteor trails give good radar echoes, but these are of no interest in present-day weather analysis.)

The frequency most used is 3 GHz, which has a wavelength of 10 cm. Good results are also obtained by 3-cm waves (frequency of 10 GHz). These, however, attenuate in the rain much more than 10-cm waves. Other frequencies are also sometimes used. Waves of 0.5 cm give echo from cirrus. The wind profiler (see Section 2-7) works with waves of over 70 cm.

Conversion between wavelength l and frequency n is

$$l = \frac{c}{n}$$

with c being the speed of light (3×10^8 m s^{-1}).

The returning echo is very weak, but the radar receiver is sufficiently sensitive to detect it. An automatic calculation gives the distance to the scatterers, based on the time difference between emission of the wave pulse until the return echo is sensed. The intensity of the echo gives some indication of the sizes and number of scatterers. Scatterers are usually raindrops, since the cloud droplets and small ice crystals (snow, dry hail) reflect disproportionately less. Snow and hail are seen well. The best reflection comes from wet hail. Aircraft are sensed well by the radar, but each aircraft occupies only one sampling volume and thus can be easily distinguished from meteorological echoes.

Radar Display

Echo processing by computers made it possible to increase the amount of data from the radar and still have a picture that can be quickly comprehended. In addition to the echo position, the information about the echo intensity can also be displayed. This is accomplished on digitally enhanced cathode-ray tube (CRT) displays, where several degrees of intensity of radar echo are displayed in various colors.

A complete rotation of the radar antenna is usually accomplished in about 10 to 20 s. In this time the information is received from about 10^4 elementary echoes. Processing of this amount of data poses some difficulties. Traditionally, the radar data were processed by analog electronic circuits and displayed on a CRT. Modern radar data are digitized and displayed on color monitors. If the display is in the form of a circular picture of the radar horizon, this is the *plan-position indicator* (PPI). It is also common to display the local polar coordinates on the screen: radii that show direction (azimuth) and circles (range markers) that show distance.

In the case of severe weather, radar images are usually stored in the computer system every few minutes. They can be played back on the screen to show the development of storms.

With repeated scanning at various elevations of the beam above the horizon, it is possible to obtain three-dimensional information about a storm. Practical techniques for fast processing of such three-dimensional data sets have been developed only recently, with the introduction of Doppler radar and with availability of powerful small computers.

An example of a PPI display is in Fig. 2-1. The circles are at 50-km intervals. The intensity of echo is explained in the stripe at the bottom of the figure. Lowest intensity is on the left, highest on the right, but the intermediate degrees of intensity are represented by various degrees of color (here shown gray).

The important echo in Fig. 2-1 is from a squall line stretching from 150 km S of the station to 110 km NE of the station. This is the dark gray area encircled by a light gray area. The darker area shows a stronger echo than the light area; those are the showers in the squall line. The squall line is traveling toward ESE. The light gray area is narrow, about 5 km, on the east side of the squall line, showing a rather quick onset of strong rain. The broad light gray area, filling most of the NW half of the field of view, shows a weak, continuous rain that typically follows the thunderstorm.

An alternate display of radar echo is on a two-dimensional vertical plane (*range-height indicator*, RHI). The information for this display comes from vertical scanning of the atmosphere with the radar beam. Both radar displays (PPI and RHI) make it possible to represent many individual echoes. Each echo corresponds to one point on the CRT display.

An example of the RHI is shown in Fig. 2-2. The station is in the lower left corner of the echo display. The lower edge of the variably gray area is at the

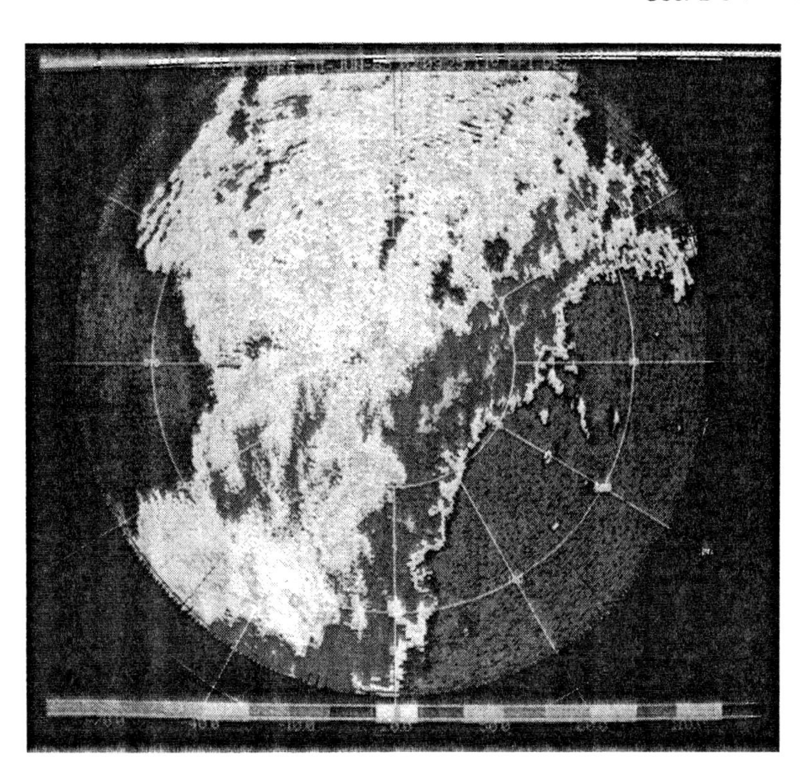

FIGURE 2-1. Radar echo from a squall line observed from Sterling, Kansas, at 0209 UTC 11 June 1985. The circles are at 50-km intervals. Photo courtesy of M. Biggerstaff and S. Veleva.

earth's surface. In this example, vertical and horizontal scales are equal, with ticks every 10 km. The radar is pointing toward the azimuth of 120°. This is the time of the PPI in Fig. 2-1. The intensity of the echo is shown in the strip at the bottom of the display. The choice here is made that lighter gray represents weaker echo, with uniform transition toward the dark of the stronger echo. The echo under a low threshold ("no echo") is represented by black. By the shape of the echo, it is possible to distinguish "no echo" from very strong echo, although both are displayed as dark. This example shows strongest rain between 10 and 35 km from the station. Comparing with Fig. 1-1, we can get a good idea about the structure of the squall line.

Echo Intensity

The intensity of radar echo is expressed in *decibels of reflectivity*, commonly abbreviated as dBZ. These units are described in specialized texts on radar tech-

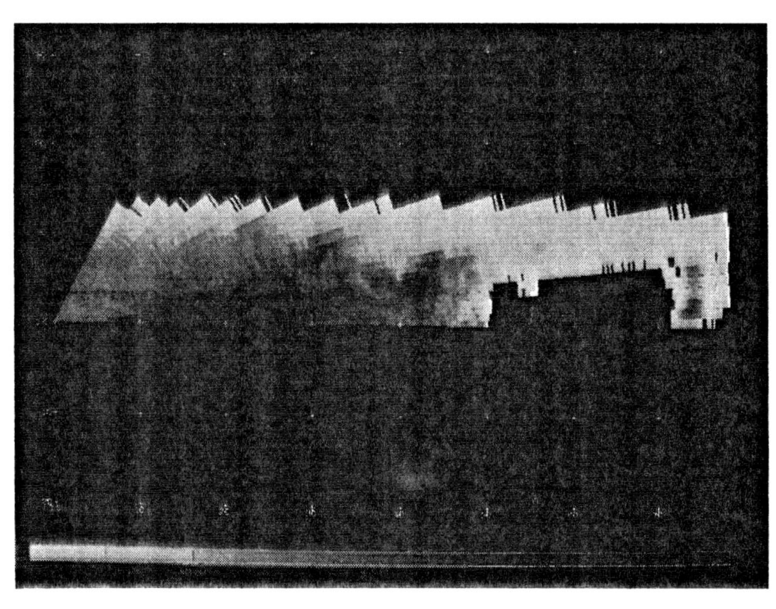

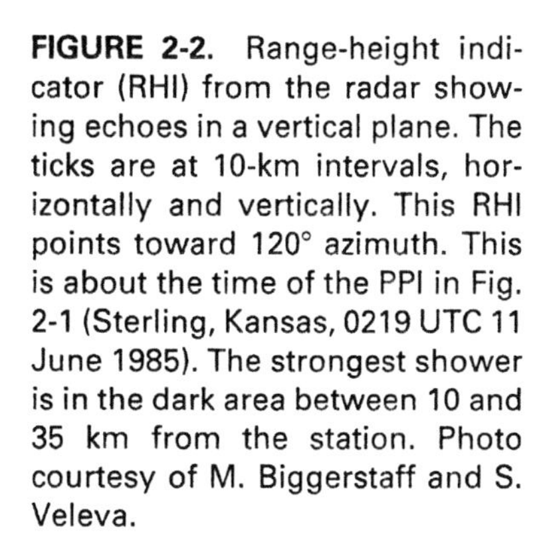

FIGURE 2-2. Range-height indicator (RHI) from the radar showing echoes in a vertical plane. The ticks are at 10-km intervals, horizontally and vertically. This RHI points toward 120° azimuth. This is about the time of the PPI in Fig. 2-1 (Sterling, Kansas, 0219 UTC 11 June 1985). The strongest shower is in the dark area between 10 and 35 km from the station. Photo courtesy of M. Biggerstaff and S. Veleva.

nology. Weak, steady rain gives echoes under 20 dBZ. Showers give echoes of 30–40 dBZ since they usually consist of larger drops and scattering from water drops is proportional to the sixth power of drop diameter. Severe storms may give echoes of 50 dBZ or more, primarily since they have large, melting hailstones. Such wet hailstones reflect like very large water drops. Dry (frozen) hail reflects about 10 times less than water drops of the same size.

2-6 DOPPLER RADAR

The radar being introduced in the weather services is *Doppler radar,* which has many advantages over "traditional" radar. The whole system of radar and associated hardware and software in the United States is known as NEXRAD (Next Generation Radar). It should be installed at about 50 stations in the United States by 1995.

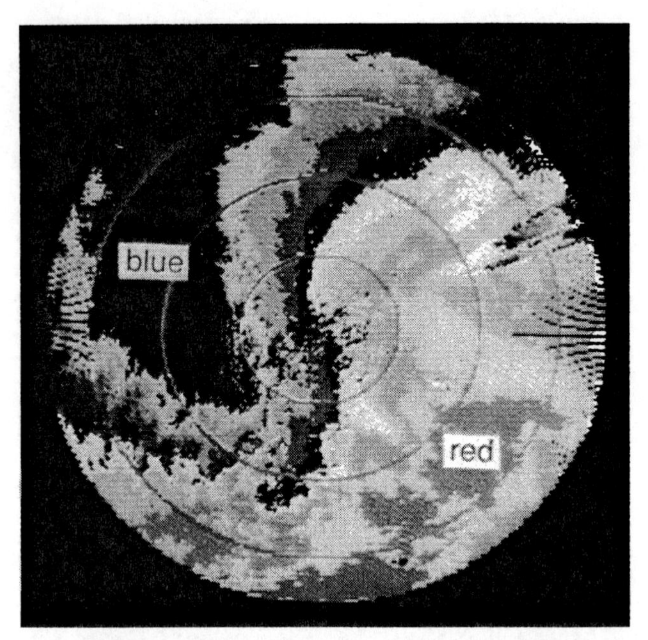

FIGURE 2-3. An example of Doppler radar display. The circles are at 20-km intervals. The dark area in the southeast (red) shows maximum receding wind speed. The dark area west of the station (blue) shows approaching wind.

Physics of Doppler Radar

The most important property of Doppler radar is that it senses the frequency shift of the reflected wave. This gives information on the radial velocity component of the scatterers. The frequency shift is the *Doppler effect,* named after the nineteenth century Austrian physicist who described this phenomenon in acoustics. This information about the speed of the scatterers comes in addition to the traditional information about the position and intensity of the echo. Sensing of frequency shift requires precisely transmitted and measured phase and frequency of the waves. The waves that satisfy this precision are called *coherent waves.* For this reason, Doppler radar is sometimes called *coherent radar.* In contrast, and when a distinction is needed, "ordinary" non-Doppler radar is sometimes called *incoherent* or *noncoherent radar.* Normally, Doppler radar provides the speed of targets (presumably the wind) and the intensity of rain. Therefore two fields need to be observed (or recorded). The display of velocity gives no information about the intensity of echo. Also, it cannot be seen whether the echo is from raindrops or from other scatterers in the "clear" air (turbulent eddies, insects). When the intensity of echo is observed, the nature of the scatterers can be determined in the reflectivity display. All signals from Doppler radar are processed in the computer that is part of the radar system.

The information about the radial speed of the target is usually represented in color on the PPI. Common choices are red for receding speed (away from the radar) and blue for approaching speed. Figure 2-3 shows receding speed of over 25 m s^{-1} in the SE quadrant of the PPI. Approaching speed of over 20 m s^{-1} can be observed in a wide band W to NW from the station. The black echoes along a curved line through the center represent zero radial speed. Along this line the wind is tangential to the radar site.

Wind Variation with Height

A tilted Doppler radar antenna receives echoes from different elevations. This yields the wind variation with height in an otherwise uniform wind field. Figure 2-4 shows several examples of the PPI under various wind conditions. Red and blue colors are simulated by solid and dashed lines, respectively. In each case the echoes from the rim of the PPI come from the highest elevation reached by the beam. The echoes near the center are from the lowest atmospheric layers. Each circle corresponds to the wind speed plotted on the top and to the wind direction plotted on the left. Straight lines imply that the wind speed is linear (or constant) with height *and* that the direction does not vary with elevation. Other combinations show that wind varies with height in different ways. When the highest radial speed is sensed within the PPI but not on the rim, a low-level jet is shown.

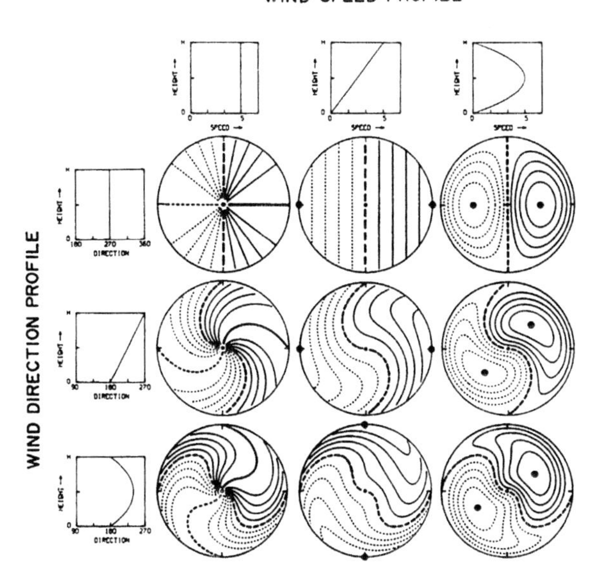

FIGURE 2-4. Several examples of Doppler radar echo in a fairly uniform wind field. The receding wind speed is shown by solid isopleths, otherwise usually simulated in red. The approaching wind speed is shown by dashed isopleths, simulating blue. The small graphs on the top indicate the wind variation with height. The graphs on the left show the variation of wind direction with height. From Wood and Brown (1986), by permission of the American Meteorological Society.

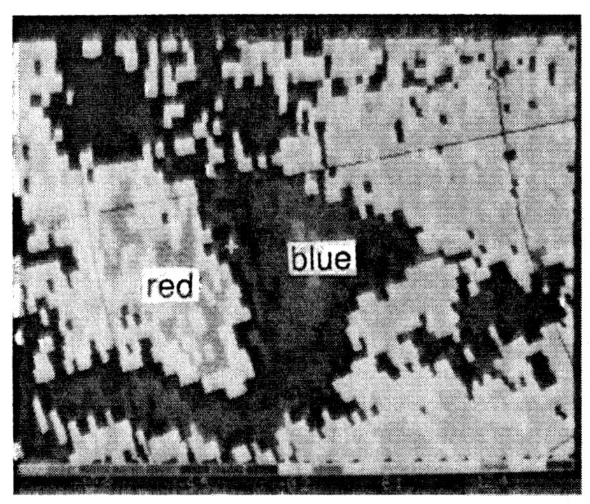

FIGURE 2-5. A part of a Doppler-radar PPI. Distance marks (circles) are at 20 and 30 km from the radar. The interval of radial speed between colors is 2.1 m s^{-1}, as explained on the color bar. The prominent red and blue patches in the center and left-hand part of the picture indicate wind divergence between them. The divergence is caused by a downburst under cumulonimbus. Photo courtesy of J. Wilson of NCAR.

Observation of Mesoscale Divergence

Figure 2-5 shows a portion of a Doppler radar PPI. Straight lines and arches show the position relative to the radar site. Small rectangles in the polar coordinate system are the *pixels* that show the average radial speed in corresponding places in the atmosphere. The radial speed is shown by color in each pixel. Maximum approaching speed is in the blue area, up to 19 m s^{-1}. Maximum receding speed is 6.1 m s^{-1} (red area). When the approaching and receding extrema lie along a radius as in Fig. 2-5, stretching in the radial direction and, consequently, divergence in the wind field are indicated. Such localized divergence appears in downbursts of cold air under thunderstorms. Detection of the downburst is of great significance for air traffic.

Critical readers will notice that the radial component of wind is not sufficient to determine divergence. The contraction in the tangential component may compensate the radial stretching. If this is the case, there is no net divergence. However, the likelihood of this occurring is negligible. It would require a hyperbolic

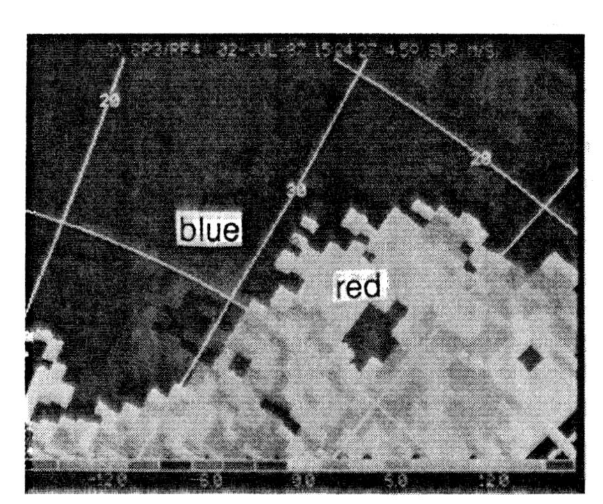

FIGURE 2-6. Doppler-radar echo, showing the radial speed in a rotating storm. The arches show distances of 18 and 20 km from the radar. The blue field shows approaching flow, toward the radar. Red shows receding flow, away from the radar. These two flows are parts of the closed circulation in a thunderstorm. Photo courtesy J. Wilson of NCAR.

wind field that is not a stable flow configuration on thunderstorm scale.

Rotating Thunderstorm

A somewhat similar situation is shown in Fig. 2-6. Again, two prominent centers of radial motion can be recognized: an approaching center in blue and a receding center in red. The increment of speed between colors in this example is 6 m s^{-1}. Now the arrangement between centers is tangential. Such echoes indicate close to circular rotation in the flow. The outgoing and incoming parts are clearly represented, and we presume that the tangential flow between the centers completes the circulation. A hyperbolic field, with the same radial components, is again an unlikely configuration since it is not stable, but a circular flow on the thunderstorm scale is a stable possibility. The atmosphere assumes stable flow patterns whenever the conditions are favorable.

Spectral Width

The return echo of Doppler radar also gives information about the *spectral width* of the returning frequencies in the echo. There is not one frequency shift that is sensed by the radar. Instead, the radial speed is determined as a prevalent or typical frequency shift within the sampling volume, from different radial velocities of scatterers within one sampling volume. Large spectral width may mean strong turbulence or large wind shear within the sampling volume. When the range of returning frequency shifts is excessive, the radial speed cannot even be determined. Such cases of indeterminate speed are shown by black pixels in Figs. 2-5 and 2-6. The few black pixels in Fig. 2-6 indicate places of enhanced turbulence or convection. The small group of black pixels in the center of the rotating thunderstorm in Fig. 2-6 indicates that a tornado may exist in the sampling volume. The variation of radial speed (relative to the radar) from one side of the tornado to the other side is too large for a typical average speed to be determined. The tornadic echo can be distinguished from a turbulent echo by its association with the rotating thunderstorm.

Echo Intensity on Doppler Radar

Compared to older incoherent radar, Doppler radar has much greater power and sensitivity. This gives it the capability to sense echoes not only from rain, hail, and snow, but also from insects, dust, and small turbulent eddies. Echoes of 10–15 dBZ can be obtained from optically clear air if turbulence or convection exist. The reflecting scatterers are the eddies with minute variations of density and refraction index. In this way, Doppler radar is useful for measurement of air currents in clear weather.

Methods have been devised to measure wind, vertical wind shear (speed and direction), and divergence. Vertical motion can be obtained by integration of divergence. Temperature gradient can be estimated from the vertical wind shear. Clear air measurements may be less available in some cooler nights when convection and insect activity are lower.

Example of Clear-Air Echoes

Figure 2-7 shows a thin line of enhanced echo going from the distance of 35 km to 50 km east of the radar. This echo comes from eddies in the air with variable density on an outflow boundary (local front) that moved from the north. The information on outflow boundaries is of great significance to the air traffic around airports. Doppler-radar information renders the prediction of outflow boundaries feasible. The movement of such lines can be predicted by extrapolation for about half an hour in advance. Another advantage of detecting outflow boundaries is the prediction of new thunderstorm cells. Such cells develop most often exactly along outflow boundaries.

For a further study of radar, the books by Doviak and Zrnić (1983) and by Rinehart (1991) are recom-

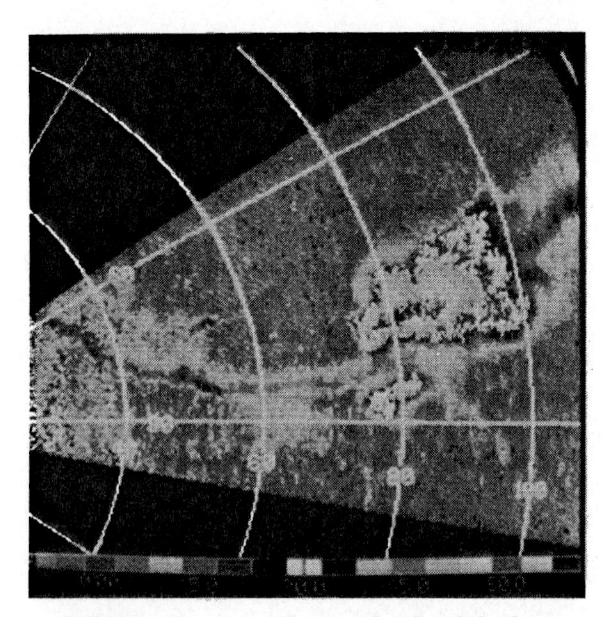

FIGURE 2-7. An example of reflectivity in a case of outflow boundary in optically clear air. This boundary caused the reflectivity band stretching radially east of the radar. The circles are at distances of 40, 60, 80, and 100 km from the radar. Photo courtesy of J. Wilson of NCAR.

mended. Rinehart's book is on an introductory level. The text by Doviak and Zrnić is written for specialists in this field.

2-7 PROFILER

Among remote-sensing instruments, very useful observations are coming from the *profiler*. This instrument, a special type of Doppler radar designed to observe the wind, provides a wind sounding from 0.5 to about 20 km elevation. Doppler radar senses reflection from irregularities in the refraction index. Contrary to other radar types, the reflected rays from rain hinder the wind observations with the profiler.

The largest part of the profiler is the antenna that consists of a horizontal net of crisscrossing wires about 0.5 m above the ground covering a field of about 12 m per side. The radar electronics are housed in a nearby shelter. The hardware of the profiler is very robust. The whole installation can be placed on a remote island where it can work for months without attention. Little animals (birds) cannot damage it, nor is the profiler harmful to the animals. Deep snow, over the level of the antenna, may cause some problem, but this is rare.

The antenna of the profiler emits radar pulses vertically (90°) and at an angle (75°). The same antenna senses the reflected signals. The slanted signal is sent through the same fixed antenna, using a minute time difference between emission of pulses from one end of the antenna to the other (and proportionally in between). Due to interference, a plane wave forms that moves in the desired direction. In this way, different directions of the beam are achieved without moving the antenna.

The slanted beam is released in two normal directions: north and east. The radial velocity along the slanted beam (75°) has a horizontal component, the observed wind.

Dispersion of echoes on turbulent eddies causes uncertainty in determination of wind. The wind in each eddy is precisely measured, but the air flow in eddies is not representative of the weather systems. It is considered that only the hourly average wind from the profiler is sufficiently representative for cyclone-scale weather systems. This information can be obtained every hour, which by far exceeds the usual 12-h intervals between radiosonde releases. The accuracy of hourly average wind from the profiler is about 1 m s^{-1}, a higher accuracy for cyclone-scale flow than normally obtained from the radiosonde.

Vertical motion is also precisely determined by the profiler, but the individual values are so variable and unrepresentative that it is believed that only 8-h averages will be of use in weather analysis.

So far 30 profilers have been deployed in the central United States. Locations were selected between the existing radiosonde stations forming a denser net of observing points.

2-8 SATELLITE OBSERVATIONS

Radiation sensors on the earth's artificial satellites provide useful information about several meteorological elements and processes. Most use is made of observations of clouds and temperature in the atmosphere and about temperature and color of the earth's surface. Current images in various wavelengths are available through standard distribution channels and should be used whenever possible in weather analysis.

Satellite images come mainly in four spectral regions: visible, infrared, water vapor channel, and radar. The images made in the visible, infrared, and water vapor parts of spectrum are routinely available on the circuits of weather data distribution.

Very informative further reading on the use of satellites in meteorology can be found in the compendium on this subject published by the National Weather Association (Parke, 1986).

Observed Weather Elements

Positioning of cloud systems is the major benefit obtained from satellite images. Many details of the atmospheric structure can be obtained from the clouds: fronts, air masses, rain areas, for example. There are severe limitations in determining those features. "Traditional" observation methods (as used at weather stations before satellites were available) give many elements much better than satellites. Nonsatellite observations, however, are restricted in space; they are available only in the points of observation. On the other hand, the geographical coverage of various features is obtained much better from satellites, although the precision may be not satisfactory. A combination of various methods gives information superior to any single set of observations.

Observation of clouds gives important information about air flow. The wind is being evaluated from cloud motion, especially over wide regions of the world where there are no direct observations of upper-level wind. The temperature of clouds gives information about the level at which the wind is evaluated. This information is routinely used in the analysis of upper-level charts.

Lifting of air can be inferred from cloud development. Since clouds form in updrafts, the formation of clouds is a fair indicator of lifting of air. On the other hand, existence (as opposed to formation) of clouds is less correlated with lifting. Often canopies of ice clouds (cirrus) cover wide regions with little or no lifting. There may be localized lifting under these canopies, but this is difficult to observe.

Sinking of air (also called *subsidence*) can also be inferred from satellite images by dissipation of clouds. Several processes in the atmosphere are characterized by sinking air. For example, the cold front often brings clear air with subsidence. Therefore, the position of the front often can be determined by noticing the edge of clear areas on satellite images.

Positions of Satellites

Most satellite observations come from the *geostationary operational environmental satellites* (GOES). These are also called *geosynchronous satellites* or *synchronous meteorological satellites*, SMS. They are positioned at the elevation of about 36,000 km above the equator. There is an equilibrium place in the earth's gravity field at that level, where the centrifugal force of the earth's rotation exactly balances the gravitational attraction. As the name implies, these satellites are stationary relative to the earth. A rather complete image of the earth ("full disk") is obtained from these satellites. Only the polar regions, latitude over 80°, cannot be seen from these satellites. Also, due to a small angle of view, high latitudes over about 60° are poorly represented. It is considered adequate to maintain five geostationary satellites, evenly spaced above the equator, to have a good image of the greatest part of the earth. Images from the satellite are obtained and broadcast about every half hour.

Successive images from geosynchronous satellites can be viewed rapidly one after the other, thereby giving good information about the motion of clouds. Methods have been developed that evaluate the wind for the places where the cloud motion is observed. The level of the cloud is determined by the cloud temperature. Wind observations thus obtained are used on weather charts, together with other observations. Current U.S. practice is to plot satellite wind observations with location indicated by an asterisk (Appendix B).

Polar regions can be seen better by images from the *polar orbiting satellites* or *polar orbiters*. These circle around the earth at an elevation of 700–1000 km, in orbits that pass close to the poles. The satellite is maintained at this elevation if it circles the earth once in about 100 min. Due to the rotation of the earth under a polar orbiting satellite, each point on the earth passes under the satellite orbit twice a day. The field of view from the polar orbiting satellite is sufficiently wide ($\approx$ 1800 km) to give an infrared image of each place on the earth twice daily. Polar orbiters do not give successive images that can be used for evaluation of wind (from cloud motion).

"Visible" Images

The rays in the visible part of the spectrum come from the tops of clouds, or from the earth's surface if clouds are absent. These rays consist of reflected sun radiation, thus this radiation can be used only in daylight. Cloud systems can be clearly seen in these images, since most surfaces on the earth are darker than clouds. Snow surfaces have the same color as clouds. Snow can normally be distinguished from clouds by the pattern of the mountains with snow and valleys without snow.

Infrared Images

Sensing of infrared (IR) radiation from various atmospheric layers and other bodies reveals the temperature of radiating bodies (Wien's law of displacement). This radiation is of about the same intensity by day and night.

The information about cloud-top temperature is a very good indicator of the cloud-top elevation. The correlation between cloud-top temperature and height is very good. Higher cloud tops are colder since they are in colder layers of the atmosphere. The information about the height of the cloud top is very useful, especially since the cloud tops cannot be observed from the surface. The weather at the surface is significantly different in the presence of tall clouds than under shallow clouds with low tops. Also, there is a fair correlation between the cloud-top height and other processes in the clouds, especially the intensity of precipitation.

Satellite observations are commonly displayed in two ways: the gray scale image and the enhanced image. The *gray scale*, or *unenhanced, image* is arranged such that the cold surfaces are bright and the warm surfaces are dark. This imitates the bright cloud tops of clouds and dark warm of the earth's surface.

The *enhanced image* gives more precise information about the horizontal extent of very cold cloud tops of high clouds. The brightness of areas in the image is selected according to *enhancement curves*. Several enhancement curves, together with examples, are described in Appendix F.

Most meteorological enhancement curves are designed to detect protruding tops of thunderstorms since these tops indicate the position of severe weather. The enhancement also helps to determine the intensity and duration of precipitation.

Low clouds (stratus) and fog are also of considerable interest. For this reason, enhancement in the low temperature range is desirable as well. Some types of enhancement are used for images designed to observe the temperature of the ocean surface. Such images are of interest to numerical modelers who need correct values of temperature at the earth's surface. Numerous other enhancement curves are used for various nonmeteorological uses.

Water Vapor Radiation

Of special interest is the radiation near 6.7 μm (4.48×10^{13} Hz) that is selectively absorbed and emitted by water vapor. The sensor that receives this radiation gives fair information about large-scale distribution of water vapor in the atmosphere without clouds. If the relative humidity is higher than about 50 percent in the whole troposphere, the radiation of 6.7 mm comes from the layers centered around 400 mb. Otherwise, when the humidity is low, this radiation comes from the earth's surface. Since the layers near 400 mb are cold, this radiation can be easily distinguished from the warm radiation of the earth's surface. Cold radiation is represented by the bright surface on the water vapor satellite image; warm radiation is shown by dark areas. Therefore, this image is a very good tool to distinguish dry and humid air masses.

There is some ambiguity with the water vapor observation. The clouds (as "black bodies") also radiate at the frequency of water vapor (6.7 μm), together with all other frequencies. Therefore the clouds may sometimes be confused with a humid layer of air since they both may radiate at the same temperature. This ambiguity is resolved when the images of the 6.7 μm are compared with images in other spectral ranges. Also, the contours of clouds are rather sharp and show small patterns. The water vapor is usually smoothly distributed.

An example of water vapor imagery is shown in Fig. 2-8. The smooth bright surfaces show humid air currents. Bright areas show radiation that comes from high tropospheric humid layers or from clouds.

Dry atmosphere lets the radiation of 6.7 μm radiate from the earth's warm surface. Such radiation is represented by dark areas. The example in Fig. 2-8 shows two vortices spiraling with bright and dark bands. These vortices are two extratropical cyclones, one over North America, the other over the Atlantic ocean. The cold front over Alabama can be recognized by the sharp transition from the humid region to the dark region where the humidity is low.

The bright regions in Fig. 2-8 are of two distinct kinds: smooth bands of water vapor radiation and lumpy spots that reveal cloud tops. Smooth water vapor bands mostly show radiation that comes from clear air with deep humid layers.

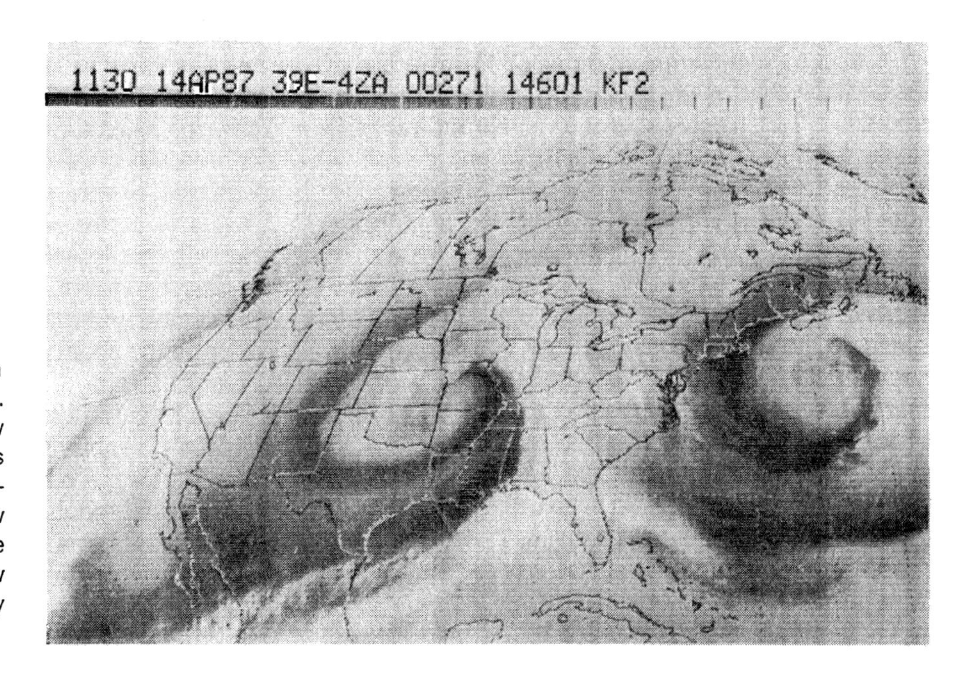

FIGURE 2-8. A satellite image in the water vapor 6.7-μm channel. The smooth bright surfaces show cold radiation coming from layers of high humidity around the 400-mb level. The dark surfaces show warm radiation originating at the earth's surface in regions of low atmospheric humidity. The lumpy bright spots are clouds.

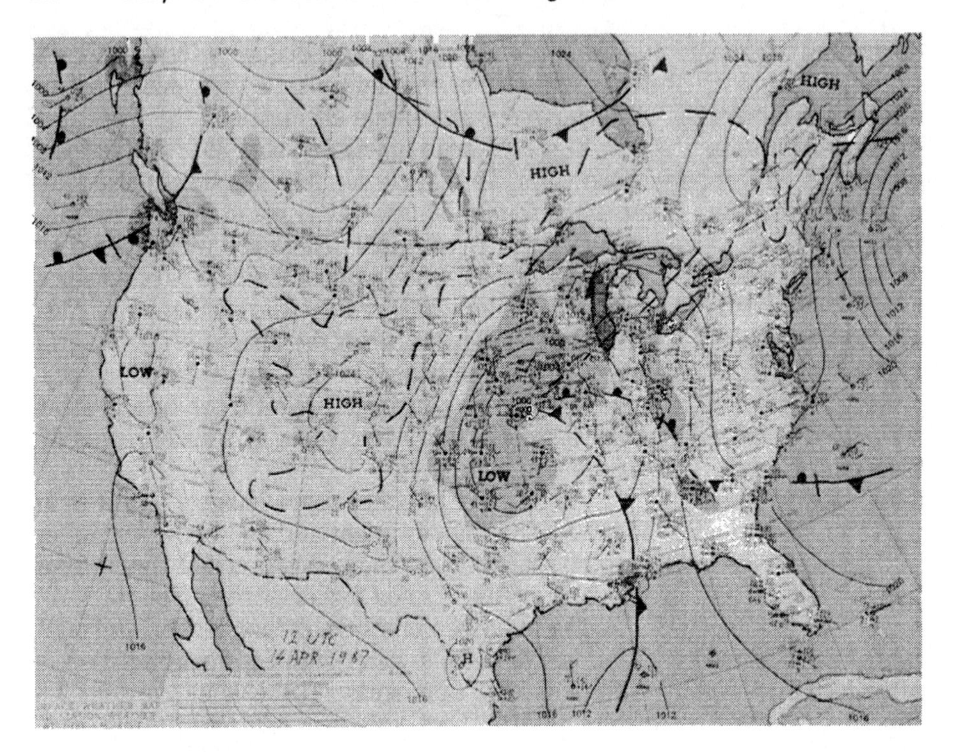

FIGURE 2-9. The surface weather chart for 1200 UTC 14 April 1987 at the time of the satellite image in Fig. 2-8. The precipitation region winds around the extratropical cyclone over the American Midwest like the humid and cloudy region in Fig. 2-8. Only part of the cyclone over the Atlantic is in the domain of this chart. From the Daily Weather Maps, published by the National Meteorological Center.

The surface weather chart for the time of Fig. 2-8 is shown in Fig. 2-9. A general correspondence of air currents and areas with precipitation (shaded) with bands in the satellite image is noticeable. The vortex over the Atlantic is only partly shown on this weather chart.

The spotty distribution of rain in Fig. 2-9 is due to an uneven distribution of clouds. The water vapor, shown by smooth bands in Fig. 2-8, is distributed much more evenly. The bright water vapor area forms elegant spiraling stripes.

A prominent feature on the satellite image and on the weather chart is the cold front over several southern states. This front is shown as an abrupt transition between bright clouds over Alabama and dark band from the western Gulf of Mexico and Missouri. This front is located in eastern Mississippi in Fig. 2-9. The discrepancy in the front's location in Figs. 2-8 and 2-9 can be attributed to the scarcity of surface observation stations. Exact location of the front could not be determined on the basis of available observations. The analyst did not have access to the satellite image when the chart was analyzed. Otherwise, the position of the front would have been corrected to better agree with the edge of the clearing in Alabama. It will be seen later that the clearing in Alabama may represent an upper front in the split-front system. There is no indication, however, of the surface cold front different from the cloud line in this case.

The dry air behind the cold front (the dark band in the satellite image) is of special interest for the dynamics of cyclogenesis. There is evidence that such dry air originates in the stratosphere and transports high potential vorticity into the cyclone. This topic is discussed in courses on dynamics of weather systems.

Microwave Radiation

A wealth of information is coming from observations in the "centimeter wave" range of the spectrum (*microwaves*, also called *radar range*, 5 to 40 × 10^9 Hz). Small amounts of these waves appear in the far infrared domain of natural radiation. Those are *background radar waves*. Since it does not involve production of electromagnetic radiation, the instrument that senses natural radar waves is also known as *passive radar*. Contrary to this, radar that emits and receives radar waves is *active radar*. The intensity of background radar waves is insignificant for all energy conversions in the atmosphere. The passive radar, however, senses the natural microwave radiation well. This gives useful information about properties of many radiating bodies, especially rain.

The earth's surface is practically a blackbody for all IR waves, but it is not a blackbody for the waves in the microwave range of the spectrum. Radar waves are transmitted by most natural surfaces on land. Therefore these waves also are not emitted from land surfaces, according to Kirchhoff's law. Microwaves have the property of being mostly absorbed and emitted from water drops of the size of raindrops and are almost unaffected by small (cloud) drops. These rays penetrate clouds, but they are absorbed on raindrops and therefore also emitted from them. In contrast with raindrops, the ocean surface absorbs and emits the microwaves only partially. This results in weak radar radiation from the ocean surface, as if this surface was very cold (perhaps near 150 K). In this way, the observations of microwaves are useful for estimation of rain over tropical oceans. If the radiating surface appears relatively warm, we may conclude that the microwaves come from the raindrops above the ocean surface. If the radiating surface over a tropical ocean appears cold, then these waves originate from the ocean. The actual evaluation of rain intensity is more complicated, since each "field of view" of the satellite sensor may contain radiation simultaneously from rain clouds and from the ocean surface.

An active radar mounted on the satellite (SASS, Seasat-A Satellite Scatterometer) was used in the 1970s with success (Black et al., 1985). It gave information about surface wind speed over the oceans, based on scattering of radar waves on capillary waves and spraying drops on breaking ocean waves. More instruments like this are under construction and data from them will be available in the future.

Significant improvement of satellite technology is planned for the years 1992-1995. New GOES (designated I through M) will be launched, as well as several NOAA polar orbiters (designated K, L, and M). All these should provide improvement in measuring the surface temperature, vertical profiles of temperature and dew point, and wind evaluation from cloud motion.

Sounding by Satellite Observation

While the clear air is practically transparent for the radiation in the visible part of the spectrum (short waves), there are numerous bands in the IR part of the spectrum in which the radiation is absorbed and therefore emitted (Kirchhoff's law). Water vapor and carbon dioxide are the most important gases that participate in the absorption and emission of IR radiation in the atmosphere. The optical depth of the atmosphere is different for various frequencies, since these frequencies are radiated by various gases. Measurement of radiation and application of the laws of radiation make it possible to compute the vertical distribution of temperature. In this way we obtain fairly good soundings of the atmosphere from the satellites. The most useful data for such soundings come from the instrument VAS (VISSR atmospheric sounder; VISSR stands for visible and infrared spin-scan radiometer) that is mounted on GOES.

The soundings obtained by VAS are very smooth in comparison with radiosonde observations. Details, as inversions, cannot be found with VAS. However, for the average temperature of 2 to 3 km thick layers, an accuracy of about 1 K can be achieved with such soundings. Since these soundings will be coming from wide regions with no other observations, their significance will be great for weather analysis.

2-9 OTHER INSTRUMENTS

A variety of new instruments that will greatly influence the procedures in weather analysis are being developed. All members of the meteorological profession will need to read the current journals and other literature to stay abreast with progress in engineering and science. A few of the instruments are described next.

Lightning Detection

Engineering methods exist that make it possible to detect every lightning stroke in a wide area. The disturbances in radio waves are *sferics*. Information about sferics is available worldwide, based on several sensing stations on the globe.

More accurate, but limited in distance, is the *direction finding technique*, which senses the magnetic disturbances at two or more sensing stations. Sensing is limited to about 400 km from the observing stations, but gives very good detail. It is possible to track thunderstorms or group of thunderstorms by these means from hour to hour. An example of application of magnetic sensing of lightning is described by Orville et al. (1983).

Work is underway to bring lightning data to the practicing meteorologists in a practical way.

Lidar

Laser rays can be used for remote sensing of the atmosphere much in the way as radar rays. This is accomplished by *lidar*, an instrument similar to radar but operating on frequencies of the order of 10^{15} Hz. The useful range of frequencies includes also the "visible" rays, that is, light. This explains the name of the instrument: It comes from light detection and ranging (similar to radar: radio detection and ranging). The laser beam consists of rays of precisely the same frequency. Thus the radiation energy is not spread out over a wide spectrum band.

The reflection of laser rays can be observed from targets in the atmosphere, as can be clouds and layers of dust, water vapor, ozone, and so forth. Vertical distribution of water vapor density is also measurable by lidar.

Lidar can be mounted on a satellite. It is expected that global fields of various lidar echoes will be available, using this technology. Use of lidar in the atmosphere also is known as LASA, *lidar atmospheric sounder and altimeter* (NASA, 1987).

Thermodynamic Profiler

Similar to satellite sounding (VAS), atmospheric radiation can be observed from the earth's surface by a thermodynamic profiler. The temperature of various layers in the atmosphere can be determined from radiation observations. The underlying mathematical theory is similar to the evaluation of soundings from satellite observations, but the possibility of moving the instrument over the whole world as on the satellite is not available here. When this instrument becomes operational, it will give observations only from points where it is installed.

Sodar

The sodar is geometrically similar to radar, but it works with acoustic (sonic) waves. It receives echoes from layers of the atmosphere above the instrument; therefore, the sound is emitted vertically. The echoes are received from inversions and other layers where stability varies. Best results are obtained in the presence of turbulent layers. Valuable observations are received in the lowest few kilometers. Sodar complements the thermodynamic profiler, adding details in the lowest kilometer where the profiler data are unreliable.

3

Patterns of Atmospheric Circulation

Weather charts are the most widespread tools for studying and presenting atmospheric processes. This chapter is an introduction to the patterns in isopleths and other lines used on weather charts. These patterns form the most common basis for the study of physical processes. We pay most attention to the geopotential of isobaric surfaces and thereby to the wind. The geopotential and wind are related by the geostrophic approximation. As an extension of wind analysis, the evaluation of trajectories is presented at the end of this chapter.

3-1 VARIABLES AND COORDINATES

Meteorological Variables

The data for weather charts come from several main observational categories:

- visual observations (for example, clouds)
- directly sensed measurements (where the sensor of the instrument is at the place of observation)
- remote sensing (radar, satellite, profiler)
- computed quantities (for example, vorticity, stability)

These observations yield a number of variables that are usually defined throughout the atmosphere, even if they are observed only at observing stations. Observed variables are the *meteorological elements*, or *weather elements*, since they constitute the weather.

There is some overlap between "observed" and "computed" variables. Not all "observed" variables are directly observed. Some evaluation or conversion is usually needed to obtain the reported value. For example, sea-level pressure and the height of isobaric surfaces are generally considered to be observed variables, although they are calculated using the hydrostatic equation. Some evaluation is also done with other observed variables, most often an automatic conversion of electric signals to the value of the reported variable. "Computed" variables have been obtained by analysts away from the observing stations. Most computed variables are evaluated as derivatives of observed variables.

Some weather elements, such as clouds and various weather phenomena (squalls, lightning, halo), cannot be treated quantitatively as continuous mathematical functions. These elements cannot be analyzed by standard computerized construction of isopleths easily. However, meteorologists must take them into account during the analysis, if possible.

Weather Charts

Most weather charts are *synoptic*, that is, they contain simultaneous weather data from a selected time. The charts are usually horizontal or isobaric, but vertical and time sections are also often used. On all charts, an important part of meteorological analysis consists of interpolation of variables away from observation points. The interpolation is often performed by manual drawing (with much erasing!) of isopleths on synoptic charts.

Routine work on interpolation is performed by computers. However, drawn lines do not represent a physical analysis. Meteorologists must apply their knowledge of weather processes to check and to improve the computer products and to introduce the physical models on the charts. Given enough time, a meteorologist is supposed to produce a better chart than any computer.

Map Projections

Geographic maps that represent meteorological variables on flat paper are usually *conformal.* They correctly represent angles and shapes of small areas. This is advantageous, since small distances in different directions that are equal on the chart are also equal on the earth's surface. For the purpose of automation, a short theory of conformal projections is presented in Appendix G. We should notice that the distance scales, printed on charts, are correct only for short distances at designated latitudes.

All horizontal coordinates suffer from the deformation of images on flat geographic charts. Therefore, spherical coordinates (latitude ϕ, longitude λ) are often used, especially in numerical weather prediction models. Spherical coordinates are always used to report the observed wind direction.

The coordinate lines of the Cartesian coordinate systems cannot coincide with geographical parallels over large distances because the geographical parallels on these charts bend differently than on the sphere. The Cartesian and spherical (geographical) coordinates coincide fairly well with each other only locally, in conveniently small regions.

The deviation of the earth's shape from a sphere is very small and is always ignored in weather analysis. Sometimes the designations x and y are used for horizontal coordinates in the spherical system. These are the curved coordinates (dimensioned as distance) exactly along the longitude λ and the latitude ϕ, respectively. Such horizontal spherical coordinates coincide well with Cartesian coordinates x and y over small distances. Next, the spherical x and y will be indicated by the subscript s, thus these will be x_s and y_s. The relationships between spherical distance coordinates x_s, y_s and angular spherical coordinates λ, ϕ are:

$$dx_s = a \cos \phi \, d\lambda$$

$$dy_s = a \, d\phi$$

where a is the radius of the earth ($a = 6.366 \times 10^6$ m). Transformation to and from spherical coordinates is shown in Appendix G. The wind direction (including sense) is observed and reported in spherical coordinates.

For various practical reasons, nonconformal charts are often used. The charts that represent the whole globe are usually nonconformal; they deform the shape of geographical features and special care is needed for measuring distances, areas, and angles. The scales are different for various directions on nonconformal charts.

Time is equal in all coordinate systems. The time axis is formally perpendicular to other coordinates, so functions varying only in space do not have derivatives in time.

The vertical spherical coordinate is the same as in any of the common Cartesian coordinate systems, as described next.

Vertical Coordinate

Commonly used coordinate systems employ one of the following four vertical coordinates: *height* (z), *pressure* (p), *sigma* (σ), or *potential temperature* (θ). The systems are named by the letters used for the vertical coordinates. The x- and y-coordinates (or corresponding spherical coordinates) in these systems are often called *horizontal,* even if they slant from the point of view of other coordinate systems. Correct expression for the x- and y-coordinates is *horizontal* for the z-system, *isobaric* for the p-system, *isosigma* for the σ-system, and *isentropic* for the θ-system.

Rigorously speaking, the horizontal part of the three-dimensional wind is different in various coordinate systems, as can be seen from a geometrical representation. On the other hand, the angle between the "horizontal" directions in various coordinate systems is regularly very small. Therefore it is common not to transform the horizontal part of the wind when the transition is evaluated from one to another coordinate system. In other words, the wind (horizontal part) is considered equal in all coordinate systems.

Horizontal derivatives are very sensitive to transformation of coordinate systems. The transformation is done with the equation

$$\frac{\partial a}{\partial x_2} = \frac{\partial a}{\partial x_1} + \frac{\partial a}{\partial z_1}\frac{\partial z_1}{\partial x_2}$$

where the subscripts 1 and 2 show the two coordinate systems.

The vertical component of motion is normally greatly different in various coordinate systems. Therefore this component requires recomputation when the transition is made from one to another coordinate system.

The right-handed Cartesian z-system (x, y, z, t) is considered (by students) the simplest for use with equations. In this coordinate system, z points up locally. It is often assumed that x points east, but this is not always the case. Frequently, we use x downwind or in some other horizontal direction.

The p-system is most often used in meteorology. This system is left-handed, since pressure increases opposite of increasing height z. However, the rule for the cross product of vectors is modified in this system so that the equations in the z- and p-systems look very much alike. The word "up" in the p-system is used to designate the direction of decreasing pressure. In this way the directions "up," and similarly "down," are approximately equal in the z- and p-systems.

There are formal difficulties in using equations at or near the earth's uneven surface. Therefore, terrain-following coordinate systems are often used. In these systems the vertical coordinate is often designated *sigma* (σ), and sometimes s. For some sigma systems the vertical coordinate increases up and for some systems it increases down, depending on whether sigma is calculated from z or p. Sigma systems are used in models of numerical weather prediction. For presentation of weather charts it is common to convert the results of numerical prediction to the p-system. Sigma coordinates are greatly preferred for models that use spectral methods for the integration of equations. All other coordinates are very cumbersome in spectral methods, since these other coordinates are not entirely continuous; they have regions with undefined variables where the lower coordinate surfaces (of non-sigma systems) intersect the earth's surface.

The use of potential temperature as a vertical coordinate (θ-system) has several advantages. The θ-system gives a better resolution of stable layers than other systems and the vertical motion in this system vanishes under adiabatic conditions. Since the logarithm of potential temperature is proportional to entropy [equation (C-40), Appendix C], the charts of the θ = const. surfaces are also called *isentropic charts*. Examples of isentropic charts are shown in Section 6-3.

3-2 GRAPHICAL MANUAL ANALYSIS

Analysis: The Process

The process of weather map analysis leads toward a synthesis of observations and models into physical systems. Bergeron (Godske et al., 1957, p. 651) commented on this process as follows:

> After careful consideration of their representativeness and reliability, all available meteorological data must be fitted into the most probable system of ideal and modified three-dimensional tropospheric models. The word "analysis" now in general use is, thus, to a certain extent misleading. What is aimed at is, in fact, more a synthesis of the weather situation than a mere analysis; the word diagnosis, also often applied, is consequently more appropriate.

This can be interpreted as a technical rule that the weather charts should contain the models of atmospheric processes. The models may be altered, as the observations demand, but they will nevertheless be identified. Using models and all technical instructions for analysis leads toward a physical concept of the processes in the atmosphere.

In light of the above, it can be seen that the process of weather analysis is entirely different from the process of isopleth drawing on weather charts. Drawing of lines is an important first step in the work, but it is mainly technical and it can be done by people who have no knowledge of meteorology. Even computers can do it with considerable success. On the other hand, the process of analysis consists of identifying and marking atmospheric processes, major currents, physical properties of air, and conceptual models that give physical insight into weather development. To be successful in analysis, one must be highly trained in meteorology.

Analysis: The Product

The term *analysis* should be used only for such charts that contain marked conceptual models of atmospheric processes. Papers that contain only isopleths should be called *charts*. There is an exception to this rule: The isopleths prepared by computers are commonly called *analysis*. It should be kept in mind that computers do not really provide a meteorological analysis; they only provide the lines. It is better to use the word *chart* for sets of isopleths and computer products. The term *analysis* should be used for a product that shows conceptual models.

Computerized processing of weather data greatly reduces the manual task of drawing isopleths. However, a meteorologist cannot passively depend on computer output. Instead, he or she must master both drawing lines and improving automatically drawn weather charts. A meteorologist is expected to prepare charts better than the computer does, even if not as fast. For example, computer output is known for the absence of discontinuities (fronts), poor resolution of mesoscale processes, and other shortcomings. Therefore, a professional meteorologist must know the classical methods of analysis, including those developed by the Norwegian and Chicago schools and numerous other investigators who advanced our knowledge of the detailed structure of the atmosphere. Thus, studying graphical manual drawing of weather charts is not obsolete in the era of automation.

Analysis in Practice

The order of topics in this book does not reflect the order in which the charts should be analyzed. Analysis progresses best when the analyst works on several charts concurrently. As soon as part of an analysis on one chart is entered, the analyst is advised to work on other charts as well, so that he or she can compare them and arrive at a unified three-dimensional concept of the atmosphere. If several analysts collaborate, charts should be exchanged among the team members and compared. During this process more data and auxiliary charts, such as satellite images, soundings, and vertical sections, may be needed. If possible, such additional data and charts should be prepared before the analysis is complete. Automatic processing of weather data by computers (drawing lines) makes it practical to prepare those additional charts.

The principal tools for analysis are colored pencils and erasers that can be used on paper. Transparent acetate sheets with appropriate markers and erasers are useful as well. These tools are also used when automated isopleths are available. During analysis, lines always need adjusting; this justifies the use of erasers. Computer products with isopleths are regularly in need of analysis and must be marked with symbols for physical processes.

3-3 ISOPLETHS OF SCALAR FIELDS

Construction of isopleths constitutes graphical interpolation. The pattern of isopleths geometrically represents a surface that presumably fits all observations. *Observations* here means "values in observation points."

Manually Prepared Isopleths

Most isopleths, or isolines, are constructed by computers. However, for the following reasons, students have to acquire the skill of drawing isopleths by hand.

- —Often a new, nonstandard chart has to be prepared.
- —The automatic isopleths from the computer may miss a smaller-scale weather pattern, requiring this pattern to be added by hand.
- —The computer may erroneously introduce a shape in the isopleths that we recognize as spurious, requiring the analyst to remedy the situation.
- —The graphical device may break down, requiring preparation of charts by hand.
- —Meteorologists are expected to improve the computer graphics routines, and they cannot do this if they are not familiar with techniques of analysis.

A worker who can design and improve computerized graphics routines must know the models of atmospheric phenomena. Other professionals, such as mathematicians or programmers, do not have enough expertise to handle the weather data in agreement with meteorological concepts.

Properties of Isopleths

The isopleths of scalar fields connect points with selected values. They do not intersect other isopleths and they do not show waves shorter than twice the distance between nearest data points. The sketched and corrected isopleths in Fig. 3-1 show probable and improbable forms of isobars. The sketched isobars in Fig. 3-1a show many short waves. The wavelength that predominates is $L = 2D$, where D is a typical distance between nearest-neighbor stations. Although it is possible that such short waves appear in nature, they are very unlikely. Therefore, the short waves

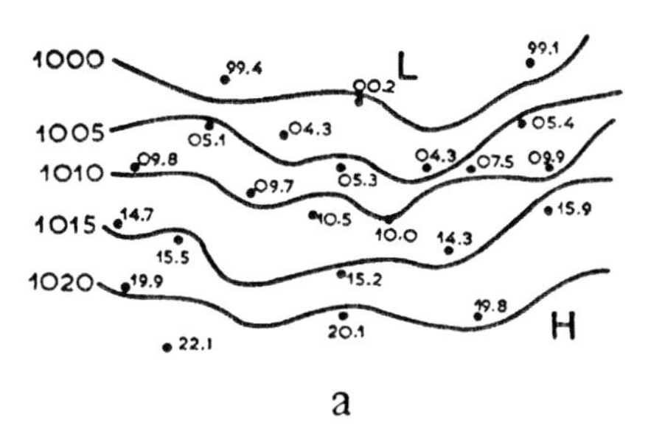

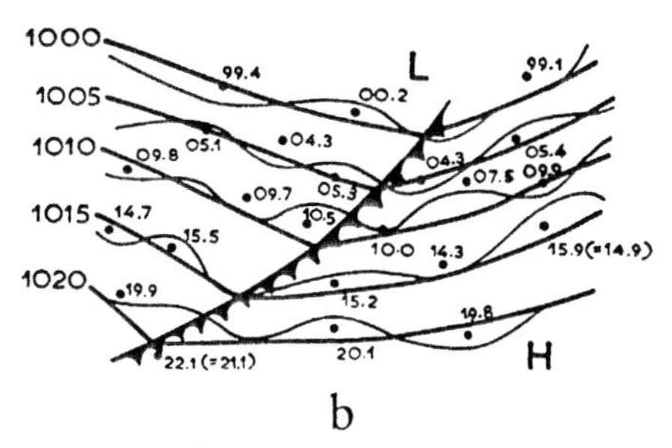

FIGURE 3-1. Improbable (a) and probable (b) forms of isobars. From Godske et al. (1957), p. 670, by permission of the American Meteorological Society.

should be eliminated from the final analysis whenever there is no physical justification for their existence.

Similar to the short waves, spacing of isopleths is sometimes improbable. This occurs when groups of two isopleths appear together, in pairs. In such cases the distance is wider between neighboring pairs of isopleths. This pattern must be recognized and eliminated. An example of such a pair of isobars is given in Fig. 3-1a. There the sketched isobars of 1005 and 1010 mb are closer to each other than to other adjacent isobars.

Short waves with length of $2D$ often appear on weather charts that are constructed by unsophisticated numerical routines. In this respect, there is a similarity between humans and machines: They both tend to produce short waves.

Smooth Isopleths versus Models

Drawing of isopleths is a graphical analog to the mathematical procedure of fitting polynomials through several data points. In this procedure we usually do not know whether a smaller structure exists between the data points; thus it is advisable to follow the rule that smooth isopleths represent the most likely case among all possibilities. Therefore, in the absence of contrary evidence, we assume that the variables are smooth between the data points. Deviation from smooth isopleths is introduced only when there is a good reason to place a known structure in the pattern. These structures are the conceptual models of weather systems: fronts, jet streams, cyclones, thunderstorms, and so forth.

An example of a structure introduced in the isobars is shown in Fig. 3-1b, where the corrected set of isobars contains a sharp trough with a cold front. The plotted data do not show the front. The information about the front in this place came from the analysis of other weather elements not shown here. The kinks are introduced in the isopleths because it is known from dynamic meteorology that isobars kink on the front. The front in Fig. 3-1b is drawn following the rules outlined in the discussion of the Margules equation (8-3). Kinks on isopleths are considered short-wave patterns, since a Fourier analysis of kinked lines reveals prominent short waves.

Names of Isopleths

Isopleths of various variables have been used so much that they obtained names such as:

isotherms, for temperature;
isentropes, for potential temperature;
isodrosotherms, for dew point;
isobars, for pressure;
isallobars, for pressure tendency;
isotachs, for wind speed; and
contours, (also *isohypses*) for the height of isobaric surfaces and for thickness.

Three- and Four-Dimensional Representation

Three-dimensional equivalents to isopleths are the equiscalar surfaces. These can be represented "three-dimensionally" on paper in perspective view, which may introduce serious misunderstanding. Therefore three-dimensional views are seldom used in manual analysis. To the contrary, computerized representation of three-dimensional variables is very successful, especially when the point of view can be changed. Four-dimensional representation shows variations in space and in time. This is displayed as an animated three-dimensional representation.

While it is of great advantage to have a good representation of all needed variables, it should not be forgotten that a complete analysis should contain physical models, not only the isopleths.

Typical Patterns

Scalar isopleths often show typical patterns. Several of these patterns, especially those referring to isobars or contours, have names. If these names are mentioned without specific reference to a variable, then it is assumed that the pressure or height of the isobaric surface is meant. Some of these names are as follows.

> *High:* a maximum in the scalar field, usually shown by at least one closed isopleth, with higher values inside.
>
> *Low:* a minimum in the scalar field, usually shown by at least one closed isopleth, with lower values inside. (A high, or a low, is not necessarily enclosed with an isopleth, if the isopleths are chosen with large intervals.)
>
> *Trough line:* a line that connects points with relative maximum of curvature on neighboring isopleths such that the isopleths bend around lower values of the scalar.
>
> *Ridge line:* a line that connects points with relative maximum of curvature on neighboring isopleths, but here the isopleths bend around higher values of the scalar.
>
> *Trough:* a region around the trough line.
>
> *Sharp trough:* a trough with kinked isobars or contours.
>
> *Ridge:* the region around the ridge line. The width of the ridge or of the trough is not mathematically defined. It is taken that the ridge (or the trough) is narrower than the space between the inflection points on either side of the ridge or the trough.
>
> *Col:* the hyperbolic point between a pair of lows and a pair of highs, also called "saddle point." In rare cases an isopleth may intersect itself in a col.
>
> *Wave:* A sinusoidal shape. Usually the space between two adjacent ridges is designated as a wave, such that the trough is in the middle of the wave.

Details in Scalar Fields

The patterns of typical isopleths are illustrated in Fig. 3-2. The locations of highs and lows are indicated by a circled X, with an appropriate letter above it (H or L) and with the central value under it. Other styles of indicating the centers are also used. For instance, if there is not enough space, just a letter L or H is acceptable; an example is the low off the New England coast in Fig. 3-2. The computers will also find highs and lows in the regions of sparse contours ("weak gradient"); such examples are the two highs over Canada in Fig. 3-2. As soon as the computer senses that a grid point has a higher (or lower) value than its immediate neighbors, it prints the letter H (or L), even if the analyst may attribute no particular significance to this point.

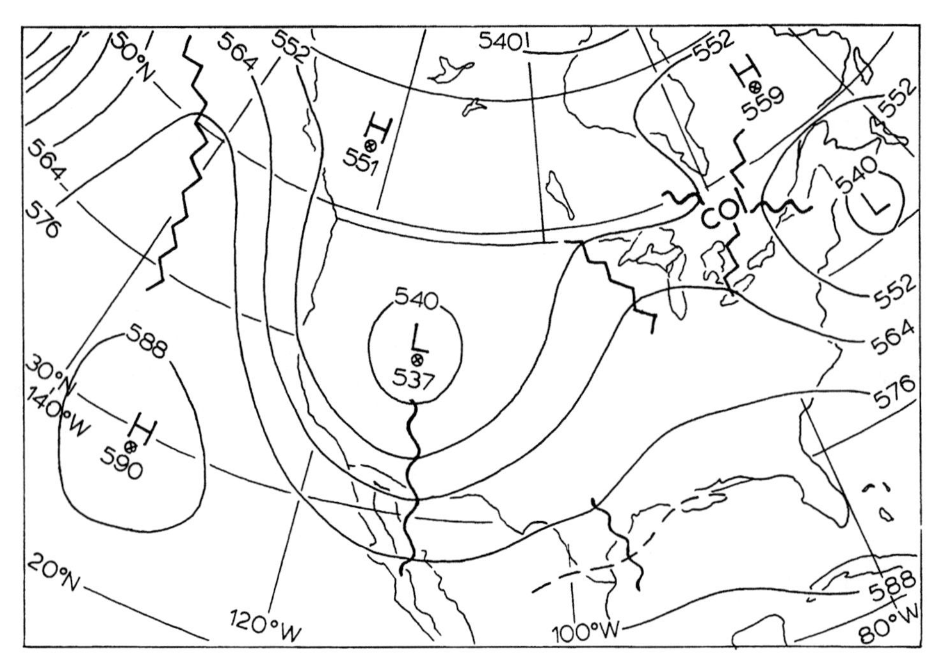

FIGURE 3-2. Typical isopleth patterns shown by 500-mb contours, intervals of 12 gp dam, for 0000 UTC 26 April 1984. Troughs and ridges are shown by wavy and kinked lines, respectively. An intermediate contour (dashed) is drawn near the Texas coast to enhance a small trough in that region.

Sometimes local weather may depend on the existence of weak troughs, for example, the one over the Texas coast in Fig. 3-2. For enhancement of important details, it is advisable to draw additional intermediate isopleths in such regions. Such additional isopleths should be drawn in a different style, perhaps with a thinner line or by using dashed lines of the same color. A different style is needed so that the computation or estimation of the gradient is not confused.

The Wave

Repetitious phenomena are usually called *waves*. Sometimes the repetition or periodicity is not easily seen. For example, an outbreak of cold air is called *cold wave* mainly due to its alternation with *heat wave*.

A wave often comes in a sinusoidal form, in space or in time. Mathematically, it is not defined how one wave should be determined in the sinusoidal pattern: for example, from the trough to the next trough, or from the inflection point to the next similar inflection point. The preferred meteorological practice is to use the segment between two adjacent ridges so that the trough is in the middle of the wave. A weather analyst's attention is usually turned more to troughs than to ridges. An approaching trough in the westerlies is a sign that the region of interest is active and that weather changes can be expected. The approach of the ridge is normally less eventful. For this reason, "wave" is often used synonymously with "trough."

A wave in tropical easterlies often contains a cyclonic eddy in the trough. Figure 3-3 shows isobars in the tropical Atlantic. The 1012-mb isobar shows several waves in the easterlies; the troughs on these waves are shown by wavy lines. The two eastern troughs have developed vortices. Each of the closed isobars in these vortices shows a tropical depression.

Not all troughs in the easterlies develop into vortices. Satellite images and measurements of cloud motion are of great help for determining whether vortices are present. Still, it is often not possible to confirm the existence of small eddies such as the tropical depressions in Fig. 3-3. If the eddies are not confirmed, sinusoidal contours are drawn. Troughs in the easterlies attract the attention of meteorologists since some troughs develop into tropical cyclones.

Relation between Wind and Contours

It is common to use the expressions *air flow* or *wind direction* when the contours direction of isobaric surfaces is considered, since the wind blows closely in the direction of contours. This is a consequence of the important property of wind to be approximately equal to the geostrophic wind. Mathematically, this rule is expressed as

$$\mathbf{V} \approx \frac{g}{f}\mathbf{k} \times \nabla z \tag{3-1}$$

or, in the form of two scalar components,

$$u \approx -\frac{g}{f}\frac{\partial z}{\partial y} \qquad v \approx \frac{g}{f}\frac{\partial z}{\partial x} \tag{3-2}$$

This property of wind is discussed in all introductory books on general and dynamic meteorology and in several places in this text. The sense of wind is the familiar low on the left, high on the right, when looking downwind in the Northern Hemisphere; it is opposite in the Southern Hemisphere. This relation between contours and wind is the *Buys-Ballot rule*.

When only the wind speed is needed, the above equations reduce to

$$V = \frac{g}{f}\left|\frac{\partial z}{\partial n}\right| \tag{3-3}$$

where V is the wind speed and n is the coordinate in the horizontal direction normal to the streamline.

The above approximate equations help to draw the contours on a weather chart. The rule is that in the

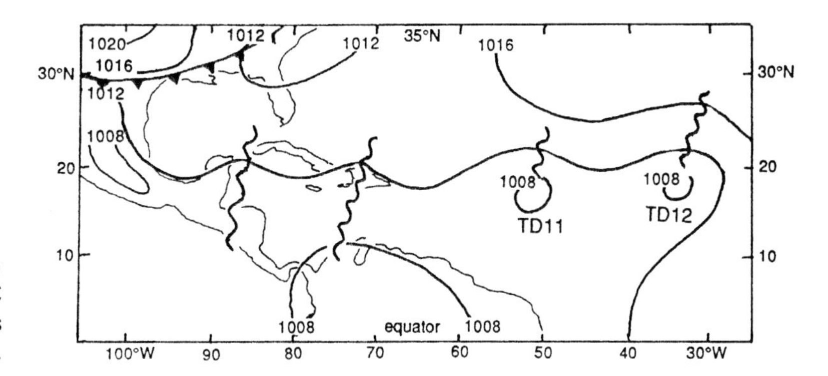

FIGURE 3-3. Sea-level isobars in the tropical Atlantic for 0000 UTC 23 September 1990. Wavy lines mark the troughs in the easterlies.

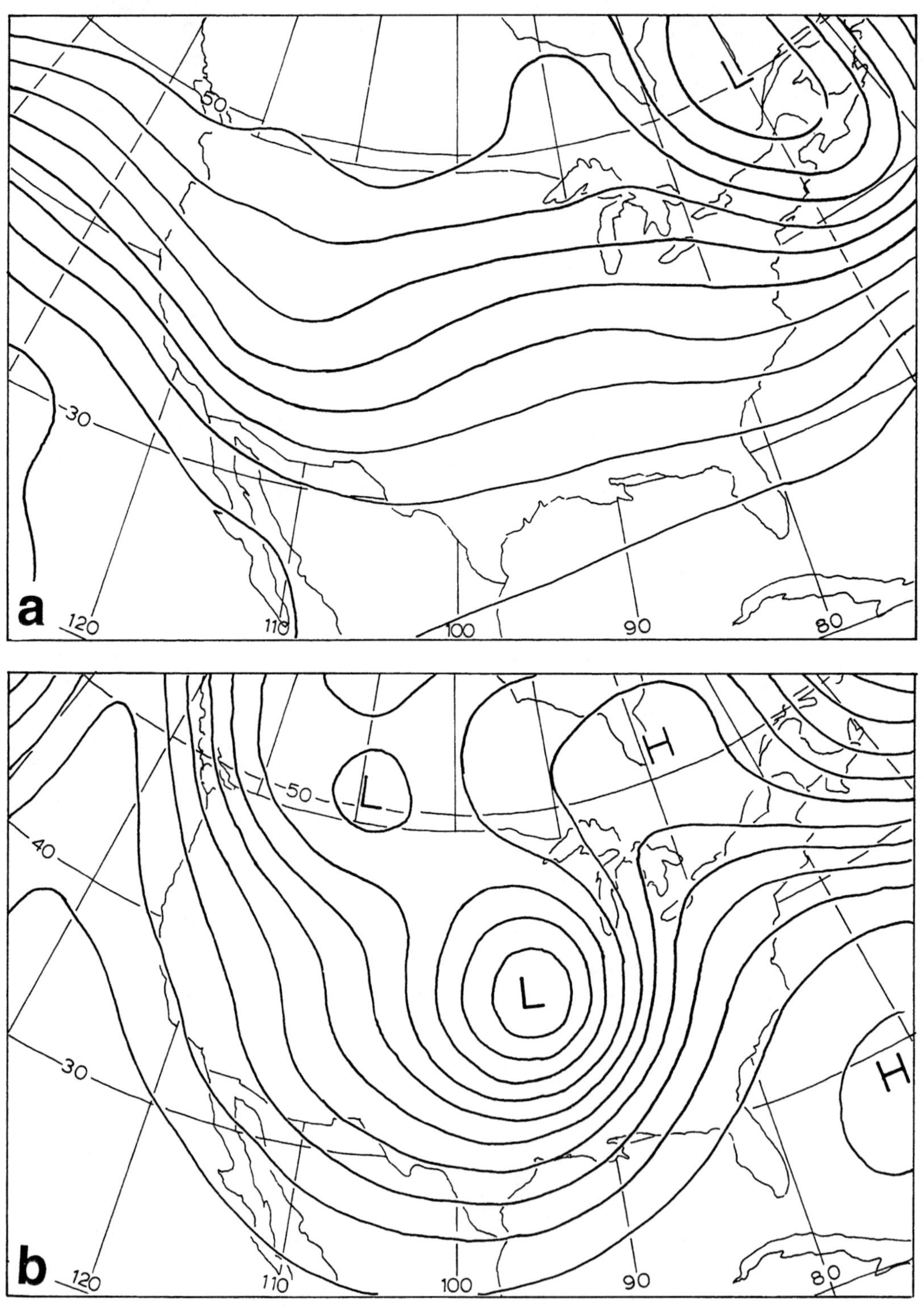

FIGURE 3-4. Examples of 500-mb contours, intervals of 60 gp dam. (a) A zonal weather situation, 1200 UTC 18 November 1983. (b) A meridional weather situation, 1200 UTC 28 November 1983.

absence of better evidence, the contours should lie approximately along the wind direction.

Jet Stream

Contours normally show elongated zones of higher concentration. If the wind (or geostrophic wind) in such a zone exceeds about 30 m s^{-1} in the upper troposphere or about 15 m s^{-1} in the lowest 2–3 km, this may well be a *jet stream*. The polar jet stream can be seen over the Pacific Ocean, along the west coast of North America in Fig. 3-2, where the contours of 552, 564, and 576 gp dam are closely spaced in a long stretch.

Zonal and Meridional Flows

Zonal patterns (westerlies and easterlies) are prominent in the climatological (time) averages of pressure, height, and temperature. The average pattern is easy to comprehend and describe since it does not have many details. Therefore, the average (zonal) pattern of isobars, contours, and isotherms is considered "normal" and all other patterns are named "disturbances." There is, of course, no natural reason for the disturbances to be something abnormal; this is only a matter of accepted terminology.

The contours on upper-level charts show more zonal configuration on some days than on others; then we talk about the *zonal weather situation* (Fig. 3-4a). A different situation occurs on days when there are large disturbances in the westerlies with prominent meridional flow (along the meridians) over large sections of synoptic charts. These are cases of a *meridional weather situation* (Fig. 3-4b). There is no strict distinction between zonal and meridional weather situations, since a variety of intermediate cases occur. For this reason we often cannot decide on the categorization. In such cases when the flow is not quite zonal and, at the same time, not as meridional as in some prominent cases, it is advisable to avoid categorization into zonal or meridional. These expressions should be used only in prominent cases, when there is little doubt about the type.

Transitions between zonal and meridional weather situations occur perhaps once every 5–20 days. Hope has been expressed many times that a regularity in these transitions would help us in forecasting beyond 3 to 5 days. Unfortunately, such a regularity has not been detected.

Blocking

The meridional situation often coincides with the *blocking* situation (Fig. 3-5). The flow is zonal over North America and the western North Atlantic. However, the zonal flow is obstructed (blocked) by a pair of centers, a high and a low, over the eastern North Atlantic. This pair constitutes the *block* for the zonal flow, hence the name *blocking* for such a weather situation. The high over the North Atlantic, centered south of Iceland, is the *blocking high*; the low off the Portuguese coast is the *blocking low*. In the space between the blocking high and blocking low is a significant area with an easterly wind; this is also typical for blocking.

Cut-off Vortices

A situation related to blocking is the appearance of *cut-off vortex*, that is, a low south of the jet stream and highs north of it. "Regular" (not cut-off) vortices are the ones in their usual positions: lows north of the polar jet and highs south of the jet. However, over time, the wave amplitude in the zonal jet stream usually increases. This is due to baroclinic development, since the strong jet stream of the middle latitudes usually is baroclinically unstable. When the amplitude of the waves increases, some ridges separate into closed anticyclonic eddies. These eddies form highs north of the jet stream. Similarly, some troughs separate into cyclonic eddies that form closed lows south of the jet stream. Such separated eddies are *cut-off highs* and *cut-off lows*.

An example of the formation of a cut-off low is shown in Fig. 3-6. At 0000 UTC 20 February 1984 (Fig. 3-6a), a large amplitude of the 500-mb contours can be seen over North America. A *deep trough* (one that extends far south, still in the westerlies) can be seen from the Great Lakes through New Mexico to Baja California. The greatest concentration of contours shows the location of the polar jet stream from British Columbia to Montana and Wisconsin, then turning south to California, going over Texas, and back north to New England. The flow near the west coast of North America shows a meridional situation, since the meridional flow is about as prominent as the zonal flow.

On the next day (Fig. 3-6b), the fast flow of the polar jet stream established itself from western Canada to the Great Lakes and New England. This new position of the jet stream may be designated as zonal. The trough moved to Texas and now contains a low center with a closed contour; this is a cut-off low. The important element in this designation is that the main jet stream is located north of the cut-off low.

A somewhat similar situation is shown in Fig. 3-7. Again there is a closed contour around a low, this time centered over Kansas. However, the polar jet is

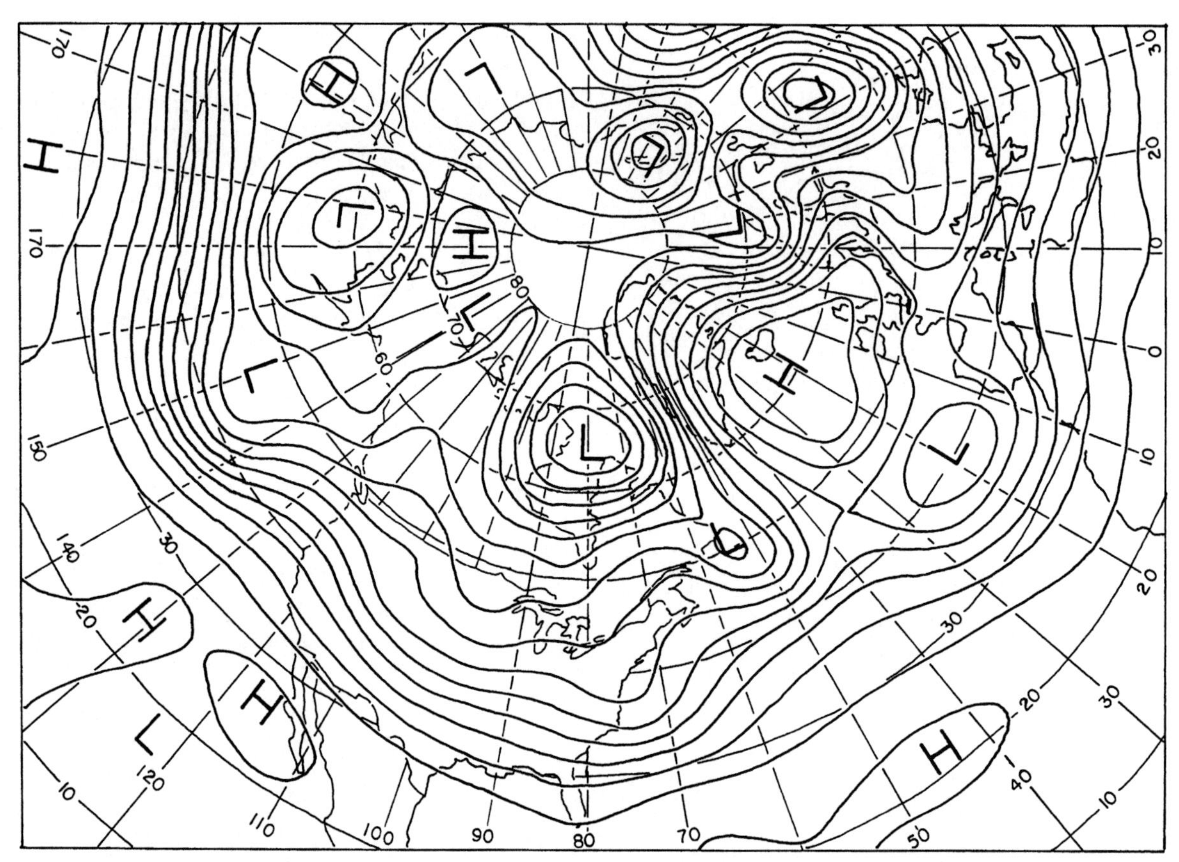

FIGURE 3-5. Blocking over the eastern North Atlantic, shown by 500-mb contours, intervals of 60 gp dam, at 1200 UTC 14 November 1983.

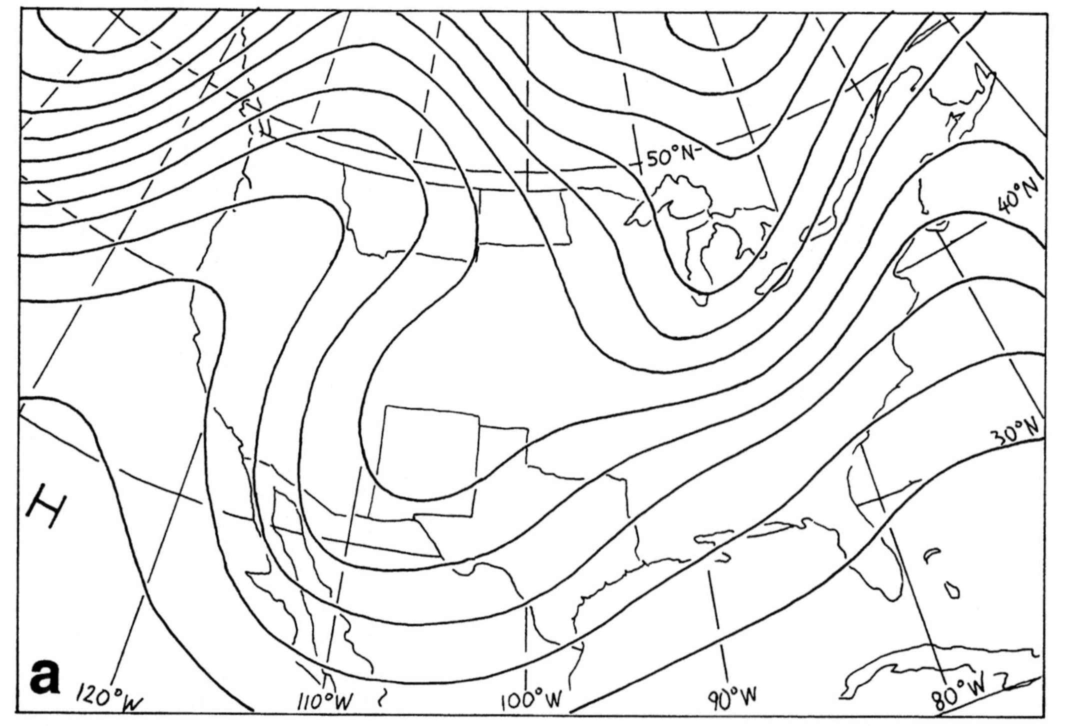

FIGURE 3-6. Formation of a cut-off low, illustrated by contours on subsequent days. (a) 0000 UTC 20 February 1984. (b) 0000 UTC 21 February 1984.

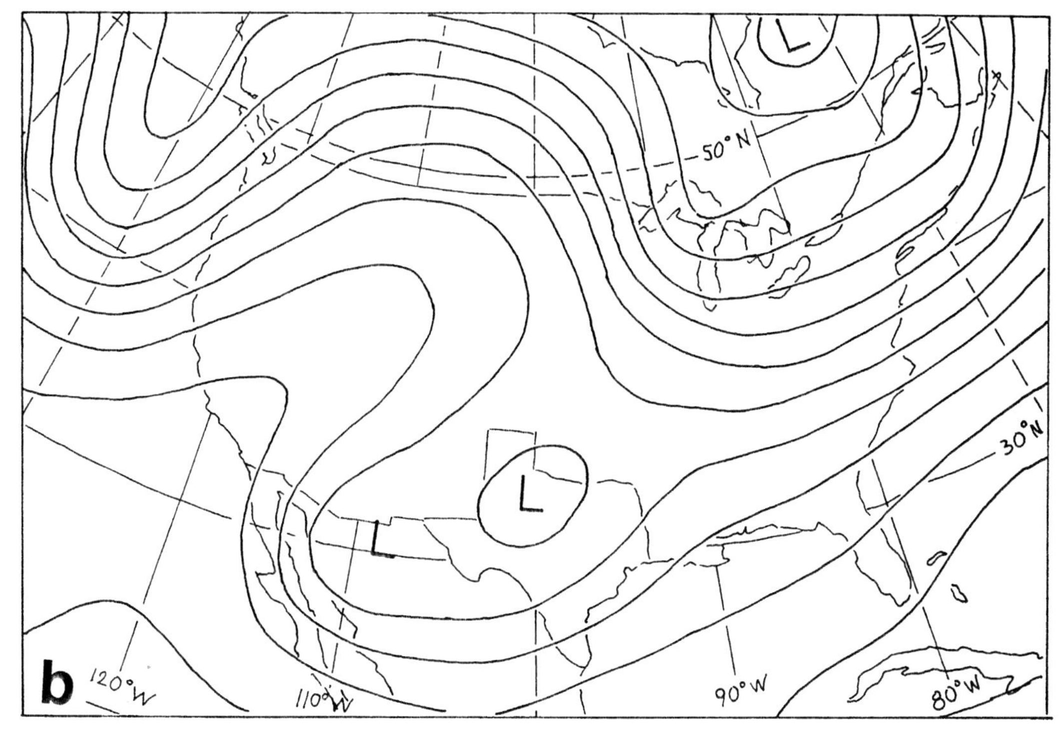

FIGURE 3-6. Cont.

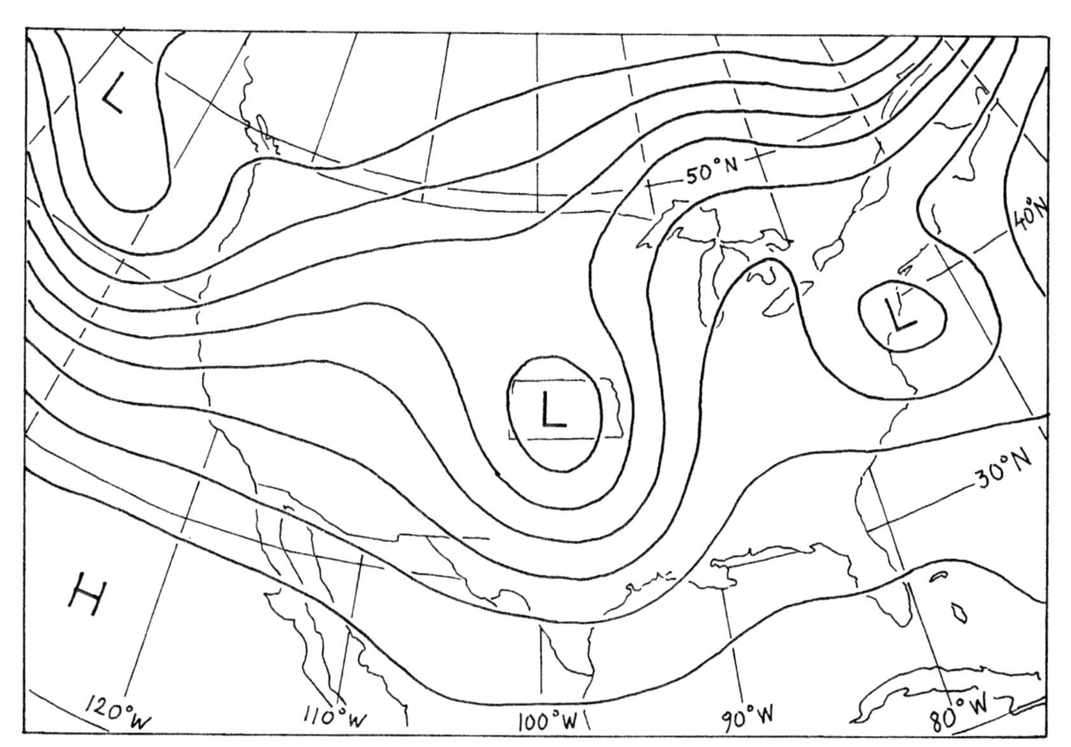

FIGURE 3-7. Examples of lows of middle latitudes: a "usual" low over Kansas and a cut-off low over the East Coast of North America. Solid lines are the 500-mb contours, intervals of 60 gp dam, at 0000 UTC 16 February 1984.

now located south of the low. This shows that the low is in a "usual" position on the polar side of the jet; therefore, this low is not a cut-off low. This is a usual low imbedded in the westerlies. The other low on the same chart, centered over the East Coast of North America, is a cut-off low since the polar jet is north of it.

3-4 NUMERICAL ANALYSIS

Weather maps are routinely produced by computers. A short description of the numerical process is given in this section so that the advantages and limitations of automatically produced isopleths can be appreciated. The grid-point representation is described next. The spectral method is described in a separate subsection.

Numerical analysis is often called *objective analysis*. This term is not appropriate, since it implies that manual analysis is not objective. Each type (numerical and manual) has both shortcomings and advantages. Educated meteorologists should take advantage of all available methods. Numerical analysis is produced in a much shorter time than manual analysis, while manual analysis offers an insight into the physical processes in the atmosphere. A practical way is to use lines prepared by numerical analysis and to supply the symbols for physical processes manually.

Grid-Point Representation

Weather maps exist as arrays of numbers in computer memory. Each element (geopotential height, temperature, a wind component, and so forth) is represented by values in regularly spaced *grid points*. The arrays of grid-point values are suitable for automatic computation of gradients and other derived quantities. Each such array is a *numerical analysis*.

Usually we do not read the numbers that constitute a numerical analysis; there are too many numbers for efficient comprehension. Instead, a graphical display of isopleths is preferred. The arrays that represent the variables in grid points lend themselves to standard graphics routines that produce isopleths. The isopleths can be viewed on computer terminals or can be printed on paper. Such automatically produced isopleths also are called *numerical analysis*.

Evaluation of Grid-Point Values

The values in grid points are obtained by interpolation between the values in the nearest observation points. Observed values in a circle around a grid point (domain of dependence) are used to find the value in that grid point. This is illustrated in Fig. 3-8, where the value of geopotential height z_G should be evaluated in the grid point G. There are several observations (station circles with wind barbs) within the circle with radius of dependence of 500 km. The value in the grid point (z_G) is then evaluated as a weighted average:

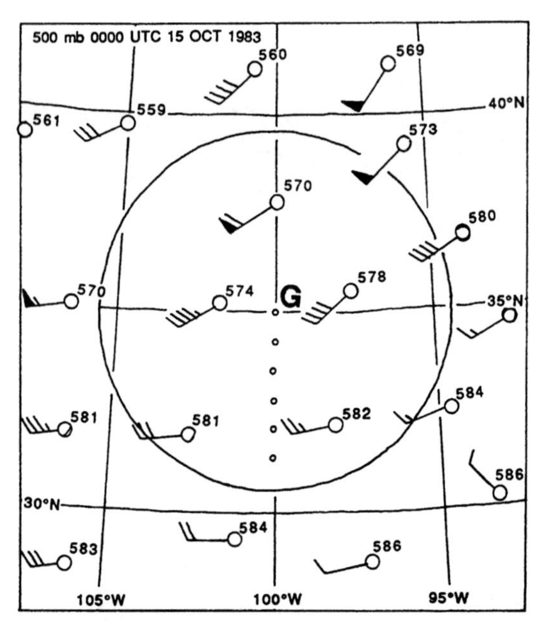

FIGURE 3-8. The domain of dependence of the value in the grid point G. The radius of dependence r = 500 km is selected. Several more grid points are shown along the meridian south of G. Other grid points come with the same intervals, but they are not drawn.

$$z_G = \frac{\sum w_i z_i}{\sum w_i} \tag{3-4}$$

where w_i are the weights allotted to the observed values z_i ("observations"). The summation is over the observations in the circle. There are five such observations in the example in Fig. 3-8. The weights are calculated from the formula

$$w_i = \exp\left(-\frac{d_i^2}{c}\right) \tag{3-5}$$

where d_i are the distances between the observation points and the grid point G, and c is a suitably determined constant that controls the effective influence of an observation on the grid-point value. As an example, the choice of $c = 1.1 \times 10^{11}$ m^2 causes the weight of about 0.1 at a distance of 500 km. Nearby observations, say closer than about 100 km, will obtain a weight in excess of 0.9. Allotting more weight to observations near the grid point imitates the manual interpolation (drawing of isopleths), when the analyst pays more attention to the nearby observations than to distant ones.

The First Guess

There are difficulties with numerical analysis, too. A grid-point value cannot be determined if there are no observations in the domain of dependence. For this reason, a numerical analysis starts with a preliminary guess of z in all grid points. This guess is usually the most recent forecast of z, valid for the time of the new observation. Then the values from (3-4) are computed one after the other and inserted into the analysis. The points that are far from all observations do not get any correction: In those points the initial guess will stay.

In the case of unavailable forecast, the value from the chart 12 h before can be used. Even the climatological average gives an acceptable first guess for numerical analysis.

Another method to handle data-sparse regions is to use a larger radius of dependence. Similarly, for greater detail, strategies have been developed that use larger and smaller radii.

Further improvement in numerical analysis is achieved when the difference between the first guess (z_{gi}) is treated as the variable that needs evaluation in the grid point. This means that the difference

$$z_i - z_{gi}$$

is used in (3-4) and (3-5) instead of the height z_i. The z_{gi} values in the observation points are obtained by interpolation of the first guess to the observation points.

Numerical analysis as described so far will yield a poor value in a grid point if the observations exist only on one side of the domain of dependence. Also, centers without observations in their middle will be underestimated. Various sophisticated methods have been developed so far to circumvent such difficulties. A detailed study of these methods is described in specialized books and scientific journals. The short example in Fig. (3-8) and in equations (3-3) and (3-4) only shows the basic idea in this process.

Use of Wind for Height Evaluation

The relation between wind and geopotential height is normally close to the geostrophic balance, as described by the approximate equations (3-1) through (3-3). Therefore the information on the wind gives another tool for calculation of height.

We may assume that in a small domain (say within the domain of influence) the contours are straight and equally spaced. Then we may consider two contours: the one through the observation point and another through the grid point. The value of height z_O in the observation point is known. The height z_G in the grid point is related to z_O by the geostrophic wind speed (3-3). Using finite differences we have

$$V \approx \frac{g}{f}\left(\frac{z_G - z_O}{\Delta n}\right)$$

where Δn is the distance between contours. This equation can be solved for z_G to give a new value of height in the grid point. Now it can be seen that other values of z_G will be obtained when another observation of wind is used, from another observation point. These values of z_G will generally be different from the value obtained by (3-4). All these values are used in the final analysis; a weighted average of the z_G values obtained from wind and from height is used.

Using of Wind Speed and Direction

A complete relation between wind and height takes care of various possibilities of geometry and wind direction. This is illustrated in Fig. 3-9.

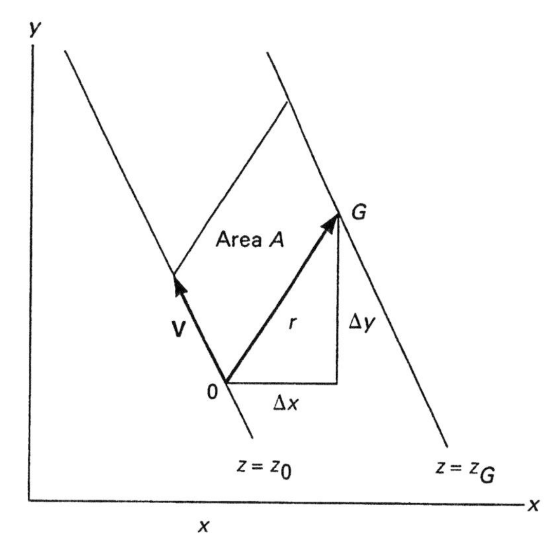

FIGURE 3-9. Evaluation of height z in the grid point G on the basis of observed height and wind in point O. Observed wind vector is **V**. Two parallel contours are drawn, through the two points G and O. The height in these two points is z_G and z_O, respectively.

The observation point is O and the grid point is G. The wind **V** in O and the radius vector **r** from O to G are

$$\mathbf{V} = \begin{pmatrix} u \\ v \\ 0 \end{pmatrix} \qquad \mathbf{r} = \begin{pmatrix} \Delta x \\ \Delta y \\ 0 \end{pmatrix}$$

The components u, v, Δx, and Δy are negative for some orientations of **V** and **r**. The angle between **V** and **r** is α. Two contours are shown, one through O, the other through G. For the depicted situation in Fig. 3-9, it is expected that

$$z_G > z_O$$

since a geostrophic balance is assumed. In other cases it may be that $z_O > z_G$, or even $z_O = z_G$, depending on the orientation of **V** and **r**.

We can also see that the difference $z_G - z_O$ is proportional to the area A of the parallelogram whose two different sides are **V** and **r**. The area A is given by any of the following expressions:

$$A = \mathbf{k} \cdot \mathbf{V} \times \mathbf{r} = u\,\Delta y - v\,\Delta x = Vr \sin\alpha \quad (3\text{-}6)$$

where V and r are the magnitudes of **V** and **r**, respectively. The area becomes formally negative for $\alpha < 0$ or $\alpha > \pi$. The reverse of sign of A shows exactly the reverse of sign of $z_G - z_O$. Now the definition of the geostrophic wind speed (3-3) can be used in finite difference form:

$$V_G = \frac{g}{f}\frac{z_G - z_O}{r \sin\alpha}$$

Solving this for $z_G - z_O$ and using (3-6) yields

$$z_G - z_O = \frac{Vfr}{g}\sin\alpha = \frac{f}{g}(u\,\Delta y - v\,\Delta x)$$

This equation yields new information about the height in the grid point G:

$$z_{iw} = z_G = z_O + \frac{f}{g}(u\,\Delta y - v\,\Delta x) \quad (3\text{-}7)$$

As mentioned in the preceding subsection, this value is also used for evaluation of z in the grid point.

Comments on Numerical Analysis

The procedures described so far are applied to all grid points on the weather chart. There are typically 10,000 points on any given level on the weather chart. As an illustration, several grid points are shown in Fig. 3-8 along the meridian south of point G, at typical intervals of 80 km. More points would clutter the figure; therefore they are not drawn. The same density of points is used also in the other direction. The work to find all grid-point values is multiplied by the number of levels needed. Contemporary models use 20 to 30 levels. The whole workload must also be multiplied by the number of weather elements that need numerical representation.

The greatest burden for computers in numerical analysis is to find which observation points are within the domain of dependence of every grid point. People see this on the first glance, but the computer has to calculate distances between all grid points and all observation points. Together with calculations in (3-4) and (3-5), we see that numerical analysis poses a great demand on computers. This is one reason why the Weather Service maintains batteries of the most powerful computers available.

Spectral Numerical Analysis

An alternative way to represent meteorological variables is by coefficients of polynomials interpolated to the variables in grid points. Each such set of coefficients is a *spectral analysis*. Spectral analysis is also displayed in the form of isopleths.

To form the isopleths, a computer uses the coefficients of the interpolating polynomials to evaluate the variables in grid points automatically and then uses a standard isopleth-producing routine. The end product is again a set of isopleths. When a set of isopleths is viewed, it cannot be recognized whether the analysis in the computer was represented spectrally or by grid-point values.

A set of grid-point values is needed to obtain the spectral coefficients. In the case of spherical harmonics, the underlying grid points are equally spaced along the geographical parallels. Spacing along the meridians is in approximately equal increments of $\sin\phi$. These are the *Gaussian points*. Such a particular arrangement of grid points is necessary for an efficient evaluation of the coefficients of spherical harmonics. Then the orthogonality rules apply, and an inversion of huge matrices is avoided.

3-5 DISCONTINUITIES

This section describes the patterns that appear on atmospheric fronts. These patterns are characterized by discontinuities on isobars and other isopleths. The physical circumstances at the fronts are described in dynamic meteorology. More detail on fronts is given in

Chapter 8 and in Appendix J. Several examples of isobars and contours on fronts are shown in this section.

Sharp Trough

Hydrostatic considerations require a sharp trough in pressure or geopotential fields along the line where the density abruptly changes. The Margules formula (Appendix J) shows a discontinuity of ∇p (or ∇z in p-coordinates) on the front, such that the isobars or contours form a kink. The pointed side of this kink is always toward the high. Such a kink does not appear on the ridge line, since that would imply an unstable stratification on the front. Statically unstable states seldom persist in the atmosphere long enough for us to observe them with standard instruments.

Practical drawing of a frontal kink is facilitated by the front usually being already in a trough. Then the preliminary isobars are already cyclonically bent and there is little doubt about how they should be kinked: The curvature of the contour should be enhanced so that a kink appears. The angle of kinking should be similar to the angle between the two wind directions on the sides of the front, recalling the approximate equality of wind and geostrophic wind. Figure 3-10 illustrates such a case at the 1012-mb isobar. This isobar was already curved cyclonically when it was first sketched, so the direction of kinking was immediately clear. Wind observations on both sides of the front also show how the isobar should kink.

Problems with Anticyclonic Curvature

The 1016 mb isobar in Fig. 3-10 is generally anticyclonically curved. This makes it difficult to construct a cyclonic kink. The drawn shape of this isobar shows an acceptable solution. An alternative way of drawing this isobar, as shown in Fig. 3-10b, cannot be accepted since the anticyclonic curvature near the front (at point A) is too strong. Dynamic meteorology shows that excessive anticyclones cannot persist in a quasi-stationary state. Therefore, it is advisable to be very cautious when drawing small anticyclones. On the other hand, there is a situation where strong anticyclonic curvature, similar to Fig. 3-10b, may appear: in or near thunderstorms. This is discussed more in Chapter 12.

For the cases where it is difficult to place a cyclonic kink on an isobar the position of the front should be checked. Inspection of the case in Fig. 3-10a reveals that the front is improperly placed. If the front is moved ahead (downwind) to the position of the dashed line in Fig. 3-10a, the difficulties with the kink disappear. Of course, we have to pay attention to other elements as well. For example, the temperature of 23°C seems high for the frontal passage at the station with pressure recorded as 155. This temperature reading may still be on the cold side of the front since the air at the leading edge of the front is not very cold. The cold air mass had traveled for a while over a warm surface until it came to this place (in Texas). During that time, the leading edge of the cold air had time to heat up. This is typical of cold fronts that travel over a warm surface. Significant local cooling after frontal passage appears only when the front moves away. By then the earth's surface has given much of its heat to the passing cold air.

Fronts on Upper Levels

Contours on upper-level fronts are usually less kinked than at the earth's surface. The higher we go in the atmosphere, so more parallel the contours and fronts are to each other. There the contours seldom intersect the front, and if they do, the kinking angle is close to 180°.

A particularly difficult situation occurs when contours intersect a front in a ridge. Then the contours are anticyclonically curved, and the cyclonic kink is difficult to find. Figure 3-11 illustrates one such case. The ridge in the middle of the chart, from northwest to

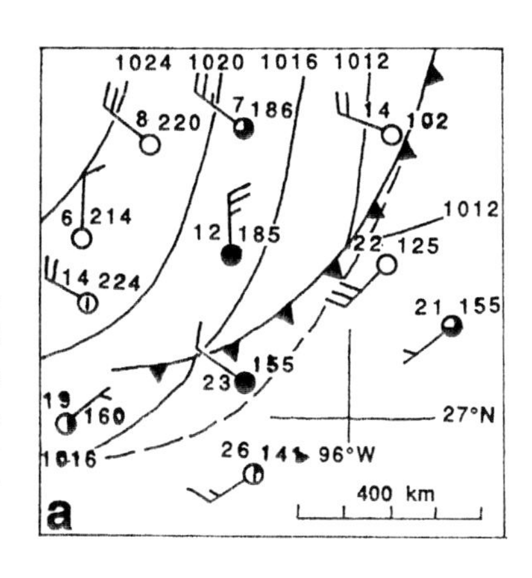

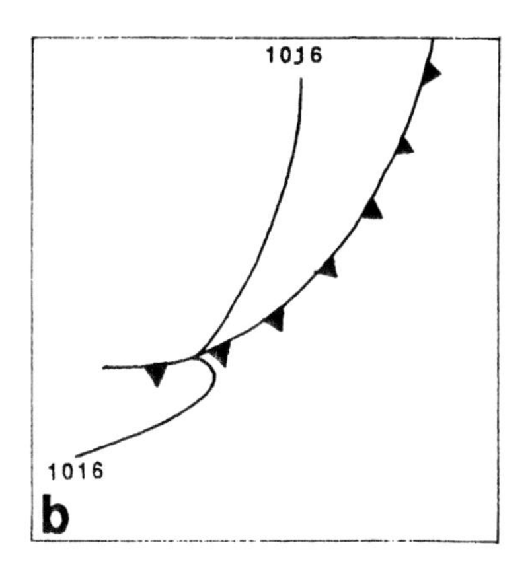

FIGURE 3-10. Examples of isobars kinking on the front (1800 UTC 26 January 1967). (a) The front with weather observations. (b) The same front, but the 1016-mb isobar has unacceptably strong anticyclonic curvature.

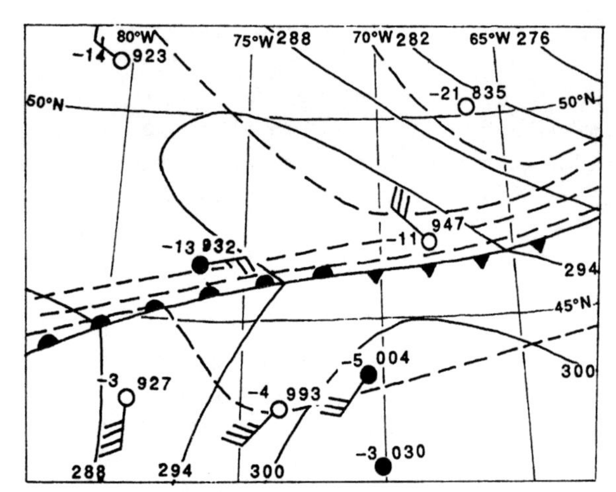

FIGURE 3-11. Contours intersecting the front in a ridge. The chart shows the contours (solid, gp dam) and isotherms (dashed, °C) at 1200 UTC 27 January 1967.

southeast, is confirmed by wind observations, following the geostrophic balance. The frontal layer intersects the isobaric surface in the zone of dense isotherms. The contours in this region would be anticyclonically curved, except at the front. Therefore, the contours show the characteristic kinks in all points where they intersect the front. Students may appreciate this situation much better if they copy the data from this figure onto a blank sheet of paper and draw the isopleths without looking at the Fig. 3-11.

3-6 STREAMLINES

Techniques of Constructing Streamlines

The wind field is often represented by streamlines, that is, by lines parallel to the wind vectors. This is especially practical in the tropics where the isobars are sparse. The technique for drawing streamlines is rather easy; the lines are drawn freehand following a quick inspection of plotted wind barbs. Analysts find it easier to draw streamlines than isobars, since the numbers do not have to be read. The basic goal is to make long lines, starting on the upstream edge of the chart and ending on the downstream edge.

The starting points of streamlines are conveniently selected, according to an analyst's judgment and need. A sufficient number of streamlines is taken to adequately represent the wind field. The purpose of a streamline chart will determine the density of streamlines. It is common to choose the density of streamlines similar to the density of isobars or contours. Contours are not drawn on the same chart with streamlines.

As with isopleths, the presented details cannot exceed what the available data offer. Wavelengths smaller than twice the distance between nearest stations cannot be correctly represented on a chart; therefore, these waves should be avoided.

Streamlines that start at the edge of a chart will not enter the regions of closed circulation within the field. In such cases new streamlines must be introduced inside the chart so that the field can be conveniently filled. In other places some streamlines will end within the boundaries of the field. This occurs in regions of convergence (that is, where $\nabla \cdot \mathbf{V} < 0$). In these regions streamlines may merge, indicating that the air disappears from the chart level. Disappearance of streamlines is also a fair indication that there is vertical motion at or near the chart level. (We can conclude only that the derivative of vertical motion is not equal to zero since $\nabla \cdot \mathbf{V} = -\partial\omega/\partial p$. However, we know that $\omega \neq 0$ near this level for the derivative $\partial\omega/\partial p$ to be different from zero).

Figure 3-12 shows streamlines drawn by hand on the basis of wind data. The lines are drawn parallel to the air flow, with occasional disregard of slow wind. It is assumed that slow wind is easily influenced by small-scale local phenomena that are not representative of the large-scale processes.

Convergence and Divergence of Streamlines

Figure 3-12 shows a weather situation with two fronts, polar and arctic, and a discontinuous wind field on each front. Frontal discontinuities are drawn by streamlines that stop or emerge at the front. It is common to draw more streamlines stopping, rather than emerging, at the front. In this way the idea is expressed that there is convergence at the front. How densely the streamlines are drawn is a subjective choice. Occasionally we emphasize faster flow by denser streamlines. The streamlines at the front show cyclonic shear in the same way as the geostrophic wind shows cyclonic shear in the contour field.

Convergence is not always discontinuous, as it is on the front. An example of smoothly convergent streamlines can be seen in the northwestern part of Fig. 3-12 where the streamlines merge into each other. This convergence is associated with an anticyclonic curvature of the flow. We know from dynamic meteorology that discontinuous anticyclonic shear cannot persist in the atmosphere. Therefore, the merging of streamlines is drawn continuously, with smooth curves.

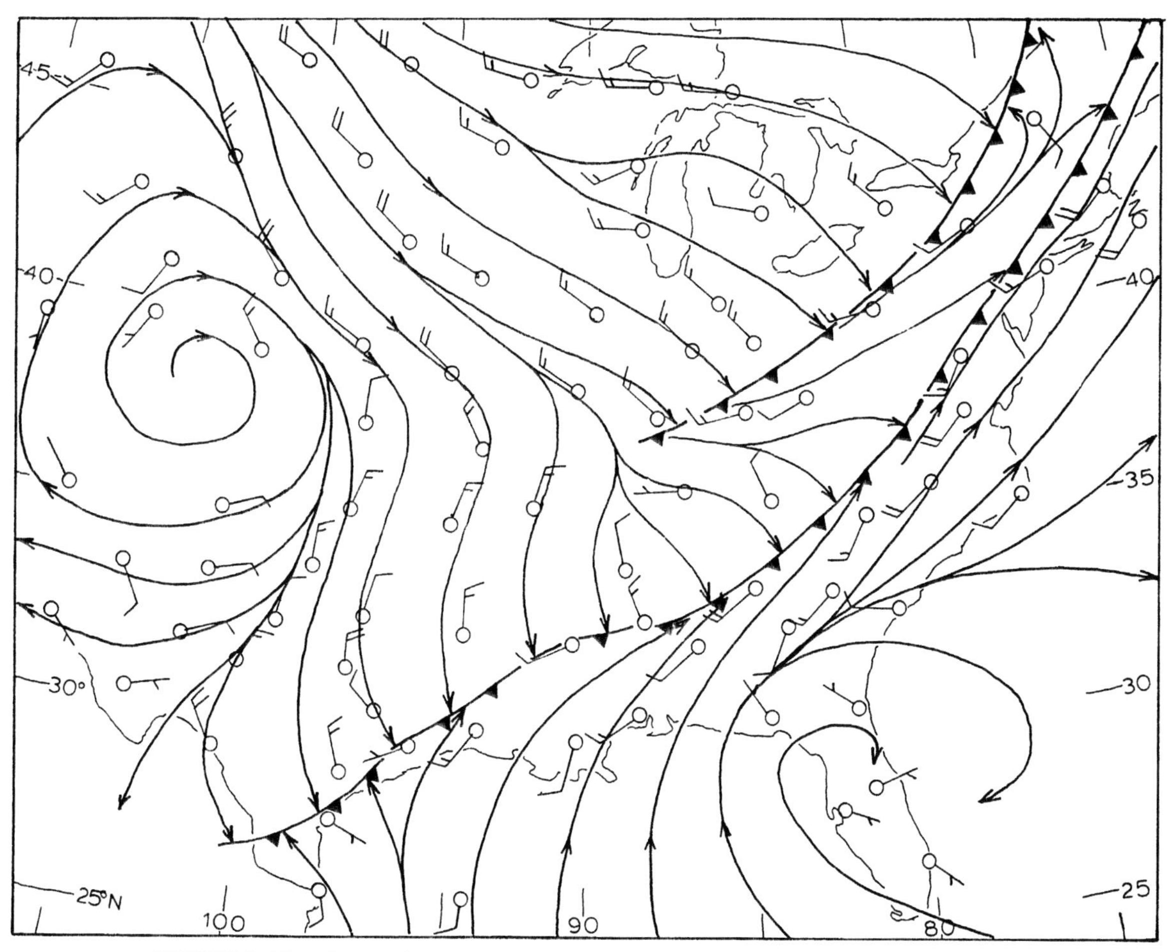

FIGURE 3-12. Streamlines at the earth's surface (1800 UTC 23 February 1988).

Streamlines are divergent in the southwestern part and over the southeastern states in Fig. 3-12. Such divergent streamlines branch out of each other or new streamlines emerge in the divergent area. Divergent streamlines are always drawn in a smooth fashion, without discontinuities as on fronts. This corresponds to our idea that the divergent flow is frontolytic, contributing to the dissolution of fronts.

Objective Construction of Streamlines

In older textbooks are detailed instructions for precise manual construction of streamlines (Saucier, 1955; Petterssen, 1956). It is doubtful that such extra effort will yield worthwhile improvements of the streamlines, especially since very good streamlines can be obtained automatically, by computers.

An example of computer-generated streamlines is shown in Fig. 3-13. Standard computer-generated streamlines do not produce discontinuities; however, several lines of convergence can be spotted where the streamlines merge. It is not difficult to change some of the streamlines manually to obtain recognized models of fronts with prominent cyclonic wind shear and convergence. For example, the streamlines that reach the Iowa–Missouri border from the southeast in Fig. 3-13 can be changed so that they continue approximately straight until they end at the front line. In this way a prominent cyclonic directional wind shear can be formed against streamlines that come from the north through Minnesota and Iowa. The final shape of streamlines can be made similar to the forms on the front shown in Fig. 3-12.

Additional Comments on Streamlines

Construction of streamlines is a very important technique in analysis of tropical weather. Due to a weak Coriolis force, the isobars (or contours on upper-level charts) are very sparse in the tropics, if judged by the middle-latitude experience. Therefore, streamlines give better information about weather systems than isobars do.

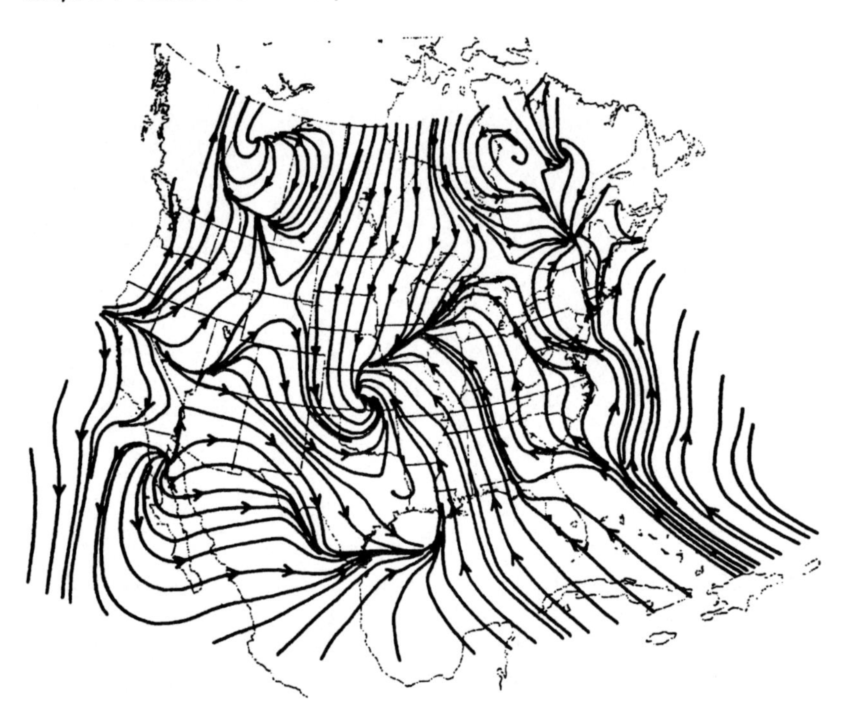

FIGURE 3-13. Computer-generated streamlines at the earth's surface at 1200 UTC 3 April 1974. From Whittaker (1977) by permission of the American Meteorological Society.

The terminology used with isobars and contours is also used on streamline charts. So, for example, a closed streamline with cyclonic curvature, or a cyclonically spiraling streamline, shows a *low*. Anticyclonically bent closed streamlines, or spiraling out of a center, show a *high*. Similarly, the *troughs* and *ridges* (also *trough* and *ridge lines*) are defined similar to such patterns on isobars or contours.

Streamlines are usually similar to the isopleths of the *nondivergent stream function*. This is a scalar function of horizontal space coordinates that exactly represents the nondivergent (rotational) part of the wind. The stream function is described in Chapter 4.

3-7 ISOTACHS

Patterns of Air Currents

Isotachs complement streamlines by showing wind speed graphically. If there is a choice of colors, the preferable color for isotachs is green. Identification of principal air currents is a major consideration in constructing isotachs. Attention should therefore be paid to the appearance of jet streams and conveyor belts, besides following the usual rules for drawing contours of scalar fields.

There is a strong tendency for wind maxima in the atmosphere to be elongated, thus showing jets. Also, the direction of the jet stream coincides fairly well with streamlines. Therefore, the jet core can be drawn like an enhanced streamline. A thick or double arrow (green or brown, if in color) is a common indication of the jet core.

In cases of fast-moving waves on the jet, there are differences in direction between the jet core and streamlines. In that case the jet core will intersect the streamlines. Due to the elongated character of fast wind regions, isotachs normally have maximum curvature on the core of the jet stream.

Figure 3-14 shows the relationship between isotachs (single lines, labeled in m s^{-1}), and the jet core (double line with arrowhead). The variation of wind speed V along the normal coordinate n (line AB) shows a maximum wind speed on the jet core, as illustrated by the isotach bending away from the normal AB. At the intersection of the isotach and jet core $\partial V/\partial n = 0$, which is the mathematical condition for maximum wind. The most common situation is for isotachs to have maximum curvature on the jet core. This results from the rule for finding the core of air currents: The jet core is preferably located in the points of maximum curvature of the isotachs. If a discrepancy appears between the jet core location and maximum curvature of isotachs, the lines should be reconsidered. The isotachs, or the jet core, or both should be moved to achieve the agreement.

Example of Isotachs

An example of isotachs and a jet core is shown in Fig. 3-15. The thin dashed lines are isotachs drawn automatically by a computer, without regard for the possi-

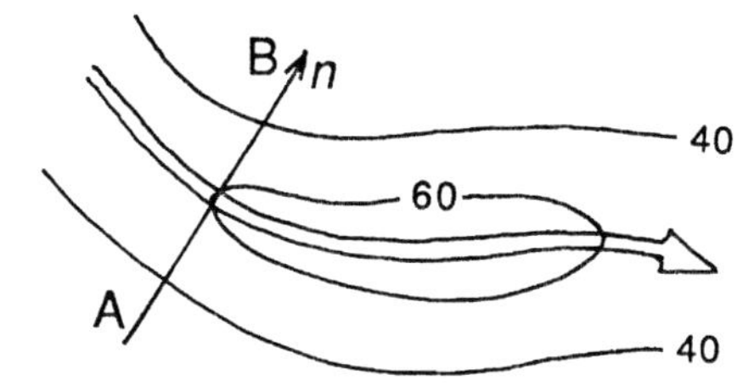

FIGURE 3-14. Relationship between isotachs (m s^{-1}) and the jet core (double line with arrowhead) on a horizontal (isobaric) chart. Line AB is the normal coordinate n at its intersection with jet core.

ble existence of a jet stream. The jet core (double line with arrowheads) is drawn by hand, along the chain of maxima in the isotachs. After the jet core is introduced, changes in the isotachs are entered. The meridional propagation of the 70-kt isotach near point X is changed to a zonal position to emphasize the elongated character of the polar jet stream. The corrected isotachs are shown by thicker dashed lines. The isotach of 90 kt northeast of X has also been changed to follow the maximum curvature to the jet core. A similar correction is made on one of the 70-kt isotachs near point Y in the western part of Fig. 3-15.

3-8 TRAJECTORIES

Frequently we need to know the origin or destination of the air, since we breathe it and since it transports thermodynamic and other properties that are of interest in our work. The information about movement of air is most practical in the form of trajectories.

Definition of Trajectory

The trajectory of an air parcel is a curve constructed by following the three-dimensional position vector

$$\mathbf{x}_3(t) = \begin{pmatrix} x(t) \\ y(t) \\ p(t) \end{pmatrix}$$

of the parcel. The subscript 3 means that the vector is three-dimensional. The trajectory is obtained as an integral of velocity. The wind velocity is defined as a total differential of the position

$$\mathbf{V}_3(t) = \frac{d\,\mathbf{x}_3(t)}{dt}$$

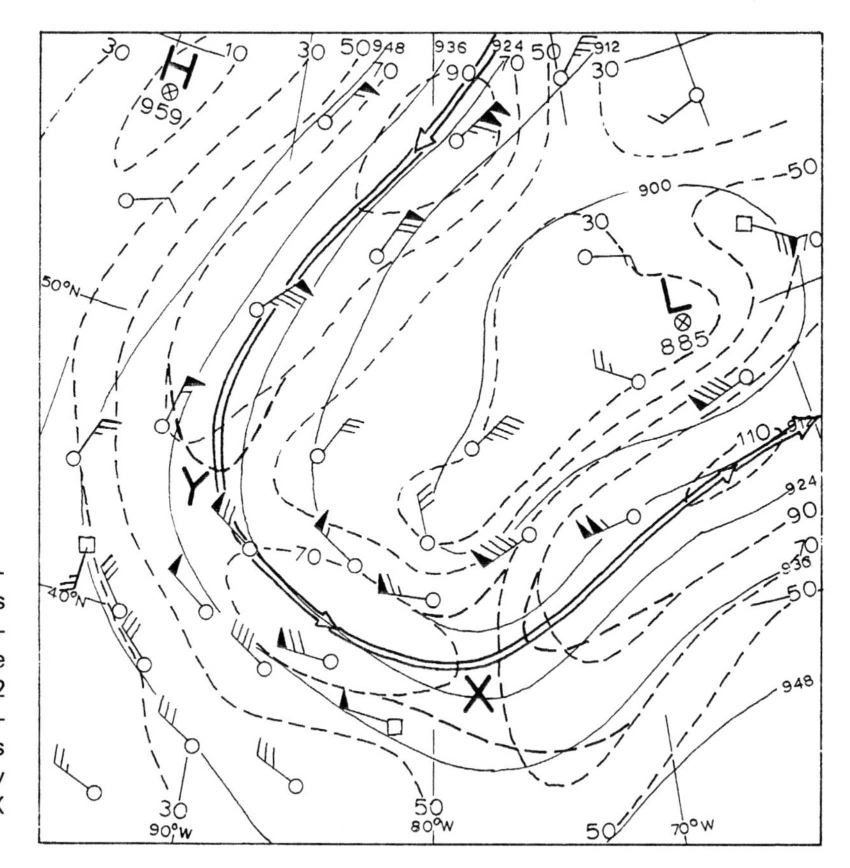

FIGURE 3-15. An example of isotachs and jet core. Dashed lines (kt) are computer-generated isotachs from the NMC facsimile chart of 300 mb for 1200 UTC 2 June 1988. Thin lines are the contours (dam). Possible corrections to the isotachs are shown by thicker dashed lines near points X and Y.

The integration along the trajectory segment yields the end point of the trajectory:

$$\mathbf{x}_3(t_1) = \mathbf{x}_3(t_0) + \int_{t_0}^{t_1} \mathbf{V}_3(t)\, dt \qquad (3\text{-}8)$$

This can also be used for intermediate points at any selected time when various choices of t_1 are used. The times t_0, t_1 are normally selected at the observation hours.

Equation (3-8) shows that knowledge of wind is necessary for evaluating the trajectory. This is a source of an implicit relationship: The wind along the trajectory cannot be determined before the trajectory is evaluated, and the trajectory cannot be evaluated before the wind is determined.

For the time being, only horizontal trajectories will be considered. The problems with vertical motion are formidable and will be commented on later. Therefore, the index 3 will be omitted in the following equations.

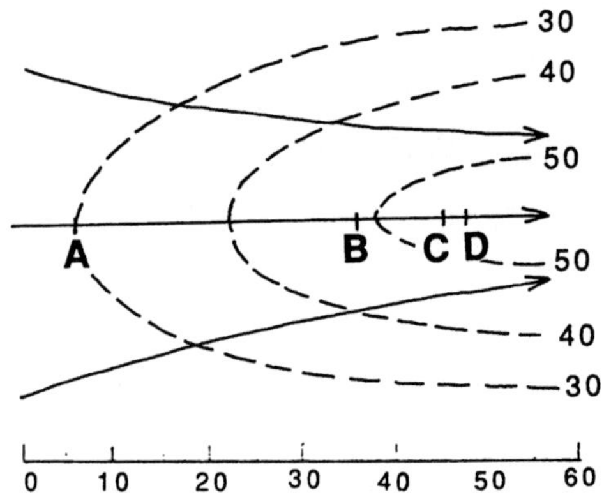

FIGURE 3-16. Construction of a straight trajectory. Arrows are streamlines and dashed lines are isotachs (m s^{-1}). The 6-h trajectory starting in point A is illustrated. Points B, C, and D show the iterative evaluation of the end point of the trajectory. The scale on the bottom shows the distance traveled in 6 h for the indicated speed in m s^{-1}.

Practical Construction of the Trajectory

The integral in (3-8) can be approximately solved as a sum of two terms (trapezoidal rule):

$$\int_{t_0}^{t_1} \mathbf{V}(t)\, dt \approx \mathbf{V}(t_0)\frac{\Delta t}{2} + \mathbf{V}(t_1)\frac{\Delta t}{2}$$

$$= \frac{\Delta t}{2}[\mathbf{V}(t_0) + \mathbf{V}(t_1)] \qquad (3\text{-}9)$$

where $\mathbf{V}(t_0)$ is the wind at the starting point, $\mathbf{V}(t_1)$ is the wind at the end point of the trajectory, and Δt is the interval of time between t_0 and t_1. The above-mentioned implicit relation can be seen here again: The wind at the end point $\mathbf{V}(t_1)$ is not available until the trajectory is determined because we still do not know where that end point is. On the other hand, as can be seen in (3-8) and (3-9), the trajectory cannot be determined until the wind $\mathbf{V}(t_1)$ in the end point is found. This implicit relationship can be resolved iteratively, as illustrated in Fig. 3-16. For the sake of simplicity, the trajectory in this example is selected along a straight streamline. It is assumed that a 6-h trajectory is needed, starting in point A. In manual work, it is useful to prepare a ruler showing the trajectory length (product of speed and time) for each wind speed. Such a ruler is shown at the bottom of Fig. 3-16. The length has been appropriately computed for the chart scale at the average geographical latitude. Here are the steps of the iterative procedure:

1. Use the speed and direction in the starting point A (here 30 m s^{-1}) to find the provisional end point (B in this example).
2. Read off the speed in point B (here 48 m s^{-1}).
3. Compute the average speed between points A and B as

   $$V = \tfrac{1}{2}(V_A + V_B) = \tfrac{1}{2}(30 + 48) = 39$$

4. Use the new speed (39 m s^{-1}) to find a better end point, again starting from A. This yields point C.
5. Read off the speed in point C (here 54 m s^{-1}).
6. Compute the average speed between points A and C as

   $$V = \tfrac{1}{2}(V_A + V_C) = \tfrac{1}{2}(30 + 54) = 42$$

   (Iteration of the procedure should be noticed by now.)
7. Use the new average speed (42 m s^{-1}) to find next better end point, again starting from A. This yields point D.

This procedure can be iterated further, but it is advisable not to use more than two or three iterations. When the last two points fall close to each other (say within $\approx$ 5% of the trajectory length), the evaluation should be stopped. The usual accuracy of observations and chart construction does not allow a more accurate trajectory. Also, the approximation in (3-9) is somewhat crude; it does not account for possible nonlinear

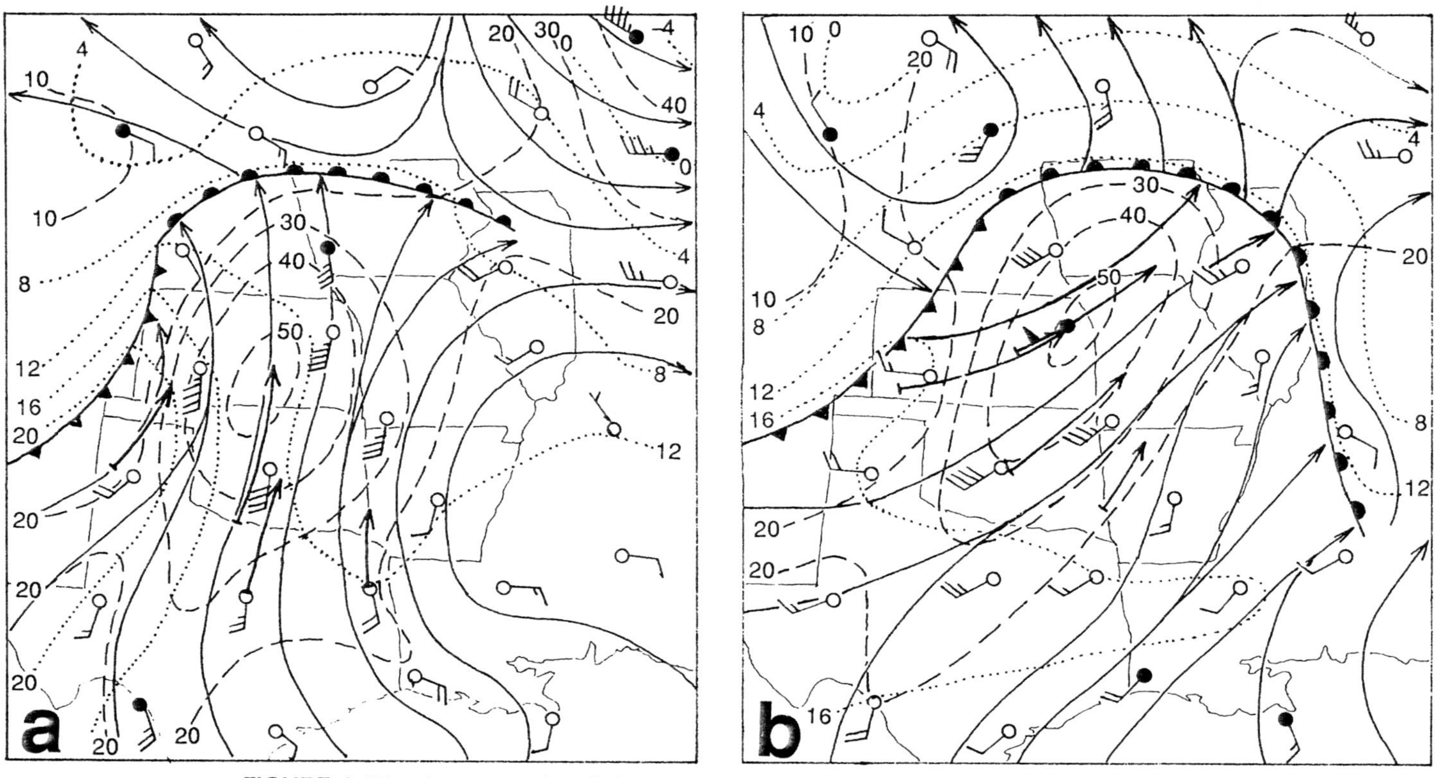

FIGURE 3-17. An example of the construction of isobaric trajectories at 850 mb between 0000 and 1200 UTC 15 October 1983. (a) and (b) are the initial and final weather charts with streamlines (solid), isotachs (dashed, kt), and isotherms (dotted, °C). The first 6-h segments of the trajectories (heavy arrows) are in (a); the last 6-h segments are in (b). Complete 12-h trajectories are in (c), with smoothed kinks (dashed). The trajectories that start in points A, B, C, and D end in points A′, B′, C′, and D′, respectively.

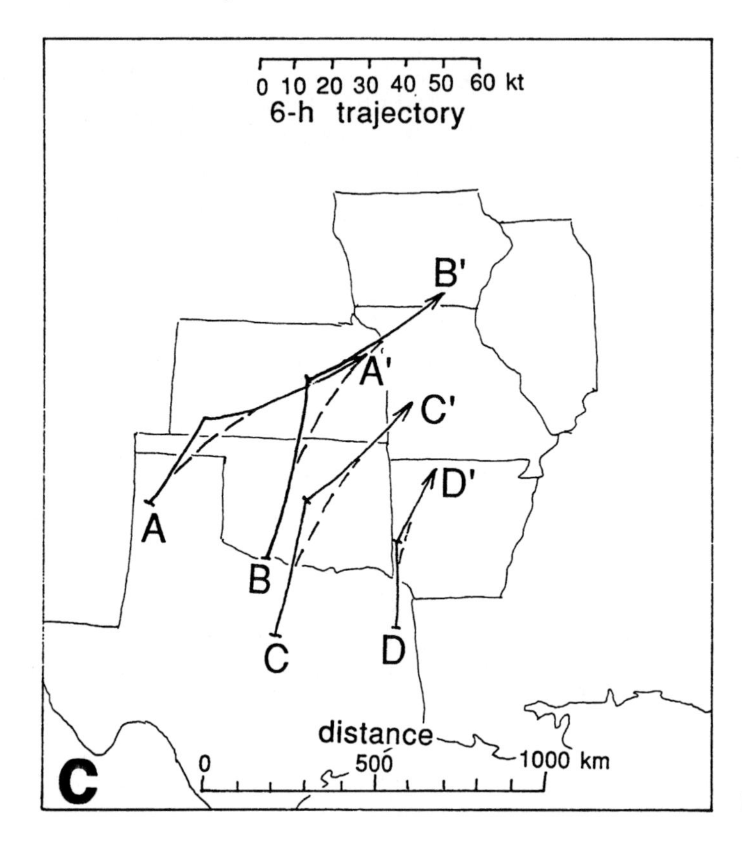

FIGURE 3-17. Cont.

wind variation between the starting and end points of the trajectory.

Variation of Direction Along the Trajectory

The above example also gives a hint of how the trajectory can be constructed in the case where the wind direction (not only the speed) also changes along the trajectory. The new position, corresponding to point C in the above example, should be computed along a vector:

$$\mathbf{V}(t_1) = \tfrac{1}{2}(\mathbf{V}_A + \mathbf{V}_B) \qquad (3\text{-}10)$$

This means that the new position (like C in Fig. 3-13) is obtained in a direction different from the direction in which point B had been obtained.

This procedure, especially with the vector averaging as in (3-5), is suitable for use on computers. The accuracy can be enhanced if the trajectories are evaluated over shorter intervals than 6 h and if the vertical motion is also taken into account. Using shorter intervals and vertical motion depends on availability of data. Computer models often provide an abundance of data at all time steps. Data at every hour give satisfactory trajectories on the scale of weather systems. A good vertical wind component is also available from the models.

A Practical Manual Procedure

When the trajectories are constructed by hand, it is not necessary to evaluate the average vector as in (3-10). It suffices to draw trajectory segments along curved streamlines. A ruler for evaluation of distances should be constructed as in the example in Fig. 3-16. The ruler for the next example is shown on the top of Fig. 3-17c.

Figure 3-17 shows several 12-h isobaric trajectories constructed by this method. The first 6-h segments are in Fig. 3-17a, and the last 6-h segments are in Fig. 3-17b. Complete 12-h trajectories are drawn in Fig. 3-17c. The constructed segments are joined by kinks at first, since the streamlines change from one chart to the next. The kinks are consequences of finite-difference solution of the integral in (3-8). Therefore, it is advisable to smooth the kinks, as shown by the dashed lines. The "correct" trajectories are likely to lie along the smoothed dashed lines.

Technical difficulties may arise. For example, overlapping of trajectories may occur, as with the segments AA′ and BB′ in Fig. 3-17c. It should be noticed that the chart in Fig. 3-17c is not synoptic; the beginning and end points of trajectories are at different times.

Common Difficulties with Trajectories

The vertical motion in the atmosphere is usually very small compared with horizontal motion. Unfortunately, small vertical displacement of air parcels may cause large errors in computed trajectories. When the parcel is displaced to another level, it moves with another wind, rather than with the wind at the level where we compute the trajectory. This error increases rapidly along the trajectory.

Another difficulty with evaluation of trajectories is that a trajectory of a "material point" is seldom needed. Usually, we are interested in motion of finite volumes of air. Then a common property of fluid flow plays an important role: Volumes of air deform drastically in fluid flow. The volumes usually elongate along the flow. Later, the stripes of fluid fold into unexpected shapes and eventually diffuse in the environment. Any computation that follows a finite number of trajectories eventually fails to capture the fate of elongating fluid bodies.

4

Kinematics

This chapter describes several meteorological elements that are derived from "observed" elements. The rules of calculus in several coordinates are used. The gradient vector is used in various forms. Most of the chapter is devoted to the evaluation of geostrophic wind, various forms of the wind shear (vorticity, divergence, deformation, and thermal wind), and advection of scalar quantities. These variables form the basis for the application of the equations of dynamic meteorology.

The word *kinematics* implies that no reference to forces is made. However, some of the quantities are naturally close to forces, as gradients of pressure and geopotential.

4-1 EVALUATION OF GEOSTROPHIC WIND

Geostrophic wind is evaluated from the height (or pressure) field, without reference to the observed wind. The relevant equations are (C-7) through (C-9) of Appendix C. Geostrophic wind is often compared with wind. Both wind and geostrophic wind are vectors. They have the same dimension and are sometimes equal.

A verbal description of geostrophic wind may be practical for description of the direction. However, the determination of speed and any use of geostrophic wind in computers should rely on the mathematical equations.

Direction

The direction of the geostrophic wind (including sense, as it is customary in meteorological usage) is tangent to the contours (or isobars), with the lower height (pressure) on the left, if traveling along the geostrophic wind vector in the Northern Hemisphere. This description is rather common and it was introduced in meteorology in the middle of the nineteenth century. It is known as the Buys-Ballot rule, after its inventor.

The direction of geostrophic wind is always expressed as the direction *from* which the corresponding equal wind blows. The opposite direction is never mentioned when the wind is referenced. Also, it is not completely correct to say that geostrophic wind "blows," since it is only a function of the height field and not of the air motion.

Speed of Geostrophic Wind

The magnitude of geostrophic wind is normally called "speed." In manual work, the speed can be obtained practically with the nomogram that is printed on many weather charts. The spacing of two neighboring contours can be measured conveniently along the edge of a piece of paper, where two ticks can be recorded. The distance between the contours can be compared with the distance in the nomogram. Then the speed of geostrophic wind can be read off the nomogram.

An example of the evaluation of geostrophic wind is shown in Fig. 4-1. There is a point A where the geostrophic wind should be evaluated. The direction is shown by the line AD. This line is tangent to an intermediate contour (not drawn) in point A. The angle between this direction and the meridian can be measured by a protractor. Here it is 296°, or 300° when rounded to the nearest 10°. Generally, the point may lie off a printed meridian on the chart. In that case we can draw the meridian at the point of interest.

To obtain the magnitude of the geostrophic wind, we can use the distance between nearest contours. The point of interest (A) will generally not lie exactly between two contours. An uncentered measure may be adequate. Here this is BC across the contours and through point A. Figure 4-1b shows a typical nomogram where the measured distance can be found and the speed can be read off. The distance B′C′ in Fig. 4-1b is equal to the distance BC in Fig. 4-1a.

Care must be taken so that the contours are taken at intervals prescribed on the nomogram. If the contours are drawn with other intervals, then the speed must be changed proportionally to reflect the difference in the geopotential height between the contours. For example, in Fig. 4-1 the nomogram is constructed for the interval of 30 gpm between the contours, and the chart is drawn with 60-gpm intervals. Therefore, the read-off speed must be doubled. In our example we read the speed of 19 kt, but due to the doubling we find that the geostrophic wind speed is 38 kt (≈ 20 m s^{-1}). The other possibility is to draw more contours to obtain the proper spacing. In a region with sparse contours, intermediate contours should be drawn for this purpose.

Height Gradient and Geostrophic Wind Speed

The distance that corresponds to speed (as B′C′ in Fig. 4-1b) is computed from the height gradient (magnitude only, $\partial z/\partial n$). The definition of the geostrophic wind (C-7) from Appendix C gives the speed $V \equiv |\mathbf{V}|$ as

$$V = \frac{g}{f}\frac{\partial z}{\partial n} \approx \frac{g}{f}\frac{\delta z}{\delta n}$$

The last expression is in finite differences, where δz is the selected interval of geopotential height between the contours. The distance between the selected contours on the chart is δn, measured in units that correspond to the distance in nature. Thus a usual value of δn is of the order of 10^5 to 10^6 m, depending on the wind speed. This value is easily distinguishable from the increment δz between chosen contour values; δz is typically 40, 60, or 80 gpm. The above equation can be solved for the distance as

$$\delta n = \frac{g}{f}\frac{\delta z}{V}$$

and this is the distance (as B′C′) that corresponds to the geostrophic wind speed. The distance δn and the Coriolis parameter f vary with geographical latitude; therefore, the construction of a diagram like the one in Fig. 4-1b takes more effort than just a straight interpretation of the above equation.

For cases where the nomogram for geostrophic wind is not provided, the height gradient ($|\nabla z|$) has to be computed and multiplication and division must be

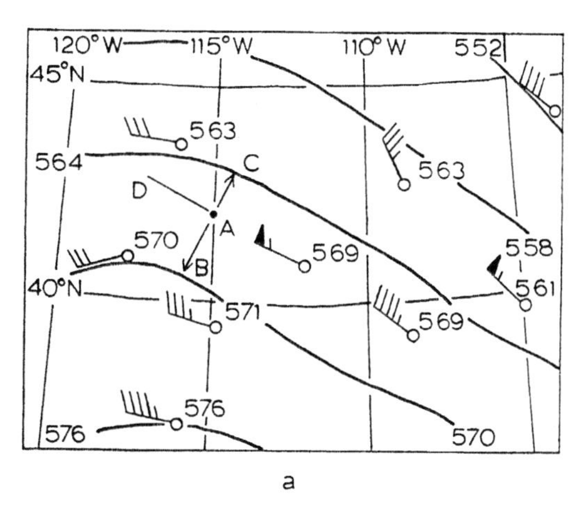

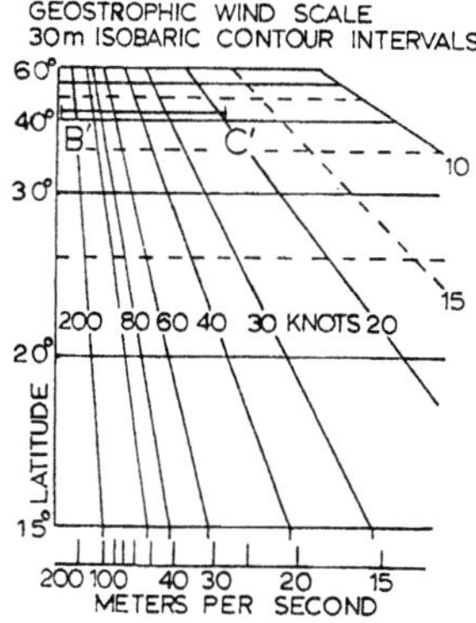

FIGURE 4-1. An example of graphical evaluation of geostrophic wind. (a) A 500-mb chart with contours in gpm. (b) The geostrophic wind scale for the chart in (a). The example shows the distance BC from (a) measured as B′C′ in (b).

performed, as required by the quoted formula for geostrophic wind speed.

It is possible to evaluate two components of geostrophic wind in a similar way; however, this procedure is twice as long as the evaluation for speed shown above. Evaluating geostrophic wind using components is suitable for computers, since computers cannot readily evaluate the direction the way people do.

The Meaning of Geostrophic Wind

Geostrophic wind is usually approximately equal to the wind vector. This can be expressed mathematically as

$$\mathbf{V} \approx \mathbf{V}_g$$

This is possibly the most important equation that the students will learn in meteorology. It is an approximate equation. It also cannot be expected to hold near the equator where the Coriolis parameter vanishes.

The equality between wind and geostrophic wind (within a desired tolerance) occurs in the case of *geostrophic flow*. For this case it can also be said that "wind is geostrophic," meaning that "wind is equal to geostrophic wind." It should not be automatically assumed that wind is equal to geostrophic wind, although this happens frequently.

A significant benefit of the concept of geostrophic wind is that geostrophic wind often may be used instead of the wind. This usage is commonly called *the geostrophic approximation*. It is used both in equations of dynamic meteorology and in practical work. It is often satisfactory to give information to pilots and other users about geostrophic wind when wind observations are not available. However, there are cases when wind is greatly different from geostrophic wind. For this reason caution should be exercised. Chapters 7 and 12 show several cases where typical "geostrophic deviation" occurs.

Some Limitations in Using Geostrophic Wind

The above procedure for evaluating geostrophic wind is complete in the sense that nothing of practical importance has been left out. For the purpose of understanding the relation of wind and geostrophic wind, it is worth noting that the procedure does not make use of several other related quantities:

friction force
acceleration
curvature of the contours
wind speed and direction

These items are irrelevant for describing geostrophic wind, since they have not been used in its determination. Therefore, when the definition of geostrophic wind is needed, those quantities should not be used either. Instead, the correct definition of geostrophic wind from equations (C-7) through (C-9) in Appendix C should be used.

The above-listed quantities have something to do with geostrophic wind, although not with the evaluation of it. These quantities are relevant when answering the question, When can we expect that wind is equal to geostrophic wind? This question requires a rather complicated answer and it is discussed Chapter 7.

4-2 WIND SHEAR AND CURVATURE OF THE FLOW

Derivative and Difference Types of Shear

Several quantities can be described as *wind shear*. These can be conveniently classified in two major categories: the *derivative type* and the *difference type*. The *derivative type of wind shear* is of the form

$$\frac{\partial V}{\partial s} \tag{4-1}$$

where V is either the wind vector, wind speed, or a component of the wind vector, and s is a direction in space. The *difference type of wind shear* is of the form

$$V_1 - V_2 \tag{4-2}$$

where V_1 and V_2 are two values of wind speed, two wind vectors, or two values of the same wind component, each in two points in space marked 1 and 2.

These two definitions of wind shear have different dimensions. The derivative-type wind shear (4-1) has the dimension of s^{-1}, whereas the difference-type wind shear (4-2) has the dimension of m s^{-1}. The derivative definition is suitable in smooth fields of wind, whereas the difference definition can be used conveniently as a measure of wind variation across a line of discontinuity. It should be remembered that the mathematical derivative of variables is not defined in points of discontinuity. For this reason it is incorrect to evaluate a derivative approximately using a finite difference across a discontinuity. In other words, the difference definition is useful at a discontinuity, but the derivative is undefined.

A relation between the two definitions of wind shear is given by the approximate definition of the first derivative (without the limit):

$$\frac{\partial V}{\partial s} \approx \frac{V_1 - V_2}{\Delta s}$$

where Δs is the distance between the points 1 and 2.

Terminology with Wind Shear

When the direction s is vertical, or when the points 1 and 2 lie one above the other, we have *vertical wind shear*. This shear will be elaborated upon further in Section 4-7.

When the direction s is horizontal and normal to the streamline, this is *normal wind shear* or *lateral wind shear*. When the line between points 1 and 2 is normal to the streamlines, we have *normal shear of difference type*. We sometimes use the term *longitudinal shear*, along the streamline. This shear is called *stretching* or *shrinking*, depending on the sign of $\partial V/\partial l$, where l is the coordinate along the streamline. In each of the above cases V may be defined as a vector or as a scalar component or wind. Thus the expression *wind shear* is rather general and requires more specification when actual values are used.

The thermal wind is a vertical wind shear, as shown in the list of formulas in Appendix C. It has two variations, derivative or difference, as shown by definitions (4-1) and (4-2).

In the case when the vertical shear of the horizontal wind is evaluated, the expressions *upshear* and *downshear* are used. In these cases V or V_1 and V_2 are horizontal vectors in (4-1) and (4-2). The resulting wind shear is also a vector. The direction along this vector is *downshear*, and the opposite direction is *upshear*. It will be seen later that thunderstorms often propagate in the downshear direction, irrespective of the environmental wind direction.

Cyclonic and Anticyclonic Shears

The designations *cyclonic* and *anticyclonic* wind shear are used when discussing the wind field on horizontal (or isobaric) charts. This refers only to the normal, horizontal shear, where the direction s in (4-1) is normal to the streamline. Such shear can be recognized by its tendency to turn fluid parcels (also floating bodies) around the vertical axis. There are two senses of rotation of parcels in the horizontal flow. For these two senses, the general agreement regarding the sign of wind shear is that when the floating parcels turn left, this shear is positive; when the floating parcels turn right, this shear is negative. Further, the positive shear is called *cyclonic* in the Northern Hemisphere and the negative shear is *anticyclonic*. These designations (cyclonic and anticyclonic) are reversed in the Southern Hemisphere. This rule about signs is so important that if one chooses the normal direction $n = -s$ in the direction that may give the opposite result, then the definition of the normal shear (4-1) is redefined as $-\partial V/\partial n$, such that the rule of signs is upheld.

Curvature

A quantity related to wind shear is *curvature* of the streamline, the contour, or the trajectory. The curvature k is defined as the inverse of the *radius of curvature* r of the tangent circle that has the same curvature:

$$k \equiv \frac{1}{r}$$

When dynamics are considered, it is often desirable to calculate the curvature of the trajectory of an air parcel. However, it is considerably easier to find the curvature of the streamline or the contour since these lines are conveniently represented on a synoptic chart. The trajectories are more difficult to find, as was shown in Section 3-8. In this chapter, only the curvature of streamlines and contours will be considered.

The sign of curvature, like the sign of shear, is positive when the streamline bends to the left for an observer who moves with the wind. This curvature is called *cyclonic* in the Northern Hemisphere. Negative curvature indicates bending to the right and is called *anticyclonic* in the Northern Hemisphere. Cyclonic or anticyclonic curvature can be recognized by the turning of fluid parcels that travel along the streamline (or contour, or trajectory). All conventions about signs apply in the Southern Hemisphere; only the designations cyclonic and anticyclonic are reversed.

4-3 VORTICITY AND DIVERGENCE

The computed variables of vorticity and divergence are of profound importance for interpreting atmospheric dynamics. By knowing the distribution of computed variables, we get an insight into the movement of weather patterns, the distribution of vertical motion, and cyclogenesis. The vertical motion, in turn, is the primary indicator of formation or dissipation of clouds and precipitation.

Several important quantities, including vorticity, divergence, and deformation, are formed by various shearing terms. Deformation is described in Section 4-6, while vorticity and divergence follow here.

Definitions of Vorticity and Divergence

In horizontal (isobaric) Cartesian coordinates, the vorticity ζ and divergence div $\mathbf{V}$ are defined as

$$\zeta = \mathbf{k} \cdot \nabla \times \mathbf{V} = \frac{\partial v}{\partial x} - \frac{\partial u}{\partial y} \tag{4-3}$$

$$\text{div } \mathbf{V} = \nabla \cdot \mathbf{V} = \frac{\partial u}{\partial x} + \frac{\partial v}{\partial y} \tag{4-4}$$

All vectors used in these definitions, like vectors in this book, are two-dimensional. Their vertical component is zero, as explained in Appendix D.

Vorticity can be described geometrically as twice the angular speed of rotation of a body that moves with the environmental fluid velocity. This moving body is best imagined as a part of the fluid itself.

Divergence can be described as a relative rate of increase of a material surface, since geometrical considerations show that

$$\nabla \cdot \mathbf{V} = \frac{1}{A} \frac{dA}{dT} \tag{4-5}$$

where A is the area of the material surface. It is assumed that A is conveniently small. More discussion on the meaning and evaluation of vorticity and divergence is given in Appendix H.

Three-dimensional vorticity and divergence are defined using three-dimensional vectors (subscript 3) as

$$\nabla_3 \times \mathbf{V}_3 \quad \text{and} \quad \nabla_3 \cdot \mathbf{V}_3$$

but they are seldom used in weather analysis.

For the purpose of numerical work in spherical coordinates, it is necessary to use the definitions

$$\zeta = \frac{\partial v}{\partial x_s} - \frac{\partial u}{\partial y_s} + u \frac{\tan \varphi}{R} \tag{4-6}$$

$$\text{div } \mathbf{V} = \frac{\partial u}{\partial x_s} + \frac{\partial v}{\partial y_s} - v \frac{\tan \varphi}{R} \tag{4-7}$$

where x_s and y_s are the spherical coordinates pointing east and north, respectively, φ is the geographical latitude, and R is the radius of the earth. A scale analysis will reveal that the additional terms with tan φ are generally very small compared with other terms. However, if these terms are omitted, a systematic error may accumulate with repetitious calculation.

Other Forms of Vorticity

For the purpose of dynamical considerations, other forms of vorticity are used:

Geostrophic vorticity:

$$z_g = \frac{g}{f} \nabla^2 z$$

This expression is obtained when the geostrophic wind (C-7) or (C-8) is inserted in the definition of vorticity (4-3). It is used when high accuracy is not needed.

Absolute vorticity:

$$\eta = \zeta + f$$

This is the sum of vorticity and the Coriolis parameter. Conservation of absolute vorticity is a very important concept in explaining planetary waves.

Potential vorticity:

$$P = \frac{z + f}{\delta h}$$

Here δh is the vertical distance between two material surfaces in the atmosphere. This quantity is of central importance in explaning cyclogenesis and upper-level frontogenesis. Vertical stretching (increase of δh) intensifies cyclonic vorticity. This is of great importance in the dynamics of motion on all scales. Alternative forms of potential vorticity are listed as (C-13) and (C-14) in Appendix C.

Other combinations are sometimes used; for example, geostrophic absolute vorticity

$$\eta_g = \frac{g}{f} \nabla^2 z + f$$

All the above equations in this subsection are essentially identities. They define various quantities. Therefore, they cannot be used as independent proofs of physical explanations. These definitions should be distinguished from the vorticity equation:

$$\frac{\partial \zeta}{\partial t} + \mathbf{V} \cdot \nabla (\zeta + f) = 0$$

or any other form of it [such as (C-43)]. It should be noticed that the vorticity equation contains a time derivative. The preceding definitions in this subsection have no time derivatives.

Natural Coordinates

With a suitable turning of x- and y-coordinates, the individual scalar terms in (4-3) and (4-4) correspond exactly to the definitions in natural coordinates (see Appendix H):

$$\zeta = \frac{V}{r} - \frac{\partial V}{\partial n} \tag{4-8}$$

$$\text{div } \mathbf{V} = \frac{\partial V}{\partial l} - V\frac{\partial \alpha}{\partial n} \tag{4-9}$$

where l is the coordinate along the streamline and n along the normal to the streamline. Both these coordinates may be curved. The term α is the wind direction expressed as the azimuth.

Since the natural coordinates are rather descriptive, the terms in (4-8) and (4-9) have names, as follows:

$\frac{V}{r}$ *curvature term* (since $1/r$ is the *curvature*)

$-\frac{\partial V}{\partial n}$ *lateral shear* or *normal shear*

$\frac{\partial V}{\partial l}$ *stretching* (*shrinking*, if the value is negative)

$-V\frac{\partial \alpha}{\partial n}$ *normal curvature term* or *diffluence term*

The derivative $\partial\alpha/\partial n$ in the last term is the *normal curvature*, that is, the curvature of the normal to the streamlines. The streamlines are not necessarily curved in the case of diffluence (or confluence, in the case of a negative value). However, the normal to the streamlines is curved in the cases of diffluence or confluence. This is discussed more with deformation in Section 4-6.

In some books different signs are used for the terms with normal derivatives in the definitions (4-8) and (4-9). There is no general agreement about the sense of the normal direction n. Therefore, the earlier discussion on determining the sign of the normal shear should be repeated here: The normal shear $-\partial V/\partial n$ or $\partial V/\partial n$ must have a positive value in the case of cyclonic shear in the Northern Hemisphere. As mentioned before, this can be achieved either by changing the sign in the definition or by choosing the other sense of the direction n.

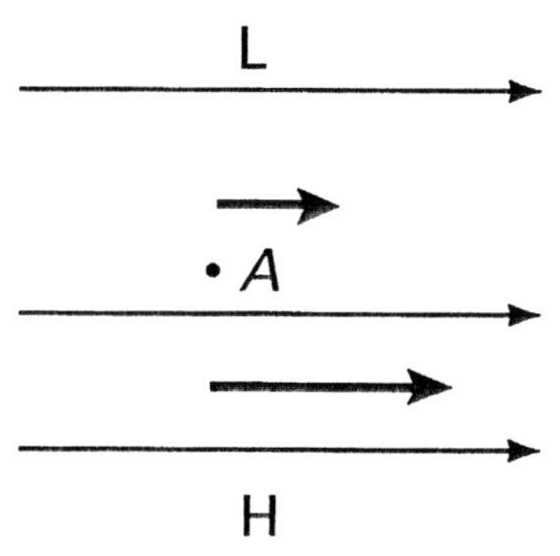

FIGURE 4-2. Wind distribution (shorter vectors) and contours (longer arrows) with positive shear (cyclonic in the Northern Hemisphere). Narrower spacing between contours shows faster wind.

Examples of Wind Shear

When analyzing weather charts, we must recognize several simple models of flow where vorticity, divergence and deformation are prominent. Several examples of flow with significant vorticity are shown in Figs. 4-2 through 4-5. Each figure covers a horizontal area of about 500 km along the short side and contains several contours. Spacing of contours shows the wind speed such that smaller distances correspond to faster wind. Use of the geostrophic approximation $\mathbf{V} = \mathbf{V}_g$ is usually adequate for the estimation of wind shear. The x- and y-coordinates may be understood as "east" and "north" directions, respectively. Vectors (arrows) show the air flow, with longer arrows indicating faster flow.

The flow is straight along the x-coordinate in Fig. 4-2. In this case the y-component of flow (v) is equal to zero everywhere. The terms on the right-hand sides of (4-3) and (4-8), differing from zero are $-\partial u/\partial y$ and $-\partial V/\partial n$. The derivative $-\partial u/\partial y$ can be recognized here

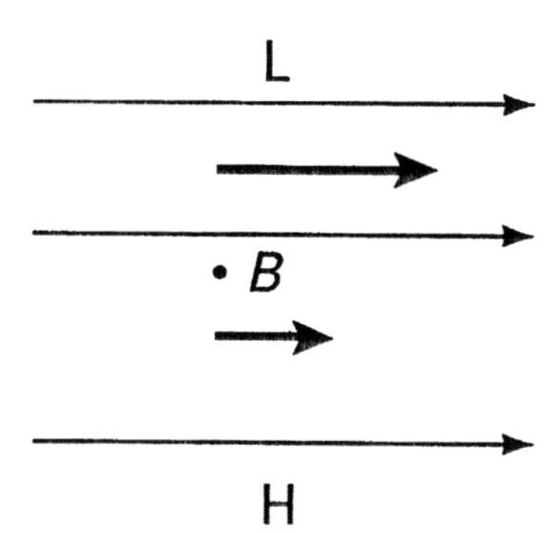

FIGURE 4-3. Wind and contours (as in Fig. 4-2) with negative shear (anticyclonic in the Northern Hemisphere).

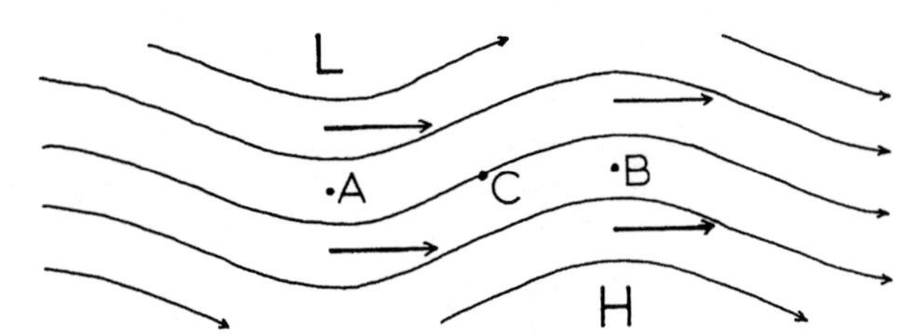

FIGURE 4-4. Wind and contours in the case with curvature without normal shear. Curvature is positive at A, negative at B.

as the normal horizontal shear, since the y-axis is normal to the flow. Therefore, the vorticity here is positive:

$$\zeta = -\frac{\partial u}{\partial y} = -\frac{\partial V}{\partial n} > 0$$

Since the only nonzero component of wind (u) decreases in the y-direction, the value $-\partial u/\partial y$ is positive and this shows that the wind shear is cyclonic. In a similar way, for Fig. 4-3 we have

$$\zeta = -\frac{\partial u}{\partial y} = -\frac{\partial V}{\partial n} < 0$$

This is the case of anticyclonic wind shear.

The above examples show that the vorticity is equal to the normal shear in the case of straight streamlines. These examples are more general than shown here, since we may rotate the coordinates in any direction. An easy practical way for estimation of shearing vorticity is to take the x-axis along the flow, as in Figs. 4-2 and 4-3. Then the existence and sign of the wind shear can be determined from the one remaining term $\partial u/\partial y$.

Individual terms on the right-hand sides of (4-3) through (4-9) should not be called components, since both vorticity and divergence are scalar. We reserve the word *component* for constituents of vectors.

The expression for vorticity in natural coordinates can be estimated in the same examples (Fig. 4-2) as

$$\zeta = -\frac{\partial V}{\partial n} > 0$$

Similarly, we estimate vorticity for point B in Fig. 4-3 as

$$\zeta = -\frac{\partial v}{\partial n} < 0$$

By comparison with (4-6), we see that the speed V corresponds to the only nonzero wind component u and that the direction (and sense!) n coincides with the y-axis. This sense of the n-direction is determined by the requirement that $\zeta > 0$ in the case of cyclonic vorticity. Therefore, the cases in Figs. 4-2 and 4-3 illustrate shearing vorticity, positive and negative (cyclonic and anticyclonic, respectively, in the Northern Hemisphere).

The situations depicted in Figs. 4-2 and 4-3 appear in nature most prominently along the flanks of jet streams. For this reason, we find expressions in the literature such as "cyclonic side of the jet stream," meaning the side with cyclonic shear. The other side is, accordingly, "anticyclonic."

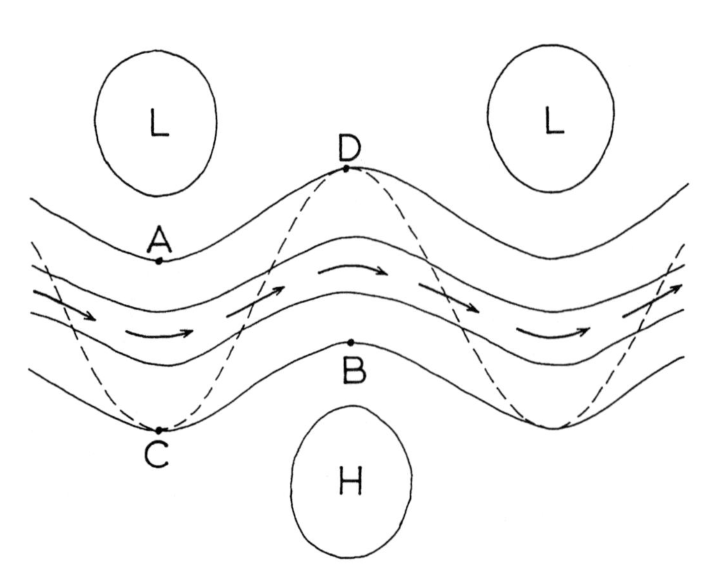

FIGURE 4-5. Schematic contours (solid lines) illustrating combined effects of shear and curvature. The isopleth $\zeta = 0$ is dashed.

Examples of Curvature

The next two examples, points A and B in Fig. 4-4, show a flow without normal shear, but with significant curvature. These are again cases with flow parallel to contours and with constant speed. In these cases the divergence vanishes (div $\mathbf{V} = 0$). For point A vorticity can be estimated as

$$\zeta = \frac{\partial v}{\partial x} > 0$$

since other terms disappear. In natural coordinates,

$$\zeta = \frac{V}{r} > 0 \tag{4-10}$$

The sign of the radius of curvature (r) here is chosen with the same logic as in choice of the normal direction in (4-8). There is no exception to the sign of curvature for the situation at point A in Fig. 4-4. Curvature $1/r$ that causes turning to the left is positive. This curvature is cyclonic in the Northern Hemisphere and anticyclonic in the Southern Hemisphere. Therefore, if (4-8) is altered so that it has a minus sign before the term V/r, then the radius must be taken appropriately negative to obtain $\zeta > 0$ in (4-10).

Analogously, at point B in Fig. 4-4 we have negative curvature (anticyclonic in the Northern Hemisphere):

$$\zeta = \frac{\partial v}{\partial x} < 0$$

and in natural coordinates

$$\zeta = \frac{V}{r} < 0$$

For this situation we must choose $r < 0$ (for anticyclonic curvature in the Northern Hemisphere) to obtain a negative result.

Combination of Shear and Curvature

Several additional cases of shear and curvature are shown in Fig. 4-5. The flow is along the contours. A few arrows show the sense. The spacing of the contours shows the wind speed. Highs and lows are indicated as in the Northern Hemisphere. There are several points with combination of normal shear and curvature in this example. At points A and B we can be certain about the sign of vorticity since the normal shear and curvature have the same sign in each point. At point A there is positive vorticity and at point B there is negative vorticity. At points C and D the situation is unclear for a qualitative estimation. There are opposite signs for the shear and curvature terms at each of these points. We cannot visually determine the sign of vorticity unless we calculate the terms quantitatively to see which predominates. The value of vorticity at C and at D is probably near zero. The isopleth of $\zeta = 0$ is also drawn in Fig. 4-5. The vorticity isopleth shows a wave with a larger amplitude than the contours. Such large-amplitude waves in the isopleths of vorticity are typical for weather charts near jet streams.

Wind Shear on a Weather Chart

Several examples of wind shear at 300 mb can be seen in Fig. 4-6. This level is close to the maximum wind level of the polar jet stream. The core of the jet stream reaches the speed of over 140 kt (or 72 m s^{-1}) over Ohio. The streamlines are not drawn to avoid clutter-

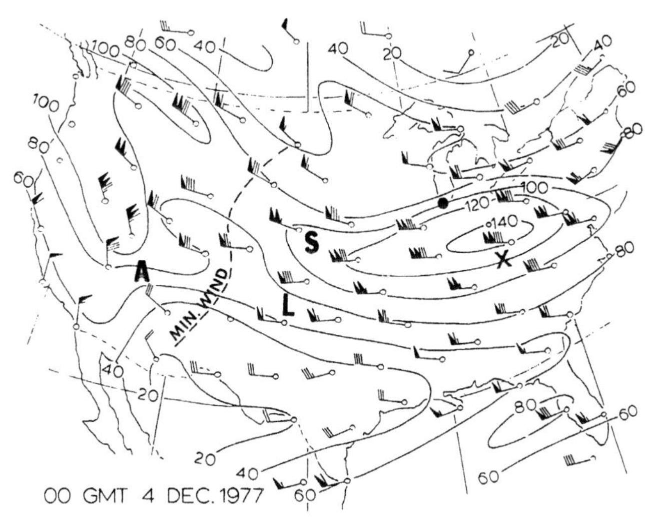

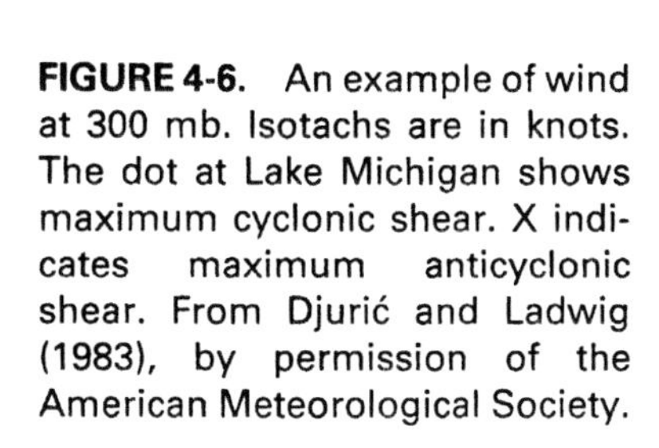
FIGURE 4-6. An example of wind at 300 mb. Isotachs are in knots. The dot at Lake Michigan shows maximum cyclonic shear. X indicates maximum anticyclonic shear. From Djurić and Ladwig (1983), by permission of the American Meteorological Society.

ing the figure, but they can be fairly well visualized on the basis of wind barbs. There are several examples of various wind shear in this figure.

(a) Cyclonic shear can be found on the left-hand flank of the jet stream. Maximum cyclonic shear is at the strongest gradient of wind speed (densest isotachs) around the southern end of Lake Michigan. The point of maximum cyclonic shear is indicated by a large dot.

(b) Anticyclonic shear exists on the right-hand flank of the jet stream. Maximum anticyclonic shear can be detected south of the maximum wind of 140 kt at point X. An important item of information from dynamic meteorology is that air currents on the rotating Earth can be stable only if the anticyclonic shear does not exceed the value of the Coriolis parameter f; that is,

$$\left|\frac{\partial V}{\partial n}\right| < f \qquad (4\text{-}11)$$

This is the criterion for dynamic stability. The measurement in Fig. 4-6 shows that the anticyclonic shear in point X reaches $10^{-4}\ s^{-1}$, which is equal to f at this latitude. In the case where the air currents exceed the above critical anticyclonic wind shear, the flow breaks down into small eddies. Then the momentum (speed times mass) mixes between fast and slow currents, until the overall shear decreases and satisfies the criterion for dynamic stability. Formation of clouds along the anticyclonic flank of a jet stream is connected with this kind of instability. Contrary to the anticyclonic shear, there is no similar restriction on cyclonic shear. Cyclonic wind shear near Lake Michigan can be measured in Fig. 4-6. It has about twice the value of the Coriolis parameter.

(c) The normal shear (both horizontal and vertical) is equal to zero in the core of the jet stream. This follows from the mathematical rule that the first derivative of a function is equal to zero at the maximum of that function. The vertical maximum of wind speed cannot be verified on the basis of an isobaric chart like this. Only from other experience do we know that the maximum wind in the polar jet stream normally appears at or near 300 mb.

(d) There is longitudinal shear at the point S in the middle of Fig. 4-6. This is a case of stretching along the streamline since the wind speed increases downwind at this point.

(e) The signs of shear and curvature do not have to coincide. At point A, in the Southwest, there is anticyclonic shear and cyclonic curvature of the flow.

Example of a Shear Line

Most fronts in the atmosphere are shear lines, that is, lines where we find discontinuities in the wind field. The discontinuous horizontal wind shear is always cy-

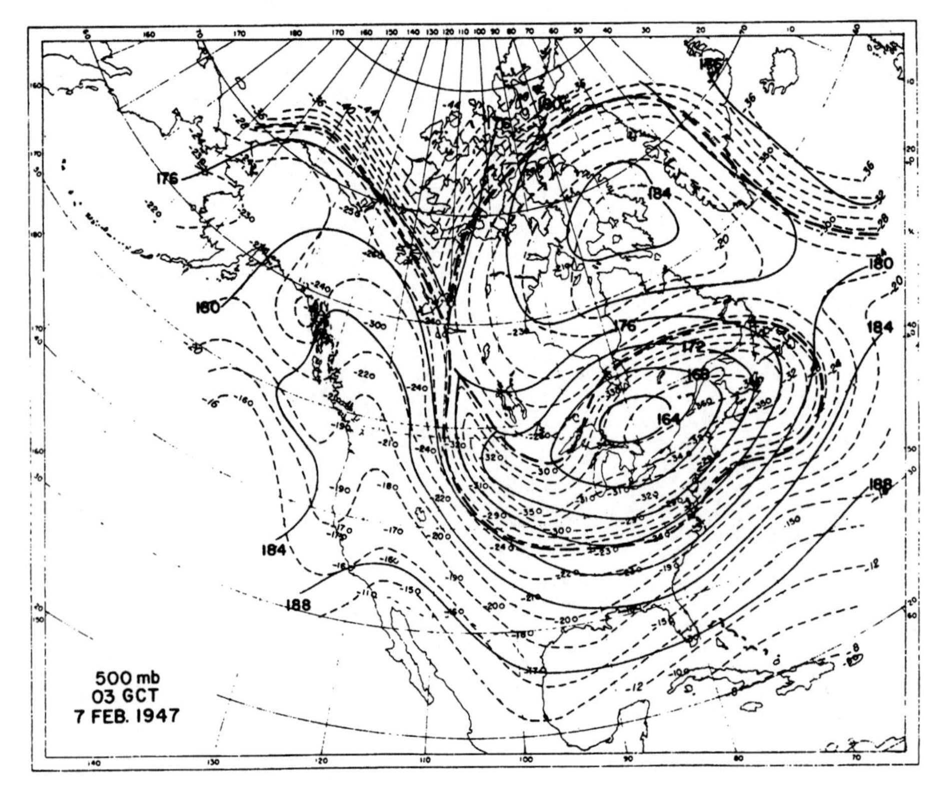

FIGURE 4-7. Example of the polar front at 500 mb. Isotherms at 2°C intervals (dashed), geopotential contours at 200-ft intervals (solid). Heavy dashed line is the warm-air boundary of the frontal layer, except at the stretch over Alberta–Saskatchewan border where the front sank under the 500-mb surface. From Palmén and Nagler (1949), by permission of the American Meteorological Society.

clonic; otherwise, the criterion for stable flow (4-11) is violated and the flow cannot maintain its shape. Shear lines may occur on isobaric weather charts also when there is no front. An example is shown in Fig. 4-7, where along the meridian of 110°W between 56°N and 60°N there is a transition between a northerly flow on the west side and southerly flow on the east side. The polar front was in this area before the observation time, but this front sank beneath 500 mb. The shear line remained.

4-4 DESCRIPTION OF DIVERGENCE AND VORTICITY USING INTEGRALS

The word *divergence* has two meanings: (1) the mathematical function defined in (4-4) and (2) positive values of div **V** (that is, when $\nabla \cdot \mathbf{V} > 0$). This second meaning is the opposite of "convergence," that is, of the case when the divergence is negative ($\nabla \cdot \mathbf{V} < 0$).

Definition of Divergence

Divergence of the flow (both meanings) can often be estimated on weather charts if its description by an integral is used. Let us assume that the average divergence $\nabla \cdot \mathbf{V}$ is evaluated over a conveniently small area A. Then the definition of the arithmetic average can be obtained from the definition of divergence from hydrodynamics:

$$\text{div}\,\mathbf{V} \equiv \frac{1}{A} \lim_{A \to 0} \iint_A \nabla \cdot \mathbf{V}\, da$$

When the limit is omitted, this gives an approximate value

$$\text{div}\,\mathbf{V} \approx \overline{\nabla \cdot \mathbf{V}} \equiv \frac{1}{A} \iint_A \nabla \cdot \mathbf{V}\, da \qquad (4\text{-}12)$$

The obtained value $\overline{\nabla \cdot \mathbf{V}}$, with overbar, is the average divergence in the area A. Green's theorem further yields

$$\text{div}\,\mathbf{V} = \frac{1}{A} \int_L v_n\, dl \qquad (4\text{-}13)$$

where L is the boundary of the area A and v_n is the normal component of the flow. The sign is such that $v_n > 0$ for outflow and $v_n < 0$ for inflow. Thus an inspection of outflow and inflow from small areas gives us a tool for estimation of divergence. If we find more outflow than inflow over the boundary of an area, we have *divergence* (or *positive divergence*, $\nabla \cdot \mathbf{V} > 0$, in accordance with the second meaning of the word). If there is more inflow than outflow, this is *convergence* (or *negative divergence*, $\nabla \cdot \mathbf{V} < 0$).

The meaning of divergence is elaborated upon in Appendix H, on evaluation of divergence from wind observations.

Figure 4-8 illustrates a case of prominent convergence near point A. An oval (dashed) is drawn, enclosing a "small" area around point A. Outflow from the oval can be expected on the northeast part of the oval where the southwesterly wind takes the air out of the oval. Similarly, inflow can be found on the southern side of the oval, where south wind on the warm side of the front brings the air into the oval. The long southeastern part of the oval has the air flow along the perimeter; therefore, there is not much outflow or inflow there. Major inflow into the oval occurs on the western side, where the air flow at OKC and GSW goes across the perimeter. The inflow from the west is not compensated by outflow elsewhere. Therefore, it is a safe (albeit qualitative) conclusion that there is a large net inflow into the oval. This shows that the integrals in (4-12) and (4-13) have negative values and thus the average divergence is negative. In other words, this is a case of prominent convergence.

An inspection of the wind distribution and other elements in Fig. 4-8 shows that the polar front is in the

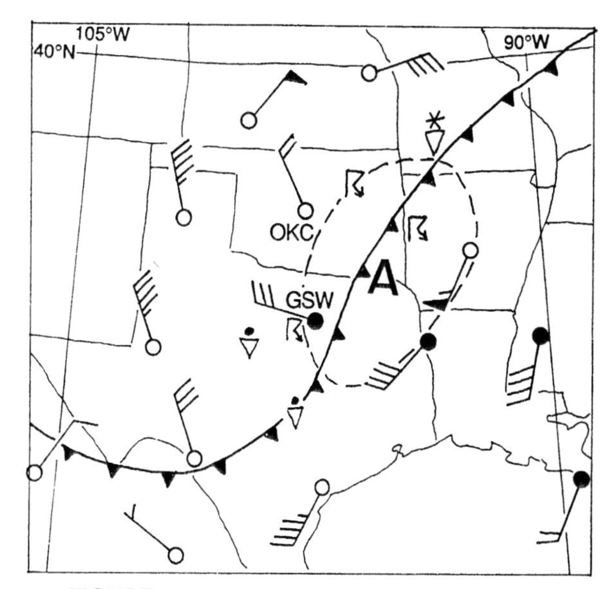

FIGURE 4-8. An illustration of convergence (negative divergence) near point A at 850 mb, 0000 UTC 27 January 1969. The oval (dashed) surrounds an area into which more flow is entering than exiting at the level. Cyclonic circulation is also prominent around the oval.

region. There is a northwesterly flow on one side of the front and a southwesterly flow on the other side. Convergence is common on fronts of this type. This case is rather instructive: The analyst is normally looking for similar cases of convergence, since they point out the existence of fronts. The front has been identified in this weather situation greatly due to the recognition of convergence.

Definition of Vorticity

There is a similar method for visualizing of vorticity on the basis of the integral over a small area A. The arithmetic average of vorticity over the domain A is

$$\bar{\zeta} = \frac{1}{A} \iint_A \mathbf{k} \cdot \nabla \times \mathbf{V} \, da$$

A limit with $A \to 0$ yields the exact definition of vorticity, similarly to the definition of divergence. Since we are interested in visual estimation, we will not apply the limit. Stokes's theorem yields

$$\bar{\zeta} = \frac{1}{A} \int_L v_t \, dl \tag{4-14}$$

where v_t is the tangential wind component along perimeter L of area A. Cyclonic circulation in the Northern Hemisphere (or anticyclonic in the Southern Hemisphere) is positive.

Turning back to the example of Fig. 4-8, we may apply (4-14) to area A. Prevalent flow along the perimeter can be found along the southeastern side; therefore, the circulation can be safely estimated as cyclonic. This illustrates the usual condition in the atmosphere: Vorticity is cyclonic at the front (positive in the Northern Hemisphere).

4-5 EVALUATION OF VERTICAL MOTION FROM DIVERGENCE

Integration of the Continuity Equation

Due to conservation of mass, divergence in the atmosphere is firmly related to vertical motion. This is expressed by the continuity equation [(C-30), Appendix C]:

$$\frac{\partial \omega}{\partial p} = -\nabla \cdot \mathbf{V} \tag{4-15}$$

The *kinematic method* of evaluation of vertical motion consists of integration of divergence in (4-15). The following steps are needed:

Integration of (4-15) between pressure levels p_1 and p_2 gives

$$\int_{p_1}^{p_2} \frac{\partial \omega}{\partial p} dp = -\int_{p_1}^{p_2} \nabla \cdot \mathbf{V} \, dp \tag{4-16}$$

A good approximation on the left-hand side is $d\omega \approx (\partial\omega/\partial p)\, dp$, since other partial derivatives are usually much smaller than $\partial\omega/\partial p$. Then the left-hand side integrates to $\omega_2 - \omega_1$, where ω_1 is the vertical motion at the level p_1 and ω_2 is at the level p_2.

The integral on the right-hand side integrates exactly to give

$$-\overline{\nabla \cdot \mathbf{V}}\,(p_2 - p_1)$$

provided the correct vertical (pressure) average divergence $\overline{\nabla \cdot \mathbf{V}}$ is selected. In the case of constant $\nabla \cdot \mathbf{V}$, or linearly varying $\nabla \cdot \mathbf{V}$, the value of $\nabla \cdot \mathbf{V}$ in the middle of the interval is equal to the average divergence $\overline{\nabla \cdot \mathbf{V}}$. With these results, (4-16) gives

$$\omega_2 - \omega_1 = -\overline{\nabla \cdot \mathbf{V}}\,(p_2 - p_1)$$

As an example, we may consider the layer between 1000 and 700 mb. Then the values ω_2 and p_2 are at 700 mb and ω_1 and p_1 are at 1000 mb:

$$\omega_2 = \omega_{700}$$

$$\omega_1 = \omega_{1000}$$

$$p_2 = 700 \text{ mb} = 7 \times 10^5 \text{ Pa}$$

$$p_1 = 1000 \text{ mb} = 10 \times 10^5 \text{ Pa}$$

For the purpose of visual estimation, the divergence at 850 mb ($\nabla \cdot \mathbf{V}_{850}$) can be assumed equal to the average divergence in the layer between 1000 and 700 mb. Another common approximation is that the vertical motion at 1000 mb is much smaller than at 700 mb. For this reason, often ω_{1000} may be neglected in comparison with ω_{700}. These approximations give

$$\omega_{700} = \nabla \cdot \mathbf{V}_{850}\, \Delta p$$

where $\Delta p = p_1 - p_2 = 3 \times 10^5$ Pa. Thus convergence ($\nabla \cdot \mathbf{V}_{850} < 0$) shows that $\omega_{700} < 0$, which is rising motion. Divergence ($\nabla \cdot \mathbf{V}_{850} > 0$) at 850 mb indicates sinking motion at 700 mb.

Utilization of Divergence on Weather Charts

The above conclusions can be used to interpret the weather situation depicted in Fig. 4-8. We already estimated that there is convergence in the region A.

Therefore we expect lifting at the 700-mb level. There are no direct observations of vertical motion to confirm the conclusion. However, weather phenomena support the conclusion on lifting. This is a region with rain and snow, and lifting is the usual cause of precipitation. The observations of precipitation are shown in the same figure, as they were observed at several weather stations, and also at stations that did not take radiosonde observations.

The wind in Fig. 4-8, and other weather elements (temperature, dew point) that are not plotted here, show that the polar front is located in the oval at A. Convergence and cyclonic circulation are commonly concentrated at the front. Also, vertical motion and precipitation are often parts of the weather system with the front. Detection of convergence and cyclonic circulation are important tools for identifying atmospheric fronts.

Comments on the Kinematic Method

Evaluating of vertical motion ω on the basis of continuity equation (4-15) is the *kinematic method* for evaluating vertical motion. This method can be used over the whole weather chart, using the methods of numerical analysis. Such evaluation of vertical motion is discussed in the course on numerical methods.

Due to natural fluctuations in the wind, the kinematic method of determination of vertical motion may yield doubtful results. Sometimes a turbulent eddy influences the radiosonde observation so that the wind observation is several meters per second off the environmental wind. Since other observing points are several hundred kilometers away, that observation will erroneously be considered representative for a large area. Therefore, the evaluation of divergence and large-scale vertical motion is susceptible to errors.

Contrary to the above conclusion, the kinematic method is very satisfactory in numerical models when the computed wind is used. In those cases, the wind does not suffer from unrepresentative observations. A sophisticated continuity equation is one of the basic equations in numerical weather prediction (Richardson equation).

Geostrophic Divergence

The evaluation of divergence in (4-15) depends entirely on a reliable estimation of ageostrophic wind. Geostrophic wind cannot be used for evaluation of divergence. Divergence of geostrophic wind depends entirely on the variation of the Coriolis parameter, and this divergence is greatly overshadowed by other processes.

The roles of geostrophic and ageostrophic wind parts can be shown as follows. Assuming that the x-axis points east so that $\partial f/\partial x = 0$, the divergence of geostrophic wind is

$$
\begin{aligned}
\text{div } \mathbf{V}_g &= \frac{\partial}{\partial x}\left(-\frac{g}{f}\frac{\partial z}{\partial y}\right) + \frac{\partial}{\partial y}\left(\frac{g}{f}\frac{\partial z}{\partial x}\right) \\
&= -\frac{g}{f}\frac{\partial z}{\partial x}\frac{1}{f}\frac{\partial}{\partial y}f \\
&= -\frac{v_g}{2\Omega \sin\varphi}\frac{\partial}{a\,\partial\varphi}(2\,\Omega \sin\varphi) \\
&= -\frac{v_g}{a}\frac{\cos\varphi}{\sin\varphi} = -\frac{v_g}{a}\cot\varphi
\end{aligned}
$$

Using typical values in the last expression, the magnitude of geostrophic divergence can be estimated as

$$\text{div } \mathbf{V}_g \approx \frac{10 \text{ m s}^{-1}}{6.4\text{E}6 \text{ m}} \approx 1.6\text{E}-6 \text{ s}^{-1}$$

This is a much smaller magnitude than the divergence normally observed in the atmosphere. Therefore, geostrophic wind is useless in the continuity equation or in other equations where the sign and value of divergence play significant roles.

Wind can be represented formally as the sum of geostrophic and ageostrophic wind:

$$\mathbf{V} = \mathbf{V}_g + \mathbf{V}_a$$

This can be introduced in the continuity equation as

$$\frac{\partial\omega}{\partial p} = -\nabla\cdot\mathbf{V} = -\nabla\cdot\mathbf{V}_g - \nabla\cdot\mathbf{V}_a$$

Divergence of geostrophic wind should be neglected in view of above arguments. The result is

$$\frac{\partial\omega}{\partial p} = -\nabla\cdot\mathbf{V}_a$$

This form of the continuity equation shows that only the divergent wind plays a role in estimation (or calculation) of vertical motion. Geostrophic wind is irrelevant for this purpose. Extending this argument another step, we conclude that the vertical motion in the atmosphere is present only if the flow is ageostrophic.

The above estimation of divergence suggests that in most practical work the variation of the Coriolis parameter should be ignored in comparison with other terms. The geostrophic flow is almost exactly nondivergent. If divergence appears in the wind field, it is due to the ageostrophic part of the wind.

Contrary to the above conclusion, the meridional variation of the Coriolis parameter $\partial f/\partial\phi$ is very important in the dynamics of the waves in the westerlies. This topic is discussed in the dynamic meteorology with Rossby and barotropic waves.

4-6 DEFORMATION

Definition

The terms on the right-hand side of (4-3) and (4-4) can be combined into two other quantities, presently called A and B:

$$A = \frac{\partial v}{\partial x} + \frac{\partial u}{\partial y}$$
$$B = \frac{\partial u}{\partial x} - \frac{\partial v}{\partial y} \tag{4-17}$$

With these, the deformation of the horizontal wind is defined as

$$\text{def } \mathbf{V} = \sqrt{A^2 + B^2} \tag{4-18}$$

Vorticity (ζ), divergence (div **V**), and deformation (def **V**) keep the same values regardless of coordinate system's orientation. The intermediate quantities A and B in (4-17) vary greatly with the rotation of the coordinate system. So do the derivatives of wind components $\partial u/\partial x$, $\partial u/\partial y$, $\partial v/\partial x$, and $\partial v/\partial y$.

Deformation (4-18) can be explained geometrically as the rate of change of shape of fluid bodies. This quantity is very important in the formation of atmospheric fronts, in the explanation of cloud shapes, and in the diffusion of materials and properties.

Pure Deformation Flow

A flow field that has no vorticity and no divergence, but has def $\mathbf{V} \neq 0$, is of a hyperbolic shape (Fig. 4-9). Individual terms on the right-hand sides of (4-17) can be interpreted easier when the coordinate system is oriented along the *stretching* (CC′, also *dilatation*) and *shrinking* (DD′, also *contraction*) axes; the proof of this is left for a student exercise. The example in Fig. 4-9 is taken with the stretching direction (also called *dilatation axis*) at an angle β to the x-axis. This choice better illustrates the evaluation of the stretching direction.

The deformation field in the atmosphere seldom appears in its pure form. It is usually superimposed on other flow patterns, as are translation, jet stream pattern, rotation, and so forth. In this section, however, for the sake of simplicity, only pure deformation is discussed. These conclusions are still applicable to all fields that contain deformation. The deformation parts A and B can be evaluated in any flow, not only in the pure deformation flow as in Fig. 4-9.

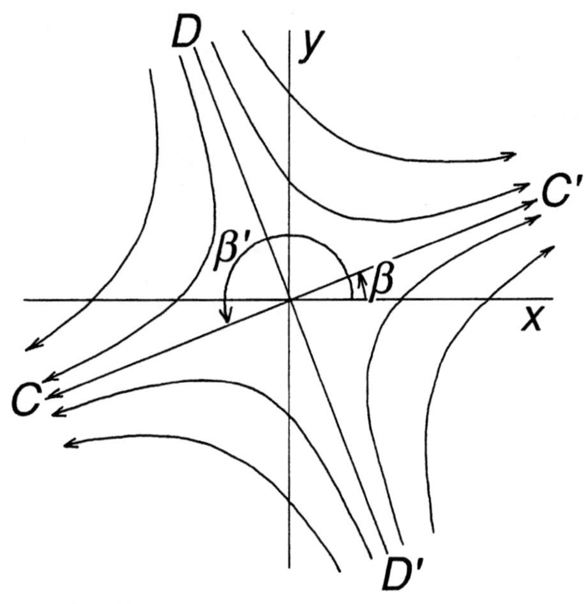

FIGURE 4-9. Pure deformation streamlines. CC′ is the dilatation (stretching) axis DD′ is the contraction (shrinking) axis. Angles β and β' show the direction of the dilatation axis.

Stretching Direction

The deformation elements A and B in (4-17) can be used to find the *direction of the dilatation axis*, that is, the line along which the material elements stretch (thus also called *stretching direction*). This is the direction CC′ shown by the angle β (or β') in Fig. 4-9:

$$\beta = \frac{1}{2} \operatorname{atan} \frac{A}{B} \tag{4-19}$$

This direction (or its opposite β') is significant in frontogenesis, as will be seen in Chapter 8. The other axis (DD′) is accordingly the *contraction* (or *shrinking*) *axis*.

Several flow patterns are characteristic of large deformation: *confluence*, *diffluence*, and *shear flow*. The shear flow was shown in Figs. 4-2 and 4-3. Assuming the x-coordinate along the flow, the principal contribution to the shear in these examples is from the term $\partial u/\partial y$. This term is equal to the term $\partial V/\partial n$ in natural coordinates when the coordinate n is oriented normal to the flow direction. In the examples in Figs.

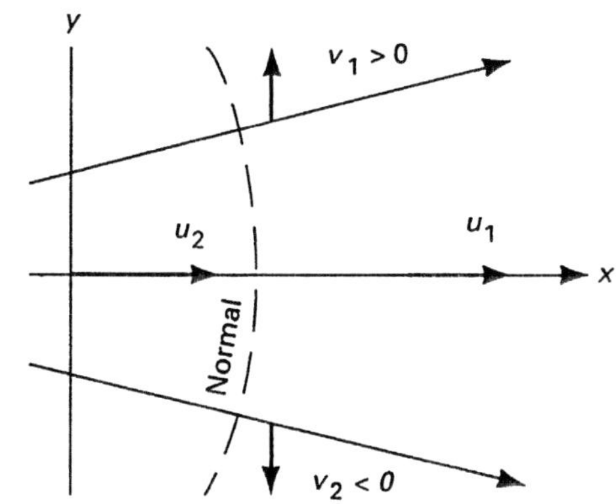

FIGURE 4-10. An illustration of diffluence with several straight streamlines (long, thin arrows). Wind components are shown as short, thick vectors **i** u and **j** v. Normal to the streamlines (dashed) is curved.

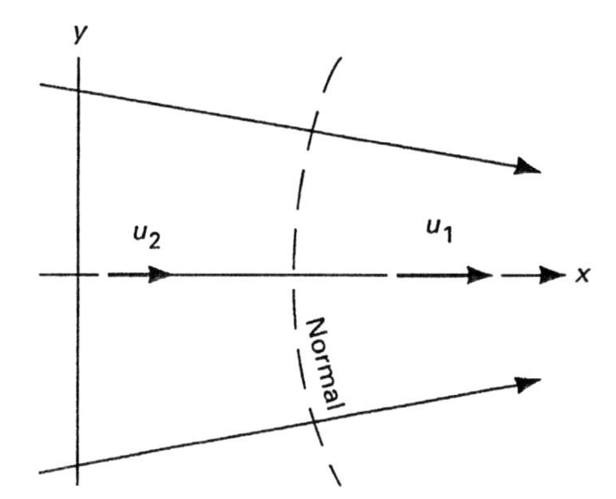

FIGURE 4-11. An illustration of confluence with several straight streamlines. Wind component u and the curved normal are as in Fig. 4-10.

4-2 and 4-3 the vorticity is also large due to the same term $\partial u/\partial y$.

Confluence (converging streamlines) in the flow can be found near points C and C′ in Fig. 4-9. *Diffluence* (diverging streamlines) is present near points D and D′. It can be said that the streamlines in these cases are convergent and divergent, respectively. However, we are using other expressions (confluent, not "convergent," and diffluent, not "divergent") to stress that we cannot easily conclude whether the flow is convergent or divergent in such cases. In the customary usage, "convergent" and "divergent" refer to the sign of $\nabla \cdot \mathbf{V}$ as in (4-4) and (4-9).

Further examples of diffluence and confluence are shown in Figs. 4-10 and 4-11. The typical quantity that shows diffluence is $\partial v/\partial y$, or its counterpart in natural coordinates $-V\ \partial\alpha/\partial n$. The relation between these terms is shown in Appendix H. The sign of $\partial v/\partial y$ is illustrated by the component v drawn in two places (as v_1 and v_2) in each of the Figs. 4-10 and 4-11. (Actually, the component v is drawn as a vector $v\,\mathbf{j}$, where $\mathbf{j}$ is the unit vector in the y-direction.) The distribution of v shows that $\partial v/\partial y$ is greater than zero near the middle of Fig. 4-10 and smaller than zero in Fig. 4-11.

Deformation of Fluid Bodies

Clouds and other fluid bodies in the atmosphere deform to elongate along the local dilatation axis. Elementary deformation of a circular body is illustrated in Fig. 4-12 for the cases of initially circular bodies in pure deformation flow.

Clouds are the most often observed fluid bodies that deform in the atmosphere. Other visually observed bodies are volumes of dry air between clouds, dust, and smoke clouds. We can observe the deformation of clouds but we cannot see the deformation of clear air since the air is transparent. Deformation of observed volumes gives a fair indication about details of air flow. We are primarily interested in the displacement of clouds since the clouds are important in weather and since cloud motion is a fair indicator of wind. The deformation of clouds gives additional details of air flow and associated weather phenomena.

Under natural conditions the elongation of fluid bodies does not go so straight as in Fig. 4-12. There is some curvature regularly present, so the elongated body bends and assumes various twisted shapes. In this way bands of clouds form. Figure 4-13 shows a

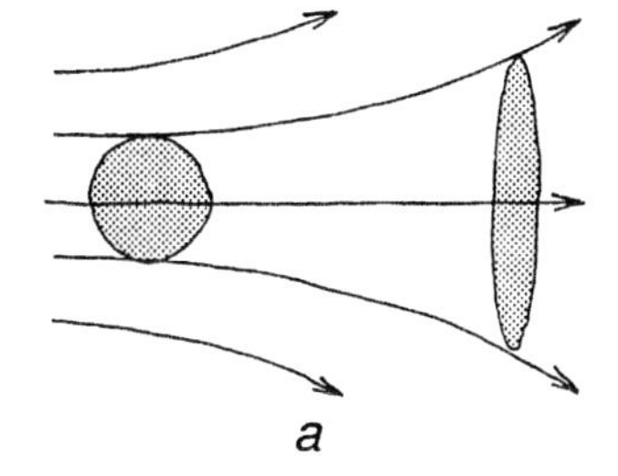

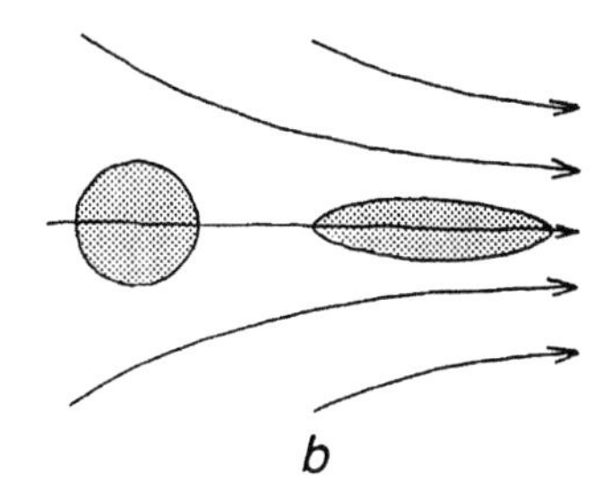

FIGURE 4-12. The change of shape of a circular fluid body in two-dimensional deformation flow. The elongation proceeds along the local dilatation axis: normal to the flow in the case of diffluence (a) and along the flow in the case of confluence (b).

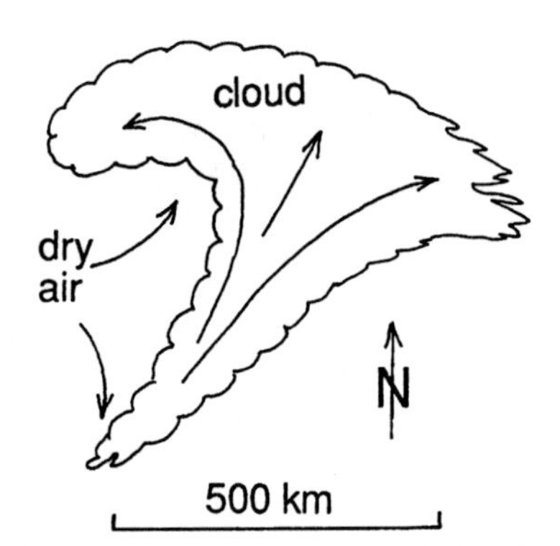

FIGURE 4-13. A hammerhead cloud as seen in satellite images. Deformation and rotation of the flow determine the shape of the cloud and of the dry air around the cloud.

hammerhead cloud shape that often appears in atmospheric vortices. Adjacent to the hammerhead cloud, dry air coming from the west also deforms in a hammerhead shape. One branch of dry air current enters under the head of the cloud "hammer."

A variation of the hammerhead cloud shape is the formation of the *mushroom shape* of buoyant bubbles. The mushroom shape forms in three-dimensional flow with vertical growth, similarly to the hammerhead, but symmetrically around the axis of growth. The mushroom shape is unfortunately best known from the cases of large explosions. However, this shape is very frequent with cumulus clouds on all scales, including the smallest lobes on the surface of cumulus congestus.

Elongation and bending of fluid bodies are intrinsic to low-viscosity fluid flows on all scales. We can see similar elongated and twisted shapes in many types of flow, from the striation of a drop of cream in a cup of coffee to the awesome spirals of galaxies.

4-7 VERTICAL WIND SHEAR

The Richardson Number

The vertical wind shear is the most frequent cause of air turbulence. Dynamic meteorology and observations show that a stably stratified fluid flow will become hydrodynamically unstable and break into a disorderly (turbulent) mode when the Richardson number is smaller than about 0.21. The Richardson number can be written in the following form that is suitable for the case when the wind direction does not change with height:

$$Ri = \frac{g \dfrac{\partial \theta}{\partial z}}{\theta \left(\dfrac{\partial V}{\partial z} \right)^2} \qquad (4\text{-}20)$$

Here the static stability is $\partial \theta / \partial z$ and the vertical shear of wind speed is $\partial V / \partial z$. The appearance of small-scale turbulence near the ground is attributed to this mechanism. The wind is slowed down due to the friction at the ground; therefore, a strong shear occurs between the wind near the ground and the wind at about 100 m above the ground. Under such circumstances, the square of the wind shear $(\partial V / \partial z)^2$ becomes large and *Ri* becomes smaller than the critical value of about 0.21. Then the turbulence develops.

Clear Air Turbulence

A similar mechanism appears in the higher layers of the atmosphere. In stable layers, with large values of $\partial \theta / \partial z$, the wind shear is often also large. Thermally stable layers with large $\partial \theta / \partial z$ promote large wind shear. Thermal stability causes weak vertical mixing. Therefore, the layers of air may glide on top of each other without significant friction between them. From (4-20) we see that the thermal stability and shear have opposing effects on the stability of the flow. With further increase in both $\partial \theta / \partial z$ and $\partial V / \partial z$, the situation eventually arises where the wind shear prevails in the competition with stability. This occurs because the wind shear is squared in (4-20) and stability is not. Then *Ri* becomes small enough for turbulence to start. This is the main mechanism of *clear-air turbulence*, which causes discomfort and occasional danger to aircraft.

In inversions of the free atmosphere, the vertical wind shear may be so large that the separation of flow occurs, with a discontinuity between superjacent layers of air. Under these conditions shear and ensuing turbulence are particularly strong.

Turbulent elements (eddies) in shearing layers have a preferable wavelength of the order of 100 m. Clouds form on these eddies on occasions of high humidity. These clouds appear as billows of altocumulus. Sometimes these waves have a shape of breaking waves on water.

4-8 ADVECTION

Terminology with Advection

Atmospheric properties and some bodies (gasses, dust, clouds) may be transported by the wind. Such transport is *advection*. Calculation or estimation of advection can be easily done for properties that are continuously distributed, that is, for properties that can be represented by isopleths. If the isopleths move due to some process other than air flow, such movement is not advection. For example, isotherms may be displaced due to heating, without air flow. Such a movement of isotherms is not advection, because it is not effected by wind. It is due to the addition of heat. Movement of isopleths as a consequence of mixing or diffusion is also not advection.

In hydrodynamics, "convection" is often used for the meteorological meaning of "advection." This text will adhere to the prevalent usage in meteorology where "convection" is used only for motions due to thermal instability in the atmosphere.

Among various books there is occasionally some disagreement about the signs in the expressions that follow, similar to the different signs in various definitions of the gradient. The usual meteorological convention will be followed here, such that positive advection contributes to the local increase of the advected quantity.

Conservative Variables

Mathematically, advection can be understood in terms of "conservative properties." A number of air properties are conserved in moving air parcels, at least under some conditions, and is expressed as

$$\frac{da}{dt} = 0 \tag{4-21}$$

where a is a variable such as potential temperature or mixing ratio. Although a is a function of space and time, it stays constant within moving air parcels (under some conditions). The distribution of this variable a can be described by a mathematical function:

$$a = a(x, y, p, t)$$

Using the Eulerian expansion [(C-22) from Appendix C], the conservation statement (4-21) can be rewritten as

$$\frac{\partial a}{\partial t} = -\mathbf{V}_3 \cdot \nabla_3 a$$

The expression on the right-hand side is the advection of a. The index 3 indicates that the vectors are three-dimensional. When the wind $\mathbf{V}$ and the del operator ∇ are split in the two-dimensional advection and vertical advection, the preceding equation becomes

$$\frac{\partial a}{\partial t} = -\mathbf{V} \cdot \nabla a - \omega \frac{\partial a}{\partial p}$$

where $\mathbf{V}$ and ∇ are two-dimensional (horizontal) vectors. In the last equation, the part

$$-\mathbf{V} \cdot \nabla a \tag{4-22}$$

is the horizontal advection and

$$-\omega \frac{\partial a}{\partial p} \tag{4-23}$$

is the vertical advection.

The advections (4-22) or (4-23) are scalar. Therefore, it is not proper to talk about "direction" of advection. The wind that enters the expression has direction; however, the advection does not.

Parts of the advection (4-22) and (4-23) are not components, since this word is reserved for components of vectors.

Alternative Formulations of Advection

When the advection is evaluated, the following other forms of advection terms may be used:

$$-\mathbf{V} \cdot \nabla a = -u \frac{\partial a}{\partial x} - v \frac{\partial a}{\partial y} \tag{4-24}$$

$$= -V \frac{\partial a}{\partial l} \tag{4-25}$$

$$= -V_n \frac{\partial a}{\partial n} \tag{4-26}$$

$$= V|\nabla a| \cos \alpha \tag{4-27}$$

In (4-24), u and v are the x- and y-components of wind, respectively. V in (4-25) and (4-27) is the wind speed, and l is the direction along the wind. In (4-26), V_n is the component of wind normal to the isopleths of a, n is the normal coordinate, and α is the angle between $\mathbf{V}$ and ∇a. Since this expression is straightforward, without ambiguity in direction and sign, expression (4-24) is suitable for advection determination with computers. Expressions (4-25) and (4-26) are practical for a manual evaluation of advection since in each of them we have fewer terms than on the right-hand side of (4-24). However, care must be taken with the normal direction (sense).

Geostrophic Advection

Advection is often estimated using geostrophic wind. All the above expressions are valid if geostrophic wind is used instead of wind ($\mathbf{V} \approx \mathbf{V}_g$). Estimation of geostrophic advection is based on the pattern formed by two sets of isopleths: contours and isopleths of a. After we determine the direction of the geostrophic wind and the direction of ∇a, we can determine the advection using the above rules for the wind and isopleths of a. Positive advection always contributes toward a local increase of a. This is shown on the charts

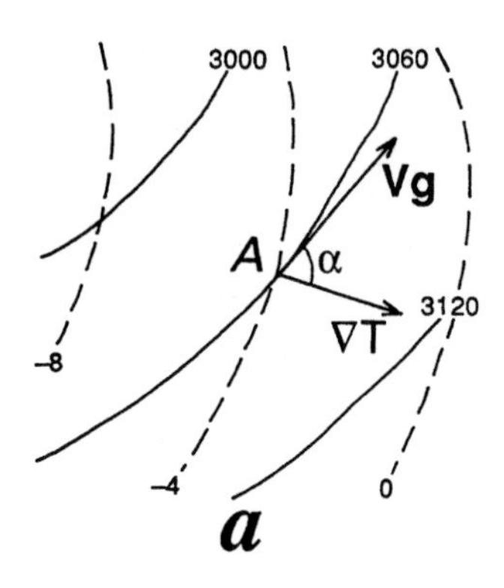

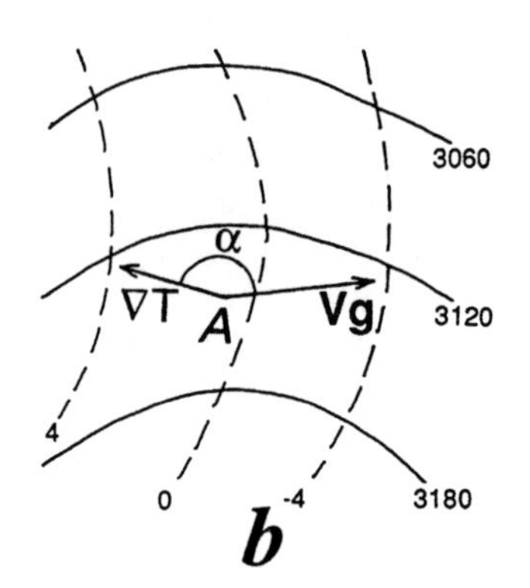

FIGURE 4-14. Examples of thermal advection (advection of temperature) at the 700-mb level. The solid lines are the contours in gpm, the dashed lines are the isotherms in °C. The angle α between the wind and ∇T determines the sign of advection. (a) Cold advection, $\alpha < 90°$; (b) warm advection, $\alpha > 90°$.

as a situation where wind (approximated by geostrophic wind) blows across the isopleths of a from the side with higher values. Similarly, negative advection is marked by wind blowing across the isopleths from the side with lower values of a. This convention about the sign of advection is very important, since in most cases we need an instantaneous estimation of this sign. In many cases we need to know only the sign of advection. Then we watch for the wind direction with respect to the isopleths of a.

The sign of advection is illustrated in Fig. 4-14 for temperature advection (also called *thermal advection*) at 700 mb. The vectors of geostrophic wind $\mathbf{V}_T$ and ∇T are drawn in point A. They constitute advection, as shown in (4-22) with $\mathbf{V} \approx \mathbf{V}_g$ and $a \approx T$. The angle α between these two vectors determines the sign of advection, as shown in (4-27). The advection is positive when $\alpha > 90°$ and the advection is negative when $\alpha < 90°$.

Mathematically, the geostrophic advection of a can be written using another set of expressions. When the expressions for geostrophic wind components [(C-8), Appendix C] are inserted for u and v in (4-24), we obtain the advection in the form of the Jacobian of z and a:

$$-\mathbf{V} \cdot \nabla a = \frac{g}{f}\left(\frac{\partial z}{\partial y}\frac{\partial a}{\partial x} - \frac{\partial z}{\partial x}\frac{\partial a}{\partial y}\right)$$

$$= \frac{g}{f} J(a, z) = -\frac{g}{f} J(z, a)$$

where J is the Jacobian operator or the functional determinant of a and z. This form of advection is often used when the advection is evaluated by computers.

Intersections between Isopleths and Contours

Geostrophic advection of a scalar quantity (as a above) is proportional to the density of intersections between the isopleths of a and height contours. The contours are streamlines of the geostrophic wind. The intersections between the contours and isopleths of a show that both the scalar product $\mathbf{V} \cdot \nabla a$. and the advection of a are different from zero. The geometry of advection is further described at the end of Appendix I. The following interpretation of advection will be given in terms of height contours on isobaric charts and isotherms.

Isotherms and contours, when they intersect, usually form quadrangular areas. The number of intersections is about equal to the number of elementary quadrangular areas between the isopleths, as illustrated in Fig. 4-15. Exceptions appear in cases of convoluted isopleths, as for example near point A where various forms of areas appear between isopleths. In any case, the intersections between contours and isotherms (or isopleths of other scalars) show the advection.

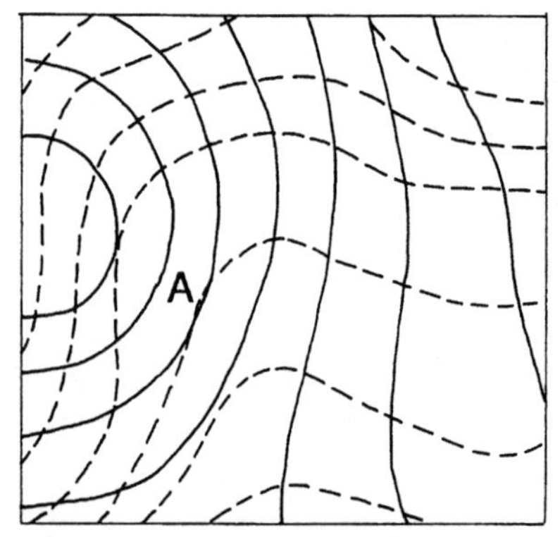

FIGURE 4-15. An illustration of advection with height contours (that is, geostrophic streamlines, solid) and isopleths of a scalar (dashed). Significant advection can be seen in regions with dense intersections between the two sets of lines. The advection is equal to zero in places where the two sets of isopleths are parallel to each other. Zero advection can be seen near point A.

Vorticity Advection

Geostrophic advection of geostrophic vorticity is a frequently evaluated or estimated quantity. Geostrophic vorticity advection can be evaluated or estimated on the basis of one set of isopleths: contours of the isobaric surface. Vorticity itself can be estimated by observing the wind shear and curvature as in Figs. 4-2 through 4-5. A rather typical case is represented in Fig. 4-5. There is only one isopleth of vorticity in that figure. The angle between the isopleth of vorticity ($\zeta = 0$) and the contours is largest at the points of inflection. Since we know that vorticity in this situation is positive on the side of the lows (L), we can see positive vorticity advection downstream from the troughs in the westerlies and negative vorticity advection upstream from the troughs. The change of sign of vorticity advection from one to the other side of the trough has profound dynamical significance, as discussed in Chapter 7.

Also important is absolute vorticity advection, defined as:

$$-\mathbf{V} \cdot \nabla(\zeta + f)$$

The contours of the Coriolis parameter f coincide with geographical parallels. When the isopleths of ζ are very sparse ("weak gradient"), the north wind represents positive advection of f. Conversely, again with a weak gradient of ζ, the south wind shows negative advection of absolute vorticity.

4-9 THE STREAM FUNCTION AND VELOCITY POTENTIAL

As any two-dimensional vector, the wind can be represented by two scalars ψ and χ using the representation

$$\mathbf{V} = \mathbf{k} \times \nabla\psi + \nabla\chi \tag{4-28}$$

or, in scalar notation,

$$u = -\frac{\partial\psi}{\partial y} + \frac{\partial\chi}{\partial x} \qquad v = \frac{\partial\psi}{\partial x} + \frac{\partial\chi}{\partial y} \tag{4-29}$$

New scalars are *stream function* ψ and *velocity potential* χ. They are presumably continuous functions of space and time, as are the wind components u and v.

The signs in (4-28) and (4-29) are different in some books on hydrodynamics. The above signs are standard in meteorology. If there are discontinuities in the wind, then the wind field must be smoothed before this theory can be used.

One advantage of the new variables ψ and χ becomes apparent when the derived variables of vorticity and divergence are computed. Differentiation of (4-36) and (4-37) yields

$$\zeta = \nabla^2\psi \qquad \text{div } \mathbf{V} = \nabla^2\chi \tag{4-30}$$

These are rather simple functions of one variable each. This is an advantage over the definitions (4-3) and (4-4) that need both wind components each.

Definitions (4-30) make it convenient to separate the rotational part and divergent parts of the wind. If one of the scalars ψ or χ is set equal to a constant (usually zero), the wind that remains can be one of the following:

(a) *nondivergent wind*, also called *rotational wind:*

$$\mathbf{V}_\psi = \mathbf{k} \times \nabla\psi \tag{4-31}$$

or, in scalar notation,

$$u_\psi = -\frac{\partial\psi}{\partial y} \qquad v_\psi = \frac{\partial\psi}{\partial x}$$

(b) *irrotational wind*, also called *divergent wind:*

$$\mathbf{V}_\chi = \nabla\chi$$

or, in scalar notation,

$$u_\chi = \frac{\partial\chi}{\partial x} \qquad v_\chi = \frac{\partial\chi}{\partial_y}$$

The upper-level wind is often closely nondivergent. In such cases, one scalar (ψ) describes the wind fairly well. The situation is very similar to the relation between geostrophic wind and the geopotential of an isobaric surface: Geostrophic wind is also completely determined by one variable, the geopotential of the isobaric surface.

A meaning of the stream function can be seen when the geostrophic approximation is used in the first equation (4-30):

$$\nabla^2\psi = \frac{g}{f}\nabla^2 z$$

This equation can be solved by the method of finite differences, using a computer and prescribing appropriate boundary conditions. However, an approximate solution is

$$\psi = \frac{g}{f}z \tag{4-32}$$

When computers were rare and expensive, the solution (4-32) was given the name *poor man's stream function*, since the multiplication of z by g/f did not require a computer; a calculating machine sufficed. The solution (4-32) shows that the stream function is proportional to z, which is the most common variable on weather charts. A further name was introduced for the solution (4-32) for the case when the Coriolis parameter f was kept constant. Such a ψ was the *very poor man's stream function*. A slide rule did the job.

The function gz/f is also called *geostrophic stream function* since the geostrophic wind can be approximately computed as

$$u_g = -\frac{\partial}{\partial y}\frac{gz}{f} \qquad v_g = \frac{\partial}{\partial x}\frac{gz}{f}$$

An instructive exercise is to evaluate the stream function numerically. This is described in the course on computational methods.

A more exhaustive discussion on kinematics in the atmosphere (besides other topics) can be found in the books by Bluestein (1992, 1993).

5

Analysis of Vertical Soundings

Vertical structure of the atmosphere is normally described by thermodynamic properties and wind. The thermodynamic part of this analysis is reviewed in this chapter. The analysis of wind is distributed among many related topics in other chapters. This chapter contains a review of basic thermodynamic quantities, but it cannot serve as a text for thermodynamics; instead, it is designed to refresh the knowledge that students acquired in dynamic meteorology. The formulas are given in explicit form to facilitate a numerical evaluation of needed quantities.

5-1 SOUNDINGS AND CHARTS

Most vertical soundings in the atmosphere are made by radiosonde. Soundings provide temperature, relative humidity, and wind as functions of pressure. From soundings we can identify several phenomena and quantities that show the state of the atmosphere. Among these are inversions, stable and unstable layers, fronts, tropopauses, air masses, humid layers, atmospheric stability, and various thermodynamic quantities. Elementary analysis techniques are shown in this chapter; however, more use of thermodynamic diagrams is made in several other chapters of this book.

New instruments, such as is VAS (VISSR atmospheric sounder, VISSR stands for visible and infrared spin-scan radiometer), are being developed on the basis of radiation measurements from satellites in various spectral bands. Another prospective instrument for atmospheric sounding is the *thermodynamic profiler*, which is based on radiation measurements like the VAS, only located at the earth's surface. VAS gives better information about the upper troposphere than the thermodynamic profiler, but the profiler is expected to yield more detail about the lower half of the troposphere. One disadvantage of instruments based on radiation is that they do not provide good resolution for inversions and other details. The main advantage of VAS is that it can economically give a great number of soundings over the surface of the earth, including soundings from vast unpopulated areas where it is not feasible to maintain radiosondes.

Manual analysis of soundings is traditionally done by drawing on the thermodynamic chart or on a vertical section that contains several soundings. Contemporary analysis of soundings is done by computers. Therefore, students must become familiar with the necessary formulas.

The earliest thermodynamic charts were constructed in nineteenth century, as theoretical thermodynamics developed. Following a proliferation of sounding devices (balloons, kites, airplanes, and, after 1928, radiosonde), numerous charts were designed by various authors. Following the tendency to standardize the procedures, only a few charts are currently used. The skew T–log p diagram is most often used in the United States. Other charts have somewhat different coordinates, but are very similar to the American skew T–log p chart.

A skilled meteorologist can find the important information from the radiosonde report in the numerically coded form. Computational interpretation of coded reports (without graphical representation) is growing in importance, since such a procedure is programmable on computers. Some equations for this task are given in Appendix C, and the ones for a moist atmosphere are given below. For instructional purposes, the graphical analysis of soundings also is shown.

A more detailed description of the skew T–log p chart can be found in the Air Weather Service manual (U.S. Air Force, 1978).

5-2 REVIEW OF THERMODYNAMIC VARIABLES IN HUMID AIR

Books on dynamic meteorology normally give the basic information on thermodynamic quantities needed for analyzing vertical soundings. Therefore, only a short review of important definitions and equations needed for calculation of thermodynamic variables is shown here. These formulas can be used in efficient calculation of thermodynamic variables.

Graphical evaluation of thermodynamic variables was standard before computers became available. Thermodynamic diagrams were developed for this purpose. Besides being practical for manual evaluation of variables, diagrams are useful for understanding thermodynamic concepts. Therefore, diagrams are shown here as well.

In many formulas, the temperature is required in °C. Such values are indicated by the lowercase letter t. Some attention should be paid to avoid confusion with the same letter used elsewhere for the time coordinate.

Saturation

One of the most important properties of water vapor is that it can be saturated. Saturation vapor pressure is a function of temperature only. The *saturation vapor pressure* e_s has been determined experimentally and is represented by

$$e_s = 6.1078 \exp \frac{19.8\, t}{273 + t} \approx 6.1 \exp 0.073\, t \qquad (5\text{-}1)$$

Here t is temperature in °C and e_s is in mb. The value 6.1078 is the saturation vapor pressure in mb at 0°C.

Saturation vapor pressure e_s is not a quantity we observe in the atmosphere. Instead, it is an intrinsic property of water vapor that has been established experimentally. Other forms of (5-1) can be found in the literature, where the numerical constants are somewhat different or where a polynomial is substituted for the exponential expression. With contemporary calculators and fast computers, there is little need to use the polynomial approximation. For estimation purposes, it is practical to notice that e_s doubles with a temperature increase of about 10°C.

Equation (5-1) gives the saturation vapor pressure over a flat water surface. The *saturation pressure over ice*, e_{si}, is lower than e_s and is given by

$$e_{si} = 6.1078 \exp \frac{22.5\, t}{273 + t} \approx 6.1 \exp 0.082\, t \qquad (5\text{-}2)$$

Saturation is different over a curved water surface (for example, over small water drops). Although important in cloud physics, this may be ignored in weather analysis.

Relative Humidity

Relative humidity r, the most frequently observed property of water vapor, can be used to find other variables. Relative humidity is practically defined as the ratio of vapor pressure to the saturation vapor pressure:

$$r = \frac{e}{e_s} \qquad (5\text{-}3)$$

Another definition of relative humidity is based on *mixing ratio* w and *saturation mixing ratio* w_s:

$$r = \frac{w}{w_s} \qquad (5\text{-}4)$$

The relative humidity r in (5-4) is slightly different from the relative humidity in (5-3), but the difference is so small that we normally neglect it. Therefore, the same letter r is used in (5-3) and (5-4).

Under normal conditions (no supersaturation), relative humidity cannot exceed unity. However, relative humidity is routinely reported in percents. Therefore, the definition (5-4) is often written as

$$r = 100 \frac{w}{w_s}$$

with the understanding that this value is expressed in percents.

Other Variables

When r is observed, (5-3) can be used to find the *vapor pressure* as

$$e = re_s \tag{5-5}$$

The relative humidity and vapor pressure in the atmosphere are regularly determined using the saturation over water (5-1).

The *mixing ratio* w is the ratio of densities of water vapor (ρ_v) and dry air (ρ_d):

$$w = \frac{\rho_v}{\rho_d} \tag{5-6}$$

These two densities (ρ_v and ρ_d) satisfy the corresponding equations of state for the two substances:

$$\rho_v = \frac{e}{R_v T} \tag{5-7}$$

$$\rho_d = \frac{p - e}{RT} \tag{5-8}$$

R_v and R are the gas constants for water vapor and dry air, respectively. The density of water vapor ρ_v from (5-7) is also known as *absolute humidity*. It is nondimensional (kg per kg), but it is often expressed in units of $g \cdot kg^{-1}$. Attention must be paid that the ratio of masses (kg per kg) must be used in calculations. The difference $p - e$ is the partial pressure of dry air.

The mixing ratio can also be expressed by dividing ρ_v by ρ_d:

$$w = \frac{eR}{(p - e)R_v} = \frac{0.622e}{p - e} \tag{5-9}$$

The constant 0.622 is the ratio of gas constants for dry air and water vapor. The difference $p - e$ is the partial pressure of dry air. Similarly, *saturation mixing ratio* is defined as

$$w_s = \frac{e_s R}{(p - e_s) R_v} = 0.622 \frac{e_s}{p - e_s} \tag{5-10}$$

The vapor pressure e and the saturation pressure e_s are much smaller than the air pressure p. Therefore, it is common to neglect e or e_s in comparison with p in the preceding two formulas. This gives us the quantities *specific humidity* q and *saturation specific humidity* q_s that are almost equal to the mixing ratio and saturation mixing ratio, respectively:

$$q = \frac{eR}{pR_v} = 0.622 \frac{e}{p} \tag{5-11}$$

$$q_s = \frac{e_s R}{pR_v} = 0.622 \frac{e_s}{p} \tag{5-12}$$

It is common to substitute the mixing ratio and specific humidity:

$$w \approx q \quad \text{and} \quad w_s \approx q_s$$

The *dew point* T_d is the temperature ("point on the temperature scale") at which the observed vapor pressure e is saturated. The dew point is evaluated using e in (5-1), instead of e_s, since the saturation may be not present. If e is used in (5-1), then t must be used instead of t_d. Solving (5-1) for t (now t_d) gives the dew point in °C as

$$t_d = 273 \frac{\ln e - \ln 6.1}{19.8 - (\ln e - \ln 6.1)} \approx 13.7 \ln \frac{e}{6.1} \tag{5-13}$$

In this expression, as well as in the next, e is still the vapor pressure (in mb) in spite of its suggestive similarity to the mathematical basis of natural logarithms. Millibars are used for e wherever the constant 6.1 is used.

The dew point is the most often reported humidity variable. The above formulas allow the evaluation of other quantities in terms of the dew point. So, for example, vapor pressure can be obtained when the shorter form of (5-13) is solved for e:

$$e = 6.1 \exp(0.073\ t_d) \qquad (5\text{-}14)$$

If this expression is inserted in (5-11), the specific humidity can be evaluated in terms of the dew point and pressure:

$$q = 3.794 \exp \frac{0.073 t_d}{p_{mb}} \qquad (5\text{-}15)$$

where p_{mb} is the pressure expressed in millibars.

Similarly, the relative humidity can be expressed as

$$r = \exp[0.073(t_d - t)] \qquad (5\text{-}16)$$

Another frequently used variable is the *dew point depression*

$$t - t_d$$

Usually, no letter is assigned to this variable; instead, this difference is used.

Lifting Condensation Level

Often the temperature and pressure at the *lifting condensation level* (LCL) are needed. Because these values are given by implicit functions of temperature and dew point, iterative procedures are needed for solving the equations. Approximation methods have been developed that give sufficiently accurate values at the lifting condensation level (subscript LCL). First, the temperature at the LCL can be evaluated using Barnes's (1968) formula:

$$t_{LCL} \approx t_d - (0.001296\ t_d + 0.1963)(t - t_d) \qquad (5\text{-}17)$$

where all values of temperature (lowercase t) are expressed in °C. Once t_{LCL} is known, the definition of potential temperature makes it possible to evaluate the pressure at the LCL. If we assume that the potential temperature stays unchanged in a parcel of air that lifts adiabatically,

$$\theta_{LCL} = \theta$$

then the definition of potential temperature (capital T for the absolute temperature in kelvins)

$$\theta \equiv T\left(\frac{p_0}{p}\right)^{2/7} \quad \text{and} \quad \theta_{LCL} \equiv T_{LCL}\left(\frac{p_0}{p_{LCL}}\right)^{2/7}$$

yield the pressure at the lifting condensation level:

$$p_{LCL} = p_0\left(\frac{T_{LCL}}{\theta}\right)^{7/2}$$

$$= p\left(\frac{T_{LCL}}{T}\right)^{7/2} \qquad (5\text{-}18)$$

The ratio 7/2 is the frequent thermodynamic ratio of the specific heat and gas constant for dry air, $c_p/R = 7/2$.

Wet-Bulb Temperature

Several quantities in the atmosphere are related to the wet-bulb thermometer. This thermometer is a part of the psychrometer. *Wet-bulb temperature* T_w is related to temperature and mixing ratio by

$$T_w = T - L\frac{w_s - w}{c_p} \qquad (5\text{-}19)$$

where c_p is the specific heat of dry air at constant pressure ($c_p \approx 1004$ J kg^{-1} K^{-1}). The latent heat of evaporation, L, which, although it varies with temperature, often may be used as a constant:

$$L = 2.5E6 \text{ J kg}^{-1} \text{ K}^{-1} \qquad (5\text{-}20)$$

Similarly to the potential temperature, the *wet-bulb potential temperature* can be computed by multiplication of T_w with the familiar expression from the definition of potential temperature:

$$\theta_w = T_w\left(\frac{p_0}{p}\right)^{R/c_p} \qquad (5\text{-}21)$$

Here p_0 is the pressure of 1000 mb if the pressure p is also used in mb. Other units of pressure can be used, if both p_0 and p have the same units. The nondimensional pressure,

$$\left(\frac{p}{p_0}\right)^{R/c_p}$$

is also known as the *Exner function*. Its inverse is used in (5-21).

Equivalent temperature can be obtained using the calculation of heat released if water vapor condenses. An approximate formula is

$$T_e = T\left(1 + \frac{Lq_s}{c_p T_{LCL}}\right) \qquad (5\text{-}22)$$

and similarly for the *equivalent potential temperature*:

$$\theta_e = \theta\left(\frac{1 + Lq_s}{c_p T_{LCL}}\right) \qquad (5\text{-}23)$$

The quantities θ_e and θ_w have the property of being constant (conservative) in air parcels during adiabatic processes with condensation or evaporation. Therefore, θ_e and θ_w are useful in air-mass identification.

Virtual temperature T_v can be used in the equation of state for evaluating density when the gas constant for dry air is used instead of the gas constant for the mixture of dry air and water vapor. A good approximation for the virtual temperature is

$$T_v = T(1 + 0.61\,w) \qquad (5\text{-}24)$$

The multiplier $1 + 0.61w$ differs from unity in warm and very humid air by only up to about 1%.

Several other thermodynamic variables are described in Section 5-8.

5-3 SKEW *T*–log *p* DIAGRAM

The Coordinates of the Thermodynamic Diagram

Most thermodynamic charts used today, including the skew *T*–log *p* diagram, are based on the energy of instability (or stability, if the sign is changed). The coordinates of such a chart (or *thermodynamic diagram*) are selected so that the area on the paper is proportional to the energy of instability:

$$E = R\int_a^b [T_L(p) - T_a(p)]\, d\ln p \qquad (5\text{-}25)$$

Here E is energy per unit mass accumulated by an air parcel that has been lifted adiabatically from level a to level b, T_L is the temperature of the adiabatically lifted parcel (*lifting temperature*), and T_a is the temperature of the environment (also called *ambient temperature*, for example, as measured by the radiosonde). The coordinates of the skew *T*–log *p* diagram are, as the name implies, temperature and logarithm of pressure. The coordinate axes are drawn at an angle, to distinguish easier the lapse rates of temperature between dry- and moist-adiabatic cases; hence the name "skew." When a parcel is "lifted" in the diagram, it starts from a point where the temperature is equal to the observed ambient temperature: $T = T_a$.

It is common to speak about *height* on the skew *T*–log *p* diagram, even if height is not a coordinate. The coordinate $\log p$ is closely proportional to $-z$, that is, to the height with the opposite sign.

Other printed lines on the diagram are families of solutions of thermodynamic equations. These lines are the adiabats (θ = const.), the pseudoadiabats (θ_w = const.), and isopleths of saturation mixing ratio (w_s = const.). Usually one schematic sounding of temperature is also printed on the graph: the U. S. standard atmosphere.

Other thermodynamic diagrams often used in various countries are the Stüwe diagram and the tephigram. The Stüwe diagram has the same coordinates (T, $\log p$) as the skew *T*–log *p* diagram, only with these coordinates normal to each other. The tephigram has the coordinates T, θ. (Different letters were used for the potential temperature in history. The name of the tephigram was selected when ϕ was fashionable.) Organization and usage of those other diagrams are practically the same as with the skew *T*–log *p* one. Also, the same auxiliary lines are printed on those diagrams. Therefore, studying the skew *T*–log *p* diagram helps understanding of most other thermodynamic diagrams as well.

Graphical Operations

The most frequently evaluated quantities on the skew *T*–log *p* diagram are illustrated in Fig. 5-1. This is a part of a complete chart, with lines selected for evaluating several of the variables in the example that follows. Usage of the graph can be demonstrated with a parcel of air having pressure p = 900 mb = 9E4 Pa, temperature T = 15°C = 288 K, and relative humidity r = 50%. These variables could be directly measured by the radiosonde. In the diagram in Fig. 5-1, p and T are represented by the doubly circled point A.

The quantities that follow can be evaluated using the diagram:

—The saturation mixing ratio w_s can be read directly from the chart using the printed isopleths of w_s. In original diagrams these lines are dashed and green. In our example we have w_s = 12 g kg^{-1}. This ratio is a function of temperature and pressure. Units of g kg^{-1} are commonly used. In equations, w_s must be used as the original ratio. In this example, w_s = 0.012.

—The mixing ratio w, the ratio of densities of water vapor and dry air, is evaluated from the definition of relative humidity (5-4):

$$w = rw_s$$

In our example w = 6 g kg^{-1}.

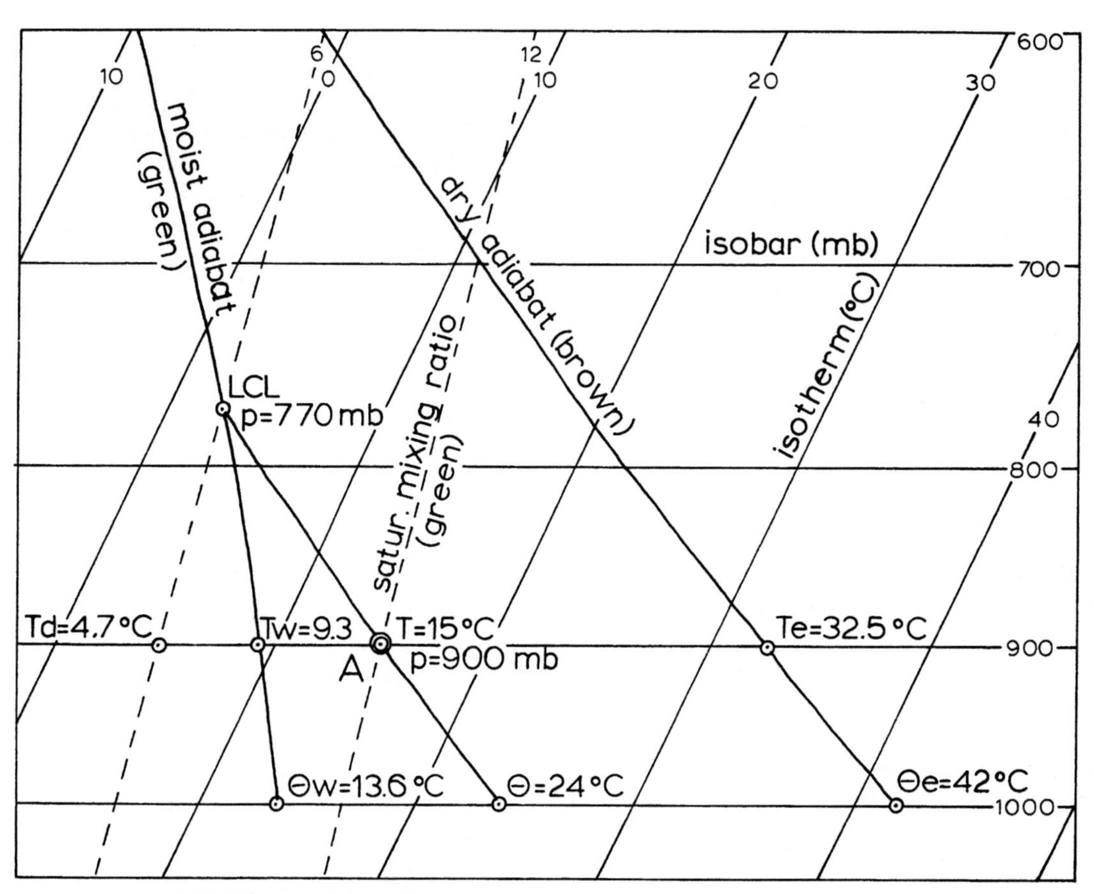

FIGURE 5-1. Evaluation of thermodynamic variables on a skew T–log p diagram. The observations of temperature and pressure are at A; other circled points show evaluated quantities.

—The dew point T_d can be found at the same pressure (900 mb), but at the isopleth $w_s = 6$ g kg^{-1}. This value is equal to 50% (observed relative humidity) of the evaluated mixing ratio. The dew point is the temperature where the current mixing ratio w becomes the saturation mixing ratio ($w_s = w$).

—The potential temperature θ is at the intersection of the dry adiabat through A and the isobar of 1000 mb. Here we have $\theta = 24$°C. Potential temperature is commonly expressed in kelvins; therefore, here also $\theta = 297$ K.

—The pressure at the lifting condensation level p_{LCL} is found at the intersection of the printed w_s isopleth for the evaluated w (that is $w = 6$ g kg^{-1} in our example) and dry adiabat through A. In our example, p_{LCL} is at 770 mb.

—The wet-bulb temperature T_w is at the intersection of the moist adiabat through the point indicated as LCL and the isobar through A; $T_w = 9.3$°C.

—The wet-bulb potential temperature θ_w is equal to T on the continuation of the moist adiabat through the LCL point to 1000 mb; $\theta_w = 13.6$°C.

—The equivalent temperature T_e can be found by following the moist adiabat through LCL until very low values of temperature are found near the upper left corner of the complete chart, outside the domain shown in Fig. 5-1. There the wet adiabat becomes tangent to a dry adiabat. From there we follow that tangent dry adiabat to the original pressure of 900 mb. In this way we find $T_e = 32.5$°C. The choice of printed adiabats and moist adiabats is such that the same values appear in both sets of lines. In this way, we may see which are the pairs of adiabats (one moist, one dry) that are tangent to each other, even if they touch outside of the printed diagram.

—The equivalent potential temperature θ_e is on the same dry adiabat as T_e, but at 1000 mb. Therefore, in Fig. 5-1 we have $\theta_e = 42$°C $= 315$ K.

Several other variables can be determined numerically, using the equations of Sec. 5-2 as follows:

—The virtual temperature T_v can be evaluated from the approximate equation (5-22), where w is given in g kg^{-1}. In our example, $T_v = 16°C$. This result is not drawn in Fig. 5-1 since it is too close to A. The virtual temperature is normally close to the temperature, since water vapor concentration (mixing ratio) is normally smaller than 0.03.

—The pressure of water vapor e can be found from the definition of specific humidity (5-11), where we may use $w \approx q$:

$$e = \frac{R_v}{R} p\, q = 1.61\, pq$$

The constant 1.61 is the ratio of gas constants R_v and R. In Fig. 5-1, $e = 1.61 \times 9E4 \times 0.006 = 869$ Pa ≈ 8.7 mb.

—The specific humidity can be expressed in terms of mixing ratio when (5-9) and (5-11) are used:

$$q = w\left(1 - \frac{e}{p}\right)$$

This formula shows again that the specific humidity is almost equal to the mixing ratio w since q is about 0% to 3% smaller than w. In our example

$$q = 0.006\left(1 - \frac{869}{9E4}\right)$$
$$= 0.00594 \approx w$$

—The density of water vapor ρ_v [also called *absolute humidity*, (5-7)] can be derived from the definition of specific humidity (5-11). The specific humidity q may be used instead of w:

$$\rho_v = q\rho = \frac{qp}{RT_v}$$

Our example yields

$$\rho_v = \frac{0.006 \times 9E4}{287 \times 289} = 6.5\mathbf{E}{-3} \text{ kg m}^{-3}$$

The above elements give useful information for the diagnosis of air masses and fronts. They provide insight into the type of precipitation (frozen or not). The likelihood of convection is another item of interest that is evaluated from the weather elements on the thermodynamic chart.

5-4 APPLICATION OF THE THERMODYNAMIC DIAGRAM

When sounding values are plotted on a thermodynamic diagram, several other atmospheric properties can be deduced. Some of the most important atmospheric features that should be recognized on the sounding graph are:

(a) the temperature of the air mass,
(b) the existence of humid layers (possibly with clouds),
(c) the type of precipitation,
(d) the stability of atmospheric layers,
(e) fronts and tropopause(s), and
(f) vertical wind distribution.

Points (a) and (e) are discussed at length in Chapters 8 and 10; points (b), (c), and (d) are discussed in this chapter. Point (f) is not a thermodynamic concept, although it is commonly drawn on the thermodynamic diagram. Analysis of wind is discussed in several chapters of the book. Thermal wind is related to thermodynamics, but it is described in Chapter 3.

Ambient Temperature and Dew Point

Figure 5-2 illustrates the elementary terminology used in analysis of the soundings. The subsections that follow describe the features seen in this figure.

Observed *ambient temperature* $T_a(p)$ and *ambient dew point* $T_d(p)$ are the most frequently drawn graphs on the chart. The ambient temperature is drawn as a solid line in Fig. 5-2. The dew point is drawn by a dashed line. The graphs of these two variables are formally functions of pressure. Several useful mathematical operations can be done with these functions as, for example, to evaluate various quantities and lapse rates. When these functions are given graphically, the mathematical operations with them are best done in the same way: graphically. Analogously, the sounding should be analyzed numerically if the data are given in a numerical form.

Wind observations are commonly drawn with wind barbs in the margin of the paper with the thermodynamic diagram. The same symbols are used as on isobaric charts. The analysis of wind is not described in this chapter. Here the emphasis is on thermodynamics and hydrostatic stability.

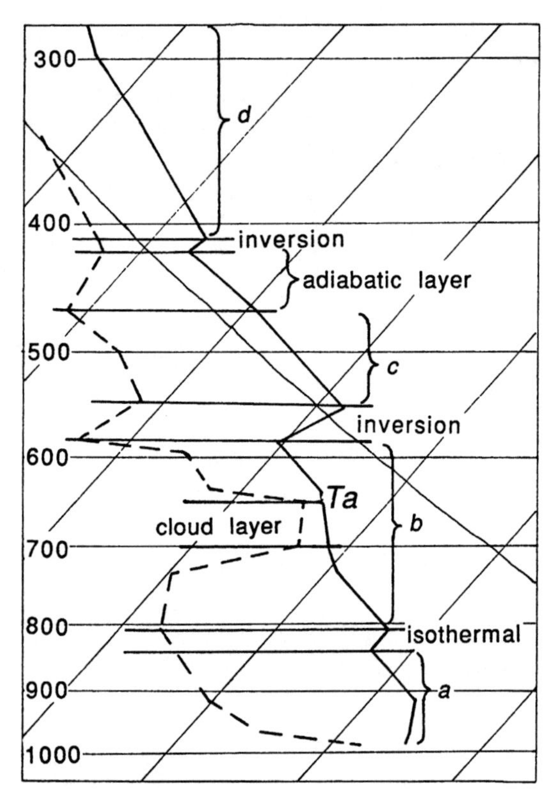

FIGURE 5-2. An example of ambient temperature $T_a(p)$ and ambient dew point $T_d(p)$ represented on the skew T–log p diagram. Several layers are characterized by the different slopes of the curves.

Lapse Rate of Temperature

The slope of the T_a curve is usually described by the *lapse rate*. The lapse rate of temperature is defined as the negative derivative $-\partial T/\partial z$. The negative sign is selected in analogy with the second definition of the gradient [(D-1) from Appendix D]. For usage in pressure coordinates the lapse rate is

$$-\frac{\partial T}{\partial z} = g\rho \frac{\partial T}{\partial p} \tag{5-26}$$

However, often only the expression $\partial T/\partial p$ is used, without $g\rho$ and without the minus sign. The value on each side of (5-26) is of the same sign.

The slope of the curves should be compared with the coordinate lines of T = const. and with the adiabats (θ = const.). Both sets of lines are printed on the diagram. The layers a, b, c, and d in Fig. 5-2 show a "normal" temperature decrease with height. Such layers are most common in the atmosphere. In these layers, the lapse rate of temperature with height is positive, that is,

$$-\frac{\partial T}{\partial z} > 0 \quad \text{and} \quad \frac{\partial T}{\partial p} > 0$$

The layer between 810 and 845 mb shows "zero lapse rate," since the ambient temperature curve coincides with the coordinate line. The derivatives in (5-26) are equal to zero. Therefore, this is an *isothermal layer*.

Temperature increases with height in the layers between 550 and 585 mb and between 412 and 420 mb. These are *inversions*. The name reflects the surprise of the first observers of inversions in the nineteenth century; they had expected that temperature always decreased with height.

The layer between 430 and 464 mb is an *adiabatic layer* since the ambient temperature distribution T_a coincides with the adiabat $\theta(p)$ = const. The adiabat of 40°C is drawn in Fig. 5-2, so its slope can be compared with the slope of the T_a curve. An adiabatic layer forms usually as a consequence of turbulent mixing. Thereby the potential temperature and mixing ratio are equalized through the layer. Similarly, the specific humidity is equalized in the adiabatic layer. The dew point in a well-mixed layer follows exactly the isopleths of saturation mixing ratio.

Humid Layers

The two curves T_a and T_d rarely touch each other. The measurements practically never show 100% relative humidity. This is due to the limited sensitivity of the hygrometer in the radiosonde. However, it has been observed that the clouds are abundant, and even continuous, when the radiosonde reading shows as low as 70% relative humidity. This corresponds to a dew-point depression of about 5°C. In agreement with this experience, the proximity of the T_a and T_d curves between 655 and 700 mb in Fig. 5-2 leads us to assume that this is a cloud layer.

The existence of clouds in a space with unsaturated water vapor can be explained by the large natural variability of relative humidity. If the measured humidity is above 70%, then it is likely that air parcels with saturated vapor are in the vicinity. Clouds develop in these parcels.

A small dew-point depression near the earth's surface may indicate fog, especially if the wind is light. If at the same time the temperature is below freezing, it may be a supercooled fog. Both fog and supercooled fog are hazardous for traffic since they obstruct visibility. The supercooled fog often causes formation of *rime* (*hoarfrost*) on solid objects. Rime forms primarily on objects that are vertical or elevated from the ground, such as tree limbs and electric wires.

Rain and Snow

Presence of thick humid layers may indicate that precipitation is falling, especially if there is indication that lifting of air is taking place. If, in addition, the temperature is below freezing, that precipitation is in form of snow. The snow may penetrate through about 300 m of air that has temperature above freezing. Such a layer cools fast due to conversion of heat to latent form during melting of snow. After a few hours of melting snow, an isothermal layer forms with a temperature of 0°C, and then all snow reaches the earth's surface. In connection with this, it is a matter of climatological interest to know that the observation of 0°C appears more often than other nearby values, due to cooling during snow melting.

More complicated cases may occur. Rain may be falling through a surface layer (that is, a layer at the earth's surface) that is colder than 0°C. If this cold layer is more than about 400 m thick, all rain freezes and falls as *ice pellets*, also called *sleet*. This is a comparatively harmless form of precipitation, except that a layer of ice pellets is slippery on highways, just as snow is.

Much more hazardous is *freezing rain*, also known as an *ice storm*. This can be recognized on the thermodynamic diagram as a thin layer of cold air below freezing and a thick humid layer with precipitation above it. In the warm layer aloft, all snow melts. Then the raindrops reach the ground in liquid form and freeze upon contact. Ice in freezing rain accumulates primarily on horizontal surfaces and much less on the sides of buildings. The solid layer of ice formed by freezing rain represents great danger for traffic. The weight of ice may do damage to tree limbs and electric wires.

Supercooled water drops of cloud or rain are of most danger to aircraft. They deposit a layer of ice on the aircraft's wings that changes the aerodynamic lift and may lead to catastrophe.

5-5 THERMAL (HYDROSTATIC) STABILITY OF THE ATMOSPHERE

Graphical Representation of Stable and Unstable Layers

The thermodynamic chart is a practical tool for representing the stability of air masses. A short review of stability categories, using the parcel method, is shown in Fig. 5-3. Each of the four frames in this figure depicts a small section of the skew T–log p diagram in Fig. 5-1. Five lines are represented in each diagram: isobar, isotherm, pseudoadiabat, dry adiabat, and ambient temperature as a function of elevation (pressure or height). The observed ambient temperature is drawn as an uneven line. It simulates the trace from a recording instrument. The four other lines are smooth, as they are the coordinates or are obtained from the appropriate equations of thermodynamics.

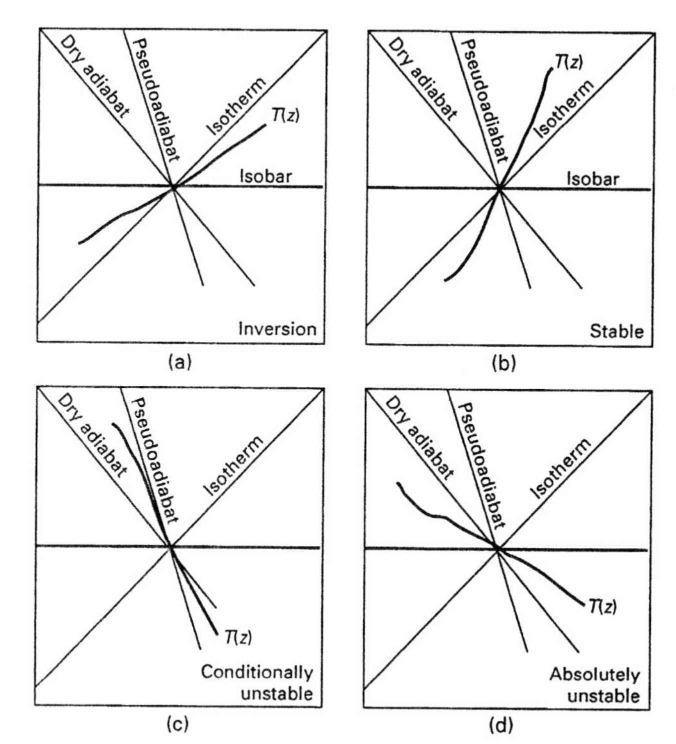

FIGURE 5-3. Four main categories of hydrostatic stability as shown on a skew T–log p diagram. The thicker, uneven line shows a possible measured ambient temperature T_a, here $T(z)$, for four important cases. The other lines (smooth, thin) are the theoretical lines that are normally printed on the diagram.

We often switch between the expressions *stability* and *instability*, since they indicate about the same concept, only with opposite signs. Instability is usually defined with the same formula as the stability, with the sign changed.

Stability in Terms of Vertical Derivative

This section describes the stability of the atmosphere that can be concluded on the basis of the *parcel method*. This method considers a parcel of air that is initially at the same temperature as the ambient atmosphere, but when it lifts its temperature decreases according to the adiabatic lapse rate. The same conclusions can be reached for a sinking parcel. To avoid encountering the earth's surface, it is common to consider a rising (not a sinking) parcel.

The four principal categories of stability are shown by the four possibilities of the slope of the ambient temperature curve against the isobar. Since the diagram has skew coordinates and since we often talk in terms of height (also in p coordinates), there are several angles that can be used as the "slope." Therefore, it is advisable to avoid the use of "slope." Instead, the stability should be expressed by the lapse rate of temperature with height. The lapse rate of observed ambient temperature with height is defined as the negative vertical derivative of temperature in (5-26).

It is also common to use the expression *lapse rate of temperature* for the expression $\partial T/\partial p$, without the minus sign. In this way, the sign of the quantity is the same as the sign of the lapse rate in z-coordinates in the definition $-\partial T/\partial z$, with the minus sign.

The equations of thermodynamics yield the definitions and approximate values of the dry (Γ_a) and moist (Γ_w) adiabatic lapse rate as

$$\Gamma_a = -\left(\frac{dT}{dz}\right)_a = \frac{g}{c_p} \approx 0.0098 \text{ K m}^{-1}$$

$$\Gamma_w = -\left(\frac{dT}{dz}\right)_{aw} = \frac{\Gamma_a}{1 + \dfrac{L}{c_p}\dfrac{dw_s}{dT}} \approx 0.07 \pm 0.03 \text{ K m}^{-1}$$

where the subscripts a and aw indicate adiabatic and pseudoadiabatic processes, respectively. The value of dw_s/dT comes from the Clausius–Clapeyron equation or from the experimentally determined w_s (5-10). The above expressions follow from the first law of thermodynamics and the hydrostatic approximation.

Meaning of Derivatives

There is a fundamental difference in meaning between the various lapse rates: Γ is an observed quantity, while Γ_a and Γ_w are theoretical values derived from the equations of thermodynamics. A possible point of confusion is that all these quantities have the same dimension of K m^{-1} and can be compared with each other.

With the above expressions, the main categories of stability can be established by a comparison of the observed lapse rate Γ with the theoretical values Γ_w and Γ_a. These categories are:

Inversion:

$$\frac{\partial T}{\partial z} > 0, \ \Gamma < 0$$

Stable, but not inversion:

$$\frac{\partial T}{\partial z} < 0, \ 0 < \Gamma < \Gamma_w$$

Conditionally unstable:

$$\Gamma_w < \Gamma < \Gamma_a$$

Absolutely unstable:

$$\Gamma > \Gamma_a$$

as they are illustrated in Fig. 5-2.

Adiabatic Layer

The observed lapse rate Γ may be equal to the dry adiabatic one ($\Gamma = \Gamma_a$). A layer with such a lapse rate is the *adiabatic layer*. Stability of the adiabatic layer is *neutral*, assuming that no condensation of water vapor takes place. Turbulent mixing of air equalizes the potential temperature and thereby produces an adiabatic stratification. Therefore, an adiabatic layer is likely to be a layer with strong turbulence. Formation of an adiabatic layer is somewhat analogous to the exchange of heat between atmospheric layers by radiation. However, the radiation alone tends to form an isothermal layer.

A turbulent adiabatic layer has a constant mixing ratio. This ratio is conservative, much like the potential temperature. The dew point in the adiabatic layer follows the printed isopleths of saturation mixing ratio.

Potential Instability

Entirely different from these four categories is the *potential instability* (sometimes called *convective instability*). Theory shows that layers within which the equivalent-potential temperature decreases may become unstable when lifted sufficiently. The potential instability is defined analogously to the ordinary static stability, but in terms of wet-bulb potential temperature,

$$\frac{\partial \theta_w}{\partial z} \begin{cases} > 0 \text{ potential stability} \\ < 0 \text{ potential instability} \end{cases}$$

Practically, this means that potentially unstable layers are humid on the bottom part and dry on the top part. A typical potentially unstable sounding superficially looks like a subsidence sounding described below. The presence of potential instability is determined with information on the vertical distribution of θ_w. A potentially unstable atmosphere becomes absolutely unstable when saturation occurs in the layer. Usually saturation is achieved by lifting of the layer, hence the designation "convective instability." However, saturation may be achieved also by evaporation of rain that falls through the layer.

Inversion

As described above, a layer where the temperature increases with height is an *inversion*. Inversions are particularly stable layers. A number of atmospheric phenomena are related to inversions. It is useful to distinguish several categories of inversions, as shown in Fig. 5-4. The main types of inversion are as follows.

Subsidence inversion. Typically, the dew point decreases with height within the inversion. The humidity at the top of the inversion is low (Fig. 5-4a).

Frontal inversion. In such an inversion the dew point increases with height. The air at the top of the inversion is humid (Fig. 5-4b).

Exceptions (of course) have been observed. The passive cold front often occurs under a subsiding warm air mass. Then the sounding looks like the subsidence model in Fig. 5-4a. Soundings in rainy weather may show "frontal inversion" even if the structure of the atmosphere shows that there are no fronts in the vicinity.

Other categories of inversions include the following.

Radiation inversion. The temperature and dew point resemble the frontal inversion, except that the bottom of the inversion is at or very close to the ground. When a radiation inversion does not dissipate during the day, it may further intensify the next night. Strong and widespread radiation inversions of the polar night develop into the arctic air. The inversion on the top of the arctic air develops into the arctic front. That is the main mechanism of generation of arctic air.

Turbulent inversion. This inversion forms on the top of an adiabatic layer. Turbulent eddies in the subjacent adiabatic layer cool the bottom of the turbulent inversion. The top of the turbulent inversion is not affected by turbulence and cooling. The subjacent adiabatic layer can be recognized by an adiabatic lapse rate and by a constant mixing ratio: The dew-point distribution follows the isopleths of constant saturation mixing ratio. An adiabatic inversion and its subjacent adiabatic layer are illustrated in Fig. 5-2 between 410 and 470 mb.

Again, as with all other models, we find many exceptions and intermediate cases when it is not easy to apply the rules. Some examples of interpretation of inversions and stable layers are shown in later sections.

5-6 INSTABILITY INDICES

For comparing stability in different places on the weather charts, it is convenient to represent instability (or stability) by a number. Such a number is usually called the *index of instability* (or *index of stability* for some indices). Other measures of instability are obtained by vertical integration of thermodynamic quan-

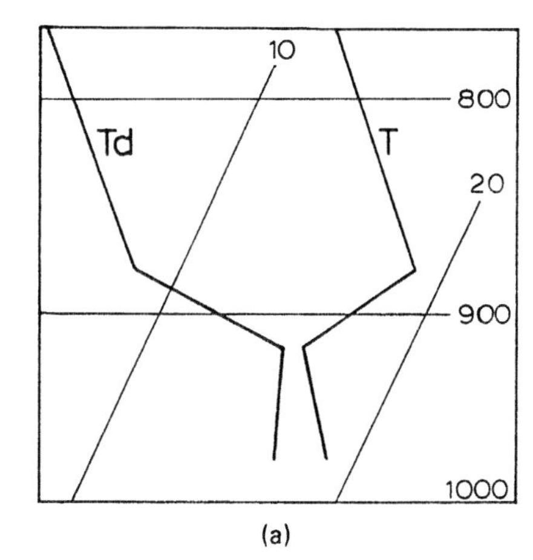

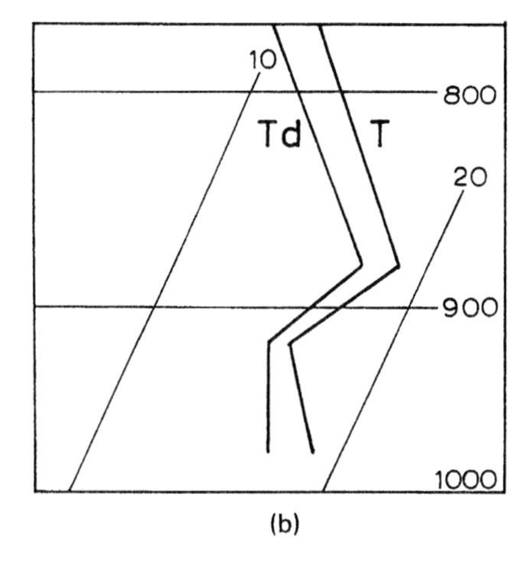

FIGURE 5-4. Vertical distribution of temperature and dew point in the cases of (a) subsidence inversion and (b) frontal inversion.

tities; these are described Section 5-8. An index of instability is usually evaluated from the observed vertical distribution of temperature and dew point. The U.S. Weather Service disseminates two such indices: the *lifted index* (described in the next section, with integration) and the *K index*. Another widely used index is the *totals index*. More detailed information about these and other indices of instability and their application can be found in the manuals by Miller (1972), U.S. Air Force (1978), Peppler (1988), and Alcorn (1990).

The three indices that follow next are all indices of instability (as opposed to indices of stability) since their values become larger in a more unstable atmosphere.

K Index

The *K index* is evaluated using the values of temperature t and dew point t_d both in °C, at several standard levels:

$$K = t_{850} - t_{500} + t_{d850} - t_{700} + t_{d700}$$

The pressure levels (in mb) of t and t_d are indicated by indices. Values of the K index over 24 indicate a likelihood of showers. Values over 30 are usually associated with severe thunderstorms.

Totals Index

The totals index

$$tt = t_{d850} + t_{850} - 2 \times t_{500}$$

has achieved a good reputation. It uses a few values from the previous formula for K. Values of tt over 44 indicate thunderstorms. Values over 50 indicate severe thunderstorms with large hail and tornadoes. If values near 55 appear, groups of tornadoes are feared.

SWEAT Index

The SWEAT (severe weather threat) index is

$$\text{SWEAT} = 12\,D + 20(tt - 49) + 2\,v_8 + v_5 + 125(S + 0.2)$$

where

$D = t_d(850 \text{ mb})$, °C; if $D < 0$, change it to $D = 0$
$v_8 = |\mathbf{V}|(850 \text{ mb})$, kt
$v_5 = |\mathbf{V}|(500 \text{ mb})$, kt
$S = \sin[\text{dd}(500 \text{ mb}) - \text{dd}(850 \text{ mb})]$
tt = totals index, °C; if $tt < 49$, change $(tt - 49)$ to 0
dd = the reported wind direction

The term $125(S + 0.2)$ should be dropped in any of the following cases:

1. when the wind direction (dd) at 850 mb is between 130 and 250° ($130° < dd_{850} < 250°$)
2. $210° < dd_{500} < 310°$
3. $dd_{500} - dd_{850} > 0$
4. $v_8 \leq 15$ kt and $v_5 \leq 15$ kt

SWEAT values of over 300 occur with severe thunderstorms. Values over 400 accompany tornadoes. The usefulness of the SWEAT index is rather low if its values are under 250 (Miller, 1972).

An important unanswered question is, Which index is best? Unfortunately, there are no conclusive results that show a definite advantage of any index of instability over others. Some forecasters, depending on their previous experience, believe that one index is better than the other. Weather analysts will have to rely on their own comparison between various indices and other tools.

5-7 VERTICAL VARIATION OF TEMPERATURE

Parcel Temperature versus Ambient Temperature

Useful conclusions about the state of the atmosphere can be made on the basis of vertical temperature variation in these circumstances:

1. in the ambient atmosphere, as observed by the radiosonde, and
2. within a parcel of air that lifts through the atmosphere.

Lifting of the parcel is a mathematical experiment that simulates a possible corresponding process in the atmosphere. The equations of thermodynamics make it possible to evaluate the temperature of the lifted parcel as a function of elevation (or pressure). Normally, an adiabatic process is considered. If the condensation of water vapor plays a role, the *pseudoadiabatic process*, the process where all liquid water (drops of cloud and rain) are promptly eliminated from the parcel, is assumed.

Besides the temperature variation with height, the variation of dew point gives also important information about condensation and stability. Therefore, the dew point is regularly plotted on the thermodynamic diagram.

The ambient temperature and the temperature of the lifting parcel both vary vertically and have the same dimension, but they have entirely different meanings. These two concepts are described in the following subsections.

Ambient Temperature

The vertical distribution of temperature, also called *ambient temperature* (T_a), is usually observed with the radiosonde. It can be formally represented as a function of pressure:

$$T = T_a(p)$$

where T_a may be given numerically as a set of values of T_a at several pressure levels. It is convenient to have equally spaced intervals between values of T_a, for example, at every 10 mb or so. Reports usually do not contain equally spaced values. Reports from the observing stations most commonly contain values at standard levels and at *significant points*. The significant points show discontinuities (usually kinks) and extrema on the T_a curve.

For the purpose of analysis, the T_a curve is usually plotted on the thermodynamic diagram with the details that come from the measuring device (radiosonde, dropsonde). A well-observed ambient temperature is usually characterized by small-scale patterns that cannot all be accounted for by the known conceptual models. The atmosphere always contains many irregularities that cannot be unambiguously explained. (The observed details in the world that do not fit in our conceptual models are called irregularities. Otherwise, they would be parts of established models.)

Lifting Temperature

Several important quantities in the atmosphere can be evaluated on the basis of the *lifting temperature* $T_l(p)$. This is the temperature *within a parcel* that is lifted pseudoadiabatically. The lifting temperature is very well determined by the laws of thermodynamics. The evaluation of lifting temperature for a parcel can be made on the basis of its initial temperature, humidity, and pressure. Since it is evaluated from comparatively simple equations, the graph of the lifting temperature is smooth. One kink on this graph occurs at the lifting condensation level, where the dry and moist regimes meet. The lifting temperature must be distinguished from the ambient temperature, since the latter is the observed (measured) temperature in the ambient atmosphere, *without the lifted parcel.*

The lifting temperature is evaluated on the basis of the adiabatic forms (dry and wet) of the first law of thermodynamics. The lifting temperature in the lower part of the troposphere is based on the first law of thermodynamics [(C-33) from Appendix C], assuming the dry-adiabatic process ($dQ = 0$):

$$\frac{dT}{dp} = \frac{\alpha}{c_p} = \frac{RT}{c_p p} \tag{5-27}$$

An integration of this equation yields the lifting temperature. This integration is performed from the surface where the pressure is p_0 until the LCL (5-18) where the pressure is p_{LCL}. The result is the lifting temperature T_l of the parcel expressed as a function of pressure:

$$T(p) = T_l(p) = T_{dry}(p)$$

$$= T_0\left(\frac{p}{p_0}\right)^{2/7} \tag{5-28}$$

as in the definition of potential temperature. T_0 and p_0 are the constants of integration that describe the initial temperature and pressure of the lifting parcel. The subscript "dry" is introduced to indicate that condensation of water vapor does not occur in the lifting parcel. The lifting temperature T_l (or T_{dry}) describes the state of the parcel only while no condensation takes place.

The boundary conditions T_0 and p_0 for the solution of (5-27) are usually the temperature and pressure at the earth's surface. Other values of temperature are sometimes used (for example, the predicted maximum temperature at the surface).

The form of (5-28) shows that the lifting temperature is a smooth mathematical function on the thermodynamic diagram. To the contrary, the observed variation of temperature T_a need not be smooth. Many small-scale processes in the atmosphere cause the observed variables to have rugged forms.

Condensation Effect

The function $T_l(p)$ in (5-28) is meaningful up to the LCL. Above the LCL, the differential of heating dq should be set equal to the differential of released latent heat of condensation:

$$dq = -L dw_s$$

where L is the latent heat of condensation of water vapor and w_s is the saturation mixing ratio. When this is introduced into the first law of thermodynamics (C-33), the equation for the moist adiabatic process (C-42) follows as

$$\frac{dT}{dp} = \frac{\alpha}{c_p} - L\frac{dw_s}{dp} \qquad (5\text{-}29)$$

This equation can be integrated to yield formally

$$T(p) = T_l(p) = T_{moist}(p) \qquad (5\text{-}30)$$

that describes the lifting temperature T_l of the parcel of air above the LCL. We are currently not interested in the exact mathematical form of T_{moist} in (5-30) since the curve can be conveniently found on the thermodynamic diagram. The boundary condition (starting point) for this curve is the lifting temperature T_l at pressure p_{LCL}, where

$$T(p_{LCL}) = T_{moist}(p_{LCL})$$

Graphical evaluation of the lifting temperature is easy on a standard thermodynamic diagram, since the graphs of general solutions of (5-27) and (5-29) are already printed on it as families of lines. These are the *dry* and *moist adiabats,* that are described for the skew T–log p diagram in Section 5-3.

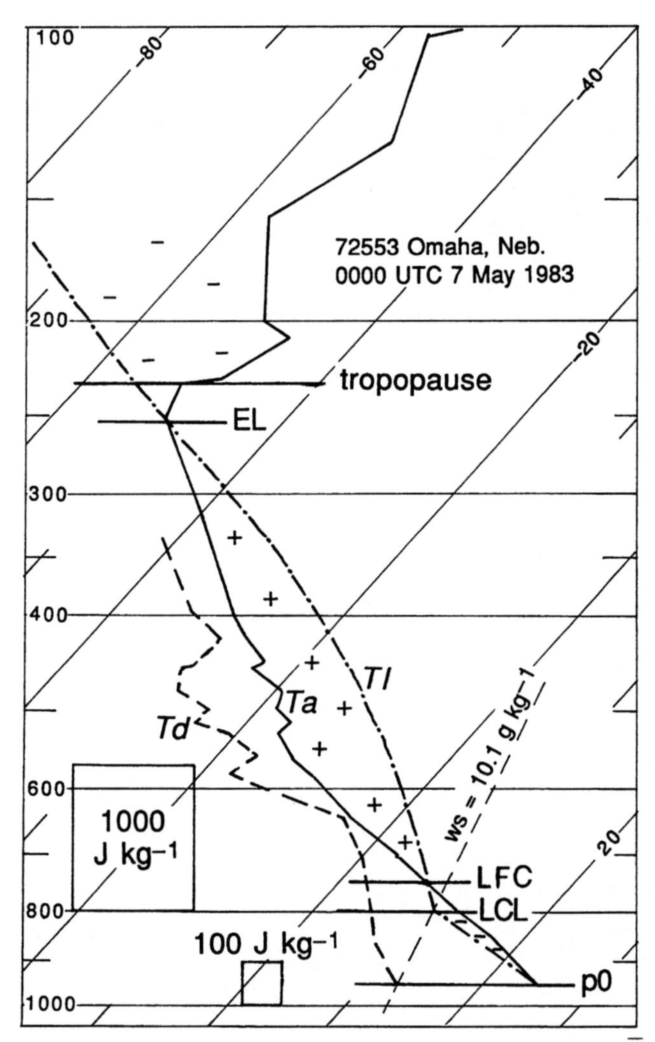

FIGURE 5-5. An example of a sounding, and lifting temperature of an air parcel that is "lifted" from the earth's surface (Omaha, Nebraska, 0000 UTC 7 May 1983). The ambient temperature T_a (solid line) and ambient dew point T_d (dashed line) represent the observations. The lifting temperature T_l (dash-dotted line) consists of the dry and moist adiabats. The dry adiabat is T_l between the surface at 954 mb and the LCL at 798 mb. T_l is drawn along the moist adiabat from the LCL into the stratosphere. The upper intersection between T_l and T_a is at the equilibrium level EL. The rectangles in the lower left portion of the graph are examples of areas of 1000 and 100 J kg^{-1}.

Example of Lifting Temperature

Figure 5-5 shows the ambient temperature (sounding) $T = T_a(p)$, the ambient dew point $T_d(p)$, and the computed lifting temperature $T_l(p)$ of the air parcel at the earth's surface. The lifting temperature T_l of the parcel consists of $T_{dry}(p)$ from p_0 to the LCL at 798 mb and of $T_{moist}(p)$ above the LCL. The lifting condensation level LCL is at the intersection of the saturation mixing ratio isopleth through the surface dew point with the lifting temperature T_l. In the figure the dew point at the surface is 12.6°C and therefore the mixing ratio is 10.1 g kg^{-1}. The w_s isopleth of 10.1 is drawn in the example in Fig. 5-5. The determination of the LCL was also described in Section 5-3, with the example in Fig. 5-1.

Determination of the lifting temperature T_l as shown in Fig. 5-5 is fairly representative of the situation in the atmosphere when lifting by some mechanism can be anticipated. The lifting may be provided by gliding flow up the sloping terrain (*orographic lifting*), general lifting in the air mass (*frontal upgliding*), or convective lifting. The example in Fig. 5-5 is an evening sounding from a thunderstorm region, when convective lifting was already present in the vicinity of the station. In such places downdrafts appear that compensate lifting. These downdrafts will mechanically stimulate the air at the surface to lift. Small stability will not be a strong obstacle to lifting. The parcels that lift from the earth's surface will be buoyant in the layer between 751 and 254 mb.

The lifting temperature of the parcel at the earth's surface may be criticized because it does not represent a realistic parcel. It is likely that the superjacent parcels also will lift, and they normally have a different dew point and temperature. Therefore, the stability of the atmosphere can be determined by a modification of the initial temperature and dew point of the lifting parcel. A more realistic lifting temperature can be obtained when average temperature and dew point are taken for a layer of about 1 km at the surface. Such a modification is described below in the evaluation of the lifted index (Section 5-7). How the initial temperature and dew point of the parcel are evaluated is not too important. It is much more important to use the same method in all cases that will be compared with each other.

Levels of Free Convection and Equilibrium

The example in Fig. 5-5 shows several more concepts that are useful in weather analysis, the intersections of the T_l and T_a lines. The lower intersection is the *level of free convection* (LFC, at 750 mb in the example). The upper intersection is the *equilibrium level* (EL, at 255 mb in the example). At these levels the parcel is not buoyant. If the lifting parcel oversteps the LFC, it will continue rising without added energy since it is buoyant above that level. It will rise easily to the EL. Above that level the parcel will be negatively buoyant (buoyant down). It may continue rising only if it has accumulated kinetic energy of vertical motion; this is described later.

The LFC is of interest only when the lifting temperature is lower than the ambient temperature in the lowest layer of the atmosphere. Otherwise, if the lifted temperature is higher than the ambient temperature in the lowest layer, the atmosphere is unstable and the LFC is at the surface.

Convective Condensation Level

The LCL may have little realistic significance if the air parcels do not lift. Therefore, it is useful to anticipate convective lifting that may be realized when the air is heated from the earth's surface, most likely during sunshine. This leads to the concept of *convective condensation level* (CCL). This is the level where the saturation mixing ratio isopleth (w_s = const.) through the surface wet-bulb temperature (T_w) intersects the ambient temperature (illustrated in Fig. 5-6). The dry adiabat through the CCL point on the ambient temperature goes through the temperature (*convective temperature*, CT in Fig. 5-6) at the level p_0. This is the temperature that the air at the surface must attain to form the first cumulus. In other words, if the air at the surface is heated to the convective temperature, then the convection will be deep enough to form clouds at the CCL.

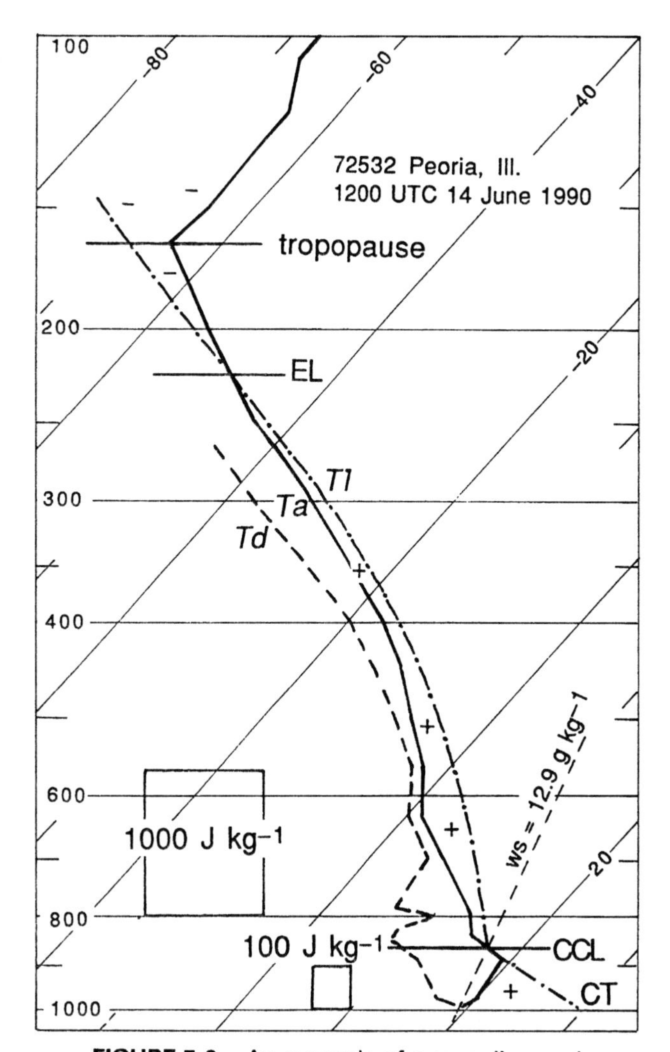

FIGURE 5-6. An example of a sounding and lifting temperature of an air parcel that reached the convection temperature (CT) at the earth's surface (Peoria, Illinois, 1200 UTC 14 June 1990). The symbols are as in Fig. 5-5, with addition of the CCL at 860 mb.

Figure 5-6 shows a morning sounding with surface temperature of 19.2°C at 992 mb. In the course of the diurnal temperature variation, the surface temperature of 29.4°C may be reached. At that time, when 29.4°C is reached, an adiabatic layer will form up to the CCL. Dry convection will be present in this layer. The parcel at the surface may still have w = 12.9 g kg^{-1}. This parcel will reach condensation at the CCL at 860 mb where 12.9 g kg^{-1} represents saturation va-

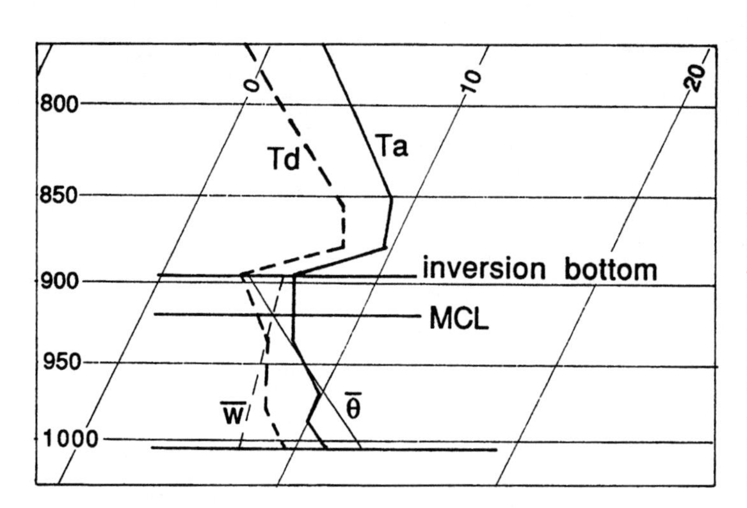

FIGURE 5-7. An example of graphical evaluation of the mixing condensation level (MCL) in the case of turbulence in a stable, humid layer under an inversion. The ambient temperature (thicker solid, T_a) and dew point (thicker dashed, T_d) have been averaged under the inversion to obtain an adiabatic layer characterized by the adiabatic lapse rate and equal mixing ratio. After mixing, the temperature distribution is adiabatic, that is, along a dry adiabat (thin, straight solid line $\bar{\theta}$. Mixed dew-point distribution is along the isopleth of constant saturation mixing ratio (dashed, thin straight line $\bar{w}$). The intersection of the $\bar{w}$ and $\bar{\theta}$ lines shows the MCL.

por mixing ratio. The LFC is at the surface. The parcel will not be negatively buoyant until the equilibrium level at 228 mb.

Mixing Condensation Level

Clouds may form in a layer that is mechanically mixed. This appears sometimes in comparatively shallow mixed layers, perhaps 1 km thick. Sufficient mixing in a layer at the earth's surface usually occurs with wind over 10 m s^{-1}, and the mixed surface layer is usually capped by an inversion. Turbulence causes the potential temperature and mixing ratio to equalize through the layer. If the humidity is high enough, the top part of the mixed layer will acquire a temperature below dew point. The condensation will start at and above the *mixing condensation level* (MCL).

Evaluation of the MCL is illustrated in Fig. 5-7. The layer under inversion is not yet mixed, but it is anticipated that the wind will be fast enough to cause mixing during the day. When the mixing is accomplished, the potential temperature and mixing ratio in the layer will be equalized. Also due to mixing, the dew point will be equalized through the layer as well.

In practical work, first the average potential temperature and average mixing ratio in the layer under the inversion are determined graphically. The dew point plays the role of the mixing ratio, since these two variables are approximately proportional to each other. The two averages are then drawn as lines on the thermodynamic chart: The ambient temperature is the adiabat and the dew point is along w_s = const. These two lines intersect at the mixing condensation level.

5-8 INTEGRATED INDICATORS OF STABILITY

Several important quantities that characterize the state of the atmosphere can be evaluated when the vertical distribution of T and T_d is integrated, as described next. For these, an integration of temperature and dew-point variations with height must be performed. The process of vertical integration requires a somewhat intricate computer program, but it can be performed rather easily graphically, on a thermodynamic chart. Also, the graphical work on the thermodynamic chart often has advantages over automatic processing on a computer since more detail can be noticed. The computers have the advantage in the speed of data processing.

Lifted Index

The *lifted index* is based on the lifting temperature at 500 mb for a parcel of air having the average mixing ratio in the lowest kilometer of the atmosphere and temperature T_p of the predicted diurnal maximum at the surface. The following difference gives the lifted index of stability S:

$$S = T_{l500} - T_{a500} \qquad (5\text{-}31)$$

T_{l500} is the lifting temperature at 500 mb and T_{a500} is the ambient temperature at 500 mb. Both may be expressed in K or °C, since the difference is used. The units (K or °C) are usually omitted for the lifted index. The predicted maximum temperature determines the selection of the dry adiabat at p_0. The LCL is found by the method shown in Section 5-3, using the moist (pseudo) adiabat through the wet-bulb temperature T_{w0} at the level p_0. Evaluation of T_{l500} constitutes integration of temperature [(C-33), with $dQ = 0$] until the level of 500 mb:

$$T_{l500} = T_p - \frac{R}{c_p} \int_{1000 \text{ mb}}^{500 \text{ mb}} T(p)\, d \ln p$$

The lifted index is positive in a stable atmosphere; therefore, it is a *stability index* (as opposed to the earlier instability indices). Negative values of −2 to −6 are rather common in regions with showers.

An earlier version of the lifted index, known after its designer as the *Showalter index*, used the parcel originating at the 850-mb level. A variation of the lifted index is the *modified lifted index*, which is obtained when the lifting parcel is taken at the level of maximum wet-bulb potential temperature, θ_w, in the lowest 300 mb. The letter S for the lifted index is selected from the name Showalter.

Energy of Convection

The energy of convection is often evaluated on the basis of a parcel of air that is being lifted through the atmosphere. The force needed to lift the parcel in a stable atmosphere need not be explicitly considered. We presume that it is made available by large-scale lifting in the atmosphere (frontal upgliding), by mechanical forcing up the mountain slope, or by some other process. What we consider below is the energy needed to lift the parcel. The energy of a lifting parcel in a stable atmosphere is considered negative. In the other case, when the atmosphere is unstable, the energy will be released by the lifted parcel and no external force is needed. In this case, the energy of convection is positive.

Energy of lifted parcel (assumed of unit mass) as described by (5-24) is the basis of several quantities that describe the stability of the atmosphere. In each case, the difference

$$T_l(p) - T_a(p) \qquad (5\text{-}32)$$

must be integrated step by step, along the coordinate p.

The area on the thermodynamic diagram between the lifting temperature and ambient temperature indicates the energy of instability (or stability, if negative). The layers in which the lifting parcel is warmer than the ambient atmosphere show the (positive) energy of instability. It is common to shade such areas between T_a and T_l in the cases when $T_l > T_a$ with the color red. The example in Fig. 5-5 shows such a large *positive area* (plus signs) between 751 and 255 mb. The top of this layer is at the equilibrium level at 255 mb, where the lifted parcel presumably reaches the same temperature and density as the ambient atmosphere.

A layer in which the parcel is colder than the ambient atmosphere shows the energy of stability (*negative area* on the chart). Such areas are usually shaded blue, or indicated with minus signs if colors are not available. The example in Fig. 5-5 shows stability for the lifted parcel between 950 mb (at the surface) and 750 mb, and again above 255 mb. The negative area from 950 to 750 mb is comparatively small. It shows that not much energy is needed to start convection in this case.

To facilitate visual estimation of the energy of instability, two rectangles are drawn within Fig. 5-5, one for the energy of 1000 J kg^{-1}, the other for 100 J kg^{-1}. The energy of instability for the selected lifted parcel (the red area between 750 and 255 mb) can be estimated to exceed 1000 J kg^{-1} since it is larger than the 1000 J kg^{-1} rectangle in the diagram. A better estimation of the energy of convection can be made by an approximate numerical evaluation of the integral in (5-25). An approximate, but easy and fast method is shown in Appendix L.

Convective Inhibition

In a stable atmosphere, the air parcels do not lift spontaneously; they are not buoyant. In cases of stable layers at the earth's surface, it is meaningful to integrate (5-25) from the earth's surface to the LFC. This yields the *convective inhibition* (CIN):

$$\text{CIN} = R \int_{p_0}^{p_{\text{LFC}}} [T_l(p) - T_a(p)]\, d \ln p \qquad (5\text{-}33)$$

This is the amount of energy needed to overcome the stability of the atmosphere for the convection to continue spontaneously. The form of the integral, with R and $d \ln p$, is analogous to the energy of instability (5-23). The beginning of integration (at p_0) is usually not taken at the earth's surface; instead, an average temperature and pressure are evaluated in the lowest 500 or 1000 m. These values determine p_0, $T_l(p_0)$, and $T_a(p_0)$ for the integration of (5-33). When these averages are evaluated, it is practical to evaluate the average mixing ratio in this layer also. This ratio is needed to find the LCL, and the LCL is needed since at that level the lifting temperature T_l changes regime to the moist adiabatic ascent.

The CIN by itself is not an indicator of possible convection. It should be accompanied with other indicators of instability. However, a significant CIN is needed for strong convection to develop. A small value of CIN, say under 10 $m^2\ s^{-2}$ on the morning sounding, indicates that the convection may start early in form of numerous smaller convective clouds. All the energy of instability will be expended on small clouds, without development of strong convection. Values of over 15 $m^2\ s^{-2}$ will prevent early development. Then the energy of instability will be concentrated on fewer, but stronger, convective cells. Strong squall lines develop with CIN over 50 $m^2\ s^{-2}$. Excessively large CIN (over 150 $m^2\ s^{-2}$) may prevent convection entirely.

Convective Available Potential Energy

When the integration of (5-32) is performed from the LFC to the equilibrium level p_e above LFC, the *convective available potential energy* (CAPE) is given:

$$\text{CAPE} = R \int_{p_{\text{LFC}}}^{p_e} [T(p) - T_a(p)]\, d \ln p \qquad (5\text{-}34)$$

The CAPE values under 1000 $m^2\ s^{-2}$ indicate a small likelihood of strong convection. CAPE may be over 2000 $m^2\ s^{-2}$ when severe storms develop. CAPE is also called *buoyant energy*.

Using the hydrostatic approximation

$$dp \approx -g\rho\, dz = -g \frac{p}{Rt}\, dz$$

and the definition of potential temperature (C-1), the expressions for CAPE and CIN can be rewritten in the form

$$g \int \frac{\theta - \theta_a}{\theta_a}\, dz \qquad (5\text{-}35)$$

where the limits must be determined in terms of height z.

The Bulk Richardson Number

A form of the Richardson number is sometimes computed with the CAPE and with the vertical wind shear over thick layers of the atmosphere. This is the *bulk Richardson number*

$$R_b = \frac{\text{CAPE}}{\frac{1}{2} U_z^2}$$

where U_z is the vertical wind shear (derivative type, magnitude only) averaged over a thick layer of the lower troposphere, perhaps the lowest 3 or 6 km. When the environment is already unstable and thunderstorms are likely, the bulk Richardson number may be useful to distinguish types of storms. It has been observed (Weisman and Klemp, 1986) that lower R_b, between about 20 and 40, favor the formation of supercells. Higher values, 50 to several hundred and even several thousand, are conducive for formation of multicell thunderstorms. This is explained by the strong shear (small R_b) needed to form the quasi-steady circulation of the supercell. Vertical convection cells prosper in low shear environment (high R_b). Then the convective cells are not torn apart by environmental wind shear.

Precipitable Water

The indicators of atmospheric stability described above all suffer from disregard of the thickness of the humid layer that is common in the lower troposphere. The lifted index and CAPE are influenced by humidity only in the lowest layer of the atmosphere. If the humid layer is 500 or 1500 m deep, it may give the same value of a stability indicator. On the other hand, thunderstorms develop more vigorously if there is an ample supply of water vapor. In that case, more latent heat can be released and more precipitation generated. A quantity suitable to complement the stability indicators is the *precipitable water* (PW), defined as the mass of water vapor over a square meter of earth's surface:

$$PW = \int_0^{\infty} q\rho\, dz = \frac{1}{g} \int_0^{p_0} q\, dp \qquad (5\text{-}36)$$

where q is the specific humidity, ρ is the density of humid air, and g is the acceleration of gravity. Unfortunately, there is no quick way to evaluate this integral graphically. Instead, a finite-difference approach must be used, whereby q is evaluated at a number of points along the vertical. In this way, the average value of q can be evaluated and this can be taken out of the integral in (5-36).

Since the precipitable water has the dimension of kg m^{-2}, it is commonly expressed in mm, like precipitation. The density of water is very exactly 10^3 kg m^{-3} (this is the original eighteenth century definition of the kilogram). Therefore, 1 kg of water, when spread over the area of 1 m^2, makes a layer of 1 mm.

Precipitable water values of 25 mm or more are generally sufficient to support showers and thunderstorms in the southern and central United States. The chances for showers are much lower if the precipitable water is smaller. However, on the high plains of Colorado and Wyoming, severe storms have occurred with precipitable water of less than 10 mm before the storms.

Maximum Updraft

An estimation of the maximum vertical motion in the cumulonimbus can be obtained with the assumption that the CAPE is converted into the kinetic energy of vertical motion. The z-coordinate system is used next since the expression for kinetic energy is shorter. It may be expected that the vertical kinetic energy E_{kin} will be equal to the CAPE:

$$E_{kin} \approx \tfrac{1}{2}w^2 \approx \text{CAPE}$$

Here w is the vertical velocity component in z-coordinates. The z-system is selected since it yields this shorter definition of kinetic energy. Solving the last equation for w gives the maximum value

$$w_{max} = \sqrt{2\ \text{CAPE}} \qquad (5\text{-}37)$$

This is supposedly the vertical motion with which the air parcels reach the equilibrium point near the tropopause. With a large value of CAPE, say 2000 $m^2\ s^{-2}$, the vertical velocity component of over 60 $m\ s^{-1}$ can be estimated using (5-37). Since mixing of parcels has not been accounted for, the above value may be an overestimate. However, appearances of similarly large vertical motion have been indirectly confirmed by the falling speed of very large hailstones. Extremely large hailstones of 10 cm in diameter fall with the speed of about 45 $m\ s^{-1}$, and they must be suspended in the cloud for about half an hour until they attain that size. Vertical motion of the above magnitude is needed to support such growing hailstones.

Growth of the cumulonimbus has been observed to progress through the tropopause and into the stable layers of the lower stratosphere. Theoretical considerations show that a vertical current of 50 $m\ s^{-1}$ may penetrate more than 2 km above the equilibrium level near the tropopause.

Overshooting Thunderstorm Tops

The kinetic energy of vertical motion can be used to lift the air parcels through stable lower layers of the stratosphere. The temperature is approximately constant with height above the tropopause. The parcels that reach the equilibrium level may have a vertical momentum that is somewhat lower than the maximum vertical speed. For the sake of simplicity, we may assume that a parcel reaches the top with maximum vertical speed as evaluated from (5-34). In the case of very strong thunderstorm, this speed may be 50 $m\ s^{-1}$. Calculation shows that the kinetic energy of this current is sufficient to penetrate 2.5 km into the stratosphere. Turrets of this size on cumulonimbus tops are regularly observed in satellite images and occasionally by radar echo. Since the estimation of overshooting of thunderstorm tops is an instructive exercise, it is given in the Appendix L.

6

Thermal Properties of the Troposphere

6-1 THERMAL STRUCTURE OF DEEP TROPOSPHERIC LAYERS

Thickness and Temperature

The thickness between two selected isobaric surfaces is a convenient tool for describing the thermal structure of the atmosphere. We can see from the hydrostatic or barometric equations that the thickness is proportional to the average temperature of a selected layer. If the layer is comparatively thin, an approximate hydrostatic equation (C-3) is

$$dz = \frac{RT}{qp}|dp| \qquad (6\text{-}1)$$

This equation can be written in finite differences when the differentials dz and dp are replaced by finite increments h and Δp, respectively. Then the thickness h becomes

$$h = \frac{R\Delta p}{gp}T = \text{const.} \times T$$

showing that the thickness $h = \Delta z$ is proportional to the temperature T. The expression $R\Delta p/(gp)$ is constant, since the pressure layer had already been selected, and the values of pressure on the top and bottom are determined. The value of p can be taken as the average pressure between the top and bottom of the layer (trapezoidal rule).

Integrated forms of the hydrostatic equation are preferable for layers with more than a 100-mb difference from bottom to top. An integrated version of the hydrostatic equation (6-1) can be taken as (C-6) or the following form:

$$h = \frac{R\overline{T}}{g}\ln\frac{p_0}{p_1}$$

where $\overline{T}$ is the temperature averaged over the pressure interval p_0, p_1. An important conclusion is that the thickness is directly proportional to the average temperature $\overline{T}$ *of the layer.*

Construction of the Thickness Chart

A systematic way to construct a thickness chart is to compute the values of thickness from the radiosonde reports, plot them on a chart, and draw isopleths. This procedure can be performed automatically by a computer. However, familiarity with the traditional graphical method of subtraction of scalar fields is useful. The isopleths in Fig. 6-1 show that the difference isopleths always intersect the other two isopleths in their intersections. A practical exercise of graphical subtraction will make this point clear. Traditionally, the thickness of the layer between 500 and 1000 mb is used most often.

Frontal Zone

The frontal zone on the thickness chart is the region of apparently denser contours, showing large baroclinicity (Fig. 6-2). This region is usually elongated along the contours, since the geostrophic balance in the atmosphere maintains the contours and isotherms roughly parallel to each other. The cold edge of the frontal zone marks the position of the 500-mb front. The 500-mb front in Fig. 6-2 is shown with two lines. The warm edge of the frontal zone marks the position of the surface front, drawn in the same place in Fig. 6-2a. The cold air at 500 and 1000 mb is northwest of the front. There is warm air at the surface at station A in the southeast.

Frontal Layer and Thickness

The vertical section in Fig. 6-3 is an example of the front that corresponds to the model in Fig. 6-2. Such a section may lie between the NW and SE corners of Fig. 6-2b. The frontal layer is situated between the frontal positions on the surface (actually at 1000 mb) and at 500 mb, in its usual sloping position. The thickness in this model does not vary horizontally in any layer away from the frontal layer, as is highlighted by several double-headed arrows. The atmosphere is, of course, more complicated than this model, but we are looking for models of this kind when analyzing the weather charts. The conclusion from Figs. 6-2 and 6-3 is that the edges of the frontal zone indicate the positions of the front at 500 and 1000 mb.

More complicated cases may occur: The frontal layer may not reach the upper level, the front may not reach the surface, or there may be multiple fronts. In any case, the analysis should not be confined to one chart only. Thickness must also be supplemented with other charts to diagnose the weather systems properly.

The frontal zone usually stretches along the polar jet since the two are structurally connected. Stronger segments of fronts are characterized by denser thickness contours. Such segments are typically 1000–3000 km long.

A weaker segment of the frontal zone (smaller gradient of thickness) often occurs downstream of the extratropical cyclone, in the region of the warm front. Accordingly, the warm front is less prominent in such a region. At the same time, the frontal zone is well prominent around the surface cold front in the cyclone.

The above-described relation between fronts and baroclinic frontal zones is rather general. It should be used only as a first approach to locating fronts. A better position of the fronts can be found when more detailed charts (surface chart and satellite images) are used.

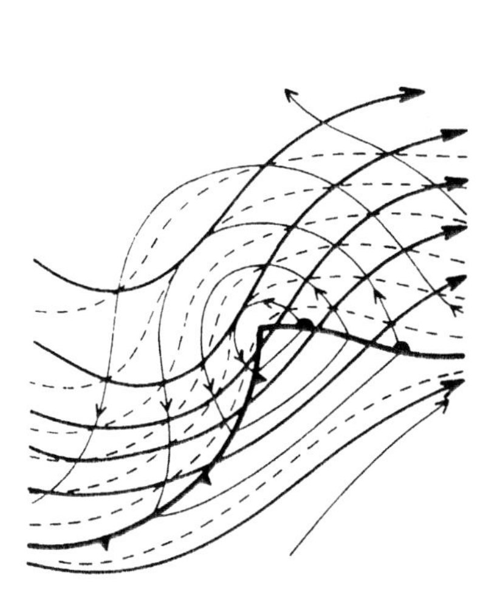

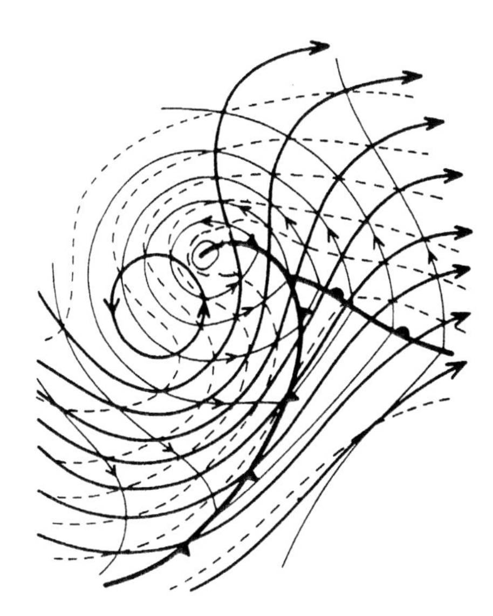

FIGURE 6-1. 500-mb contours (heavy, solid), 1000 mb contours (thin), and 1000/500-mb thickness (dashed) in two stages of cyclone development. From Palmén and Newton (1969), p. 326, by permission of the Academic Press.

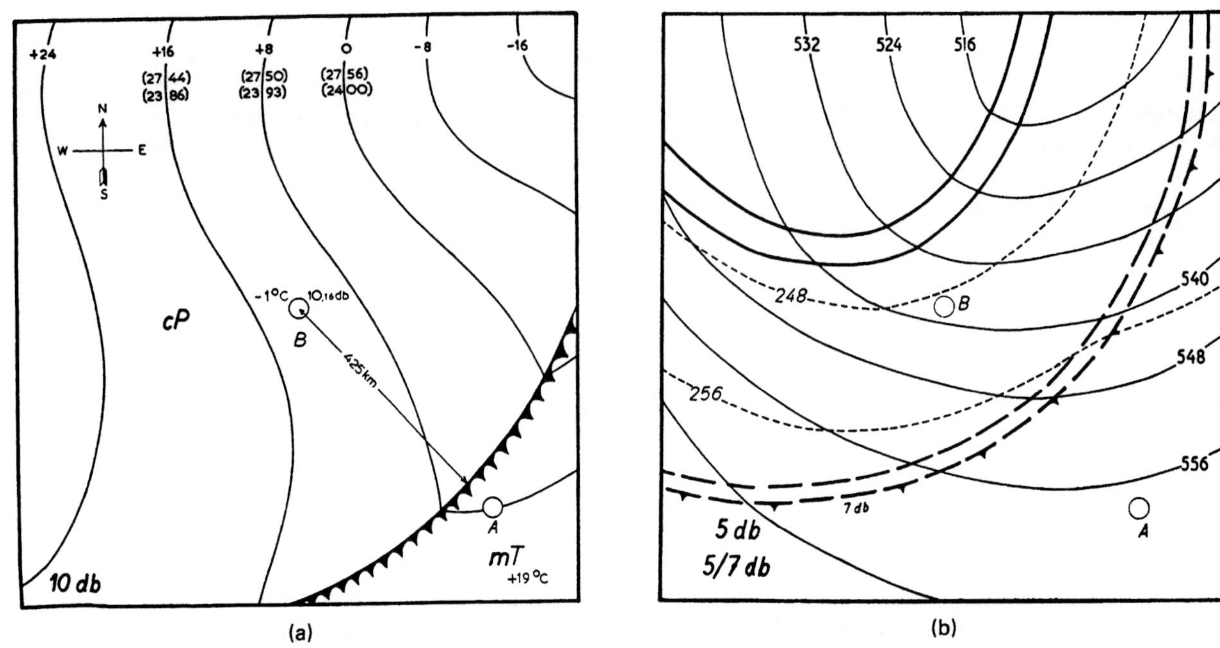

FIGURE 6-2. Relation between thickness and the front at 1000 and 500 mb. All isopleths are in dam; 1 db = 100 mb = 10^4 Pa. (a) The front at 1000 mb and contours of the 1000-mb surface. (b) 500-mb front (two thicker lines), 500-mb contours (solid), front at 700 mb [same location as in (a)], and thickness contours (dashed). The frontal zone is the region between the surface front and the 500-mb front. From Godske et al. (1957), p. 695, by permission of the American Meteorological Society.

An Example of Frontal Zones

Figures 6-4 and 6-5 illustrate the relationship of the thickness of the 500- to 1000-mb layer with the polar front. The main frontal zones in Fig. 6-4 appear as regions with denser thickness contours. These zones have steep fronts. Three such frontal zones are outlined by thick dashed lines: one over the Pacific, one between Nevada and east Texas, and one over eastern Canada.

Let us consider the Nevada–Texas zone first. Going by the simple rule illustrated in Fig. 6-2, we expect the surface (1000-mb) front in Fig. 6-4 to be somewhere near the Mexican border and along the Texas coast. The 500-mb front may be near a line from central Utah to Oklahoma. Data from other charts, primarily surface, indeed show the surface cold front in east and south Texas. The part of the front that is expected along the Mexican border dissipated due to divergence in the low-level flow in this region (frontolysis).

Final locating of the front is not done with the thickness chart alone. The thickness provides only a large-scale outline about frontal zones. Detailed information on the location of the front must be found on the charts with all available observations.

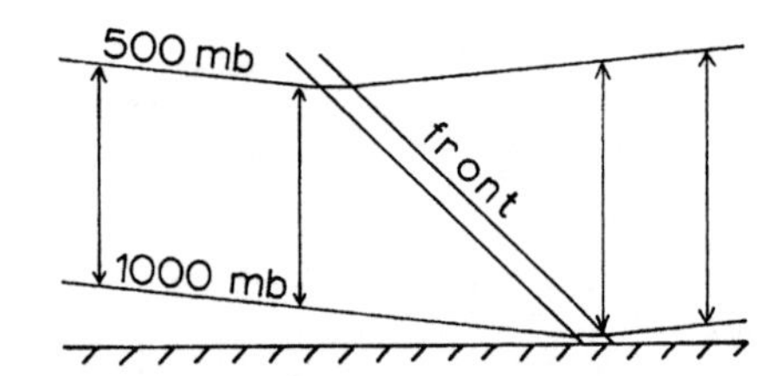

FIGURE 6-3. A schematic vertical section through the front. Thin lines are the isobars, and double-headed arrows show the thickness.

Use of Isotherms at 500 mb

The position of the front at 500 mb can best be determined using the isotherms at this level. Figure 6-5 shows the isotherms at 500 mb at the time of the thickness chart of Fig. 6-4. The 500 mb front, taken as the southern edge of the frontal layer, can be located in west Texas and southern New Mexico. The northern edge of the frontal layer is near the northern border of

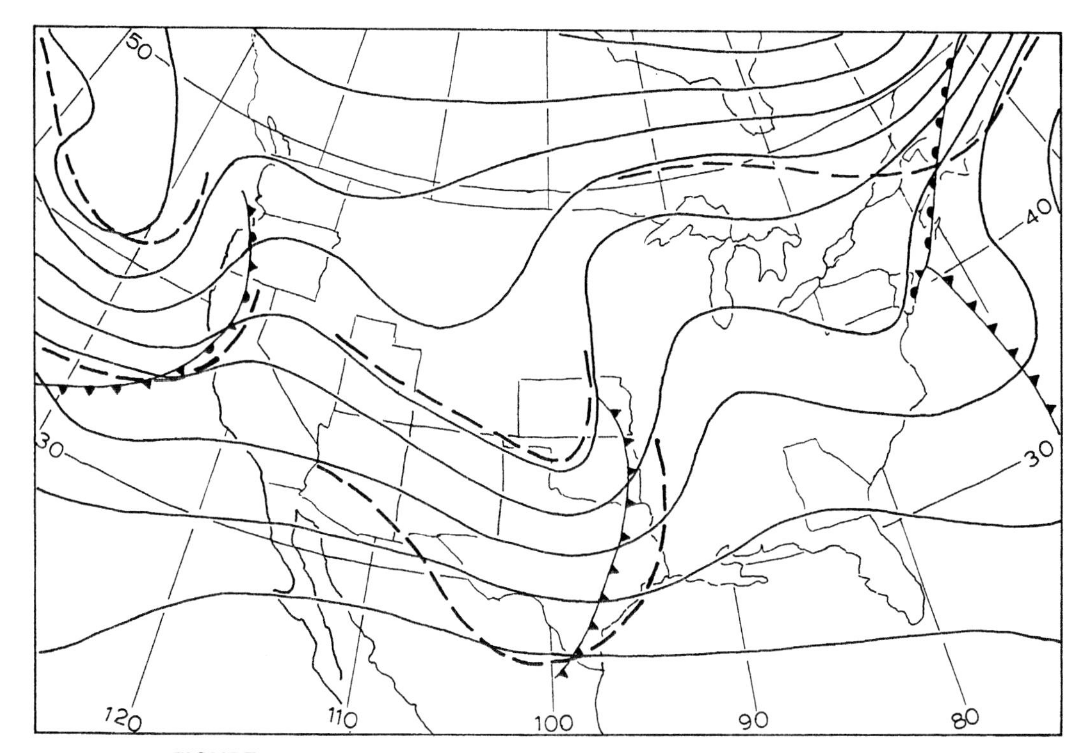

FIGURE 6-4. Thickness of the 500 to 1000-mb layer for 0000 GMT 16 February 1984, intervals of 60 dam. The frontal zones are delineated by thick dashed lines. The surface front (from the National Meteorological Center) is shown by conventional symbols. The states mentioned in the text are outlined.

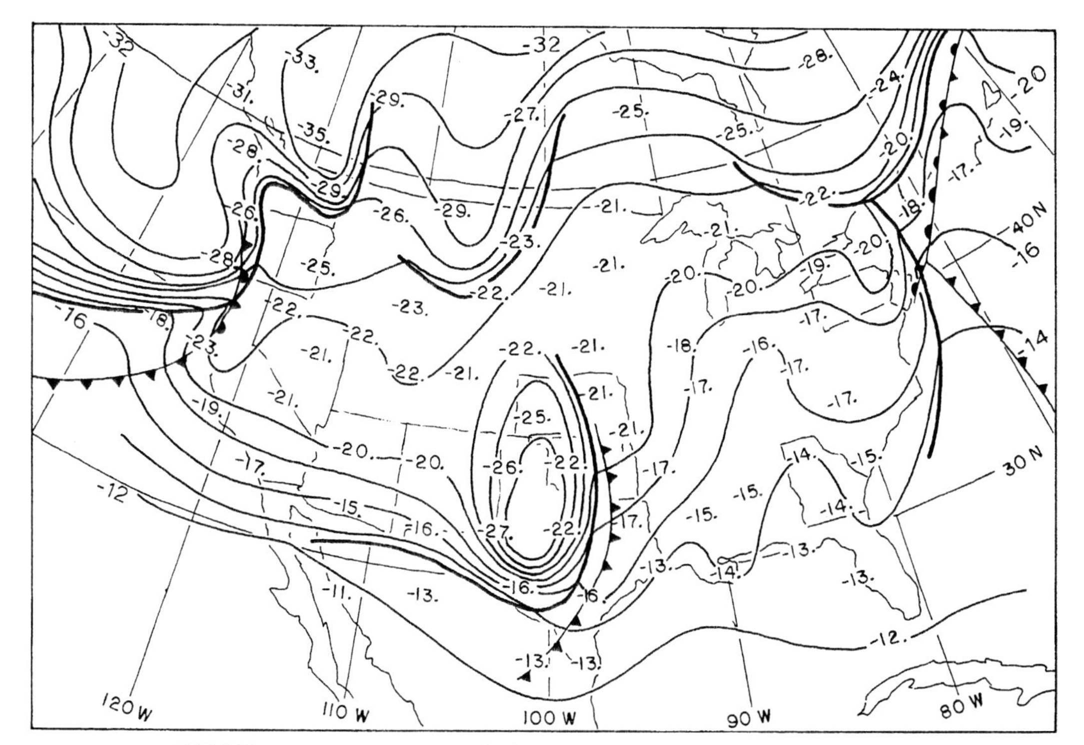

FIGURE 6-5. Temperature (°C) at 500 mb, 0000 UTC 16 February 1984. The reporting stations are indicated by points in place of decimal points. The isotherms (intervals of 2°C) are labeled by numbers without decimal points.

New Mexico, not too far from the northern edge of the frontal zone in Fig. 6-4.

Weak Warm Front

A warm front may be expected along the coast of Texas and Louisiana, since there is south wind ahead of the cold front, blowing across the thickness contours (the charts for this are not shown here). However, this warm front was so ill-defined at the surface that it could not be determined with confidence. Therefore, the warm front was not drawn. Such poor prominence of warm fronts is very common.

A similar situation occurs with the frontal zone over the Pacific at the West Coast, where the thickness contours are again diffluent and curved cyclonically (Fig. 6-4). The locations of the surface and upper-level fronts are similar to the configuration in Texas. The surface warm front is again indeterminate and therefore is not drawn. As the examples in Figs. 6-4 and 6-5 show, the surface warm front is often so weak that it has to be left off the charts.

Uncertainty of Analysis

The positions of the front in the Oregon–Washington area seem to defy the Margules equation by sloping toward the warm side. This is not necessarily so, since the surface front may be an occluded front and the upper-level cold front may travel ahead of the surface front. The other explanation of this situation is that this is a split front (Section 7-9), where the upper-level front progresses ahead of the surface front.

Another uncertain analysis is around the front over eastern Canada (Fig. 6-4). That frontal zone does not show the location of the surface front well. A significant part of the baroclinic frontal zone can be found along the coasts of Maine and Nova Scotia on the warm side of the surface front.

The front that can be identified over the Atlantic off the East Coast (Figs. 6-4 and 6-5) is not accompanied by a clearly defined frontal zone on the thickness chart. The polar air behind this front, covering the East Coast states from New Jersey to Georgia and neighboring regions, is not very cold. It was transformed by heating from below and by adiabatic warming due to subsidence. The contrast in temperature between this polar air and middle-latitude and tropical air masses is already small. Also, the front dissipated around the southern edge of the polar air. This is an area of frontolysis. The polar front is being regenerated near the northern border of the United States. Several segments of the polar front (north of the Great Lakes and two segments in the northwestern United States) are already distinguishable at 500 mb (Fig. 6-5).

Looping of the Front

Moving from the pole to the equator, the polar front is sometimes encountered more than once. At least two such examples can be found in Fig. 6-5. If we draw a line from Montana to the Texas coast, we see that this line crosses the polar front in two points. We also see two positions of the polar front along the line from northern Ontario, south to Pennsylvania and farther southeast out to the Atlantic. This repeated appearance of the polar front is a consequence of looping of the front and the formation of cut-off vortices. Most separate parts of the polar front can be traced to the same continuous front, if we follow the analysis back in time 2 to 4 days.

Looping of the polar front is generalized in Fig. 6-6. This example shows the behavior of the polar jet and front in the upper troposphere during formation of a cut-off low. This model is typical for the 300-mb level, where the jet core and the front normally are separated but remain close to each other. A similar picture appears at other levels, differing only in relation between the front and the jet. At 500 mb, the front and the jet core are in about the same geographical position.

Figure 6-6 illustrates how the amplitude of baroclinically unstable waves increases (Fig. 6-6a). The deepening frontal wave cuts off a part of the cold air (Fig. 6-6b). Now the front is already separated, but the jet stream is still in the form of a large-amplitude wave. The dashed line north of the frontal loop is not a front anymore; this is a shear line that separates rather fast air currents on both sides. Such shear lines always show cyclonic shear, since the anticyclonic shear lines are unstable and therefore cannot persist in fluids on rotating planets. The example in Fig. 4-7 shows such a shear line.

The cutting off is complete in Fig. 6-6c. There is a looping segment of jet stream around a low, with an accompanying front. The geopotential of the isobaric surface in the loop is low, due to the quasi-geostrophic balance in the atmospheric flow. The front often dissipates in the cut-off low, since the cold air in the loop warms adiabatically as it sinks. Therefore, the difference in temperature across the front decreases. The large-scale confluence in this region may not be

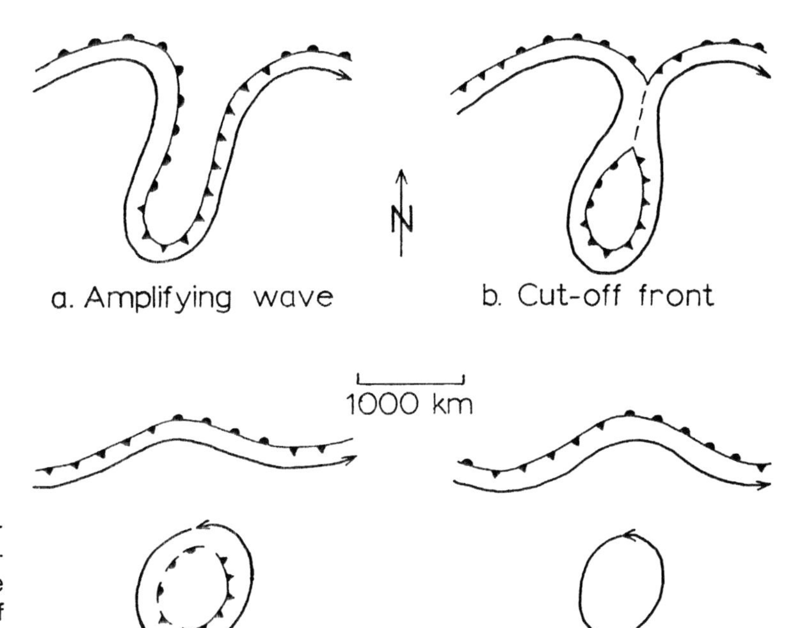

FIGURE 6-6. Four stages of cutting off the polar front in the upper troposphere. The solid line with the arrowhead is the core of the jet stream. The front at the jet level is shown by conventional symbols.

present to regenerate the front. The disappearing polar front is indicated by a dashed line on the northwestern part of the loop in Fig. 6-6c.

The last stage of a cut-off low is shown in Fig. 6-6d, where the cold air mass in the loop sank under the pressure surface. Now the loop is hardly colder than the surrounding air. The tropopause is usually lower above such loops. As the loop weakens, the tropopause gradually lifts to its usual middle-latitude position near 11 km.

Large-Scale Mixing

Separate segments of the polar front in Figs. 6-4 and 6-5 and the development of the loop in Fig. 6-6 show an important mechanism of transformation of air masses. A large volume of cold polar air travels south and cuts off the main polar air mass in the north. The cut-off mass slowly warms up until it cannot be distinguished from the ambient middle-latitude or tropical air. The warming proceeds by heating from the ground and adiabatically during sinking.

A similar cutting off occurs with the warm air mass in the higher parts of the troposphere. There the warm sectors of the cyclones expand toward the north, cut off into warm anticyclonic vortices north of the polar front, and eventually cool by radiation.

6-2 THERMAL WIND

Definitions

Thermal wind, like every other case of wind shear, is defined in two ways: as a difference,

$$\mathbf{V}_T \equiv \mathbf{V}_1 - \mathbf{V}_2 \tag{6-2}$$

where the wind vectors are at two points (1 and 2) above each other; or as a derivative,

$$\frac{\partial \mathbf{V}_g}{\partial z} \tag{6-3}$$

Sometimes the pressure derivative $\partial \mathbf{V}_g/\partial p$ is used in the second definition, with an appropriate change in sign. The points 1 and 2 in the first definition are always chosen such that point 1 is higher (farther away from the earth's surface) than point 2. Geostrophic wind is normally used for the vectors $\mathbf{V}_1$ and $\mathbf{V}_2$. All conclusions in this section will be based on the geostrophic approximation $\mathbf{V} \approx \mathbf{V}_g$. The two definitions (6-2) and (6-3) are greatly different from each other, as discussed in the beginning of Section 4.2 on wind shear.

The following expressions can be used for an interpretation of thermal wind. The definition of geostrophic wind [(C-8) from Appendix C] when inserted in (6-2) yields the difference form as

$$\mathbf{V}_T = \frac{g}{f}\mathbf{k} \times \nabla h \tag{6-4}$$

where h is the thickness of the layer between the levels 1 and 2 where the wind ($\mathbf{V}_1$ and $\mathbf{V}_2$) was observed. This equation is easy to memorize if the definition of geostrophic wind has been memorized, since both appear in the same form. Practically, (6-4) shows that thermal wind is geostrophic wind that is evaluated from a thickness chart. Therefore, the two sides of the thermal wind vector can be described as the *cold side* (left flank, in the Northern Hemisphere) and the *warm side* (right flank). Analogously, each thermal contour has a cold side and a warm side.

Thermal Wind Equation

The derivative definition of thermal wind (6-3) can be found to be equal to the following expressions (see Appendix I):

$$\frac{\partial \mathbf{V}_g}{\partial z} = \frac{g}{fT}\mathbf{k} \times \nabla T \tag{6-5}$$

and in pressure coordinates

$$\frac{\partial \mathbf{V}_g}{\partial p} = -\frac{R}{fp}\mathbf{k} \times \nabla T \tag{6-6}$$

An important property of these expressions is that the horizontal and vertical derivatives, albeit of different variables, can be evaluated from each other. This property is used in several important theories of dynamic meteorology. Several more applications of the thermal wind will be shown next in schematic examples.

Thickness Advection

Equations (6-1) and (C-6) show that the average temperature of a layer is proportional to thickness h:

$$T = \text{const.} \times h$$

where the constant is $(gp)/(R\Delta p)$. This allows us to substitute temperature for thickness (or vice versa) in the expressions for thermal advection of type (4-24) through (4-27) and this justifies using the expression *thermal advection* for thickness advection. Also, the thickness chart is sometimes called the *thermal chart*. Further, if we want only to see the advection qualitatively, then the conversion between thickness and temperature becomes unnecessary.

A particular advantage of the thickness advection appears when it is evaluated in terms of wind at two levels (see Appendix I). Then the thickness advection is given by

$$-\mathbf{V} \cdot \nabla h \approx \frac{f}{g}\mathbf{k} \cdot (\mathbf{V}_1 \times \mathbf{V}_2) \tag{6-7}$$

where $\mathbf{V}$ is the average wind in the layer. The sign of advection is determined by

$$-\mathbf{V} \cdot \nabla h \approx \begin{cases} > 0 \text{ warm advection} \\ < 0 \text{ cold advection} \end{cases} \tag{6-8}$$

Expressed in words, (6-7) and (6-8) show the warm advection when the wind turns to the right with height. Conversely, there is cold advection when the wind turns to the left with height (all in the Northern Hemisphere).

Traditional marine expressions for turning ships are often used for wind turning with height: Turning of wind to the right is *veering*; turning to the left is *backing*. When using these expressions, it should be added that the turning (veering or backing) is with height, since turning of wind (veering or backing) may also occur with time.

Turning of wind with height (or with pressure) is very well expressed by the *helicity*

$$\mathbf{V} \cdot \nabla_3 \times \mathbf{V}$$

or *relative helicity*

$$\frac{(\mathbf{V} - \mathbf{c}) \cdot \nabla_3 \times (\mathbf{V} - \mathbf{c})}{|(\mathbf{V} - \mathbf{c}) \cdot \nabla_3 \times (\mathbf{V} - \mathbf{c})|}$$

of the flow, as described in Appendix I. With mild approximation, it can be shown [(I-16) from Appendix I] that the thermal advection is proportional to the helicity:

$$-\mathbf{V} \cdot \nabla h \approx \frac{f\Delta z}{g}\mathbf{V} \cdot \nabla_3 \times \mathbf{V} \tag{6-9}$$

The helicity has been shown to be an important signature of likelihood of thunderstorms. From (6-9) it follows that warm advection also provides an environment favorable for thunderstorms. Warm advection usually is strongest in the lower atmosphere. This destabilizes the atmosphere: Lower layers are becoming warmer, whereas the upper layers do not change much. Warmer lower layers enhance thermal instability in the atmosphere.

Graphical Examples of Wind Turning with Height

An elementary relationship between wind at two levels and thermal wind is illustrated in Fig. 6-7. The zonal (west) wind component is u, and the meridional (south) component is v. The wind speed is proportional to the length of drawn vectors. The difference vector is drawn twice, in the coordinate origin and as a dashed line between the tips of wind vectors. The difference vector ($\mathbf{V}_T$) intersects both wind vectors ($\mathbf{V}_1$ and $\mathbf{V}_2$) from the same (warm) side. Concerning the warm and cold sides, the situation in the Northern Hemisphere is depicted. Cold and warm sides should be interchanged for application in the Southern Hemisphere.

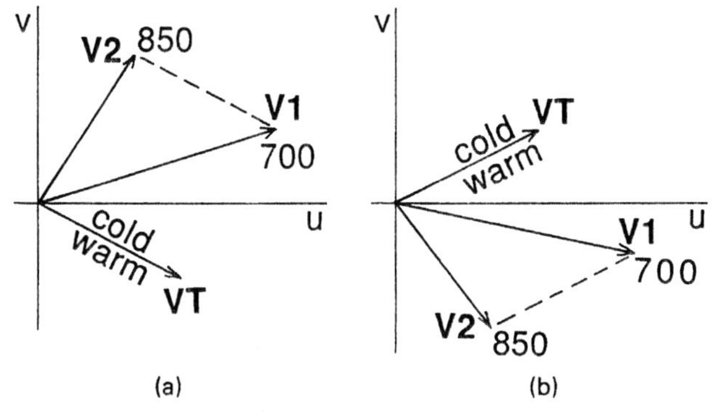

FIGURE 6-7. The relationship of wind at two levels ($\mathbf{V}_1$ at 700 mb and $\mathbf{V}_2$ at 850 mb) with thermal wind ($\mathbf{V}_T$). The sides of the thermal wind vector are marked "warm" and "cold," according to the rule that the thermal wind is parallel to the thickness contours with the cold side on the left. The dashed line is the segment of the hodograph. Warm advection is in *a*; cold advection is in *b*.

The case of warm advection is shown in Fig. 6-7a, where the wind at 700 mb ($\mathbf{V}_{700}$) is from 260° and the wind at 850 mb ($\mathbf{V}_{850}$) is from 220°. The arrangement of wind vectors and the cold and warm sides of thickness contours shows that the wind at both levels blows across the thermal wind from the warm side. Therefore, this is the case of warm advection.

The example in Fig. 6-7b represents the other case: cold advection. The main difference from Fig. 6-7a is that here the wind at both levels blows across the thermal wind vector from the cold side.

Hodograph

A complete wind sounding can be represented on a diagram where the coordinates are wind components. A point is entered for each level where a wind observation is available. The curve that connects the points is the *hodograph*. An example of the hodograph is shown in Fig. 6-8. In this case, the wind at the surface is 2.5 m s^{-1} from 220°, at 850 mb it is 7.5 m s^{-1} from 50°, at 700 mb it is 15 m s^{-1} from 340°, and so forth. Dashed lines in Figs. 6-7a and 6-7b also are segments of corresponding hodographs.

When interpreting a hodograph, it is useful to keep in mind the following schematic models, all based on thermal wind:

(a) A hodograph may consists of one point. In such an atmosphere, the wind is equal at all levels. This is a very unlikely case of a *barotropic atmosphere* where there are no isotherms on isobaric surfaces. This follows from thermal wind: The case of no thermal gradient ($\nabla T = 0$) implies no change of wind with

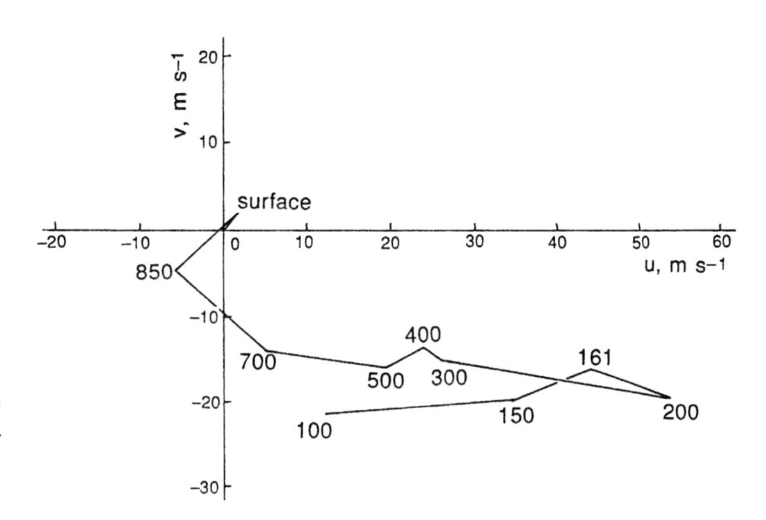

FIGURE 6-8. Hodograph for 1200 UTC 5 December 1969 at Huntington, West Virginia. Wind level is marked in mb.

height. In view of its simplicity, a barotropic atmosphere is used in some mathematical models.

(b) The hodograph is close to a radial direction. This is the case of weak or zero thermal advection, as between the surface and 850 mb in Fig. 6-8. This means that the wind is parallel to thermal wind. Also, the isotherms are parallel to the contours. Such an atmosphere is baroclinic, as is attested to by the variation of geostrophic wind with height. This case is sometimes called *equivalent barotropic* after a classical mathematical model of the atmosphere. For such a model, it has been shown that the flow at the level of zero divergence develops like in a barotropic model. Baroclinic development (cyclogenesis) does not take place in such an atmosphere. The development occurs when the isotherms intersect the contours in a special way (the thermal wave lagging the geopotential wave).

(c) The hodograph is significantly different from the radial direction. This is the case of prominent thermal advection. According to the rule in (6-7), strong cold advection can be seen between 850 and 400 mb in Fig. 6-8. The cold advection can be recognized by the hodograph that turns to the left (backs) with height. The cases in Fig. 6-7 are examples of hodographs with prominent thermal advection. Baroclinic development occurs in weather situations with significant thermal advection.

More complicated cases may occur. Of interest are the cases where warm and cold advection occur above each other. The sign of advection depends on the tangential components of hodograph segments, according to the cases described under (c). In the case of warm advection in lower layers and cold advection aloft, there is a destabilization of the atmosphere and an increased likelihood of thunderstorms.

Thermal Wind and Isotherms

Another useful application of thermal wind is determining the direction of isotherms at observation stations that are far from other stations. The isotherms in data-sparse regions can be drawn more reliably when the wind is available at several levels at a station. The thermal wind equation [(C-18) or (C-19) in Appendix C] when integrated yields:

$$\mathbf{V}_T \equiv \mathbf{V}_1 - \mathbf{V}_2 = (\mathbf{k} \times \nabla \overline{T}) \cdot \text{const.}$$

This equation shows that the isotherms are parallel to the thermal wind (or that the temperature gradient is normal to the thermal wind). Therefore, the vertical difference in wind should be found graphically or numerically and the isotherms should be drawn parallel to the difference vector $\mathbf{V}_T$. This rule is analogous to the rule that near widely separated stations the contours should be drawn parallel to the wind vector.

Advection with Thermal Wind

The concept of advection can be extended to advection with thermal wind. In this case, the meaning of advection is somewhat different, since the thermal wind is a wind shear and the patterns of isopleths need not move with the thermal wind vector. However, the rules of signs and the expressions are the same as the rules of "ordinary" advection with the wind or geostrophic wind. The vorticity advection with thermal wind will be discussed in Section 7-6, with estimation of vertical motion.

Baroclinicity

The intersections between contours and isotherms should not be confused with *baroclinicity* (also called *baroclinity*). Baroclinicity is recognized by the existence of isotherms on isobaric charts, without respect to possible intersections between the lines. It is possible to have a baroclinic situation without thermal advection, where the isotherms and contours do not intersect. That is the *equivalent barotropic* situation that has been described in the section on the hodograph. When the isotherms and contours intersect, then there is also advection of temperature, besides baroclinicity.

6-3 ISENTROPIC CHARTS

Using the horizontal coordinates along isentropic surfaces (θ-system) offers several advantages in the detection of atmospheric processes. The main advantage comes from the absence of vertical motion under adiabatic conditions. In this way, the air parcels stay on the same "horizontal" (x, y) coordinate surface. Therefore, the considerations about air trajectories and advection of various quantities are simplified. The trajectories on θ surfaces give a better depiction of air motion than isobaric or isohypsic trajectories. The methods described for isobaric trajectories (in Section 3-7) apply also to the trajectories in θ-system. The word *isentropic* practically means "constant potential temperature" since the potential temperature can be obtained from entropy, as implied in (C-40) in Appendix C.

Disadvantages of the θ-system mostly come from intersections of horizontal (isentropic) surfaces with the earth's surface. These intersections appear in the θ-system more frequently than in z- or p-systems. Another formal disadvantage of the θ-system is that the data in a θ-system are currently not readily available. It is hoped that the development of computerized techniques will alleviate this problem.

Excellent reviews of using the isentropic charts are by Uccellini (1976) and Moore (1988).

Manual Construction of Isentropic Charts

Before a more complete isentropic chart is described, it is instructive to learn the following fast and easy method for manual construction of isentropic charts. Suppose that we want to construct the isobars on the isentropic chart of 290 K. Also suppose that we have three standard charts at 850, 700, and 500 mb. These charts will yield three isobars since the isotherms of 3.8°C at 850 mb, −11.1°C at 700 mb, and −35.1°C at 500 mb are the isobars of the desired 290 K isentropic surface. The underlying theory for this is in the definition of the potential temperature (C-1):

$$T = \theta \left(\frac{p}{p_0}\right)^{\kappa} \qquad (6\text{-}10)$$

From this definition it can be seen that, for a fixed θ, the temperature and pressure are uniquely determined from each other. The above values of isotherms (3.8, −11.1, and −35.1) have been computed from this equation after inserting $\theta = 290$ and the three values of pressure (850, 700, and 500). The selected isotherms can be copied on a chart. Then they become the isobars of the 290 K isentropic surface.

The isentropic chart can be completed by plotting wind and some humidity variable on the same chart. The needed values can be obtained from radiosonde reports by interpolation from the nearest reported levels. Interpolation from the standard charts may yield acceptable values, if they come from levels not too far from the θ-level.

It is not necessary to use the above values of θ, temperature, and pressure. Equation (6-10) allows any other suitable values of T and θ. It is advisable to select θ that corresponds to an isotherm in the warmest region at 850 mb. This isotherm will likely be in the warm sector of an extratropical cyclone. The corresponding isentrope will go to higher levels (lower pressure) in the region of the polar front.

The described method cannot compete with thoroughly computerized methods that yield many isobars. As can be seen, the three underlying isobaric charts (850, 700, and 500 mb) yield only three isobars. The advantage of the shown manual method is primarily instructional.

Structure of an Isentropic Chart

Figure 6-9 shows the isentropic chart of the 305 K surface for 0000 UTC 2 April 1982. The elements for this chart have been obtained from standard radiosonde reports, using interpolation between reported data. The values plotted are as follows.

—Wind, using standard wind barbs, with values in m s^{-1} (short feather for 5 m s^{-1}, long feather for 10 m s^{-1}, pennant for 50 m s^{-1}). This plotting is different from the usual plotting in knots.

—Pressure, in millibars, plotted on the upper left (northwestern) side of the observing stations.

—Montgomery stream function, in 10 $m^2 s^{-2}$ less 3×10^5, plotted on the upper right (northeastern) side of the stations.

—Pressure difference between the 305 and 300 K surfaces (on the left).

—Last digit of the wind speed (m s^{-1}, under the station point).

—Mixing ratio (10^{-3}, on the right).

Standards for plotting arrangement of weather elements on isentropic charts have not been established. Other ways of plotting also are used in the literature.

Isopleths in Fig. 6-9 are drawn only for pressure (solid isobars, intervals of 100 mb) and for the Montgomery stream function (dashed lines, every 1000 $m^2 s^{-2}$).

Interpretation of the Isobars

Continuing the interpretation of Fig. 6-9, we can see that the elevation of the 305 K surface can be inferred from the isobars. The potential temperature in the atmosphere between northern Mexico and Nebraska is lower than 305 K. Therefore, no data are plotted in this region. This situation can be interpreted as if "the isentropic surface is under ground level."

An example of an almost horizontal isentropic surface can be seen in Fig. 6-9 where the 305 K surface slopes little in the region from California to Canada. The pressure in this region is between 300 and 400 mb. Since this region is on the polar side of the polar jet stream, it is likely that the 305 K surface is in the stratosphere.

Dense isobars indicate a steep isentropic surface, and this is likely to occur in a slanted polar frontal layer. Such is the case in a belt between Arizona and South Dakota where the isobars of 600 and 800 mb are

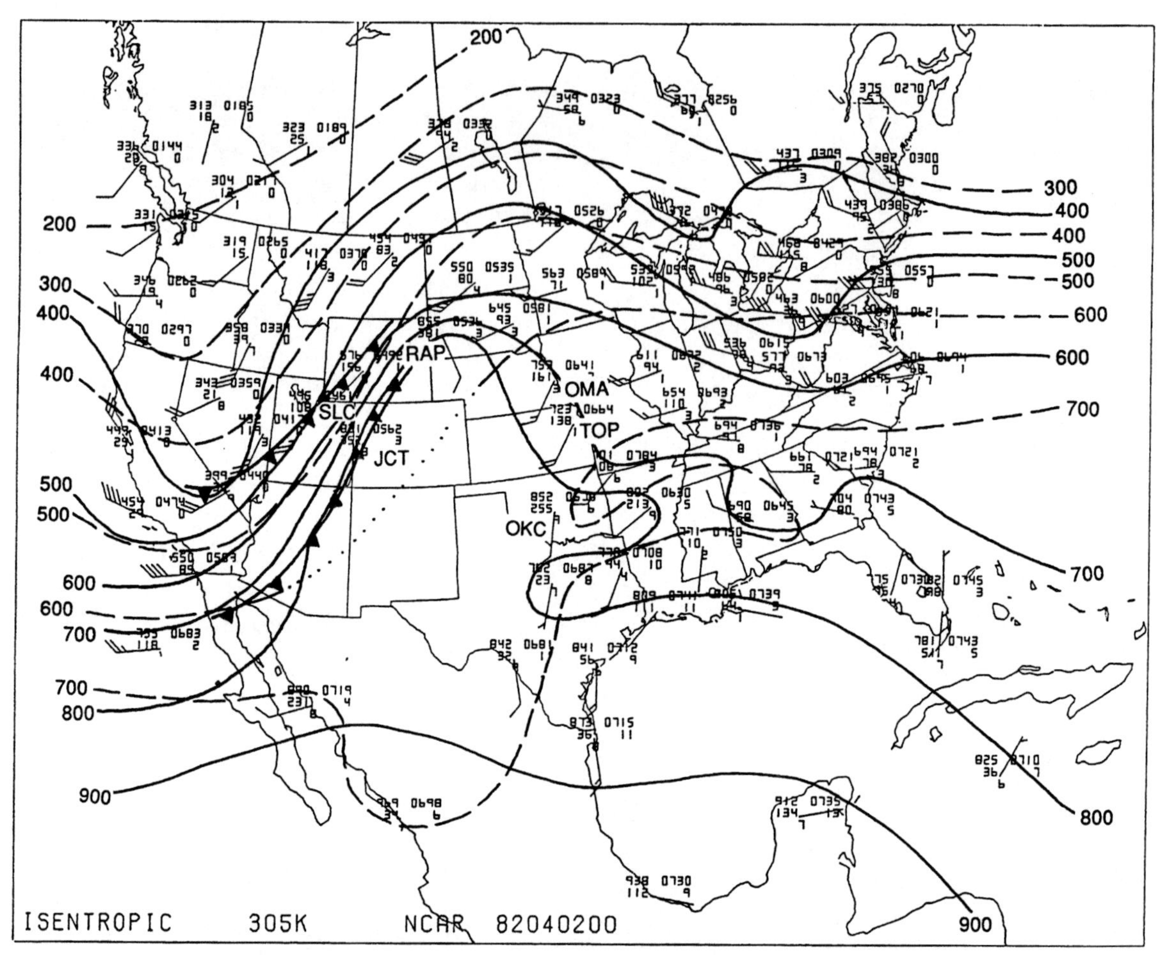

FIGURE 6-9. 305 K isentropic surface for 0000 UTC 2 April 1982. Isobars (mb, solid, plotted in the upper left position for each station) and the Montgomery stream function (10 m^2 s^{-2}, first digit omitted, dashed, plotted in the upper right position). Other plotted variables are pressure difference between 305 and 300 K surfaces (on the left), last digit of the wind speed (m s^{-1}, under the station point), and mixing ratio (10^{-3}, on the right). No data are plotted where the potential temperature at the earth's surface is higher than 305 K. The data for this chart were evaluated, processed, and kindly supplied by P. Pauley.

near each other. The strong pressure gradient in this belt signifies the polar front. The selected isentropic surface of 305 K descends from about 6 km (446 mb) above Salt Lake City (SLC) to the surface (near 850 mb in Colorado). This steep isentropic surface lies in the polar frontal layer.

If colors are available, the front should be marked by two blue lines, one on each side of the belt with dense isobars. The front in Fig. 6-9 is marked by the conventional symbols for monochromatic charts. The pattern of the frontal layer is somewhat similar to the style of front with two lines in Fig. 8-18. The frontal layer in the isentropic surface is typically much wider than on isobaric charts. The frontal layer in Fig. 6-9 occupies the whole frontal zone between the surface and the upper-level position of the front.

An overall estimate of temperature at ground level can be obtained from the isobars. Low pressure (high elevation) indicates that the cold, presumably polar, air occupies the lower troposphere. High pressure (low elevation), or even missing data due to the surface below ground, indicate a warm air mass. The example in Fig. 6-9 shows a wide warm sector around the ridge in the westerlies from Minnesota to Louisiana. The western part of the warm sector is characterized by the absence of surface data near the southern Rocky Mountains. The earth's surface in the eastern part of the warm sector is still covered by a comparatively shallow layer of cold air, so the 305 K surface is raised to about 700 mb.

Temperature

Isobars on the isentropic charts are also isotherms. Once the potential temperature and pressure are given, the definition of potential temperature

$$T = \theta\left(\frac{p}{p_0}\right)^{\kappa}$$

determines the value of temperature. For this reason, the isobars can be renumbered to become isotherms. This renumbering is usually not done. It is sufficient to notice that low pressure on isentropic charts also shows cold regions. This is in agreement with a general decrease of temperature with height (or with decreasing pressure) in the troposphere.

Wind

The wind barbs in Fig. 6-9 show the wind in m s^{-1}. This is different from the usual representation in knots. It is the hope of the scientific community that the metric units (such as m s^{-1}) will be used more in the future. The representation with long and short feathers is somewhat imprecise; the values that end in digits other than 0 and 5 cannot be represented correctly. Since precise values are routinely observed and reported, the last digit of the wind speed is separately plotted under the station point. A precision higher than 1 m s^{-1} is neither reported nor can be used in view of natural turbulent fluctuations that exist in the wind.

The wind on an isentropic surface is regularly assumed to be equal to the wind on the corresponding isobaric surface. Only the vertical component of the wind is greatly different in θ- and p-systems.

The Montgomery Stream Function

Figure 6-9 contains the Montgomery stream function

$$M = c_p T + gz \tag{6-11}$$

from (C-27) in Appendix C. The last digit is omitted, in effect meaning that the units are 10 m^2 s^{-2}. The first digit is also omitted; it is usually 3, and omitting it means that 3×10^5 is subtracted. For example, the number 0736 (at Nashville, Tennessee, BNA) stands for 307360 m^2 s^{-2}. On other charts values under 3×10^5 appear sometimes. Then the leading 2 is omitted.

Definition (6-11) implies that the Montgomery stream function can be evaluated from vertical soundings of the atmosphere using the standard vertical integration to obtain the height z at the desired level. Unfortunately, a vertical numerical integration of height cannot be performed with the accuracy needed for a later evaluation of the horizontal gradient of M. The two terms in the definition ($c_p T$ and gz) are large, and small inaccuracies in evaluating z yield comparatively large errors. Sensitivity of M can be appreciated also from the circumstance that the term $c_p T$ decreases with elevation, whereas gz increases. In this way, M changes much less with elevation than its two constituents.

A practical way to evaluate M accurately is to find a value near the earth's surface where z is small and, consequently, accurate. This makes the error in gz small. Then the vertical numerical integration of the hydrostatic equation in θ-coordinates

$$\frac{\partial M}{\partial \theta} = c_p \frac{\partial T}{\partial \theta} = c_p\left(\frac{p}{p_0}\right)^{R/c_p} \tag{6-12}$$

yields much more accurate values of M on upper levels. An integration of (6-12) yields the following expression for evaluating of M over comparatively small intervals in the vertical:

$$M_2 = M_1 + c_p \bar{T} \ln\left(\frac{\theta_2}{\theta_1}\right)$$

where $\bar{T}$ is the average temperature in the layer between θ-levels marked 1 and 2. This formula should be used to evaluate M at all levels above the earth's surface. A sequential procedure should be applied, one layer at a time.

Geostrophic Wind

The isopleths of M lie along the wind in geostrophically balanced flow. This is analogous to the isobaric surfaces, where contours represent the geostrophic streamlines. The geostrophic wind is given by the gradient of the stream function as

$$\mathbf{V}_g = \frac{1}{f}\mathbf{k} \times \nabla M$$

$$u_g = -\frac{1}{f}\frac{\partial M}{\partial y} \qquad v_g = \frac{1}{f}\frac{\partial M}{\partial x}$$

The variable factor f^{-1} in these equations shows that M is not exactly a stream function as is ψ in (4-34). The designation "stream function" should imply that f is constant. Since f varies only slowly in the horizontal, the usage of "stream function" is greatly justified. For this reason, the designation "stream function" is now generally accepted. With a similar justification, it is common to use the expression *streamlines* for the M isopleths.

There is a degree of redundancy in representing the stream function and wind since the air generally flows along the isopleths of the stream function. However, there is a purpose for this redundancy. As on other charts, we are paying the most attention to the comparatively rare cases where the wind is different from geostrophic wind. Plotting the wind and the stream function on the same chart may show the regions of strong weather development where the geostrophic balance is not attained.

Geostrophic deviation can be detected when the wind observations show a different direction than the streamlines. In such cases we expect that the balanced divergence and vertical motion can be detected, as discussed in Chapters 4 and 7. Some changing patterns of ageostrophic flow have prominent local divergence or convergence that show mesoscale weather processes. Other cases may reveal that the flow pattern is undergoing changes or oscillations that characterize the geostrophic adjustment.

Contrary to the wind direction, the deviation of wind speed from geostrophic wind speed cannot be easily detected by visual inspection of the charts. Computerized analysis of wind and geostrophic wind is needed for a complete assessment of geostrophic deviation.

Streamlines (isopleths of M) in Fig. 6-9 follow the wind direction very well in most parts of the chart. A prominent deviation from the geostrophic wind can be noticed at Rapid City, South Dakota (RAP). This is near the center of a developing cyclone in Montana. The wind in this place has a large isallobaric part. The isallobaric wind has about the same main property on the M chart as on the isobaric chart: It has a component of air flow toward falling M.

Another place with ageostrophic flow is the Arkansas-Mississippi-Alabama area where the warm air recently started spreading northeastward. The stream function shows a weak gradient, and the wind speed is accordingly low. The geostrophic adjustment has not yet been accomplished in this region.

Thermal Wind

Isobars practically are streamlines of thermal wind, except for comparatively small variations of f and p in definition (I-5) from Appendix I:

$$\frac{\partial \mathbf{v}_g}{\partial \theta} = \frac{R}{f} \frac{p^{\kappa-1}}{p_0^{\kappa}} \mathbf{k} \times \nabla p \qquad (6\text{-}13)$$

The most common conclusion from this equation is that thermal wind points along isobars, analogously to thermal wind pointing along isotherms in p- and z-systems. This analogy is understandable, since the isobars are also isotherms on the isentropic charts.

Equation (6-13) also shows that a large pressure gradient (∇p) implies a large vertical wind shear. The same condition of large ∇p was shown above to indicate the possible presence of a sloping frontal layer. These two conclusions confirm the familiar concept that large thermal wind and a sloping frontal layer usually appear together. An important advantage of the isentropic chart (over isobaric chart) is that both phenomena (large thermal wind and a sloping frontal layer) can be detected from one element (pressure).

The thermal wind equation requires that the wind turn closer to the direction of the isobars at levels above the depicted isentropic surface. For example, in Fig. 6-9, a southwesterly wind over Topeka (TOP) and Omaha (OMA) blows at a large angle across the isobars. The thermal wind equation (6-13) and the direction of isobars help us to conclude that the wind is more westerly above the chart level than at this level. This is confirmed by the observation that the polar jet stream is westerly at 300 mb (this 300-mb chart is not shown here).

Thermal Advection

Wind blowing across the isobars indicates thermal advection since the isobars are isotherms as well. This conclusion can be also made from the streamlines (M = const.) that intersect the isobars. This latter case shows geostrophic thermal advection. The sign of advection can be concluded from the familiar rules that also hold in other systems:

> the veering or backing of wind with height implies warm or cold thermal advection, respectively; and
>
> wind coming from the warm side of the isobars shows warm (positive) thermal advection. Wind from the cold side shows cold (negative) advection.

The example in Fig. 6-9 shows prominent warm advection from Kansas and Nebraska until Illinois, where the wind blows at a considerable angle across the isobars. This is a region of warm advection since the wind blows from the side of higher pressure and since the thermal wind direction is to the right of wind direction.

Stability

The difference in pressure Δp between neighboring isentropic surfaces of 305 and 300 K is a fair measure of stability of the atmosphere. Namely, the stability s

can be expressed as the vertical derivative of potential temperature:

$$s = \left|\frac{\partial \theta}{\partial p}\right| \approx \frac{\Delta \theta}{\Delta p}$$

Since the thermal difference $\Delta\theta$ has been selected as 5 K, the stability is inversely proportional to the pressure difference Δp:

$$s = \frac{\text{const.}}{\Delta p}$$

Thus, as Δp is larger, the stability is smaller. Small stability may mean conditional instability, and this is one of the favorable conditions for development of convective clouds and showers.

Small stability (large Δp) can be noticed in the active region of the jet stream over the Rocky Mountains in Fig. 6-9. The largest value of 352 mb can be found at Grand Junction, Colorado (JCT). Other charts (not shown here) show that thunderstorms are in progress in this region.

Another region of small stability is at the warm front from eastern Nebraska to Oklahoma and Arkansas. Values of 255 mb at Oklahoma City (OKC) and 213 mb at Little Rock, Arkansas (LIT), are also comparatively large and indicative of conditional instability. The release of potential instability and formation of showers in this region is also aided by large-scale lifting, as explained below.

Large stability (Δp of 12 to 18 mb) over Washington and British Columbia are in the lower stratosphere. Values of 60–90 mb in eastern United States show a typically stable cold air in the lower troposphere. This is the region of subsidence in the polar air under an inactive region of the polar jet stream.

Humidity

The selected element in the example in Fig. 6-9 is the mixing ratio. The units are 10^{-3}, also popularly known as g kg^{-1}. As shown in Chapter 5, high humidity is an important contributing factor in potential instability. Other measures of humidity (vapor pressure, relative humidity, and so forth) can be obtained from the mixing ratio using the equations from Chapter 5.

Mixing ratio of over 7×10^{-3} (or over 7 g kg^{-1}) that can be seen in Fig. 6-9 in Texas and neighboring states is comparatively large and sufficient to support strong rain. Such large values can be found only in the lower troposphere where high temperature allows a higher saturation mixing ratio. The example in Fig. 6-9 shows that the 305 K isentrope is in the lower troposphere in Texas, since the pressure is higher than 750 mb. Also, high values of mixing ratio (8 to 10×10^{-3}) are usually associated with southerly wind that brings humid air from a tropical ocean.

Vertical Motion

The logic behind introducing isentropic coordinates is that the vertical motion does not exist in this coordinate system, as long as the processes are adiabatic. The definition of vertical velocity in θ coordinates

$$\dot{\theta} \equiv \frac{d\theta}{dt}$$

coincides with the main expression in the adiabatic form of the first law of thermodynamics (C-39):

$$\frac{d\theta}{dt} = 0$$

If students are not yet familiar with the meaning of vertical motion in θ-coordinates, it may be useful to try to solve the following easy puzzle: How can we prevent lifting in the atmosphere by a mathematical transformation of the coordinate system? The answer is that lifting still exists in other coordinate systems (p, z) whereas the θ-coordinates move together with the air. When the coordinates move with air parcels, the air parcels do not have motion in those moving coordinates. It is the lifting in the more "realistic" coordinates (p, z) that can be inferred from the wind on isentropic surfaces.

"Real" vertical motion, say $\omega = dp/dt$ in the p-system, can be estimated in θ-coordinates as

$$\omega = \frac{\partial p}{\partial t} + \mathbf{V} \cdot \nabla p + \dot{\theta}\frac{\partial p}{\partial \theta} \qquad (6\text{-}14)$$

Interpretation of vertical motion ω can be made in several schematic cases, depending on the terms on the right-hand side of (6-14). First conclusions may be done by ignoring the local time derivative $\partial p/\partial t$. Such conclusions are justified in the cases when the wind is "much" faster than the movement of weather patterns. In these cases, the troughs and ridges move slower than the wind and the air parcels traverse through the patterns.

The last term with $\dot{\theta}$ disappears under adiabatic conditions. The flow in the atmosphere is most closely adiabatic in the case of subsidence. In such a case there is no release of latent heat of condensation. Otherwise, condensation of water vapor is a major source of heat for weather development.

The pressure advection $-\mathbf{V} \cdot \nabla p$, which appears in (6-14), is an important term that needs interpretation. Its effect can be studied when (6-14) is rewritten without the first and third term on the right-hand side. The case of $-\mathbf{V} \cdot \nabla p > 0$ in the adiabatic stationary flow corresponds to lifting:

$$w \propto -\omega = -\mathbf{V} \cdot \nabla p > 0 \qquad (6\text{-}15)$$

and the other case

$$w \propto -\omega = -\mathbf{V} \cdot \nabla p < 0 \qquad (6\text{-}16)$$

shows sinking.

It should be noticed that $-\mathbf{V} \cdot \nabla p$ represents also the thermal advection since the isobars are isotherms on isentropic surfaces. Therefore, the conclusions from (6-15) and (6-16) correspond to our old experience (Section 6-6) that warm advection usually is accompanied by lifting.

Using the above conclusion in interpretation of Fig. 6-9, we can find lifting in eastern Kansas and eastern Nebraska where the wind points at a large angle across the isobars. With very small stability, it can be understandable that potential instability is released. This explains the showers in this region. This area was also mentioned above as an area of warm advection.

Subsidence is noticeable in Georgia and Florida, where the wind barbs show air flowing down the sloping isentropic surface. The polar air subsides and spreads out near the earth's surface. This region also shows cold advection, as mentioned before.

Vertical Motion in Fast-Moving Patterns

The time derivative in (6-14) is important in the cases when the weather patterns move about with the wind velocity. The influence of the time derivative can be seen in the two following typical cases, one with no wind and the other when the pattern moves with the wind. Both cases will be discussed only in the adiabatic form.

Case (a): No wind. Equation (6-4) becomes

$$\omega = \frac{\partial p}{\partial t} \qquad (6\text{-}17)$$

In this case, the vertical motion ω is equal to the local pressure tendency on the isentropic surface. Such cases are rather rare in the atmosphere. For example, subsidence in the center of an anticyclone may show lowering of isentropic surfaces; there, $\partial p/\partial t > 0$ and there will be sinking with $\omega > 0$.

Case (b): The weather pattern moves with wind speed. This case can be schematically represented as in Fig. 6-10. The isentropic surface of 290 K is shown in two positions, perhaps 6 h apart. Several isobars are shown as dots on the isentropic surface. The pressure on the isentropic surface above A has increased during the 6 h; thus $\partial p/\partial t > 0$. At the same time the wind (presumably along the isentrope) crosses the isobars from the side of higher values; thus $-\mathbf{V} \cdot \nabla p > 0$. This weather situation may be described by the condition

$$\omega = 0 = \frac{\partial p}{\partial t} + \mathbf{V} \cdot \nabla p \qquad (6\text{-}18)$$

where each of the two terms on the right-hand side is large, but they balance each other. There is no vertical motion ω.

The practical problem is to distinguish when the pressure advection $-\mathbf{V} \cdot \nabla p$ represents vertical motion as in (6-15) and (6-16), and when the vertical motion is

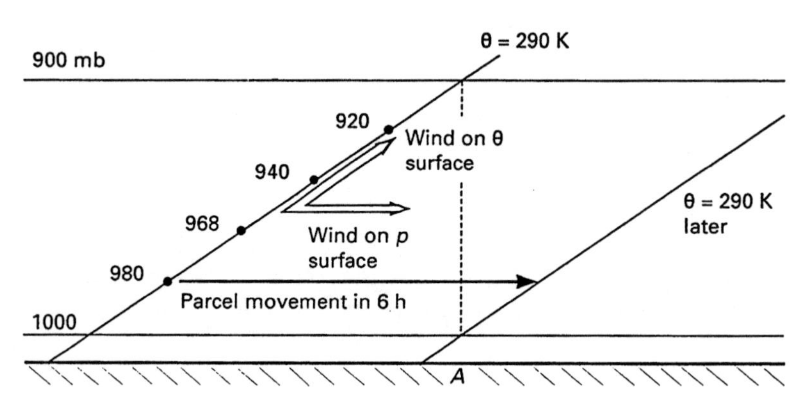

FIGURE 6-10. Displacement of an isentropic surface (290 K) in the presence of strong pressure advection and large local pressure tendency. Dots along the 290 K surface show the pressure. Conditions above point A show that the pressure is 900 mb on the 290 K isentropic surface initially. After a 6-hour displacement, the pressure is 1000 mb on the same 290 K surface over point A, indicating a positive pressure tendency ($\partial p/\partial t > 0$). Pressure advection is positive ($-\mathbf{V} \cdot \nabla p > 0$) since the wind blows from the higher pressure values.

equal to zero, as in (6-18). In each case, $-\mathbf{V} \cdot \nabla p \neq 0$, but the vertical motion is entirely different. The answer is that the motion of the system should be considered. Two cases are typical.

1. If the system is stationary, the vertical motion is correctly represented by (6-15) and (6-16).
2. The weather systems in the air current may move with the wind velocity. This is the case with short waves. Then the pressure tendency $\partial p/\partial t$ may counteract the conclusion from pressure advection.

The two terms on the right-hand side of (6-18) do not necessarily balance each other. Many other cases may occur where one of the terms predominates. The above examples are only idealized cases that show the basic circumstances with the quantities $\partial p/\partial t$ and $-\mathbf{V} \cdot \nabla p$.

Another small puzzle awaits the reader: Why are there two wind vectors in Fig. 6-10? The answer is that the "horizontal" (or isobaric or isentropic) flow is assumed equal in all these coordinate systems; only the vertical component is different. The "horizontal" wind vector in Fig. 6-10 is drawn twice, one for each coordinate system. If the flow is adiabatic, then the air moves along the isentropes. The flow in this example is also assumed "horizontal" (isobaric), so it displaces its isentropic surface, as shown. In this case, the parcel stayed on the same isobar *and* on the same isentrope. Still, for finding the pressure advection, the wind had to be assumed "slanted" along the isentrope. The angle α between the two drawn wind vectors is assumed small and $\cos \alpha \approx 1$.

Vertical Motion under Diabatic Conditions

There is hardly a need to explain the importance of diabatic processes. Differential heating is the ultimate cause of all atmospheric motions. Most heating for the processes that last less than 3 days comes from the release of latent heat during lifting. For this reason, the cases with lifting are regularly more interesting in weather development than the cases of subsidence. Precipitation and several other phenomena occur due to lifting.

When heating due to condensation of water vapor appears, the above-mentioned conclusions about vertical motion are less correct since the term with $\dot{\theta}$ was ignored. The influence of heating due to the release of latent heat can be seen next in several cases where diabatic influences appear. The time derivative will be ignored in each case.

Case (a): No wind. Vertical motion is different from zero in the case of heating, even if there is no vertical motion in the p-system. Naively we may think that (6-14) reduces to

$$w \propto -\omega = -\dot{\theta}\,\frac{\partial p}{\partial \theta}$$

However, with heating, the isentropic surfaces move from air parcel to air parcel, rushing toward new high values of θ. In this way, the isentropic surfaces approach the earth's surface. They descend from the standpoint of the p- or z-systems. In these cases the local pressure derivative is significantly different from zero and must be retained in consideration. Therefore, the relevant equation is

$$w \propto -\omega = -\frac{\partial p}{\partial t} - \dot{\theta}\,\frac{\partial p}{\partial \theta} \qquad (6\text{-}19)$$

The first term $-\partial p/\partial t$ is negative since the isentropic surfaces move toward higher pressure. The second term is positive since $\partial p/\partial \theta < 0$. These two terms essentially counteract each other. This situation is well described by

$$\frac{\partial p}{\partial t} + \dot{\theta}\,\frac{\partial p}{\partial \theta} \approx 0$$

As can be expected, heating without air movement is not necessarily associated with appreciable vertical motion. Heating only causes small vertical displacement (from the standpoint of z-system), showing an increase of thickness between neigboring isobaric surfaces.

Case (b): Heating occurs together with pressure advection. If the time derivative is temporarily ignored, (6-14) yields

$$w \propto -\omega = -\mathbf{V} \cdot \nabla p - \dot{\theta}\,\frac{\partial p}{\partial \theta} \qquad (6\text{-}20)$$

The heating term with $\dot{\theta}$ is close to impossible to estimate from weather charts. (This term, however, may be available from models of numerical weather prediction.) Fortunately, it is rather easy to estimate the sign of $\dot{\theta}$. This sign is positive for the case of lifting and condensation of water vapor. Therefore, a qualitative estimation of lifting vertical motion is still reliable. If the estimation based on pressure advection indicates lifting, then the lifting is enhanced by the contribution of the heating (diabatic) term.

The situation of case (b) can be also described in the following way, by an estimation of the sign of the terms in (6-20). In the case of lifting, $\dot{\theta}$ can be positive (if condensation occurs) and $\dot{\theta}\ \partial p/\partial\theta$ is negative (since $\partial p/\partial\theta$ is negative). The term $\mathbf{V} \cdot \nabla p$ is negative with lifting too. Here both terms on the right-hand side of (6-20) have the same sign. The estimation of lifting using pressure advection only $(-\mathbf{V} \cdot \nabla p)$ will yield an underestimation of lifting. This is useful in visual inspection of weather charts: The lifting can be inferred from advection, and we know that our conclusion will not be refuted by diabatic processes. The vertical motion will be likely stronger due to the term $\dot{\theta}$ than concluded from the advection term only.

In the case of sinking, $\dot{\theta}$ is zero, and this is the adiabatic case described in the preceding subsection.

7

Relation of Wind and Forces in the Atmosphere

The quasi-geostrophic balance between wind and pressure distribution is among the strongest physical relationships between meteorological elements. This relationship is of great help in weather analysis. Therefore, this chapter gives a review of the balanced motions of geostrophic and ageostrophic characters. Balanced vertical motion is probably the most important deviation from geostrophic balance; therefore, it is included.

Ageostrophic wind is important to numerous dynamic processes in the atmosphere, such as vertical motion, cyclogenesis, and the movement of weather systems. Unfortunately, weather observations are seldom adequate for accurately evaluating ageostrophic wind. To circumvent the difficulties with observations, various indirect methods for estimating geostrophic deviation are used. These represent the main topics of discussion in this chapter. In particular, ageostrophic wind will be discussed in connection with friction, pressure tendency, curvature of the flow, advections of vorticity and temperature, and streaks in the jet stream.

7-1 WHY IS THE WIND ABOUT EQUAL TO GEOSTROPHIC WIND?

Definitions

Air flow is characterized by the wind being approximately equal to geostrophic wind (except in equatorial regions). Often, when the wind observations are not available, geostrophic wind can be used instead of wind.

Geostrophic wind is the vector given by the equations (C-7) and (C-8) from Appendix C:

$$\mathbf{V}_g = \frac{g}{f}\mathbf{k} \times \nabla z$$
$$u_g = -\frac{g}{f}\frac{\partial z}{\partial y} \tag{7-1}$$
$$v_g = \frac{g}{f}\frac{\partial z}{\partial x}$$

Geostrophic wind is $\mathbf{V}_g$, and its components are u_g and v_g. Evaluation of geostrophic wind is described in Section 4-1. The above definitions show that geostrophic wind conveniently describes the gradient of the geopotential and the pressure force. The wind is not included in these definitions.

The approximate equality of wind ($\mathbf{V}$) and geostrophic wind ($\mathbf{V}_g$) can be expressed as

$$\mathbf{V} \approx \mathbf{V}_g \tag{7-2}$$

Geostrophic flow is a flow in which (7-2) is well satisfied, so that the equal sign can be used:

$$\mathbf{V} = \mathbf{V}_g$$

Since the observations of wind and pressure field often are available independently, the vector difference

$$\mathbf{V}_a = \mathbf{V} - \mathbf{V}_g$$

is *ageostrophic wind.* The component of $\mathbf{V}_a$ across the contours is often observed on weather charts and it is appropriately known as the *cross-contour component.* There is also an *along-contour component,* but this cannot be easily observed on weather charts; it takes calculation of geostrophic wind speed.

Stability of Geostrophic Flow

Geostrophic flow is usually stable for perturbations (waves) shorter than several hundred kilometers. For this reason, wind and height fields assume the familiar pattern where the streamlines closely follow the contours.

On a larger scale, with perturbations longer than 1000 km, geostrophic flow may be baroclinically unstable. Then the flow changes its pattern rather slowly, however. It may take 2 or 3 days to form a cut-off low or high. The details of such an unstable long wave maintain a fairly geostrophic balance.

Geostrophic Adjustment

Geostrophic flow is laminar (or nonturbulent), like several other stable stationary flows. A flow pattern can be recognized as stable when the disturbances on it are dampened. The flow is stationary when its pattern does not change. A slowly changing pattern, therefore, is *quasi-stationary.* Possible deviations from the laminar stable state behave like dampening waves and, after a few oscillations, the flow returns to the stationary mode. The conditions in the atmosphere (at least for shorter waves) are normally stable for geostrophic flow. Therefore, the atmospheric flow adjusts toward geostrophically balanced flow. The oscillations that lead to geostrophic flow constitute the process of *geostrophic adjustment.* This explains why the wind is equal to geostrophic wind. The next examples illustrate the adjustment process in more detail.

Stable and Unstable Flows

A stable, stationary flow is normally *laminar* and can be described using streamlines. *Turbulent* flow, however, is disorderly and the details of flow are hard to observe and describe. When a laminar flow persists for some time, this flow is usually *stable.* When a laminar flow cannot persist, it breaks down into turbulent flow with numerous eddies. Such laminar flow is already *unstable* before it breaks up into vortices.

The distinction between stable and unstable laminar flows is illustrated in Figs. 7-1 and 7-2. The flow in both examples is laminar upstream from the point of disturbance (A or B). The behavior of these two flows is different after they are disturbed. Figure 7-1 represents the trajectories in a stable fluid flow. This flow is characterized by straight parallel trajectories (or streamlines) until point A. A disturbance occurs at A. The trajectory through A continues through several oscillations before it assumes the earlier straight shape. The oscillations were stable, therefore they dampened. Neighboring parcels were also disturbed, but their oscillations dampened out in the same way. The disturbance in the flow is not necessarily small. If the disturbance is large, it may take a longer time to dampen the oscillations, but the process is essentially the same.

Figure 7-2 represents an unstable flow. A relatively small disturbance at B creates waves whose amplitude increases and the laminar flow breaks down. Disorderly (turbulent) flow results.

The flow that is considered in Sections 7-2 through 7-6 is of the balanced stable type as illustrated around point F in Fig. 7-1. It is assumed that the oscillations of geostrophic adjustment have already dampened out and a quasi-stationary balanced flow has been established.

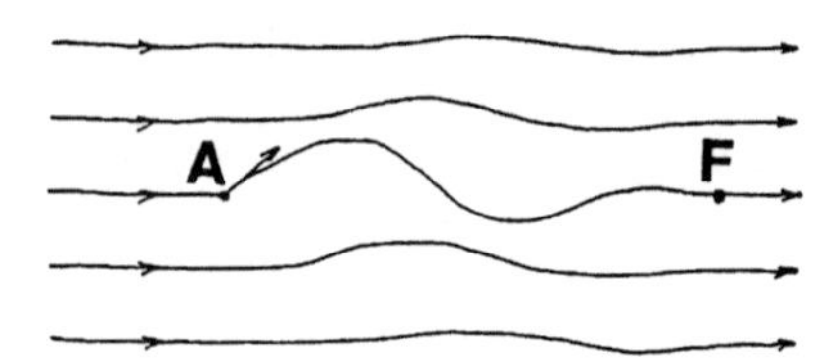

FIGURE 7-1. Schematic streamlines or trajectories in stable fluid flow. This flow is disturbed at A, but resumes a straight shape after several oscillations downstream from A. This flow stays laminar.

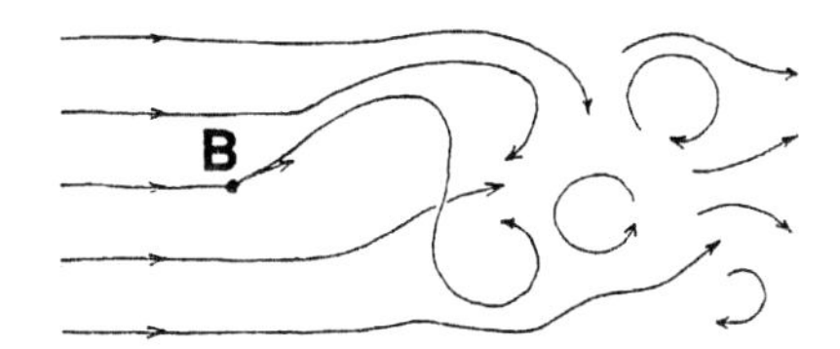

FIGURE 7-2. Schematic streamlines or trajectories in unstable flow. The flow is disturbed at B. The flow is laminar (also unstable) upstream of B and turbulent downstream.

General Considerations on Stable Flow

Figures 7-1 and 7-2 were deliberately drawn without coordinates and scales to illustrate generalized disturbances on any scale and in any direction: horizontal, vertical, or slanted. For example, the flow most familiar to students is the flow in an atmosphere that may be thermally stable or unstable. Then Fig. 7-1 represents a vertical section and the waves are similar to the lee waves behind the mountains or other obstacles. Such waves are often visible if clouds form on them. If the stratification is unstable, then every disturbance causes the flow to break down into eddies as in Fig. 7-2.

Other stable and unstable flows are similar. Figures 7-1 and 7-2 may represent large-scale horizontal flow as the westerlies over a continent. The first example (Fig. 7-1) will be a stable zonal flow. This zonal flow may become baroclinically unstable and break down in eddies, as in Fig. 7-2. The eddies in this case represent cyclones and anticyclones of the middle latitudes.

Other types of instability or stability are superficially similar to the cases in Figs. 7-1 and 7-2. Some types of instability are known as *dynamic, shearing, symmetric, baroclinic,* or *barotropic* instabilities. The stability of any flow is determined by various hydrodynamic criteria, usually by the horizontal or vertical wind shear or combinations of thermal stability and wind shear. With each kind of instability, the flow follows either the stable or the unstable pattern, as in these figures.

An important property of fluid flow is that it naturally assumes stable, stationary patterns, if only such states exist. The geostrophic flow is a stable, stationary possibility, and the atmosphere tends toward that state through oscillations. This is the reason we find geostrophically balanced flow in the atmosphere.

Experience with waves and the theory of fluid dynamics, as in the above examples, allows us to consider separately two kinds of flow: (a) stationary balanced flow, as the flow upstream from points A or B in Figs. 7-1 and 7-2, and (b) transient (disturbed) flow, as the flow downstream from A or B. The transient flow, in turn, may behave in two different ways, as in Figs. 7-1 and 7-2. A great part of fluid mechanics is dedicated to the study of various types of transient flow. However, for the purpose of weather analysis, it is useful first to study the undisturbed stationary flow (the flow upstream of A or B or far downstream from A).

Stationary flow can be conveniently studied with the equations of motion in which the time derivative can be set equal to zero. Thereby the equations for the flow reduce to rather simple forms. The technique of setting the time derivative equal to zero will be used in several sections of this chapter.

All examples that follow in this chapter are of the balanced type, as in Fig. 7-1 far downstream of A. That balance is not necessarily geostrophic. It may be of several related kinds (gradient wind balance, isallobaric wind, and so forth). Since these cases are close to geostrophic balance, they are often said to be in *quasi-geostrophic balance*.

An Example of Stationary Flow

An example of flow will be considered next where the equations for a stationary flow can be applied. This is the flow which is governed by the equations of motion [(C-28) from Appendix C]:

$$\frac{du}{dt} = -g\frac{\partial z}{\partial x} + fv$$
$$\frac{dv}{dt} = -g\frac{\partial z}{\partial y} - fu$$

This is a closed set of equations for wind components u and v, if the geopotential field z is given. An interesting theoretical problem here is to search for stationary solutions. This can be done by setting the derivatives du/dt and dv/dt equal to zero. (Another tacit approximation used is that the advection terms of the type $u\,\partial u/\partial x$ are negligible.) Then the equations reduce to

$$u = -\frac{g}{f}\frac{\partial z}{\partial y}$$
$$v = \frac{g}{f}\frac{\partial z}{\partial x}$$

which can be recognized as the familiar geostrophic wind (7-1). This kind of flow is often observed in the atmosphere and is stable under fairly wide ranges of vertical and horizontal wind shears in a statically stable atmosphere. This explains the frequent occurrence of wind that is about equal to geostrophic wind.

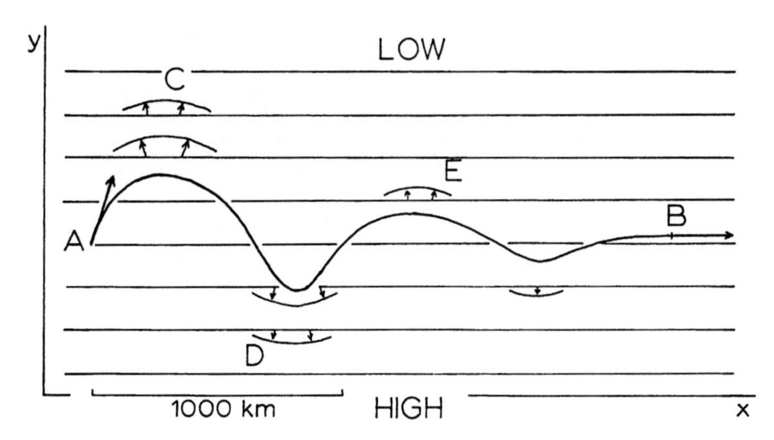

FIGURE 7-3. An idealized example of geostrophic adjustment. The trajectory of a parcel is the heavy curve AB. Contours are the thin lines. Gravity-inertia waves are illustrated by arches at C, D, and E.

Damping of Oscillations

Figure 7-3 further illustrates the process by which a geostrophic equilibrium of forces is achieved. This example is based on theoretical work of Rossby, Cahn, Obukhov and other researchers. The pressure field in this figure is described by a few straight contours. Most parcels move exactly along the contours, that is, they are in geostrophic balance.

One parcel is disturbed, and it moves across the contours at point A. The parcel turns to the right under the influence of the Coriolis force. After this, the parcel oscillates several times until it emerges at B, where it assumes almost exactly the geostrophic balance $\mathbf{V} = \mathbf{V}_g$. Damping of oscillations is due to "radiation" of energy of oscillating air parcels. Our moving parcel interacts with other parcels through the pressure force, and it literally pushes them out of their presumed geostrophic balance. These other parcels in turn push their neighbors, and soon a field of oscillating parcels is generated. These generated waves move away from their sources. These waves are sketched in Fig. 7-3 as arches at C, D, and E, which show temporarily deformed contours. The waves sketched near C, D, and E propagate with velocities of compression waves and external and internal gravity-inertia waves. The described process is another schematic example of geostrophic adjustment.

Both fields of wind and pressure change during geostrophic adjustment. It cannot be correctly claimed that the wind adjusts to the pressure field or, conversely, that the pressure distribution adjusts to the wind. Instead, both adjust to each other. There is, though, a dependence on scale. For short waves ($L \leq$ 1000 km), there is relatively little adjustment of wind and strong adjustment of pressure. For longer waves (say $L \geq 3000$ km), the wind adjusts more, while pressure changes little.

The theoretical examples in the literature whose abstract is given in Fig. 7-3 have all been computed without friction. The damping is due to dispersion of energy to the environment and not due to friction.

For the geostrophic adjustment, we do not need parcels that travel across the isobars or contours as in Fig. 7-3. Any situation with imbalance can cause this adjustment. An imbalance exists, for example, when the wind blows along the contours with a speed different from a balanced value.

In the atmosphere, we rarely have strong impulses that suddenly cause a large ageostrophic flow. Instead, disturbances and adjustment processes occur gradually; we seldom observe processes similar to the waves in Fig. 7-3. The processes that cause geostrophic imbalance and subsequent adjustment are those that generate weather processes, such as heating, deflection of flow over mountains, or bursts of instability of various kinds.

Since the adjustment applies also to other stationary solutions of the equations of motion as well– such as the gradient wind, motion in the frictional planetary boundary layer, and a few other cases of "balanced" flow–the above described oscillations are also called *quasi-geostrophic adjustment*.

7-2 FLOW IN THE PLANETARY BOUNDARY LAYER

Guldberg-Mohn Wind Approximation

Ageostrophic wind is commonly observed as the cross-isobaric flow at the surface. This is attributed to surface friction and can be explained by the equations of motion in which a friction force is added on the right-hand side. A usable form of the friction force is with the second derivative of the wind components on the right-hand side:

$$\frac{du}{dt} = fv_g + fv + K\frac{\partial^2 u}{\partial z^2}$$
$$\frac{dv}{dt} = fu_g - fu + K\frac{\partial^2 v}{\partial z^2} \qquad (7\text{-}3)$$

This set of equations is closed if the pressure force, expressed by u_g and v_g, is prescribed. The situation is analogous to the evaluation of geostrophic wind: The wind is evaluated on the basis of the pressure force.

A stationary and nonaccelerating flow is described with $du/dt = dv/dt = 0$. With this, (7-3) becomes a set of differential equations for u, v whose solution yields the Ekman spiral. However, if we are not interested in accurate solutions and consider one layer only, we can see the meaning of the friction force as first shown by C. M. Guldberg and H. Mohn in 1883. The wind profile near the earth's surface is usually convex, if represented on a graph with wind speed versus height. The wind speed increases from zero at the earth's surface to significant values near the anemometer level, and further increase of wind speed with height continues at a slower rate. In this situation, the second derivatives of u and v may be assumed proportional to $-u$ and $-v$, respectively:

$$K\frac{\partial^2 u}{\partial z^2} = -C\,u \qquad K\frac{\partial^2 v}{\partial z^2} = -C\,v$$

The parameter C is a dimensional constant of proportionality. It is also called "friction" because it shows the magnitude of the friction force. Further, we may drop the acceleration terms in (7-2) since we are interested in stationary, straight flow. The equations of motion now assume the form

$$0 = -fv_g + fv - Cu$$
$$0 = fu_g - fu - Cv \qquad (7\text{-}4)$$

The equations (7-4) are conveniently linear for wind components. If the pressure field is known, then u_g and v_g are known as well [given by (7-1)]. If we turn the coordinate system so that v_g vanishes, equations (7-4) can be solved for the wind components as

$$u = \frac{u_g}{1 + \left(\frac{C}{f}\right)^2} \qquad v = \frac{C}{f}\frac{u_g}{1 + \left(\frac{C}{f}\right)^2} \qquad (7\text{-}5)$$

The quantities in these equations are illustrated in Fig. 7-4. The wind that is evaluated from the pressure gradient using (7-5) is the *Guldberg-Mohn wind*.

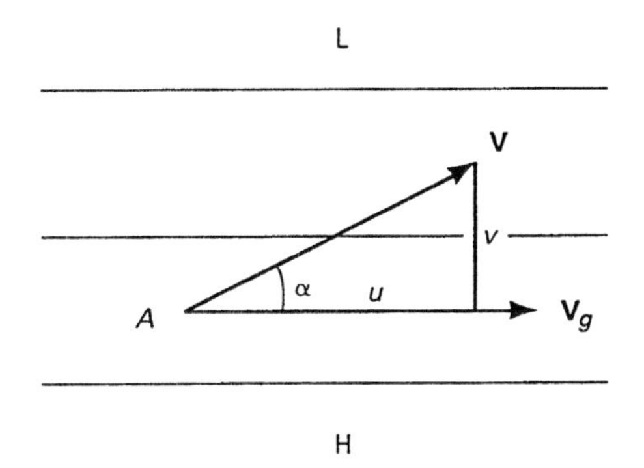

FIGURE 7-4. Wind **V** in the presence of friction force. Contours of an isobaric surface are thin solid lines, wind components are u and v, geostrophic wind is V_g, and the cross-contour angle is α.

The angle of the cross-isobaric flow with the isobar is given by

$$\tan\alpha = \frac{v}{u} = \frac{C}{f}$$

This equation shows that the cross-isobaric angle α is larger for a larger friction C. When C is equal to f, the angle α is 45°. The angle α is smaller for smaller C. This explains a nearly geostrophic flow over frictionless, smooth surfaces, as over oceans. The deviations from geostrophic wind over rough terrain (as over mountains) are large. Then a larger value of C is appropriate, of the order of the Coriolis parameter.

Antitriptic Wind

In some cases, the Coriolis force is weaker than the friction force. This may occur over rough terrain or in a thin layer (≈1 km high) of a cold air mass. This is illustrated next for the extreme case when the Coriolis force can be ignored in comparison with friction force. If the x-axis is taken along the wind, the first equation of motion from (7-4) reduces to

$$0 = -fv_g - Cu$$

Inserting the pressure force in the $f\,v_g$ term,

$$0 = -g\frac{\partial z}{\partial x} - Cu$$

This gives *antitriptic wind* as

$$u = -\frac{g}{C}\frac{\partial z}{\partial x} \qquad (7\text{-}6)$$

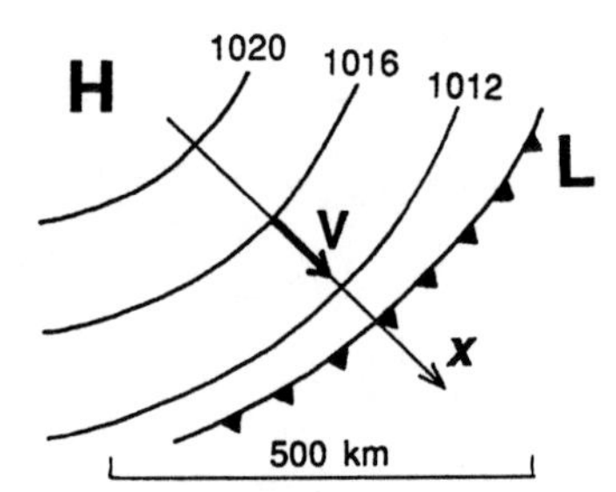

FIGURE 7-5. Antitriptic wind on a surface chart with several isobars. Wind like this occurs in cases when the Coriolis force can be ignored compared with the friction force.

The other component v is equal to zero due to the orientation of the coordinate system.

The antitriptic wind vector points across the contours (or isobars), such that the air fills the low. This is illustrated in Fig. 7-5 for the case of a shallow cold air mass moving behind the arctic front. The movement of the front itself is governed by the dynamics of the passive front (Chapter 8 and Appendix K).

Cold outflow under thunderstorms is superficially similar to the antitriptic wind by its relation to the isobars. The thunderstorm outflow is not a genuine antitriptic wind, since the flow under a thunderstorm is not balanced. This flow changes rapidly, with strong acceleration $d\mathbf{V}/dt$ and with insignificant friction. Then (7-6) does not hold. The thunderstorm case is more similar to the *passive front* that is discussed in Chapter 8 and in Appendix K.

The cross-isobaric frictional flow is of limited dynamical significance, since it is restricted to the lowest several hundred meters. Convergence at the passive front may be caused by the antitriptic wind on the cold side. This may trigger showers, but it will not be sufficient to explain the powerful upgliding that occurs in extratropical cyclones.

7-3 ISALLOBARIC WIND

Often the ageostrophic flow can be attributed to the gradient of the pressure tendency. Since the isopleths of tendency on the surface chart are called isallobars, the related form of the ageostrophic wind is called *isallobaric wind.* If this approximation is used on a constant-pressure (isobaric) chart, then it is called *isallohypsic wind.* When this approximation was first derived in the nineteenth century, ∇p was used instead of ∇z. The expression "isallobaric wind" is still widely used also with isobaric charts.

There are serious practical problems in interpreting isallobars, since they often depend on small-scale processes related to gravity waves. Such small-scale processes are far from the usual quasi-geostrophic balance and generally do not fit into the synoptic scale models. However, assuming that representative isallobars are available, an approximation of the isallobaric wind on a large scale can be obtained from the equations of motion in the form

$$\frac{\partial u}{\partial t} = -f(v_g - v)$$

$$\frac{\partial v}{\partial t} = f(u_g - u) \qquad (7\text{-}7)$$

This is a rather crude approximation since the advection and friction terms have been ignored. This does not mean those terms are unimportant; they have been deleted to avoid formal complications with additional terms in the equations.

Equations (7-7) constitute a set of two differential equations for two variables u and v. It is comparatively easy to solve the equations since the absence of advection terms makes the equations linear. However, another approximation can be made using geostrophic approximation in the time derivative. Thereby it is assumed that the synoptic-scale wind varies with time like geostrophic wind. With this approximation,

$$\frac{\partial u}{\partial t} \approx \frac{\partial u_g}{\partial t} = -\frac{\partial}{\partial t}\frac{g}{f}\frac{\partial z}{\partial y} = -\frac{g}{f}\frac{\partial}{\partial y}\frac{\partial z}{\partial t}$$

$$\frac{\partial v}{\partial t} \approx \frac{\partial v_g}{\partial t} = \frac{\partial}{\partial t}\frac{g}{f}\frac{\partial z}{\partial x} = \frac{g}{f}\frac{\partial}{\partial x}\frac{\partial z}{\partial t}$$

When these expressions are introduced on the left-hand side of (7-7), the equations can be solved for the wind components as

$$u = u_g - \frac{g}{f^2}\frac{\partial}{\partial x}\frac{\partial z}{\partial t}$$

$$v = v_g - \frac{g}{f^2}\frac{\partial}{\partial y}\frac{\partial z}{\partial t}$$

These two scalar equations can be rewritten as one vector equation:

$$\mathbf{V} = \mathbf{V}_g - \frac{g}{f^2}\nabla z_t$$

where z_t is the height tendency $\partial z/\partial t$. The last equation shows that the vector $-(g/f^2)\,\nabla z_t$ is added to geostrophic wind vector. This other vector points "down the gradient" of the height tendency z_t. Thus we expect the wind to have a component across the isallobars, coming from high z_t.

There are several difficulties in the identification of the isallobaric wind. It is rather difficult to construct

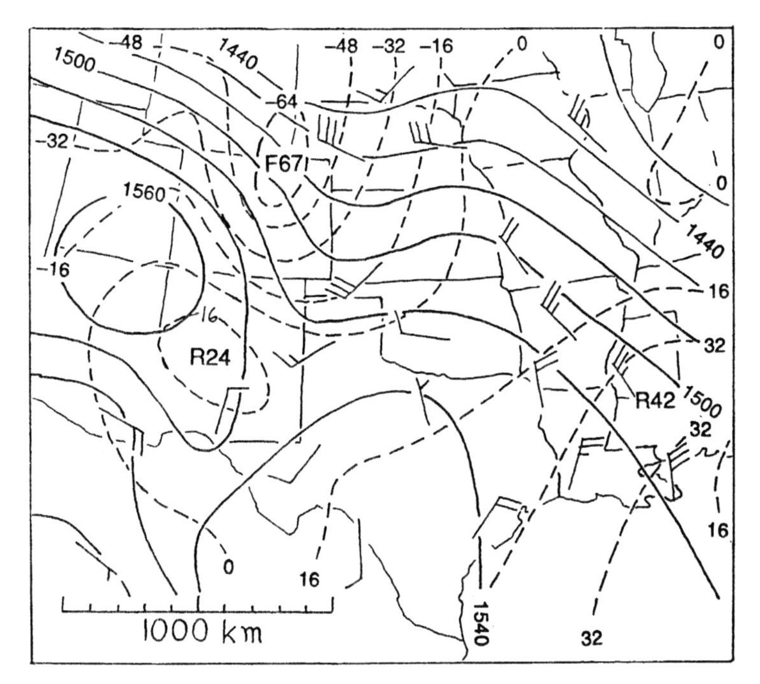

FIGURE 7-6. An example of isallobaric deviation from geostrophic wind at 850 mb for 1200 UTC, 4 February 1974. Wind is in knots (long feather = 10 kt), geopotential height (solid, gpm), and height change over past 12 h [dashed, gpm (12 h)$^{-1}$]. From Djurić and Damiani (1980).

isallobars that are representative of synoptic-scale processes. Also, if the wind deviates from geostrophic wind, we may not spot it unless it blows across the contours.

One example of isallobaric deviation from geostrophic wind is offered in Fig. 7-6. This is an 850 mb chart, with contours, wind barbs, and isopleths of height change in the previous 12 h. The 12-h height change may not represent the tendency at the hour of the chart, but we cannot obtain a better time variation on the upper-level charts. Two regions of significant cross-contour wind can be seen, one in east Texas and Louisiana, the other from northwest Texas to Kansas. In each case, the wind deviates from the contours so that the air moves toward the height fall, that is, "down the gradient" of tendency. The word *gradient* is taken in the traditional meteorological sense: toward low values.

7-4 GRADIENT WIND

Inclusion of Centrifugal Force

One of most widely known deviations of the wind from geostrophic wind occurs with the *gradient-wind balance*. The stationary flow may occur as frictionless flow along the contours (or isobars) in the case of significant curvature. One of the equations of motion, for example, the second one [(C-28), Appendix C], can be written as

$$\frac{dv}{dt} = fu_g - fu \tag{7-8}$$

The coordinate system can be conveniently rotated so that the normal component v vanishes at the point of interest A (Fig. 7-7). The acceleration dv/dt, however, does not vanish at point A. It should be noticed that $v \neq 0$ on the contour on either side of A. The difference between the forces on the right-hand side of (7-8) gives a radial centripetal acceleration. This acceleration is balanced by the centrifugal force of inertia in the amount

$$\frac{dv}{dt} = \frac{u^2}{r} \tag{7-9}$$

The radius of curvature is r. Positive values of r ($r > 0$) indicate positive curvature of geostrophic wind and a low on the concave side of the contour (for the Northern Hemisphere). Conversely, $r < 0$ indicates the high side of the contour is concave. Now (7-8) becomes

$$\frac{u^2}{r} + fu - fu_g = 0$$

This equation is algebraic (not differential) of the second degree. A practical way to find its two roots is to solve it for $1/u$ and then invert. The solution is known

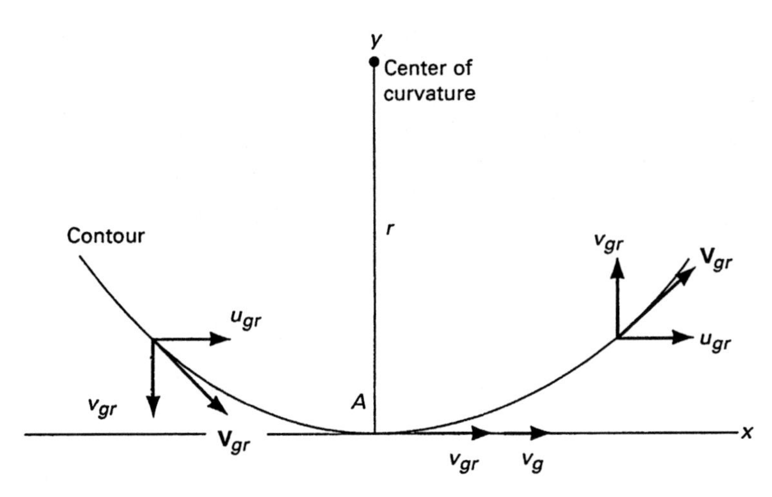

FIGURE 7-7. Gradient wind on a cyclonically curved contour. Bold letters show vectors, regular letters show components. The components are drawn as vectors, each multiplied by its respective unit vector. Geostrophic wind $\mathbf{v}_g$ is larger than the gradient wind $\mathbf{v}_{gr}$.

as the *gradient wind speed*

$$u_{gr} = \frac{u_g}{\frac{1}{2} \pm \sqrt{\frac{1}{4} + \frac{u_g}{rf}}} \tag{7-10}$$

The gradient wind vector

$$\mathbf{V}_{gr} = \begin{pmatrix} u_{gr} \\ v_{gr} \end{pmatrix} \tag{7-11}$$

points along the contours exactly like geostrophic wind. The second component of $\mathbf{V}_{gr}$ is equal to zero when the coordinate system is turned as in Fig. 7-7 (example in point A). The component v_{gr} may not be zero for other orientations of the coordinate system. Analogous to geostrophic approximation, using (7-10) and (7-11) for the wind represents the *gradient-wind approximation*.

The physics with the centrifugal force described by (7-9) may be criticized since the flow seldom takes place along stationary streamlines. More commonly the streamlines themselves move, albeit slower than the wind. Then the appropriate r in (7-9) is the radius of the trajectory. The equations for moving trajectories and appropriate gradient wind have been developed, but these equations are rather complicated and, therefore, impractical for use during manual analysis of weather charts. The theory espoused in this section is much easier, but it applies correctly only when the weather patterns move significantly slower than the wind speed.

Four cases of u_{gr} in (7-10) that need mention are for plus and minus signs before the square root and for positive and negative radius of curvature r. In each of the four cases, the wind speed deviates from geostrophic wind speed. Each case has been observed in the atmosphere. In the following, the coordinate system will be oriented so that u_g is positive ($u_g > 0$).

Circulation around Lows

The two cases of $r > 0$ are for circulation around lows. In these cases, both roots of the radical in (7-10) are real, meaning that a stationary flow exists with any choice of r, u_g, and f. The absolute value of the square root in (7-10) is greater than 0.5. This shows that one value of u_{gr} is positive and the other negative. The positive root indicates that gradient wind has the same sense (frequently called "direction") as geostrophic wind. This is the case observed in all tropical and extratropical cyclones.

The other root of (7-10) is negative. For it we have $u_{gr} < 0$ and, very importantly, $|u_{gr}| > u_g$. This is the cyclone with opposite, anticyclonic circulation. Lows with opposite circulation have been observed on the thunderstorm scale, with diameters of about 10 km, and on the smaller, tornado scale. The evidence indicates that those anomalously rotating storms develop higher wind speed than normal storms, thus agreeing with (7-8). Lows with opposite rotation have never been observed on the scale of tropical cyclones or larger. This is a fortunate experience, since such opposite-rotating cyclones should possess much faster wind than normal cyclones.

Anticyclonic Circulation

Cases with $r < 0$ are for circulation around highs. Real roots of (7-10) exist only for

$$|r| > \frac{4u_g}{f} \tag{7-12}$$

meaning that stationary flow ("stable flow") can exist only if the radius of curvature $|r|$ is large enough or if the pressure gradient, expressed by u_g, is small enough. Consequently, the observed pressure gradient (expressed by u_g) in highs is regularly weaker than in lows. Too high values of u_g (for a preset radius r) violate (7-12). In these cases, there are no real solutions of (7-10). The flow and the pressure distribution cannot persist in such patterns and therefore both break down, or change patterns, presumably until some stable shape develops.

Both anticyclonic solutions of (7-10) are faster than geostrophic wind. They are known also as the *normal* and *anomalous* gradient wind. The slower solution is considered normal, since it is observed more often. The anomalous solution sometimes appears in the ridges of the jet streams. Although there is always some doubt in the interpretation of upper-level data, some cases show very fast wind that can be interpreted as the anomalous gradient wind (Fig. 7-8). The air parcels at the jet-stream level sometimes undergo oscillations between the normal and anomalous balance. This is important when analyzing weather charts since the simple idea about the wind being in geostrophic balance may be greatly in error.

Divergence of the Gradient Wind

It may be thought that the flow balanced as in the gradient-wind equations (7-10) and (7-11) cannot produce divergence and convergence since the flow goes along circular contours. However, the dynamical significance of the gradient-wind balance predominates when the atmospheric currents assume a sinusoidal pattern. In such a wavy pattern, we find that the wind [balanced as in (7-10)] is slower in the troughs and faster in the ridge if the spacing between the contours is equal. This effect creates divergence and convergence in deep layers of the atmosphere and becomes very important in the dynamics of the waves in the westerlies.

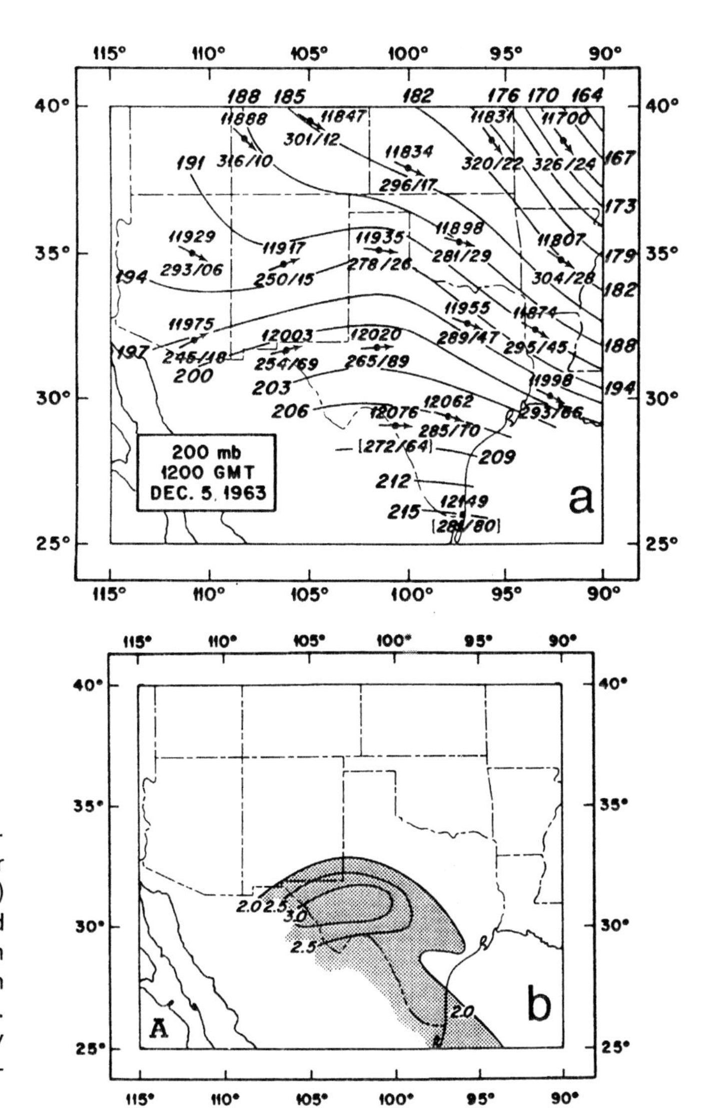

FIGURE 7-8. Example of anomalous gradient wind at 200 mb for 1200 UTC 5 December 1963. (a) Observed height (m) and wind (direction in degrees/speed in m s^{-1}). (b) Ratio V/V_g; the region with values over 2 is shaded. From Mogil and Holle (1972), by permission of the American Meteorological Society.

7-5 CYCLOSTROPHIC WIND

A frequent stable flow configuration occurs on small scales when the Coriolis parameter can be neglected. This is the circular flow that can be observed in small eddies in air and in water. In these cases, a balance between the centrifugal and pressure forces is expressed as

$$\frac{u^2}{r} = -g \frac{\partial z}{\partial n} \tag{7-13}$$

where u is the fluid speed, r is the radius of curvature at the circular streamline, and the right-hand side is the pressure force that balances the centrifugal force. This equation again represents a stationary solution of the equations of motion, as (7-10) represents gradient wind. This stationary solution is also often observed, justifying the assumptions made (vanishing time derivative, no friction, no Coriolis force).

Small circular eddies are very frequent on the turbulent scale, but they appear also on the scale of dust devils, tornadoes, and smaller thunderstorms. Since the Coriolis force is insignificant at such small eddies, there is no preference of these eddies to rotate one way or the other. This refutes the popular but erroneous belief that all eddies in the Northern Hemisphere turn the same way.

7-6 ESTIMATION OF VERTICAL MOTION

Scale of Vertical Motion

Vertical motion is one of the most important variables in the atmosphere, since many weather phenomena are caused by lifting or sinking. The clouds preferably develop in regions of lifting. The clouds may develop in stable, layered forms; however, if the lapse rate of temperature exceeds the conditional instability criterion, convective clouds develop. Lifting also supports cyclogenesis since the lifting is associated with convergence in the lower layers and this, in turn, intensifies cyclonic vorticity. Another effect of lifting and convergence is frontogenesis. Sinking, or *subsidence*, signifies fair (cloudless) weather, where anticyclones intensify and fronts dissipate.

The patterns of vertical motion appear on two distinct scales:

1. Small scale: This is lifting in horizontal domains of $1 - 10^4$ m, mostly due to turbulence, convection, and mountain waves. The vertical component of flow is typically $1 - 10$ m s^{-1}, thus easily measured by instruments mounted on aircraft. A narrow frontal cloud band is of this scale.
2. Large scale: Lifting of this scale occurs in the active regions in the westerlies or in large parts of extratropical cyclones. The vertical motion is typically $0.01 - 0.1$ m s^{-1}. It is unfortunate that there are no practical instruments that can reliably measure such small air speed. However, even if a reliable instrument was developed, the large-scale vertical speed could hardly be detected amidst muffling by small-scale eddies.

The patterns of small-scale vertical motion (up to the horizontal dimension of about 100 km) are not normally represented on weather charts. The amount of detail in the fields of turbulence and convection exceeds the capabilities of weather analysis using standard-scale synoptic charts. Therefore, the appearance of convective lifting is represented statistically: as a region of convection, without detail. This is described in Chapter 9. Some larger convective structures (squall lines, tropical cyclones, and so forth) that exceed about 100 km in length can be traced on synoptic charts. Such structures are described in Chapter 12.

The vertical motion of large scale is possibly the most important weather element: It shows where it rains and where cyclones develop. As mentioned above, there is no hope to acquire observations of this element. However, the theory provides reliable means to evaluate the vertical motion that occurs in quasi-geostrophically balanced flow. The rest of this section is devoted to the practical application of this theory.

The Omega Equation

Large-scale lifting and sinking are another deviation from geostrophic flow that can be fairly easily evaluated or estimated from the weather charts. The formula for estimating vertical motion comes from the omega equation (Appendix M), which can be used in the following form:

$$\sigma \nabla^2 \omega + f^2 \omega_{pp} = -2\, h \nabla \cdot \mathbf{Q} \tag{7-14}$$

The parameters σ, f^2, and h are positive in a thermally stable atmosphere. The right-hand side of (7-14) contains the divergence of the $\mathbf{Q}$ vector. This vector is defined as

$$\mathbf{Q} = -\begin{pmatrix} u_x\theta_x + v_x\theta_y \\ u_y\theta_x + v_y\theta_y \end{pmatrix} \tag{7-15}$$

Indices x, y, and p indicate partial derivatives in respective directions. A derivation of (7-14) and the explanation of constants are shown in Appendix M. As implied by (7-14), a knowledge of **Q** is useful for evaluating or estimating vertical motion. When the variables u, v,, and θ are given as continuous numerical fields, as in grid points, then (7-14) can be solved for the vertical motion ω. Mathematically, (7-14) is an elliptic equation where on the left-hand side is a scaled three-dimensional Laplacian of ω. However, (7-14) can also be used for practical estimation of vertical motion because of reasonably good approximations for wavelike functions that vary around the value of zero:

$$\nabla^2\omega \propto -\omega \quad \text{and} \quad \omega_{pp} \propto -\omega \tag{7-16}$$

Therefore, (7-14) can be rewritten as a proportionality:

$$\omega \propto \nabla \cdot \mathbf{Q} \tag{7-17}$$

From this proportionality we can see that the knowledge of **Q** or div **Q** gives a fair indication of vertical motion. Several examples follow that can be used as models for a variety of weather situations.

Simplification of the Q Vector

An important property of the definition of **Q** in (7-15) is that it can be evaluated reliably using geostrophic approximation for the derivatives of wind components (u_x, u_y, v_x, v_y). Unlike the evaluation of wind divergence, where divergence is very sensitive to small inaccuracies in wind data, **Q** is not prone to large errors due to small observation inaccuracies.

The **Q** vector, as defined in (7-15), can be easily evaluated using small personal computers. Data are needed only at one pressure level. There are no vertical derivatives. For this reason, **Q** is being computed in numerous weather offices and is readily available to meteorologists.

The meaning of the **Q** vector can be understood if the local coordinate system is oriented so that $\partial\theta/\partial x$ disappears. This happens when the x-axis points along the isotherms. Then there is

$$\mathbf{Q} = -\begin{pmatrix} v_x \theta_y \\ v_y \theta_y \end{pmatrix} \tag{7-18}$$

Vertical Motion around the Trough

For visual estimations of vertical motion, (7-18) can be illustrated in several schematic examples. The case in Fig. 7-9 is with $v_y = 0$. If north is taken in the y-direction, this figure also illustrates shearing deformation between the northerly flow on the west side and southerly flow on the east side. The **Q** vector has only one component, along the x-axis:

$$\mathbf{Q} = -\begin{pmatrix} Q_1 \\ 0 \end{pmatrix} \qquad Q_1 = v_x \theta_y \tag{7-19}$$

In Fig. 7-9a the **Q** vector is drawn only in places where its magnitude is relatively large and not in places where **Q** is small. Since the intensity of **Q** is proportional to v_x, the largest **Q** is found in the region of strongest shear v_x, along the trough. Also, since $Q_1 > 0$ (Q_1 is the first, that is, the x-component), **Q** points downstream along the x-axis. Further, we find convergence of **Q** in the places the **Q** vectors point to, and divergence in the places from which the **Q** vectors point away.

The rules about divergence and convergence of **Q** can be applied on this flow pattern. The omega equation in form (7-17) indicates upward vertical motion in the region of convergence of **Q** (indicated conv.

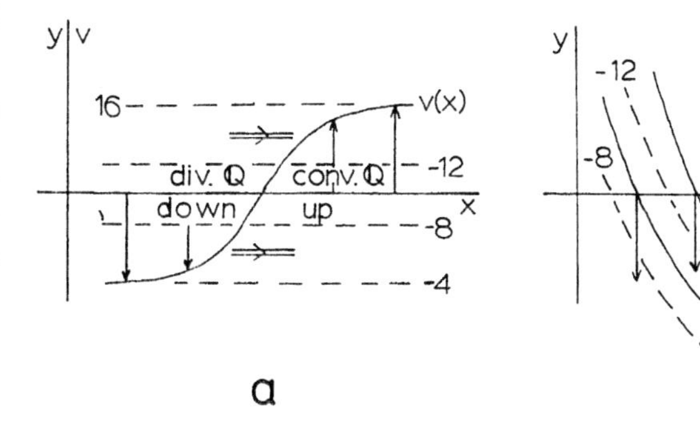

FIGURE 7-9. Estimation of vertical motion in a deformation flow with northerly wind in the west and southerly wind in the east. The meridional component of wind is shown by thin vectors. The temperature distribution is shown by dashed isotherms (°C). (a) **Q** vector (double, with arrowhead in the middle) and its convergence and divergence. (b) A trough with the same distribution of the meridional wind component and of **Q**. The solid lines are the geopotential contours (gpm).

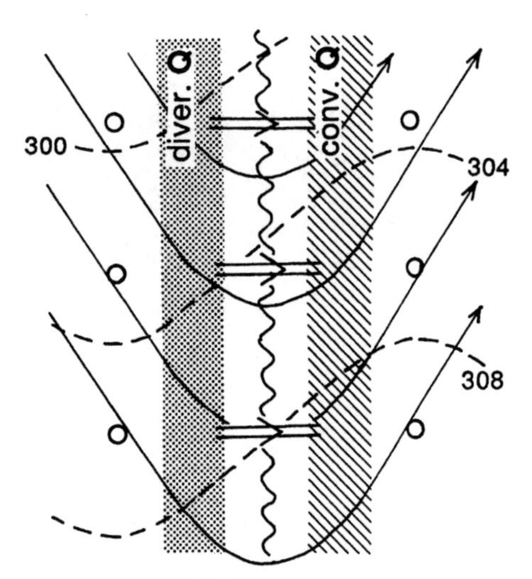

FIGURE 7-10. Distribution of **Q** (double vectors, with arrowhead in the middle) around the trough (wavy line) in the westerlies. Zero values of **Q** are shown by small circles. The streamlines are straight and equidistant near zero values of **Q**. Regions of prominent convergence and divergence of **Q** are shaded. Streamlines are solid lines; isentropes (K, for potential temperature) are dashed. Convergence of **Q** is located downstream from the trough, in the active region. In this region, lifting of air maintains a quasistationary balanced state. Similarly, there is divergence of **Q** and sinking in the inactive region upstream from the trough.

Q in Fig. 7-9a), and downward motion in the region of divergence (div. **Q** in Fig. 7-9a). The definition of convergence is div. $\mathbf{Q} < 0$. In this sense of the word, the divergence (div., with a period in Fig. 7-9a) is div $\mathbf{Q} > 0$. The markings in Fig. 7-9a indicate two regions of interest: one with div. **Q** and sinking ("down") and another with conv. **Q** and lifting ("up").

Figure 7-9b represents a trough in which the component v has the same distribution as in Fig. 7-9a. A positive constant u is added to the wind from Fig. 7-9a and the isotherms are placed along the streamlines-contours. Now the contours show a trough in the westerlies. The addition of constant u does not alter **Q**, since **Q** consists only of derivatives and the derivatives of constants vanish.

The isotherms need not be parallel with contours to reach the conclusion about lifting and sinking around a trough. The example in Fig. 7-10 is a more realistic model of a trough in the westerlies. The isentropes show a phase shift against the streamlines, as is typical for developing baroclinic waves. The simplifications include absence of a jet stream; it is assumed there is no meridional shear in this example. Also, the streamlines are straight at a distance from the trough. In those regions, **Q** vanishes due to zero wind shear. Zero values of **Q** are shown by small circles.

Figures 7-9 and 7-10 illustrate the important conclusion that lifting is located downstream of troughs in the westerlies. Analogous examples (not shown here) yield the conclusion that sinking prevails upstream of troughs and downstream of ridges in the westerlies. Both these conclusions blend together in the case of a train of waves in the westerlies. Maximum vertical motion is expected near inflection points, between troughs and ridges. Lifting prevails downstream of troughs; sinking prevails upstream of troughs. Lifting and sinking in the wave pattern is the basis for identification of active regions in the westerlies.

Thermal Advection and Vertical Motion

The balanced flow with thermal advection must involve vertical motion. An elementary situation of this kind is shown in Fig. 7-11. This figure depicts a jet stream that crosses straight isotherms. The **Q** vector in this case is again of the form (7-19). It points toward the jet core from both sides. Lifting is necessary in the jet core in such a balanced flow. This coincides with one of the old rules used in meteorology: The warm advection is favorable for lifting.

Analogously, the opposite rule can be constructed: The cold advection favors sinking of balanced flow.

A physical explanation of this situation can be given as follows: Increase of temperature by advection

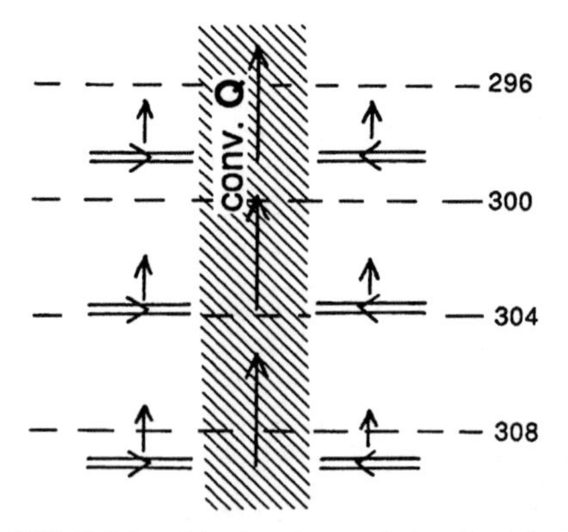

FIGURE 7-11. Plane view of the jet stream crossing isentropes (K). Longer arrows show higher wind speed. Thus the advection is largest in the jet core. The **Q** vector (double) has prominent convergence along the jet axis (shaded area). This is the situation when the conveyor belt lifts in the warm sector of the extratropical cyclone.

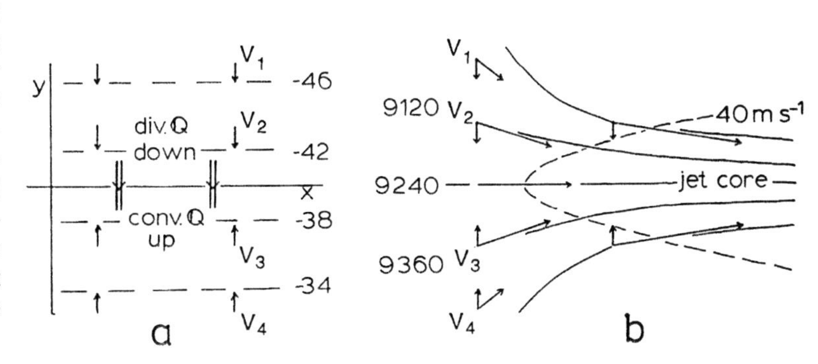

FIGURE 7-12. The y-component of wind (v) in the case of shrinking along the y-axis and stretching along the x-axis. (a) **Q** vector (double, with arrowhead in the middle), its divergence and convergence, and an assumed distribution of temperature (°C). (b) Streamlines with the same distribution of v and temperature when a constant zonal wind component is added. This is now the entrance to the jet streak.

is exactly counterbalanced by adiabatic cooling due to lifting. In this way, the flow can be stationary. If either of these two processes (advection or adiabatic cooling) prevails, the situation would cease being stationary. A similar, but more complicated, explanation can be formulated for other cases where the vertical motion is estimated from the Q vector.

Entrance to the Jet Streak

Another example of **Q** vector is shown in Fig. 7-12. This is a typical flow pattern at the entrance to a fast segment of the jet stream, the entrance to a *jet streak*. In this case we have $v_x = 0$ and $\theta_x = 0$. Figure 7-12a contains only the y-component of flow (v) and the corresponding **Q** vector. In this example, we have $v_y < 0$. Superposition of a positive x-component u (Fig. 7-12b) shows that this is a typical case of confluence. An examination of derivatives from (7-15) shows that the largest negative y-component of **Q** occurs on the x-axis, as shown in Fig. 7-12a by two **Q** vectors. In this situation, it is also assumed that v does not increase indefinitely away from the x-axis. The values of v_1 and v_2 in Fig. 7-12a are about equal, whereas v_2 and v_3 are of opposite signs. Then again, v_3 and v_4 are about equal. We find convergence of **Q** and lifting "south" of the x-axis, and divergence and sinking "north" of the axis.

A corresponding situation exists at the exit of the jet streak. There we find lifting on the left-hand flank and sinking on the right-hand flank. This situation is further elaborated upon as a part of several following examples.

Ageostrophic Circulation on the Vertical Section

Vertical circulation around the jet stream is illustrated in Fig. 7-13. The sense of circulation is shown by oval streamlines. The **Q** vector at the jet level is represented by a double vector. The jet stream in both cases *a* and *b* flows into the plane of drawing. The circled X is an arrow that points into the paper. The distribution of temperature is the same in both parts a and b, with cold air on the "northern" side (N in Fig. 7-13) and with warm air on the "southern" side (S). The only

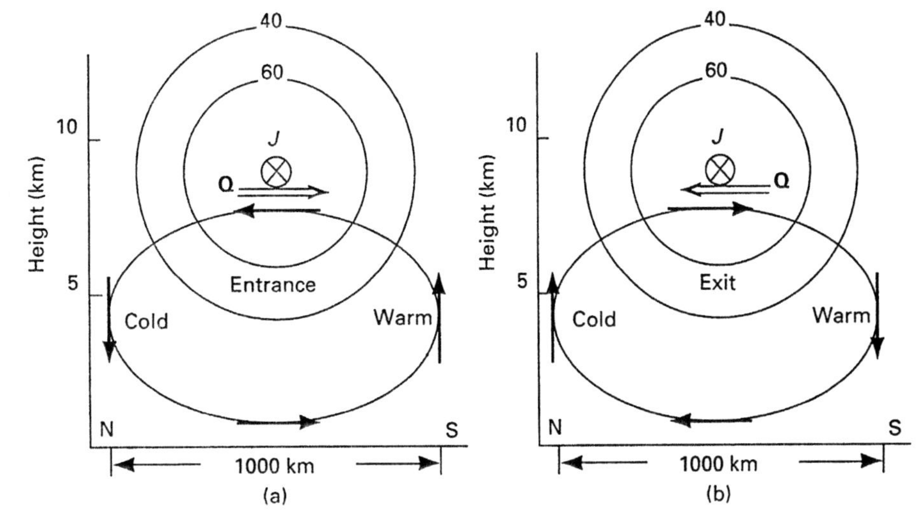

FIGURE 7-13. Vertical circulation across the jet stream. Isotachs are the circles, labeled 40 and 60 m s^{-1}. The oval with arrows is a streamline of ageostrophic flow. (a) Direct circulation at the entrance of the jet streak where the jet accelerates downstream. (b) Indirect circulation at the exit, where the flow decelerates downstream.

difference between the cases is that one of them (a) is for the entrance into a jet streak and the other (b) is for the exit.

Ageostrophic flow is expected again such that the low level flow goes in the direction of the **Q** vector at the jet level. The vertical motion on both sides of the jet is compensatory; it completes the ageostrophic circulation. Lifting in each case is under the region toward which **Q** points.

An important difference in circulation between Fig. 7-13a and b is that part (a) shows a direct circulation, that is, such that cold air descends and warm air rises. This circulation generates kinetic energy, and this is in agreement with the acceleration of air particles that enter the jet streak ("entrance"). Part (b) shows indirect circulation, with cold air on the "northern" side (N in Fig. 7-13) rising and warm air descending on the "southern" side (S).

A physical description can be formulated for the exit case as follows: The flow is diffluent and it causes a local weakening of the thermal gradient. Indirect circulation explains cooling of cold air and warming of warm air, that is, it shows intensification of the local thermal gradient. These adiabatic effects exactly counterbalance the effect of diffluence and make the situation stationary. The kinetic energy of the jet stream is expended to maintain the thermally indirect circulation.

A similar explanation can be formulated for the entrance case, as well as for other cases. This should be an instructive exercise in dynamic meteorology.

Cross-Contour Flow

The findings about lifting and sinking near the entrance and exit of jet streaks lead also to the explanation of frequently observed cross-contour flow. Balanced flow with lifting and sinking in various places requires compensating currents that maintain the continuity of mass. At the jet level, the air flows away from the place of lifting. Also at this level, the air converges at the place of sinking. This explains the cross-contour flow from the high side at the entrance to the jet streak and, similarly, explains the flow from the low side at the exit from the jet streak.

Compensating currents in the lower atmosphere develop opposite from the jet-level ageostrophic flow. This explains the low-level "northerly" flow under the entrance to the jet streak and the "southerly" flow under the exit. These low-level flows often concentrate in shallow (1–2 km) layers of the atmosphere, thus forming the low-level jet.

Figure 7-14 represents a generalization of ageostrophic flow near the jet streak. This is a model pattern at about 40 kPa, under the maximum wind in the polar jet stream. The depicted jet streak is chosen conveniently straight. A curved jet streak is actually more common. The straight example was selected to illustrate several important points. The distribution of temperature is assumed with straight isotherms, parallel to the jet core and with cold air on the "north" side.

Using the example in Fig. 7-14, the ageostrophic circulation can be summarized as follows.

At the exit: An ageostrophic component of wind "up the pressure gradient," that is, from the low ("northern") side, appears in this region at the jet level. At the same time, in the low levels under the exit, an opposite ("southerly") flow occurs, moving the air toward the left flank of the jet stream. These currents compensate for the vertical motion that shows lifting on the left-hand flank and sinking on the right-hand flank. Since it is colder on the left-hand side of the jet stream than on the right-hand side, this circulation is thermally indirect, with the cold air lifting and warm air lowering. This circulation consumes kinetic energy of the jet stream, but it steepens the front and supports the circulation of extratropical cyclogenesis. The southerly flow in the lower layers of the troposphere usually forms the conveyor belt of the extratropical cyclone and often assumes the form of the low-level jet stream.

At the entrance: In this part of the jet streak, there is downgradient ("southerly") flow at the upper-tropospheric jet level and opposite flow ("northerly") in the lower levels. This lower-level flow normally comes in the cold air, behind a cold front. The accom-

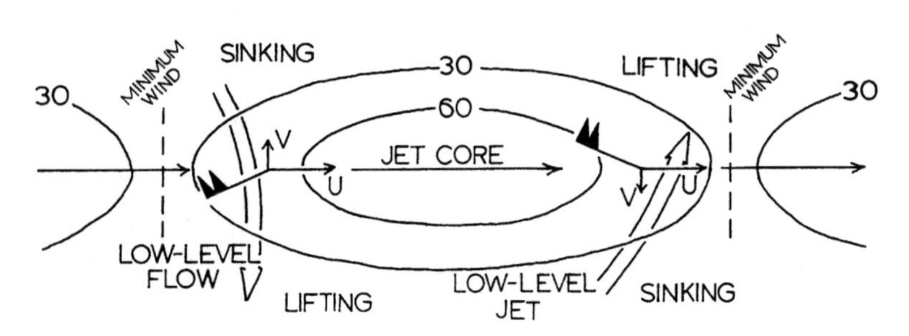

FIGURE 7-14. Distribution of wind around a straight upper-tropospheric streak of the jet stream. Isotachs are in m s^{-1}.

panying lifting on the right flank (that is, in the warm air) and sinking on the left flank (in the cold air) show a direct thermal circulation. This is the weather situation when the polar air spreads far south, behind the cyclone on the polar front.

Classical Omega Equation

An older form of the omega equation, used since the 1950s, reads

$$\sigma \nabla^2 \omega + f^2 \omega_{pp} = -f \frac{\partial}{\partial p} [-\mathbf{V} \cdot \nabla(\zeta + f)] + h \nabla^2(-\mathbf{V} \cdot \nabla \theta) \qquad (7\text{-}20)$$

The stability $\sigma = -h\ (\partial\theta/\partial p)$ is necessarily positive, which implies that all conclusions apply in a thermally stable atmosphere. Stability is required in all quasi-geostrophic and hydrostatic considerations. Equation (7-20) originates from the equations of vorticity and the first law of thermodynamics, as shown by Holton (1992, pp. 166–67).

An interpretation of (7-20) can be made with the approximations used in (7-16). The advection of vorticity $[-\mathbf{V} \cdot \nabla(\zeta + f)]$ usually increases with height under the maximum wind level. Also, a decrease of a variable with pressure means an increase of the same variable with height. Then, if the term

$$-f \frac{\partial}{\partial p} [-\mathbf{V} \cdot \nabla(\zeta + f)]$$

is positive, it indicates positive vorticity advection above that point (for example, at the jet-stream level). Conversely, if this term is negative, it indicates negative vorticity advection aloft.

A similar argument as given in (7-14) for $\nabla^2 \omega \propto -\omega$ leads to the conclusion that a prominent negative value of

$$\nabla^2(-\mathbf{V} \cdot \nabla \theta)$$

indicates a warm advection maximum. Therefore, we have an approximate symbolic form of the omega equation:

$$\text{lifting} \propto \text{vorticity advection aloft} + \text{warm advection} \qquad (7\text{-}21)$$

The old approximate rules about synoptic-scale lifting were formulated in view of (7-21) such that lifting in the atmosphere was associated with

(a) cyclonic vorticity advection in the upper troposphere, and
(b) prominent warm advection.

Since these two quantities can be detected by visual inspection, we can make conclusions about the vertical motion, which are also good for estimating the sinking motion: Sinking is accompanied by an anticyclonic advection of vorticity and cold advection.

The simplifications used in points (a) and (b) above need some restriction: Warm advection by itself is not really a contributing factor. The Laplacian of warm advection plays a role in (7-20). Therefore, a differential warm advection is important, indicating a space variation of advection that gives rise to the Laplacian. This is the explanation for the use of the word "prominent" in the rule (b) above.

The above rules are limited in their usefulness because they can be applied with confidence only when the cyclonic advection of vorticity appears together with warm advection. In this case, we are fairly certain that there is lifting. Contrary to the classical form (7-20), the rules about estimation of vertical motion based on the **Q** vector (7-15) are shorter, fewer, and safer. They do not require additional considerations with estimation of Laplacians and vertical derivatives.

The mathematical form of (7-20) [as well as (7-14)] shows that this is an elliptic equation in which the terms on the right-hand side constitute the *forcing function*. For this reason, we sometimes say that "vorticity advection forces the vertical motion" or similarly that "thermal advection forces the vertical motion." It should be understood that there is no power in nature about vorticity advection to really "force" the other variables. All these variables are mutually dependent and the expression *forcing* refers only to the mathematical form of the equation. In this sense we sometimes compute omega using only the thermal advection terms in (7-20), omitting the other term on the right-hand side. Then we say that we obtained the part of omega that was "forced" by the thermal advection. This does not mean that thermal advection causes vertical motion. Advection only shows the vertical motion diagnostically. The arrangement of the flow is such that the wind deviates from geostrophic wind by having a significant vertical component.

Simultaneous positive advections of vorticity and temperature occur most prominently in the regions downstream from the troughs in the westerlies. This explains prevalence of lifting in the active region downstream of the upper-level trough. By the same rules, sinking motion can be found upstream of troughs. Thus meteorologists are greatly concerned about the positions of upper-level troughs, both on current charts and forecast charts. Therefore, it is useful to identify the active regions on the charts between the troughs and downstream ridges.

The concern about troughs in the upper troposphere is sometimes misinterpreted because the troughs themselves are not in the centers of lifting or sinking. Instead, prominent vertical motion occurs primarily near inflection points between troughs and ridges. In a similar way, it is incorrect to say that the trough "causes" some type of weather. A better expression is "an approaching trough signifies lifting in the atmosphere."

Interpretation of Weather Charts Using the Omega Equation

The conditions with advections of vorticity and temperature are illustrated in the example of Newton (1956) in Fig. 7-15.

The 300-mb chart (Fig. 7-15a) shows a low over the central United States, with a trough from Missouri to Texas. Positive vorticity advection can be seen in the region between this trough and the ridge over the Great Lakes. For a qualitative recognition of vorticity advection, we do not really need the isopleths of vorticity. We know that vorticity is exported from troughs. In any event, the example in Fig. 7-15a con-

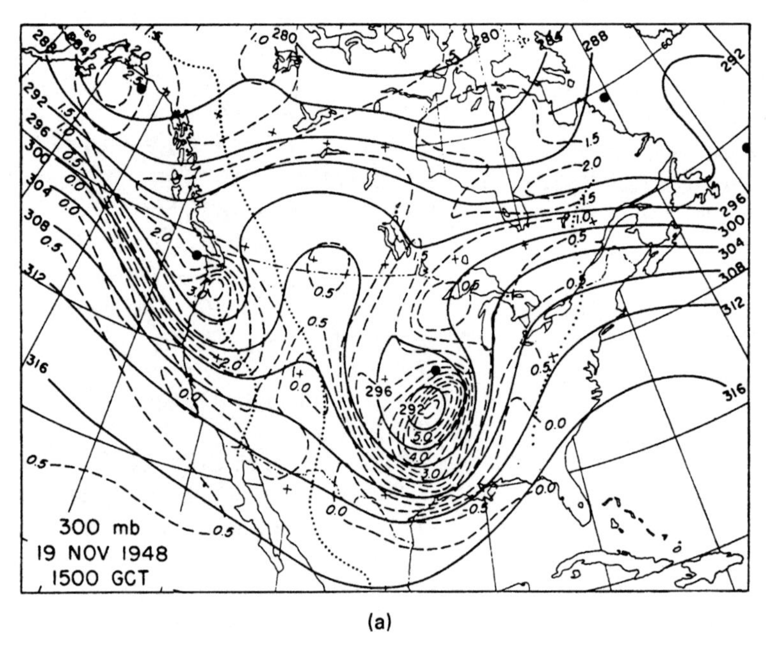

(a)

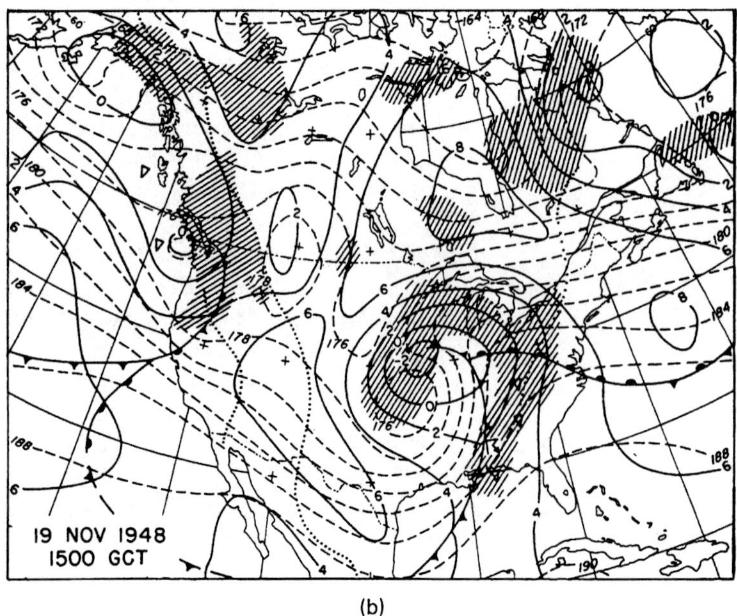

(b)

FIGURE 7-15. Weather charts for 1500 UTC 17 November 1948. (a) 300-mb contours (400-ft intervals) and geostrophic absolute vorticity (dashed, 10^{-4} s^{-1}). The heavy dots show cyclone centers. (b) Surface front and 1000-mb contours (solid, 200-ft intervals) and 500/1000-mb thickness (dashed, 200-ft intervals). Precipitation areas are shaded. Principal mountain ridges are shown by dotted lines. From Newton (1956).

tains the isopleths of absolute geostrophic vorticity. Therefore, the estimation of vorticity advection can also be seen from the characteristic intersections of the isopleths of vorticity and contours (geostrophic advection of geostrophic absolute vorticity). Judging from the vorticity advection, we may expect lifting in this region. However, the cold front is advancing in the same region and behind it there is vigorous cold advection, at least in the lower troposphere. The temperature advection can be recognized in Fig. 7-15b by the intersections of thickness and 1000-mb contours. Therefore, vorticity advection aloft and prominent warm advection in the space east of the cold front and northeast of the surface cyclone coincide until the trough over the Great Lakes. Widespread precipitation (depicted by shading in Fig. 7-15b) verifies the rule of estimation of vertical motion.

Behind the cold front, over Louisiana to Tennessee, there is positive vorticity advection aloft and cold advection in the lower layers. Therefore, we cannot conclude with certainty whether there is lifting or sinking in this area. The absence of precipitation suggests sinking.

The upper-level low in Fig. 7-15a is largely symmetrical about the north–south line along the trough from Missouri to Texas. Contrary to this symmetry, the weather phenomena are dissimilar on the western and eastern sides of this upper-level low. To the east are thunderstorms and rain, and to the west is an anticyclone with clear skies. This is a typical example. Such troughs march over North America about 50 times each year.

Trenberth's Form of the Omega Equation

An improvement of the classical omega equation (7-20) was made by Trenberth (1978) who, with mathematical manipulation, succeeded in eliminating the canceling parts of the terms on the right-hand side of (7-20). With some approximation, Trenberth showed that a consistent form of the right-hand side of (7-20) is in the form of advection of absolute vorticity by the thermal wind (see Appendix M). Therefore, the omega equation yields the following practical form:

$$w \propto -\frac{\partial \mathbf{V}}{\partial z} \cdot \nabla(\zeta + f) \qquad (7\text{-}22)$$

This proportionality indicates that lifting exists in the regions of cyclonic vorticity advection if this advection is estimated using thermal wind $\partial \mathbf{V}/\partial z$ instead of wind $\mathbf{V}$. When rather thick layers of air are considered—for example, 500/1000 mb—rule (7-22) is not much different from the part of rule (7-21) that mentions the advection of vorticity. That is, the contours of thickness are usually similar to the contours of isobaric surfaces in the upper troposphere. The example in Fig. 7-15 shows a typical case. It can be seen that the thickness contours in Fig. 7-15b are similar to the 300-mb contours in Fig. 7-15a.

Historical note: The conclusions about the cross-contour flow, as in Fig. 7-14, were already formulated by R. Scherhag in Germany in late 1930s, but they were published only after World War II. Scherhag observed the systematic upper-level flow toward high pressure in diffluent contours. This was the case of the exit from the jet streak. Scherhag called this situation *delta*, an analogy to rivers that spread wide before they enter the sea. Scherhag also attributed the circulation in the delta to cyclogenesis. However, this part of the theory was more consistently formulated by R. C. Sutcliffe in England. It was unfortunate that scientists such as Scherhag and Sutcliffe could not communicate during the war, and the development of science was greatly hindered.

8

Fronts

Fronts are discontinuities that form in strongly baroclinic zones of the atmosphere. Many significant weather events occur on or near atmospheric fronts. It is useful to mark fronts on weather charts since fronts reveal the location of a variety of weather phenomena. Various instabilities occur at fronts, and these instabilities give rise to some of the most active weather events. For this reason, a significant part of a weather analyst's work is devoted to the search for fronts.

8-1 FRONTOGENESIS

The processes that generate fronts frequently also generate clouds and precipitation, sudden wind shifts, and other changes in local weather. Therefore, we are interested in the generation of fronts (*frontogenesis*). If we can recognize frontogenesis, we also can recognize development of weather. Moreover, fronts usually exist in regions where frontogenesis is active. In this way, the identification of fronts is usually an identification of processes that cause weather development. The opposite process of *frontolysis* (dissipation of fronts) receives much less attention in meteorology since the weather activity is diminished in the presence of frontolysis.

Formal Expressions for Frontogenesis

Frontogenesis is characterized by deformation in the flow, by convergence, or by unequal heating. All these processes contribute to the intensification of the horizontal gradient of temperature. The frontogenesis in continuous flow can be expressed as

$$\frac{d}{dt}\frac{\partial\theta}{\partial y} = -\frac{\partial v}{\partial y}\frac{\partial\theta}{\partial y} - \frac{\partial\omega}{\partial y}\frac{\partial\theta}{\partial p} + \frac{\partial}{\partial y}\frac{d\theta}{dt} \qquad (8\text{-}1)$$

(see Appendix N). Here θ is the potential temperature, y is the direction of the normal to the isentropes, v is the y-component of wind, and ω is the vertical motion. The total (individual) derivative $d\theta/dt$ represents the heating of air. The three terms on the right-hand side of (8-1) express the three main processes that contribute to frontogenesis: (a) configuration of the horizontal flow, (b) tilting of stable layers, and (c) differential heating. These processes will be explained next in several schematic examples.

Confluence

The flow configuration with confluence, as in Fig. 8-1, is a typical case of confluence. The flow is along the x-axis, accelerating downstream. The term $-(\partial v/\partial y)(\partial\theta/$

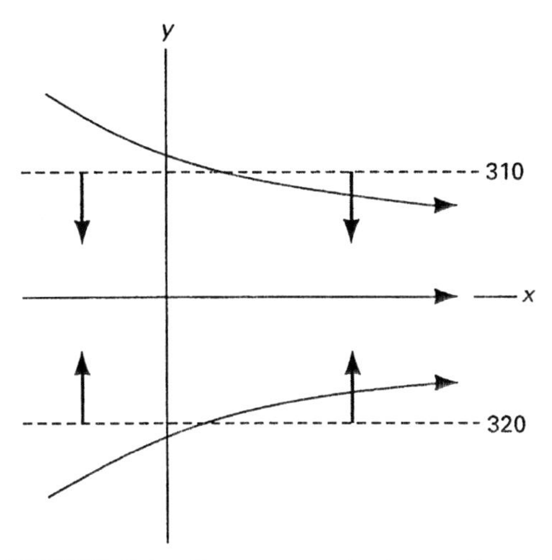

FIGURE 8-1. Frontogenesis in confluence. Natural horizontal coordinates are l and n. Long arrows are streamlines. Isentropes (K) are dashed. Short vectors show the component of flow along the normal.

∂y) in (8-1) indicates frontogenesis. In this example, with $\partial v/\partial y < 0$ and $(\partial \theta/\partial y) < 0$, the first term causes the already negative value of $\partial \theta/\partial y$ to grow more negative (the gradient $|\partial \theta/\partial y|$ intensifies, meaning frontogenesis). The streamlines in Fig. 8-1 are about equal to the example of confluence in Fig. 4-11. The isentropes are selected such that they are roughly parallel to the flow. The normal wind component v causes the warm and cold air parcels to approach each other. In this way, the gradient of potential temperature $|\nabla \theta| = |\partial \theta/\partial y|$ increases, and this is frontogenesis. In this example, the term $\partial v/\partial y$ plays an important role as a measure of deformation. This term represents the shrinking in y-direction, as in Fig. 4-9.

The appearance of minus signs several times in preceding paragraph tends to confuse students. Therefore, a visual inspection of converging isotherms in Fig. 8-1 is an easier way to make conclusions on frontogenesis. In short, confluence signifies frontogenesis and diffluence signifies frontolysis.

The term $\partial v/\partial y$ is also a part of divergence, as seen in definition (4-4). Therefore, very similar effects on frontogenesis occur due to convergence (negative divergence) in the flow.

Overturning

The term $-(\partial \omega/\partial y)(\partial \theta/\partial p)$ contributes positively to frontogenesis in the case of favorable overturning of isentropes. This process is illustrated in Fig. 8-2. In Fig. 8-2a, only one isentrope intersects the isobar of 300 mb along the drawn stretch of 500 km. The vertical motion is indicated by the vector $\omega \mathbf{k}$. Since the potential temperature θ is conserved under adiabatic conditions, the isentropes rotate with the flow and assume a steeper position at a later time (Fig. 8-2b). At that later time (Fig. 8-2b), three isentropes are shown to intersect the same stretch of the 300-mb isobar. This illustrates the intensification of the horizontal gradient of temperature in the case of overturning with $-(\partial \omega/\partial y) > 0$.

Formation of fronts by overturning is most prominent at the tropopause. Frequently, vertical circulation develops near the upper-tropospheric jet streams such that the tropopause sinks several kilometers and forms a frontal layer.

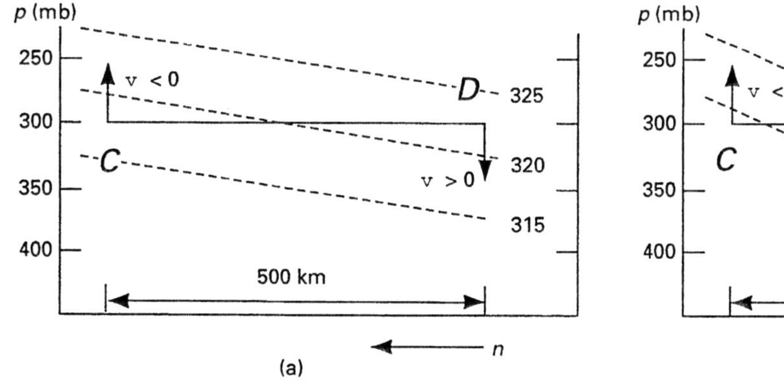

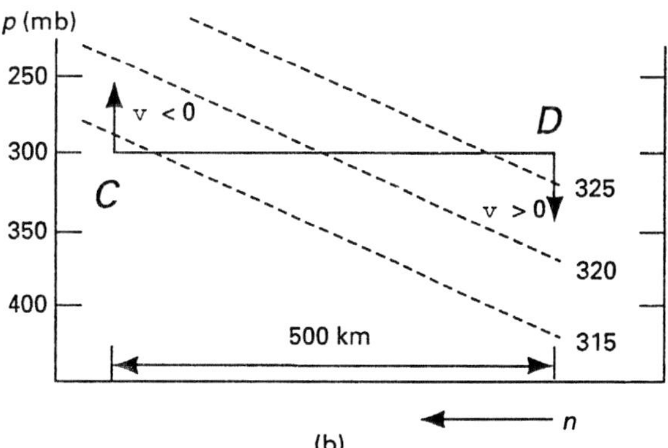

FIGURE 8-2. Vertical n,p-plane with frontogenesis due to favorable overturning of isentropes (dashed, K). Short vectors indicate vertical motion. The first situation (a) shows one isentrope intersecting the isobar of 300 mb between points C and D. At a later time (b), three isentropes intersect the same isobar, thus showing a stronger horizontal gradient of potential temperature.

Differential Heating

The third term on the right-hand side of (8-1), $(\partial/\partial y)(d\theta/dt)$, shows that contrasts in temperature (frontogenesis) may occur due to differential heating of air. Such cases occur most often when the earth's surface has widely different thermal properties in adjacent regions. During winter there are large differences between the temperature of snow surfaces and adjacent water bodies (ocean, large lakes). In such regions, local fronts may occur due to heating or cooling from below.

Another case of differential heating occurs when a cloud cover prevents parts of the earth's surface from heating during the day. This situation may generate a front between heated clear air and cold air under the cloud cover.

Deformation

Frontogenesis due to horizontal (isobaric) flow can be generalized in the relationship between the dilatation axis (Sec. 4-6) and the direction of isotherms. The two examples in Fig. 8-3 show the same schematic hyperbolic flow, but with two sets of isotherms. In Fig. 8-3a the isotherms are at an angle β smaller than 45° from the dilatation axis CC′. The isotherms are transported toward each other, indicating frontogenesis. The example in Fig. 8-3b shows the isotherms at an angle greater than 45°. The transport by wind separates the isotherms. This is *frontolysis*. The angle β is the smaller of the two angles formed by the isotherm and the local dilatation axis.

Given enough time, all isotherms in Fig. 8-3b will be brought into an orientation close to the dilatation axis, and frontogenesis will set in. Of course, universal frontogenesis will not continue indefinitely since it will be counteracted by several other physical processes, such as changing of flow patterns to other shapes, mixing of air masses that diminishes contrasts, and heating or cooling that may act to equalize the temperature.

An Example of Frontogenesis Implied by Deformation

An example of weather analysis using information about deformation and the dilatation axis is shown in the next two figures. Double-headed arrows in Fig. 8-4a and b show the local dilatation axis. The isotherms are shown on the same charts so that conclusions about frontogenesis can be made.

As stated with deformation in Fig. 8-3, frontogenesis is strong when the dilatation axis points along the isentropes. This is the case along the coasts of Virginia and North Carolina at 850 mb (Fig. 8-4a) and in Kentucky and Virginia at 500 mb (Fig. 8-4b). The colinearity of dilatation and isentropes is more significant in regions with higher values of deformation. Therefore, the amount of deformation is also supplied in Figs. 8-4a and b. As can be noticed, the deformation is relatively high over Virginia at both levels where colinearity is high. The deformation over Virginia amounts to about $4 \times 10^{-5}\ s^{-1}$ at 850 mb and over $12 \times 10^{-5}\ s^{-1}$ at 500 mb. This shows prominent frontogenesis.

Strong frontolysis can be spotted at 500 mb along the Atlantic coast from Delaware to Vermont (Fig. 8-4b). The angle β between the isentropes and dilatation axis is greater than 45°. Moreover, the dilatation axis in this region is almost perpendicular to the isentropes.

Geostrophic Frontogenesis

Frontogenesis due to horizontal (geostrophic) flow can also be expressed using the **Q** vector. Frontogenesis can be written as $(d/dt)|\nabla\theta|^2$. Assumptions of adiabatic and geostrophic (horizontal) flow yield frontogenesis

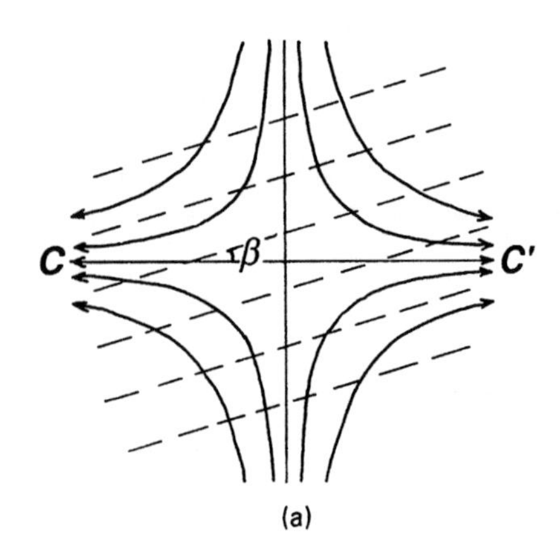

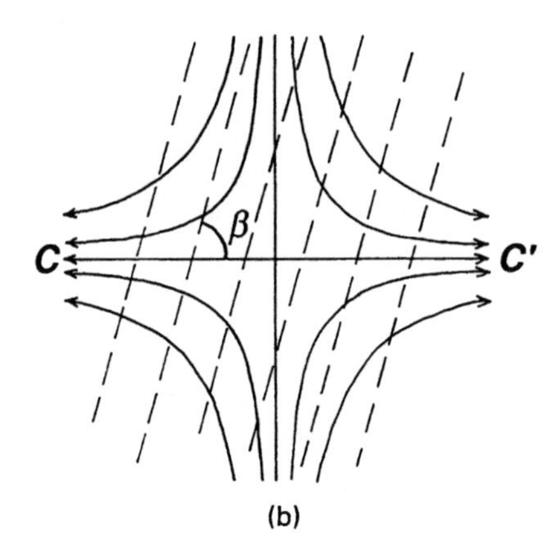

FIGURE 8-3. Schematic streamlines (solid, with arrowhead) and isotherms (dashed) in the cases of frontogenesis (a) and frontolysis (b). The angle β between the isentropes and the dilatation axis shows frontogenesis or frontolysis when it is smaller or greater than 45°, respectively.

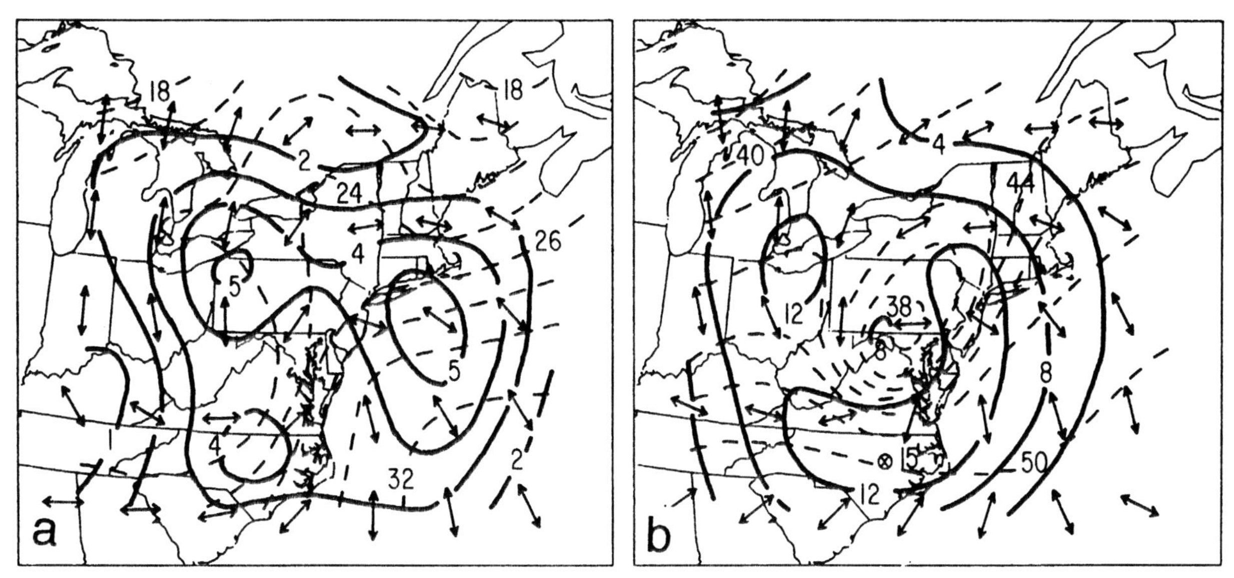

FIGURE 8-4. Geostrophic deformation (solid, 10^{-5} s^{-1}), potential temperature (dashed, °C), and local dilatation axis (double-headed arrow) at 850 mb (a) and at 500 mb (b) for 1200 UTC 3 October 1979. Strong frontogenesis can be found in places where the local dilatation axis lies along the isotherms and where the deformation is large. From Riley and Bosart (1987), by permission of the American Meteorological Society.

in terms of thermal gradient and **Q** vector as

$$\frac{d}{dt}|\nabla\theta|^2 = \mathbf{Q} \cdot \nabla\theta \qquad (8\text{-}2)$$

(The details are in Appendix N.) We conclude that frontogenesis is strongest in the places where **Q** and $\nabla\theta$ coincide. If they point in opposite directions, this shows frontolysis. Since the **Q** vector is used quite often, a chart with the **Q** vector and isotherms gives a rather easy tool to assess frontogenesis. Besides, (8-2) gives a compact equation for evaluation of frontogenesis by computers.

Interpretation of the expression $(d/dt)|\nabla\theta|^2$ from (8-2) is not essentially different from the interpretation of the corresponding expression $(d/dt)(\partial\theta/\partial y)$ in (8-1) since $|\nabla\theta| = |\partial\theta/\partial y|$. To avoid mathematical difficulties with first-order discontinuity around the points where $(\partial\theta/\partial y) = 0$, the square of the thermal gradient is conveniently introduced.

An Example of Frontogenesis Implied from the Q Vector

Frontogenesis in Fig. 8-5a and b has been evaluated from the **Q** vector and $\nabla\theta$, similar to (8-2). A somewhat different formula for frontogenesis F_g has been used:

$$F_g = 2\left(\frac{p_0}{p}\right)^{\kappa} \frac{p}{R}\,\mathbf{Q} \cdot \nabla\theta$$

This also shows that the standards of frontogenesis have not yet been established. Different authors use different formulas.

Figure 8-5, showing the same weather situation as Fig. 8-4, illustrates the comparison between the two methods to show frontogenesis. There is strong frontolysis at 500 mb (Fig. 8-5b) along the Atlantic coast from Delaware to New York.

Formation of Discontinuities

Intensification of the thermal gradient ($|\nabla\theta|$ or $|\nabla\theta|^2$) by the process where $\partial v/\partial y$ is significant is usually sufficient to show frontogenesis on a large scale. This process is effective over distances over 1000 km and time of 2 to 5 days. This consideration is essentially geostrophic: It is assumed that the flow progresses along the contours. However, this process is too slow to explain the formation of discontinuities on a typical frontal scale where a sharp trough forms in less than 1

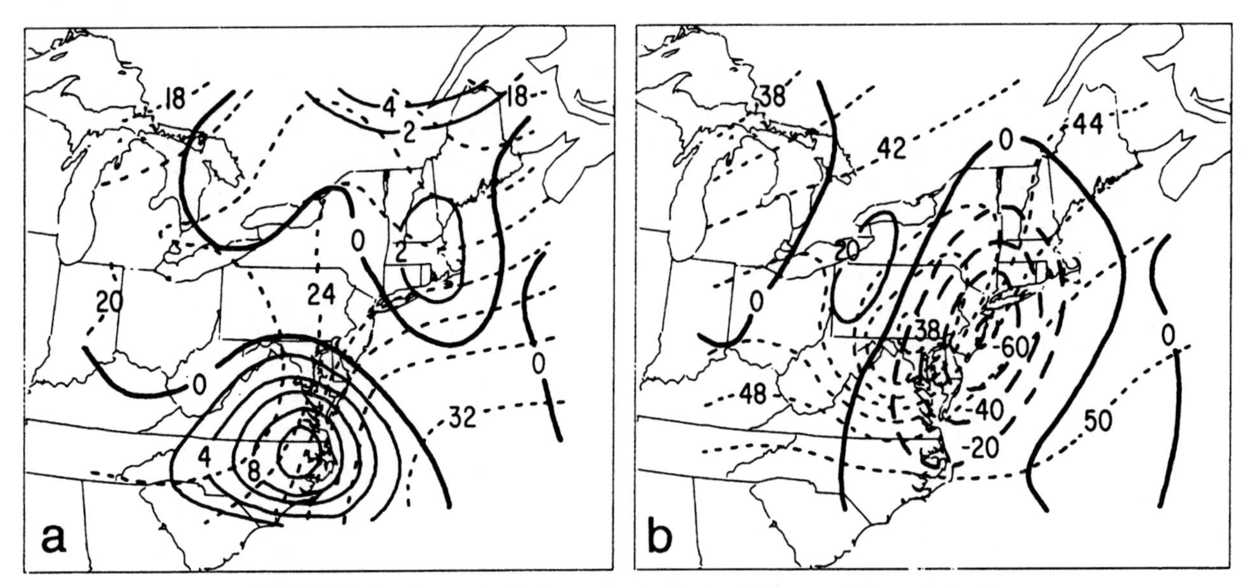

FIGURE 8-5. Geostrophic frontogenesis (solid for positive, dashed for negative values, °C m^{-2} s^{-1}) and potential temperature (dashed, °C) for 1200 UTC 3 October 1979 at 850 mb (a) and at 500 mb (b). (a) Strong frontogenesis is centered in North Carolina at 850 mb. (b) Strong frontolysis is centered over New Jersey at 500 mb. From Riley and Bosart (1987), by permission of the American Meteorological Society.

day. Sluggishness of frontogenesis by deformation is shown in Appendix N. Moreover, the described process of confluence does not form a complete zero-order discontinuity. Additional mechanisms are needed to explain the formation of a complete discontinuity with *separation of flow*. Separation of flow, a flow configuration where two neighboring parcels of air move with different velocities because they are on opposite sides of the front, is common on discontinuous fronts. A theoretical result from dynamic meteorology may be recalled here without proof: Baroclinic zones collapse into discontinuities in the presence of suitable ageostrophic flow. This process can be understood physically if the convergent flow is considered. This flow is ageostrophic, as shown in Sec. 3-3. Such flow may remove the air from an isobaric level and cause two formerly separate isotherms to touch each other. In this way, a zero-order discontinuity in temperature forms. Removal of air from a level does not occur in geostrophic flow (when $\mathbf{V} = \mathbf{V}_g$) since the geostrophic flow does not have a vertical component. Theory of frontogenesis also shows that the zero-order discontinuities appear most readily at the earth's surface.

Originating in different regions, the parcels of air on opposite sides of a front usually have different thermodynamic properties. In this way, the familiar phenomenon of abrupt weather change with frontal passage can be explained.

Frontolysis

Fronts dissipate as often as they develop. Dissipation of fronts is *frontolysis*. Deformation in the flow causes frontolysis when the isotherms are oriented differently than in the examples of frontogenesis. However, more often the fronts dissipate in the regions of divergence, frequently observed in the domain of the anticyclone that develops under the inactive region of the jet stream, between a ridge and nearest downstream trough. Frontolysis has received much less attention in the literature than frontogenesis since it occurs in the regions of descending motion and few clouds.

8-2 PHYSICS OF FRONTS

Margules Equation

Physical conditions required for quasi-stationary inclined frontal surfaces are described by the Margules equation. This equation gives the relationship between the frontal slope and the distribution of wind and temperature. Dynamic meteorology gives us several practical forms of this equation. Some forms are suitable for sharp discontinuities of zero order in temperature or density. Other forms are suitable for first-order discontinuities, where only the gradient of temperature varies discontinuously across a surface.

Physically understanding how a sloping interface may exist between fluid parts of different density and not spread out until it becomes horizontal may be difficult. A slanted surface of discontinuity can exist in a stationary state on a rotating planet where the geostrophic approximation adequately describes the flow. The circumstances at a slanted quasi-stationary front are described by the Margules equation. One form of this equation is

$$s = B \frac{\left(\frac{\partial z}{\partial x}\right)_w - \left(\frac{\partial z}{\partial x}\right)_c}{T_w - T_c} \quad (8\text{-}3)$$

where s is the slope of the front with respect to the isobaric surface and B is a positive constant (see Appendix J). The subscript w indicates the warm side of the front, and c indicates the cold side. The horizontal coordinate x is normal to the front. The slope s is positive when the cold air is under the slanting front; this is the stable case. The height derivatives $(\partial z/\partial x)_w$ and $(\partial z/\partial x)_c$ show the geostrophic wind on the two sides of the front, parallel to the front. Since the geostrophic balance can be maintained as a stationary state, the slopes $(\partial z/\partial x)_w$ and $(\partial z/\partial x)_c$ can stay stationary as well.

The Margules equation (8-3) is explicit for the frontal slope. However, this equation cannot really be used for evaluation of frontal slope. The inaccuracies of measurement of quantities on the right-hand side are too large for an acceptable evaluation of the slope s. The utility of the Margules equation is evident only in physical justification of our analysis methods. We are able to develop correct physical concepts about the atmospheric processes when we know the circumstances described by the Margules equation. Some of the concepts follow.

Geometry of Isopleths at the Front

Equation (8-3) is suitable for an ideal case of sharp discontinuity in temperature (of zero order), when $T_w \neq T_c$. The main conclusions drawn from the Margules equation are:

1. There is a sharp trough in the pressure field (or field of geopotential height) that coincides with the abrupt change of temperature at the front. The expression *sharp trough* practically means that the contours are kinked on the front.
2. The kinks on contours and the associated discontinuous wind shear are always cyclonic. An anticyclonic kink on the contours (also an anticyclonic wind shear) cannot persist. Such a configuration is unstable.
3. The front is steeper for smaller thermal contrast.

Frontal Layer

A frontal layer of thickness δ is bounded with two surfaces on which the temperature gradient is discontinuous. The physical circumstances at such frontal layers are described in Appendix J. The following theoretical conclusions are important for further discussion about such frontal layers:

4. Maximum cyclonic wind shear is situated within the frontal layer. (This shear is always cyclonic!)
5. For a vertical front, where the thermal contrast vanishes, the frontal layer is characterized by a strong wind shear.

The reason that an "infinitely thin" front cannot exist for a prolonged time is that turbulence mixes the air around fronts into layers of finite thickness. The turbulence at the front originates in the same way as described for clear air turbulence with wind shear in Section 4-8. Large stability at a thin front initially prevents mixing, and the two air masses glide on top of each other. An absence of friction enhances the thermal contrast between the air masses on the two sides of the front. In the same way, the difference in wind across the front increases. After a while the wind difference is higher than is dynamically allowed in a stable flow. The instability is reached when the Richardson number (4-20) decreases below its critical value of about 0.21. Then turbulence breaks out and the air masses mix until a finite transitional layer (*the frontal layer*) develops. Under typical atmospheric conditions, the thickness of the frontal layer is of the order of 500 m, as is shown in Appendix J. If a thinner frontal layer appears for a given wind shear and stability, turbulent mixing occurs and creates a thicker layer. The frontal layer is therefore sometimes called a "transition zone" between two air masses. The name "frontal zone" is sometimes used for the frontal layer. In this book, *frontal zone* is used for the baroclinic zone under the jet stream that contains the front at the surface and the upper-level front. Thereby the distinction is made between frontal layer and frontal zone.

The concept of zero-order discontinuity is still used on weather charts. It is easier to draw one line on weather charts than to draw a transition area. Also, a single line serves as a good approximation on smaller-scale charts. On such charts we often do not attempt to identify both sides of the frontal layer.

8-3 RELATION BETWEEN FRONT AND JET STREAM

Thermal Wind in the Frontal Zone

Baroclinic zones in the troposphere of 5–10 km depth are normally associated with upper-tropospheric jet streams and fronts. This relationship of fronts and jets is due to the strong tendency in the atmosphere to assume the thermal-wind balance. This balance can be shown by the first component of Eq. (I-3) from Appendix I:

$$\frac{\partial u_g}{\partial z} = -\frac{g}{fT}\frac{\partial T}{\partial y} \qquad (8\text{-}4)$$

The coordinate system is oriented along the wind (or geostrophic wind). Since $v_g = 0$, u_g is the (geostrophic) wind speed in this case. Equation (8-4) shows that the wind varies (usually increases) with height in baroclinic zones. This equation also implies that the wind increases with height until the level where the temperature contrast vanishes. Above that level the wind decreases, and the temperature gradient is reversed. The sign of $\partial T/\partial y$ changes above the level of maximum wind.

An Illustration of the Frontal Zone

The relationship of the jet stream and temperature gradient is illustrated in Fig. 8-6. The slope of isotherms is idealized, and only the main features of observed frontal zones are shown. The sloping isotherms are concentrated in the central portion of the scheme under the jet core (J) and above the jet core. These are the regions of thermal contrast where the cold and warm air masses can be identified side by side. The terms *warm* and *cold* are used for horizontal comparison (not vertical.

The warm and cold sides are different under and above the jet core. This can be described as a reversal of the thermal gradient. The reversal is noticed by the opposite slopes of the isotherms above and below the jet in Fig. 8-6. One isotherm that does not slope across the jet stream is at the level of maximum wind where $(\partial u/\partial z)$ and $(\partial T/\partial y)$ both are equal to zero. This horizontal isotherm is seldom a member of the set of drawn isotherms; instead, it is usually one of an intermediate value and therefore not readily found. The horizontal isotherm in Fig. 8-5 is intentionally selected as a member of the set of drawn isotherms. It is conveniently chosen at −50°C.

The quadrangles formed by intersecting isobars and isotherms in the vertical section are *thermodynamic solenoids*. The existence of thermodynamic solenoids reveals the baroclinicity of the atmosphere and a possible front. There is an overturning acceleration in the presence of thermodynamic solenoids (*acceleration of circulation*) even if the air does not always move along this acceleration.

The above description of solenoids applies only to vertical sections, not to isobaric ("horizontal") charts. As mentioned in Sec. 4-8, the solenoids on isobaric charts are identified by the existence of iso-

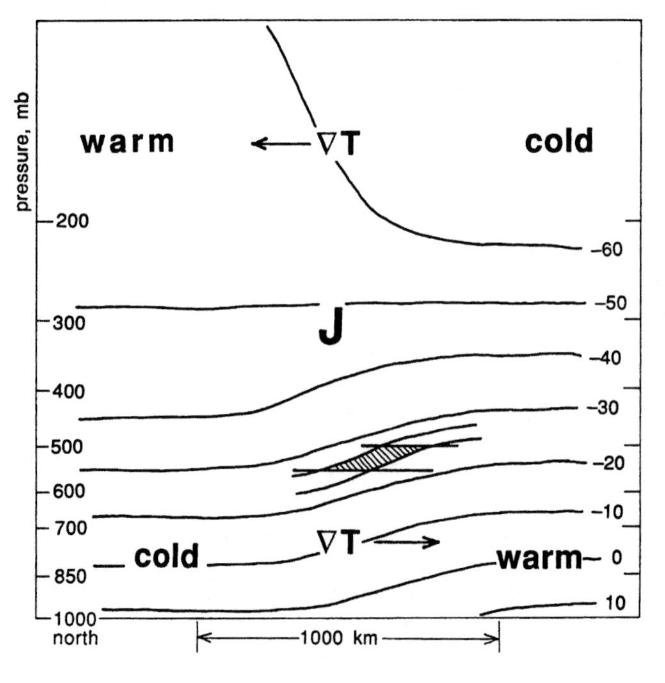

FIGURE 8-6. Schematic vertical section with isotherms (°C) near the jet stream (core at J). A thermodynamic solenoid (hatched) is enclosed between two intermediate isotherms and two isobars. The vectors show the horizontal part of ∇T.

therms, irrespective of their intersecting the contours. When the intersections of isotherms and contours exist (on isobaric charts), then there is advection of temperature, in addition to baroclinicity.

The troposphere and the stratosphere are recognizable in Fig. 8-6. The troposphere is under about 400 mb on the "northern" side and under 200 mb on the "southern" side. It is characterized by numerous, predominantly horizontal, isotherms. The region on top, with almost no isotherms, is the stratosphere. The troposphere on the equatorial ("southern") side is higher than the troposphere on the polar side due to stronger convection in warmer regions. The rather abrupt transition of the tropopause height near the jet developed due to confluence in the large-scale westerly flow. Baroclinicity developed also due to confluence.

Discontinuous Front in Vertical Section

The isotherms in a baroclinic frontal zone seldom stay in the form shown in Fig. 8-6. Frontogenetic processes near the jet stream usually act in such a way that a situation similar to Fig. 8-7 occurs. This figure shows a slanted frontal layer that often extends through the troposphere and a few other normally observed features at the polar front. Tropopauses and the frontal layer are indicated by heavy lines. The isotherms are kinked, as required by the Margules equation. The frontal layer is vertical at the level of maximum wind, as required by the thermal wind equation. The vertical front is also in agreement with the Margules equation. This equation shows that $s \to \infty$ when the thermal contrast across the front vanishes. It is hardly appropriate to use the expression *layer* for a vertical front, but this is now an accepted term. A strong (always cyclonic!) wind shear is the main indicator of the vertical front, since the thermal contrast vanishes. The vertical front at the jet level stands between the troposphere on the equatorial side of the front and the polar stratosphere on the polar side.

A slanted frontal layer in the troposphere is characterized by large thermal (static) stability, as indicated by steep, or even inverted, isotherms within the frontal layer. Inverted isotherms are characteristic of inversions. Vertical, or very sparse isotherms, are characteristic of high stability in the stratosphere.

The tropopause normally assumes a strongly deformed or discontinuous shape in the region of the jet stream. The model in Fig. 8-7 is drawn with the assumption that the tropopause is discontinuous. Some other shapes of the tropopause near the jet will be shown below.

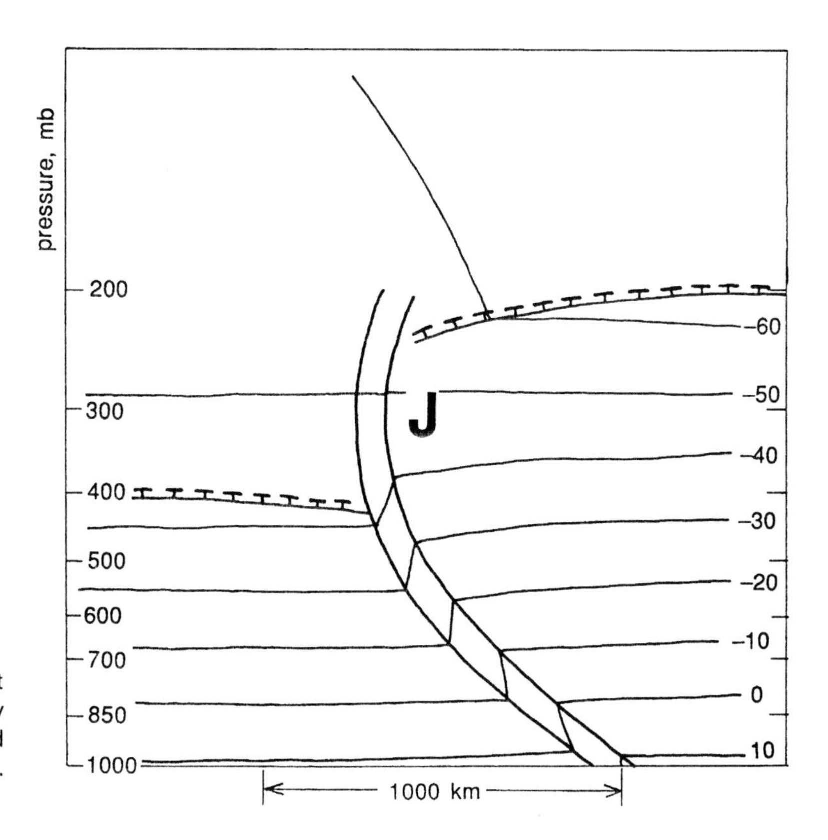

FIGURE 8-7. Model of front (heavy lines), tropopause (heavy lines with T marks), and modified isotherms (thin lines, °C) from Fig. 8-6.

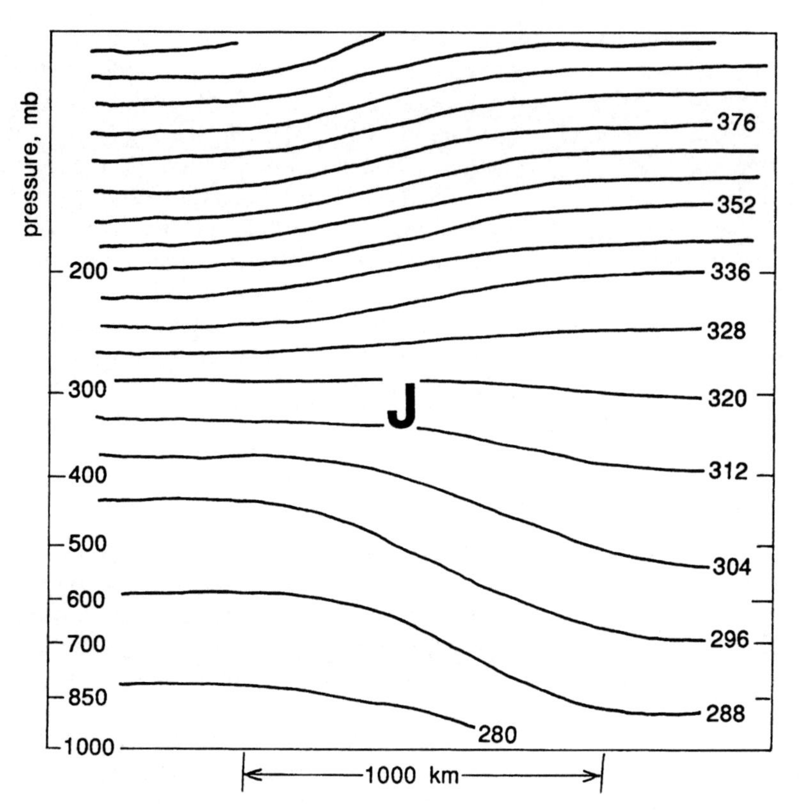

FIGURE 8-8. Vertical section with potential temperature (K) computed from temperature and pressure in Fig. 8-6.

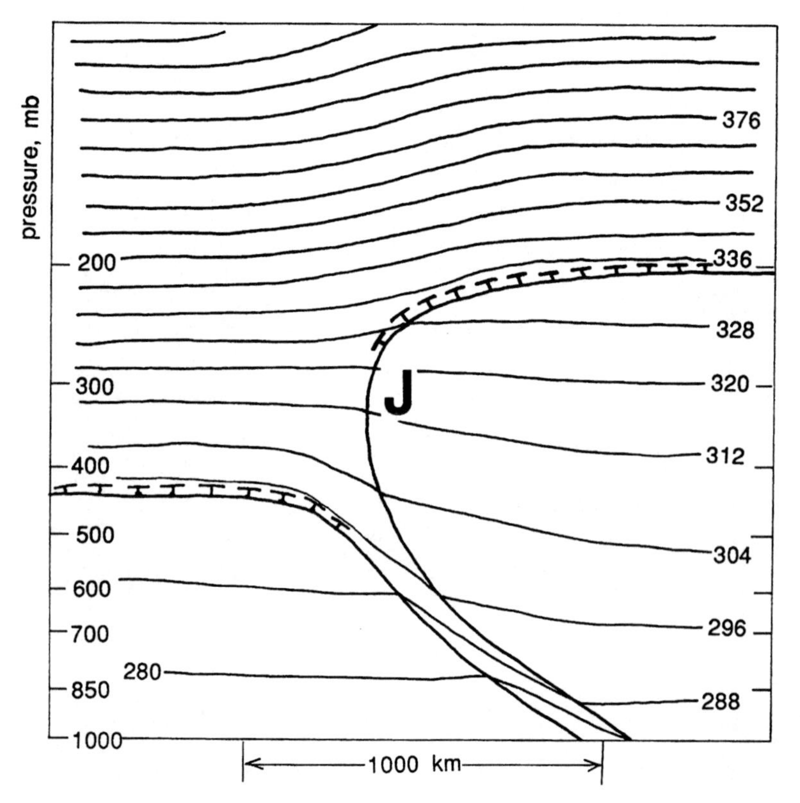

FIGURE 8-9. Model of front and tropopause (lines with T's) with modified isentropes (K) from Fig. 8-8.

Vertical Sections with Isentropes

The examples in Figs. 8-6 and 8-7 are repeated in Figs. 8-8 and 8-9 using the isentropes, based on the temperature and pressure in Fig. 8-6. Stable layers, especially the stratosphere, are characterized by dense isentropes. Less densely spaced isentropes in the troposphere indicate lower stability. Slanted isentropes under the jet stream in Fig. 8-8 are indicative of a possible front. The isentropes in Fig. 8-8 slant the other way than the isotherms in Fig. 8-6 due to the opposite signs of the vertical derivatives of T and θ. T usually decreases with height, while θ increases. The front and the tropopauses in Fig. 8-9 are in about the same place as in Fig. 8-7. The only difference is the shape of the tropopause near the top part of the front. Stratospheric isentropes that are funneled into the frontal layer suggest movement of air from the stratosphere to the troposphere. Under adiabatic conditions, the air moves along the isentropes. Therefore, we conclude that stratospheric air may be entering the frontal layer in the break between two tropopauses. For this reason, the upper part of the front in Fig. 8-9 is open toward the stratosphere, suggesting the entry of stratospheric air.

When the front model is applied, the isentropes in the frontal layer are drawn close together. This emphasizes that the frontal layer is more stable than the surrounding atmosphere. Continuous packed isentropes entering the frontal layer from the top suggest an extension of a slice of the stable stratosphere into the frontal layer. This style of drawing the connection between the front and the tropopause can also be used when isotherms are drawn, although it is more common with isentropes.

Conceptual Model with Isotherms

A model of the front-jet system in the usual style with isotherms is summarized in Fig. 8-10. The wind shear replaces the thermal contrast in the vertical frontal layer near 300 mb, as required in (8-3) for the case $T_w = T_c$. The distinction is made between the terms *frontal zone* and *frontal layer*. As described in Sec.

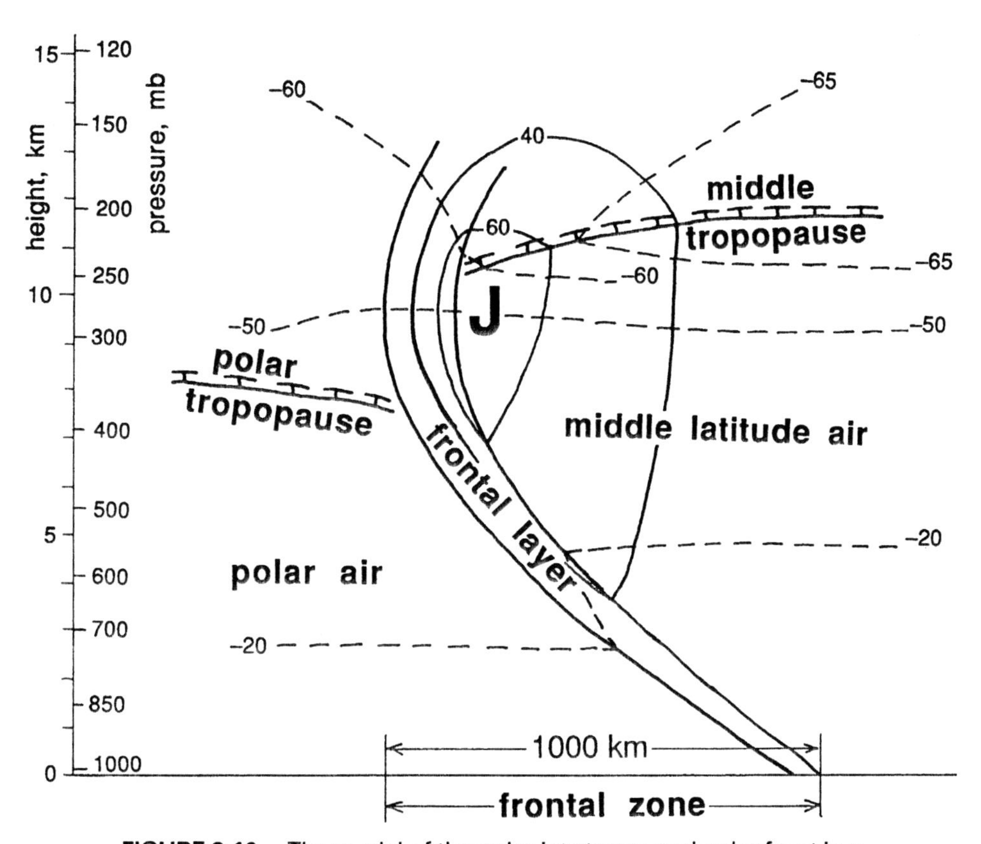

FIGURE 8-10. The model of the polar jet stream and polar front in a vertical section with isotachs (thin, m s^{-1}) and isotherms (dashed, °C). The break in the tropopause is associated with the vertical frontal layer.

8-2, the frontal zone is the wide region that contains the jet stream and the most baroclinic part of the troposphere. The frontal layer is a small-scale, prominent, slanted, stable layer that typically accompanies the jet stream.

8-4 TROPOPAUSE FOLDING AT THE JET STREAM

Vertical Circulation

A discontinuous transition between the tropopause levels above various air masses can be explained by circulation that occurs around jet streams. The vertical motion that develops around the entrance to a jet streak (Section 5-5) lowers the stratospheric air on the polar side of the jet. The tropopause is lowered during this process as well. Similarly, lifting of air lifts the tropopause on the equatorial side of the jet stream. The folding of the tropopause is illustrated in Fig. 8-11. This is a variation of Fig. 7-13a, with direct circulation near the entrance to a jet streak. The tropopause is being lowered on the polar side of the jet stream. The lowered stratospheric air at *S* is still rather stable and has started forming a frontal layer. It is possible (although not yet definitely shown) that such a frontal layer may develop all the way to the earth's surface, along the dashed lines in Fig. 8-11. This point is also discussed with the example in the next subsection.

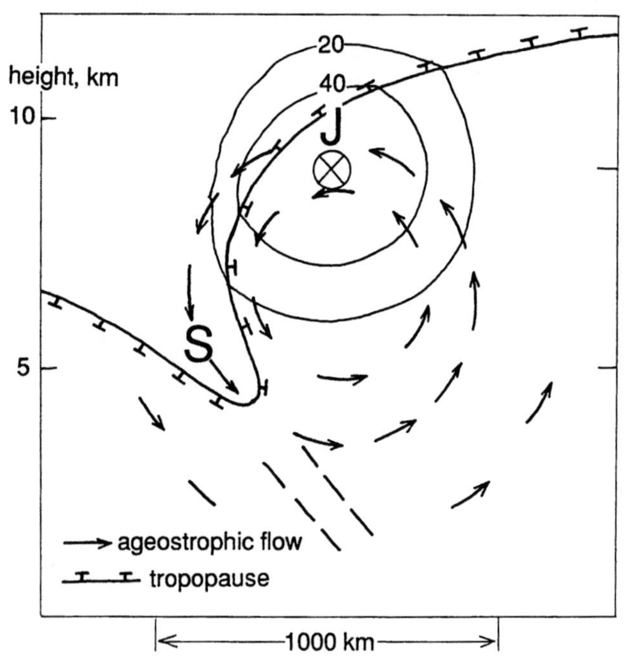

FIGURE 8-11. Folding of the tropopause with direct ageostrophic circulation around the jet stream (entrance to a jet streak). Stratospheric air descends at *S*. The T's on the tropopause show the direction of tropopause motion.

An Example of a Folding Tropopause

A discontinuous tropopause is illustrated in Figs. 8-12 and 8-13, as observed and analyzed by Shapiro (1978). Dense isentropes in Fig. 8-12 show that the polar stratosphere reaches down to the tropopause at 410 mb above MFR (Medford, Oregon). A middle tropopause is near 250 mb on the edge of the drawing south of OAK (Oakland, California). The polar jet stream is imbedded in the middle-latitude air, as in Figs. 8-7, 8-9, and 8-10. The wind speed is higher than 85 m s^{-1} in the core. Isentropes show a strong frontal layer under the polar jet. The isentropes between 300 and 310 K strongly suggest a descent of stratospheric air into the frontal layer since the air flow in the atmosphere closely follows the isentropes.

In this weather situation, detailed observations of meteorological elements and ozone concentration were made from an aircraft that traversed the jet-front system several times. Some of the results are shown in Fig. 8-13. The associated distribution of potential vorticity [(C-14) from Appendix C] is shown in Fig. 8-13a. Potential vorticity of over 100×10^{-7} K mb^{-1} s^{-1} is normally characteristic of the lower stratosphere, and this is well confirmed in this case. A characteristic feature is the fold of the isopleth of 100×10^{-7} K mb^{-1} s^{-1} down in the frontal layer, allowing stratospheric air to enter the troposphere. Stratospheric air with high potential vorticity forms a stable upper-level frontal layer.

The downward displacement of stratospheric air in Fig. 8-12 is also documented by measurements of ozone concentration. Figure 8-13b shows that the isopleth of 8 pphm vol^{-1} (parts per hundred million, per volume) coincides well with the isopleth of 100 units of potential vorticity. High values of ozone are typical for stratospheric air. A complete agreement between distributions of ozone and potential vorticity is still not present due to significant generation and dissipation of potential vorticity near strong jet streams.

The process of folding of the tropopause and forming of the front from above was also confirmed by calculation of vertical motion. Descending motion occurred within the folded tropopause. Similar cases reported by other investigators also confirmed that tropopause folding is the most important process for the formation of upper-level fronts.

A connection between the formation of the upper-level front and surface front may occur by devel-

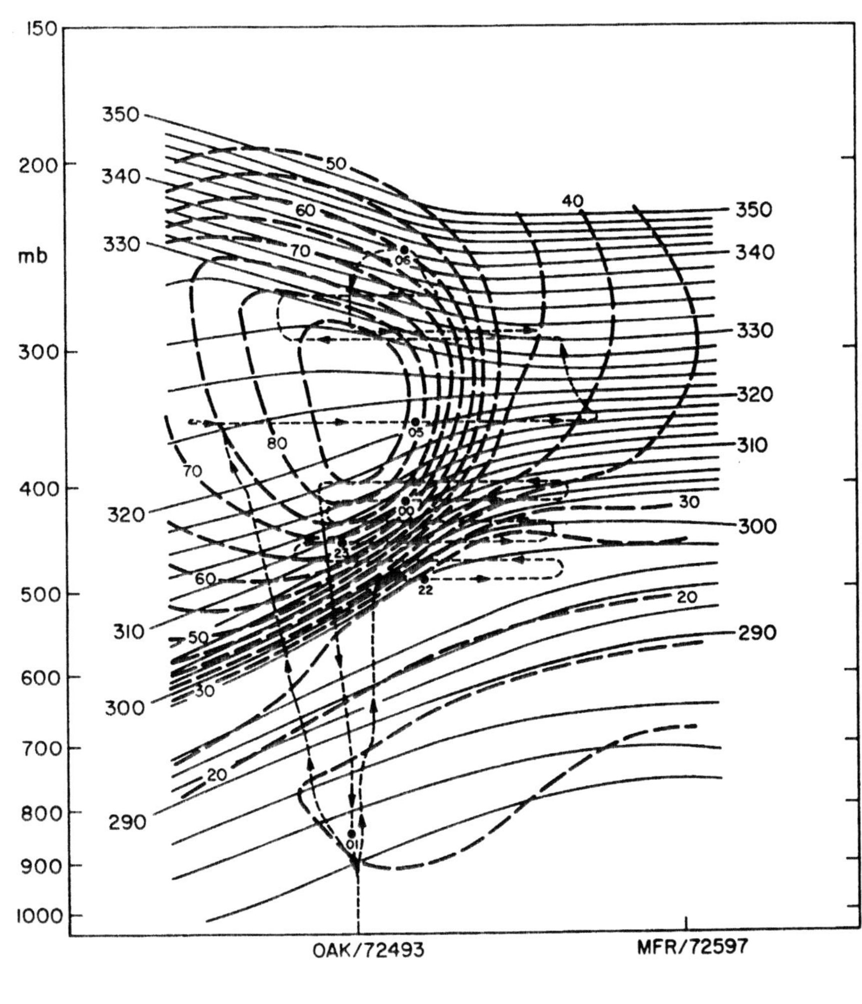

FIGURE 8-12. Vertical section based on aircraft and radiosonde observations for 0000 UTC 16 April 1976 between OAK (Oakland, California) and MFR (Medford, Oregon). Heavy dashed lines are isotachs (m s^{-1}). Isentropes (K) are solid. Flight path is the light dashed line with observation hour at dots. Distance between OAK and MFR is about 500 km. From Shapiro (1978).

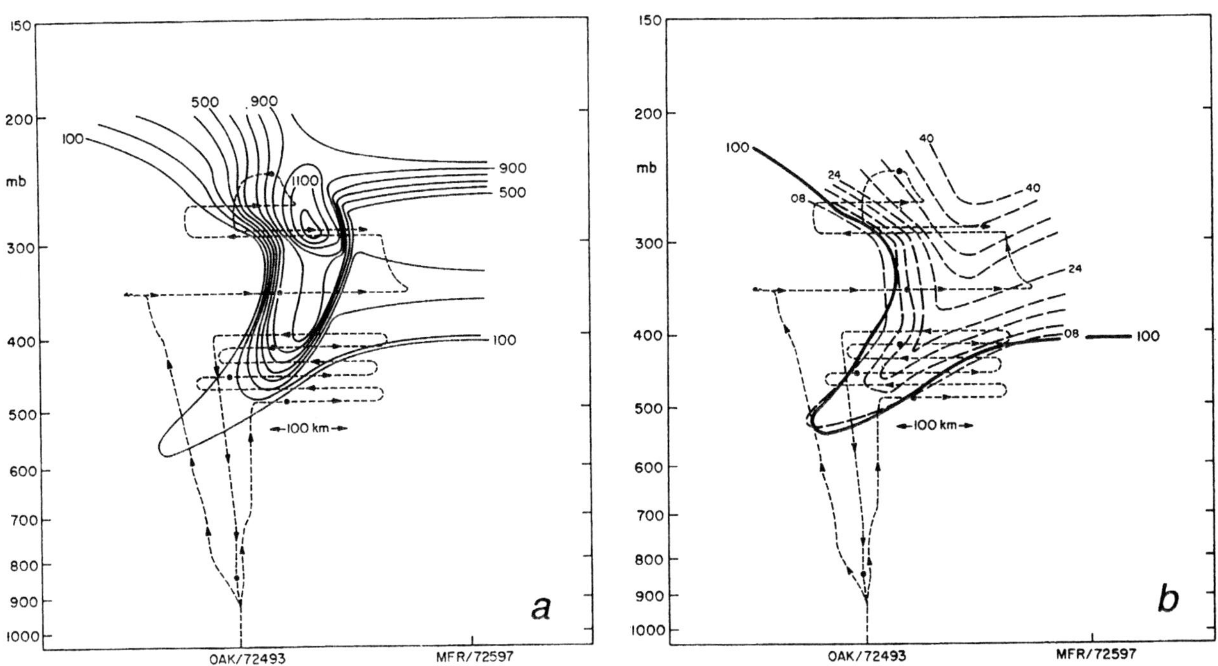

FIGURE 8-13. Isopleths for the section in Fig. 8-12. (a) Potential vorticity (10^{-7} K mb^{-1} s^{-1}). (b) Ozone concentration (dashed, pphm vol^{-1}) and potential vorticity isopleth of 100×10^{-7} K mb^{-1} s^{-1} (solid) that is indicative of the tropopause. From Shapiro (1978).

opment and descent of the upper front, until it reaches the ground. This process is suggested by Keyser (1986) and is hinted at in Fig. 8-11. The upper-level tropopause fold preferably forms in the northwesterly segment of the jet stream. This is the region with convergence at the jet level and divergence near the ground (inactive region). The upper-level front gradually penetrates deeper down into the troposphere and is transported downwind toward the upper-level trough. The front reaches the earth's surface downstream from the upper-ievel trough, in the active region, and then assumes its usual position in the extratropical cyclone that develops in the active region.

The upper-level frontogenesis, as described above, is not a perpetual process. It occurs only when suitable downward circulation near the jet appears. There are also situations somewhat similar to those described, but with opposite circulation and, consequently, with dissipation of the front.

Indirect circulation occurs in the exits of jet streaks, which tends to equalize the tropopause level on the two sides of the jet stream. However, it seems that the large-scale deformation in the horizontal flow counteracts the equalization of the tropopause and the split-level tropopause is maintained as a fairly stable state. The indirect circulation with the jet exit often has another effect on the front: It causes the front to split. This is illustrated in Section 8-9.

8-5 FRONT TYPES

Overview of Fronts

Chapter 1's review of atmospheric circulation patterns shows that the largest front is the polar front. The polar front is associated with the polar jet stream, with a break (or fold) of the tropopause, and often can be identified at the earth's surface.

The subtropical front is related to the subtropical jet in the same way as the polar front is related to the polar jet. The subtropical jet is described in greater detail in Chapter 11, with the associated break between the middle and tropical tropopauses. As mentioned in Chapter 1, the subtropical front is confined to the upper troposphere.

In addition to fronts that are structurally connected with the jet streams, several other fronts may develop in some weather situations. The most prominent fronts not associated with jet streams are the intertropical front, various mesoscale features as the breeze front and outflow boundary with thunderstorms, and the arctic front.

Intertropical Front

A line of clouds is regularly observed near the equator, where the trade winds from the northern and southern hemispheres meet. This is the *intertropical convergence zone* (ITCZ). In some cases, the ITCZ may become narrow, especially when it moves away from the equator; the ITCZ can then be interpretted as a front along the leading edge of the equatorial air. The existence of the intertropical front between the easterlies and equatorial westerlies is sometimes doubted. Conclusions on its existence and sharpness are largely made from the sudden onset of monsoon in the affected regions.

Mesoscale Fronts

Some fronts appear on a small scale, of the order of 100 km in length. These are outflow boundaries near thunderstorms and various local discontinuities near mountains and water bodies. The discontinuity at the leading edge of the sea breeze (or lake breeze) is the *breeze front*. The *land breeze* can be observed over water at night. Most of smaller-scale discontinuities will not be considered in this chapter; instead, attention will be paid to synoptic-scale fronts that are most often associated with upper-tropospheric jet streams. A review of mesoscale fronts is given by Young and Fritsch (1989). These fronts are also described briefly in Section 1-4.

A phenomenon that may be classified with fronts is the *dry line* that often develops in the warm sector of the extratropical cyclone. This line is described in Chapter 10.

Arctic Front and Arctic Jet Stream

Most often the arctic front is a widespread inversion that covers the lower part of the polar air mass. The motion of the front at the surface is usually governed by the passive front dynamics, as described in Section 8-10.

Sometimes the arctic front is associated with strong cyclonic vortices poleward of the polar jet stream. In some of these vortices, another fold (or break) in the tropopause develops, in the domain of the arctic air. This fold is similar to the fold described at the polar and subtropical jets. The arctic front may extend higher than the usual 1–2 km; in those cases, jet stream segments also appeared, so that the physical thermal wind relation was satisfied as in the polar front model in Figs. 8-6, 8-8 and 8-9. Prominent segments of jet stream that accompany the arctic front may be designated as the *arctic jet stream*.

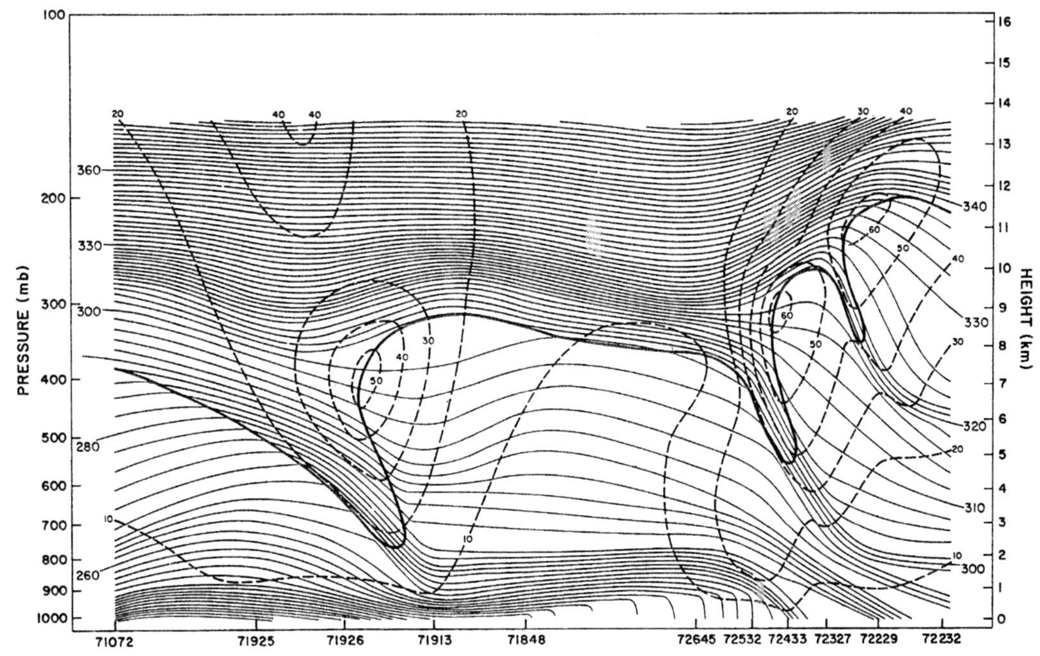

FIGURE 8-14. Vertical section of potential temperature (K, thin solid lines) and wind speed (dashed, m s^{-1}) for 1200 UTC 6 March 1979 between Resolute, Canada (71072) and Boothville, Louisiana (72232). The isopleth of potential vorticity of 100×10^{-7} K s^{-1} mb^{-1} (heavy solid) is shown as a fair representation of the tropopause. From Shapiro, Hampel, and Krueger (1987).

An example of a tall arctic front with a loop of the jet stream is shown in Fig. 8-14. Polar and subtropical jets are close to each other in the southern part of the section, each with a tropopause break. A third jet stream and a corresponding break can be seen near station 71926 in southern Canada. This shows the tropopause fold at an unusually well-developed arctic front. The isopleth of 100×10^{-7} K s^{-1} mb^{-1} folds at each jet stream, suggesting a descent of stratospheric air along the frontal layer. That jet stream—the *arctic jet stream*—can be seen on upper-level charts as a closed low imbedded in the polar air mass.

Classification of Fronts

Fronts are formally classified in two ways: structurally, by their relation to the jet streams, air masses, and other atmospheric structures, or by their motion. Figure 1-3 serves as a guideline for the structural identification of several major fronts. The polar front is on the polar edge of the polar jet stream, between the polar air and middle-latitude air. The tropopause above the polar front has a characteristic abrupt transition from one side of the front to the other side. As indicated in Fig. 1-3, a similar relationship exists between the subtropical front and the subtropical jet. The most apparent difference between the subtropical and polar fronts is that the polar front often descends to the earth's surface, but the subtropical front is confined to the upper half of the troposphere. Only a few cases when the subtropical front could be discerned at sea level were reported in the literature.

By its movement, a front can be *cold, warm,* or *stationary*. Such classification of fronts is normally determined by their movement over the preceding 3 or 6 h. However, for slow fronts, the normal direction of the geostrophic wind on either side of the front is used. This means that the geostrophic movement of the front is determined by the sense in which the isobars or contours intersect the front.

The front is stationary if there is no appreciable movement and there are no isobars crossing it on a longer stretch. This means isobars at every 4 mb and a stretch of more than 500 km.

There are cases when the movement of the front progresses opposite from the direction (sense) of geostrophic wind. Then the front's type is determined by its movement. As with air masses, there is no "cool" front in the classification.

8-6 THE POLAR FRONT

Weather Elements at the Front

The polar front is usually a slanting, narrow transitional layer associated with the baroclinic zone under the polar jet stream. This layer is anywhere between zero and several hundred kilometers wide. At its narrowest, this front transforms into a discontinuity surface with the variables changing abruptly from one side of the surface to the other. The passage of such a sharp front can be easily observed. A sudden wind shift and an abrupt change of temperature and humidity with the passage of the front are an impressive event for every observer in the field, especially when that observer is a trained meteorologist who knows and expects this phenomenon.

The frontal zone is usually identified as the most baroclinic region on a weather chart. This region is characterized by a strong temperature gradient, as can be seen from closely spaced isotherms. Even when there is a complete zero-order discontinuity, we see a region of tightly spaced isotherms, since the observation stations are not very closely spaced. The cases of zero-order discontinuity were observed by continuously recording instruments when the front passed over the stations, or when the instruments were carried through the front.

The polar front is physically connected with the polar jet stream. The front and jet are formed in an elongated shape by frontogenetic processes. Narrow jets appear in agreement with the thermal wind relation, and narrow frontal layers form due to confluence and convergence in the baroclinic regions. Consequently, the polar jet stream is always accompanied with a polar front, even if this front is weak and cannot be found at all levels. Conversely, the polar front is always associated with the polar jet stream, even if its lower part wanders more than a thousand kilometers away from the jet.

Discontinuities in weather elements are the first signs that show the front. Most prominent are the cyclonic wind shift and abrupt changes in temperature and humidity.

Precipitation often accompanies the front. Clouds and rain (or snow) usually originate in the warm air near the front at the surface. However, the distribution of precipitation is very variable.

Limitations in Interpretation of Discontinuities

Larger and smaller variations are common on weather charts. Many of these variations can be interpreted as fronts. We must be careful, though, not to interpret every observed discontinuity and every baroclinic zone as the polar front. Only a complete structural analysis of the atmosphere will show which discontinuity is the polar front.

Another difficulty often encountered in polar-front analysis is that sometimes this front is very inconspicuous. The polar front may be only a weak shear line without contrast in temperature, and it may intensify later. Along this polar front, convection or even cyclogenesis may start. On some fronts the temperature difference may even be reversed, as when there is strong sunshine in clear polar air and low dense clouds in tropical air on the warm side of the front.

Sometimes the polar front penetrates deep into the tropics. Such a front (the *passive front*, described in Section 8-10) is usually much less prominent than the front of higher latitudes. The polar front in the tropics, which is practically reduced to a shear line, often reaches Venezuela, about 10°N. Some fronts have been spotted in Brazil at the equator. In Brazil, the polar front has been observed coming from the Southern Hemisphere as well, on rare occasions with freezing weather.

Vertical Structure of the Polar Front

The polar front takes various shapes, depending on environmental conditions. Accordingly, various models have been proposed, differing in the way the surface front is placed in relation to the upper-level front and the jet stream. It is possible to classify the models in three main categories (Fig. 8-15). Each model in Fig. 8-15 has polar air on the left-hand side and middle-latitude (or tropical) air on the right-hand side. There is a subtropical trade-wind inversion near 900 mb in each case.

The middle frame (Fig. 8-15b) shows the "classical" tall model. The polar front assumes a continuous, fairly steep position from the tropopause to the surface.

In Fig. 8-15a, the surface front has progressed far from the polar jet stream toward the tropical air. If there is a front in the middle troposphere, it is distant, together with the jet stream. A front of this form occurs when slowly moving and subsiding polar air passively spreads along the earth's surface. The movement of this front is fairly well described as the *gravity current* (Section 8-10).

Figure 8-15c shows a fast-moving upper-level front that moves ahead of the surface front, with a break in the front in the vertical. In this situation, the hydrostatic stability is lowered due to cold air aloft appearing ahead of the surface cold front. The model

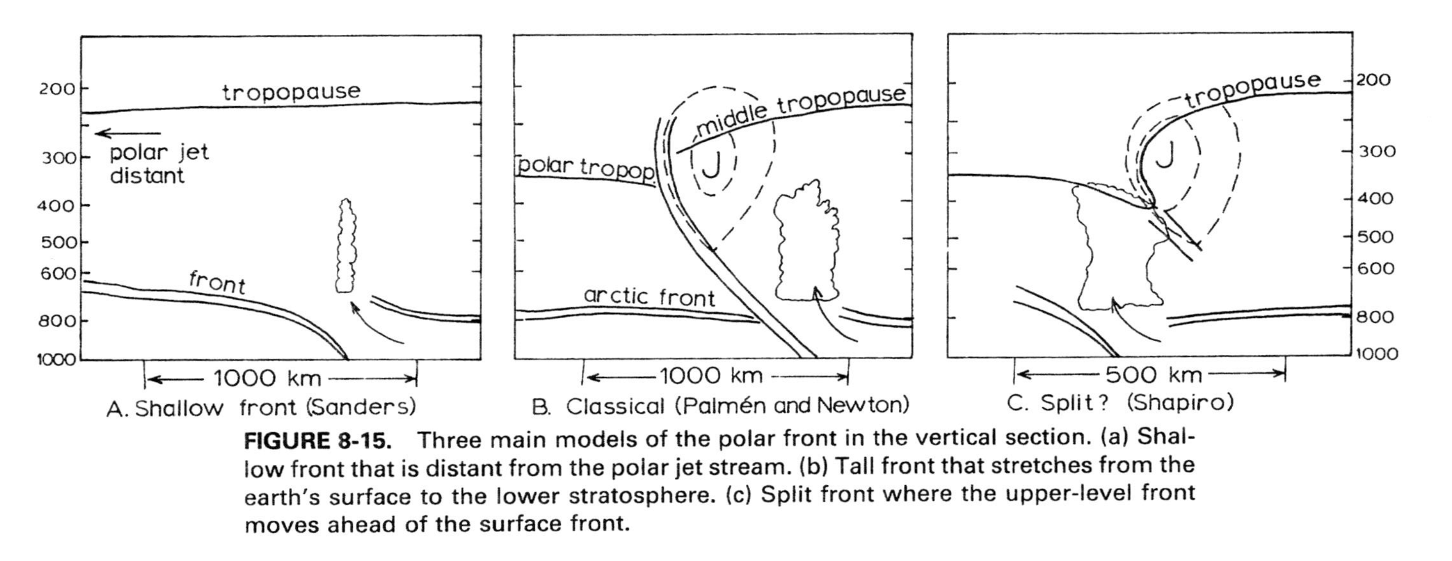

FIGURE 8-15. Three main models of the polar front in the vertical section. (a) Shallow front that is distant from the polar jet stream. (b) Tall front that stretches from the earth's surface to the lower stratosphere. (c) Split front where the upper-level front moves ahead of the surface front.

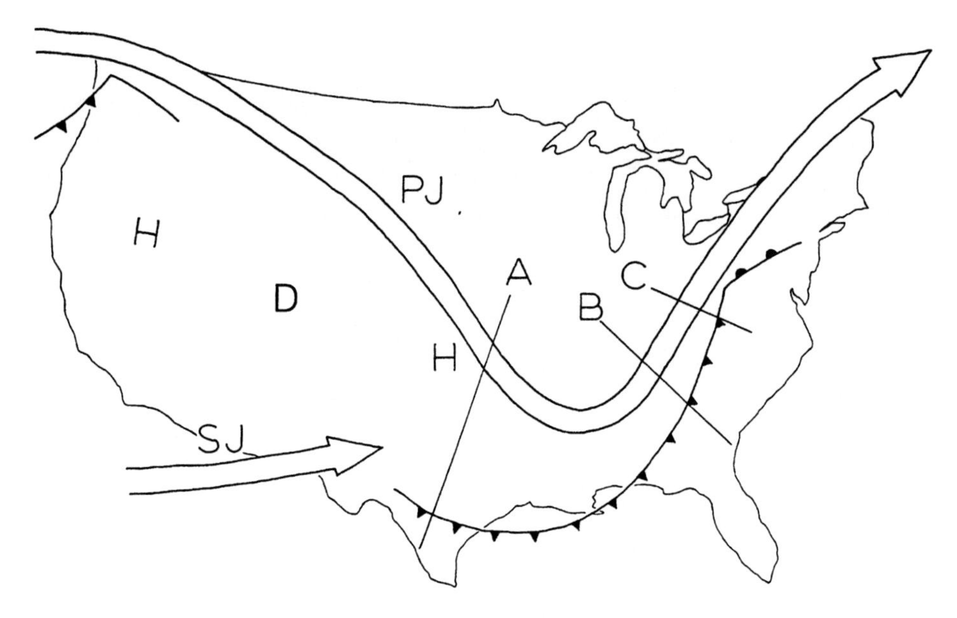

FIGURE 8-16. The polar jet with a trough. The sections A, B, and C show where the shapes of the polar front from Fig. 8-15 typically appear.

in Fig. 8-15c sometimes appears as a *split front*, described in Section 8-6.

Shapes of the Front in the Jet-Stream Wave

The shape of the polar front depends greatly on the conditions of vertical motion, frontogenesis, and frontolysis. These processes, in turn, vary with position in the wave on the polar jet. Figure 8-16 shows a trough in the middle-latitude westerlies, with its typical polar front at the surface. The upper-level polar front is not shown, but it is normally close to the jet core such that it reaches the tropopause 200–400 km poleward from the jet core. The lines A, B, and C are the sections along which the corresponding models from Fig. 8-15 can be found. Subsidence and anticyclogenesis are widespread upstream of the trough (at A). In this region, the model of Fig. 8-15a is typical. On the other side of the trough (downstream from it, at B), the steep front is common, as is the one in Fig. 8-15b. In the region of fast westerlies, sometimes the upper-level jet and front overtake the surface front; then the model from Fig. 8-15c applies.

Frontolysis and anticyclogenesis are typical at the earth's surface in the region near D in Fig. 8-16. The surface front dissipates in this region, and contrasts between air masses diminish. Often there is a trough in the surface anticyclone, where the transition between the air masses can be suspected, but the contrasts of a front cannot be detected. At the same time, the upper-level front may be prominent in this region.

An inspection of isobaric charts is often not sufficient to determine the shape of the polar front. It is possible that the shapes of the tropopause and front will be different in cases when the waves in the jet stream look similar. We should be aware of the various models and use the appropriate one as it fits the available observations. We also have to watch for patterns that have not yet been described in the literature. We cannot afford to be constricted by book models. One purpose of having models from the book is to appreciate the variety that may occur in nature. Many other phenomena exist in the atmosphere that have not yet been described. When students start practicing weather analysis, they will be expected to find those new, presently unknown, models.

8-7 FRONTAL PATTERNS ON CHARTS

Baroclinic Frontal Layer

The front is primarily detected by dense isotherms on a constant-pressure level. This is the baroclinic (sometimes hyperbaroclinic) zone of the frontal layer that intersects the isobaric surface. When the front is detected on a chart, with help of other charts, the analyst should adjust the isotherms to resemble the model.

Ignoring data should be avoided. Only when the data cannot be interpreted at all, can they be ignored. The ignored items must be marked on weather charts by crossing them out. The crossed items must stay legible because there is a possibility that a better inter-

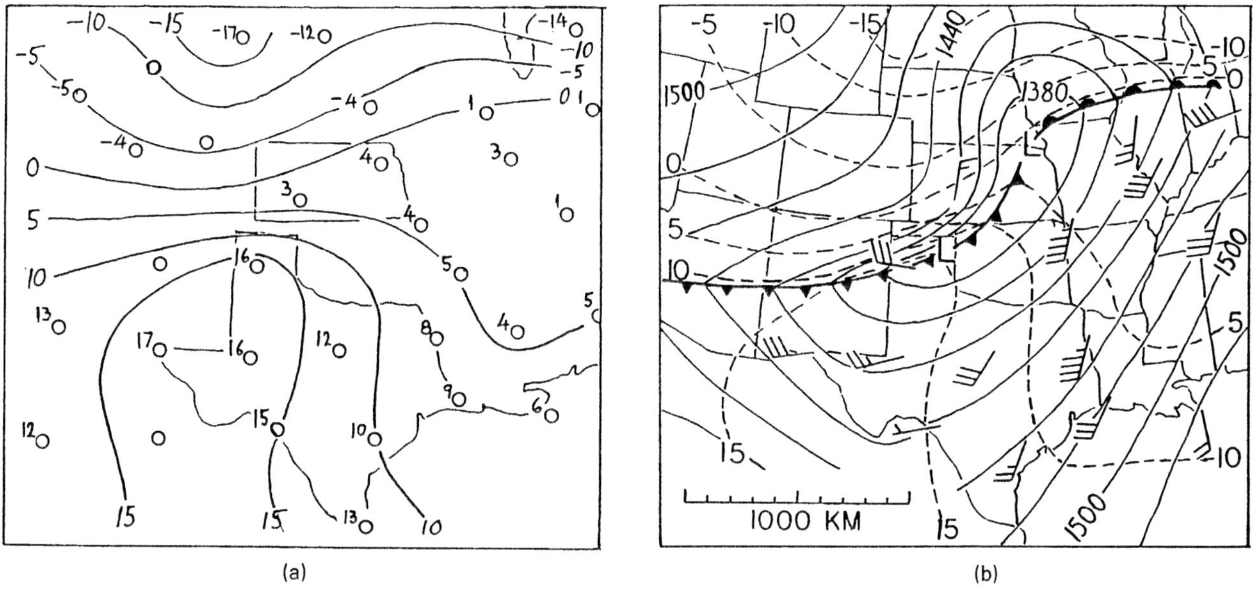

FIGURE 8-17. An example of isotherms (°C) at 850 mb for 0000 UTC 6 February 1976. (a) Sketch. (b) Final isotherms (dashed), style with one line depicting the front. Contours (solid, gpm), wind (kt), and front are added to illustrate the analysis.

pretation will be discovered and the seemingly incorrect data will offer valuable information.

Isotherms

Determining the polar front is normally easiest when isotherms or isentropes are available on isobaric charts and vertical sections. The first sketch of isotherms can be obtained by hand or from computers. Further steps in analysis can be done by visual inspection and by hand, since judgment must be made on the similarity with known conceptual models. Computer-made charts usually do not show discontinuities well; therefore, they should be corrected by hand in the places where the front needs enhancement. Unfortunately, it is not always possible to redo computer products; then at least the interpretation of charts should follow the models described below.

It is convenient to draw isotherms every 4°C or other 2^n intervals (n = integer), making it easy to halve each interval by visual inspection. The World Meteorological Organization recommends 5, 2.5, and 1°C intervals. Intermediate isotherms are drawn when the standard isotherms do not show some important detail. When intermediate isotherms are drawn, a different style of line (dashed or thinner continuous) should be used, so that the estimation of gradient is not easily confused. Red is the standard color for isotherms in manual analysis. On monochromatic (black-and-white) charts, isotherms are most often represented by dashed lines.

Styles of Front Drawing

There are two main styles for depicting fronts on the upper-level charts: (a) with the front drawn as a line on the warm side of the zone of dense isotherms (the style prevalent in the United States) and (b) with two lines, one on each side of the zone of dense isotherms.

The standard color for fronts is blue, regardless of the motion of the front. One-color (monochromatic) charts show upper-level fronts by a thicker solid line. Symbols from the surface chart (Section 9-3) are also often used, especially in black-and-white figures where it is difficult to distinguish lines.

Figure 8-17a shows the isotherms at 850 mb drawn as a preliminary sketch, when it is not yet clear that the front is in the region. An analyst with some experience will draw the isotherms free of short waves already at this stage of analysis, as shown in this figure. Dense isotherms in the western and northeastern parts of the chart indicate that the front may be there. When the other levels are consulted and the front is determined to be in the position indicated in Fig. 8-17b, the isotherms are adjusted to enhance the front. The style of analysis in Fig. 8-17b is with one line on the warm side of the frontal layer.

There are also physical reasons for drawing the front along the warm side of packed isotherms. It has been shown in theoretical models that frontogenesis near the earth's surface causes the temperature gradient to stay rather weak on the warm side and stronger on the cold side of the front. Also, the temperature gradient gradually decreases from the front toward the cold air. This is represented by gradually increased spacing of isotherms on the front's cold side. Such isotherms are frequent in the lower atmosphere: at the earth's surface and at 850 mb.

The other style of depicting fronts on isobaric charts is shown in Fig. 8-18, a 500-mb chart where the isotherms are concentrated in the frontal layer between two thicker continuous lines. The line on the warm side of the frontal layer is the one that would be drawn in the style with one line. Even when only one line is drawn, the analyst should keep this other style in mind and adjust the isotherms similarly to the style with two lines.

The frontal zone does not consist of the same isotherms over a large domain. The isotherms often separate from the group of dense isotherms. This is the case with the isotherms of 5, 10, and 15°C in Fig. 8-17b and −16 and −20°C in Fig. 8-18.

Computer-produced charts do not routinely represent discontinuities. Therefore, trained meteorologists are expected to detect the front using available "continuous" fields given by the standard graphical or numerical displays.

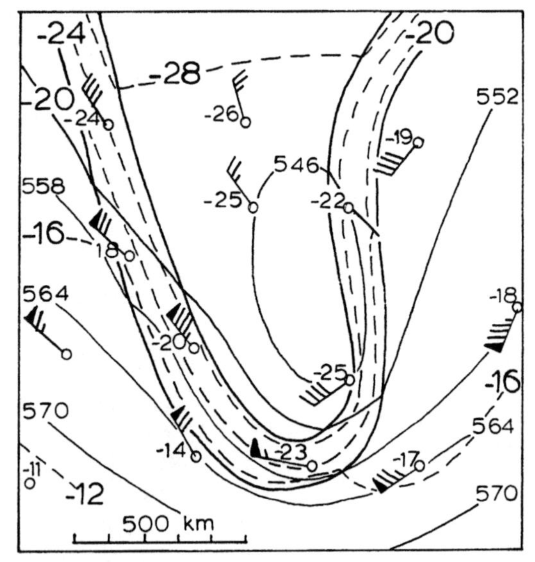

FIGURE 8-18. An example of the polar front in the style with two boundaries of the frontal layer (thicker solid lines) at 500 mb for 0000 UTC 27 January 1967 over the central United States. The contours (gp dam) are the thinner solid lines, the isotherms (°C) are dashed, and the wind is in knots.

8-8 VERTICAL SECTIONS WITH THE POLAR FRONT

Use of Vertical Sections

Vertical sections of the atmosphere can be used effectively for identification of fronts. The usual choice is to select the section across the frontal zone, since the baroclinicity (sloping isotherms or isentropes) is detected most easily. The goal is to find the phenomena depicted in the schematic vertical sections through the polar front in Fig. 8-9 or 8-13.

Isopleths are the basis for analysis, as on practically all charts. The usual choice is to draw isotherms (or isentropes) and isotachs. If the isopleths are drawn in color, the preferable choice is red for isotherms and green for isotachs. Other variables can be used for special purposes. Variables that show humidity are frequently chosen. The momentum ($v + fx$) is sometimes used since it may show symmetric instability. A discussion of this element will be postponed to a course on dynamics. Drawing of isopleths can be accomplished by computers.

Frontal Characteristics of an Example

The basic technique for analyzing vertical sections is illustrated in Fig. 8-19, showing the vertical section from Sault Ste. Marie, Michigan (SSM) to Hatteras, North Carolina (HAT). The ordinate is the logarithmic pressure scale, with the top at 90 mb. The section is selected across the polar jet stream at 1200 UTC 28 August 1986.

The isotherms are drawn at 8°C increments in dashed lines. The isotachs are solid lines, every 40 kt. The isopleths in the first frame (Fig. 8-19a) are drawn on the basis of observed data, without respect to conceptual models. Computer contouring yields such isopleths. The other frame (Fig. 8-19b) contains the finished analysis. To keep the picture clear, values of temperature are not plotted in this example.

When the isopleths are available (or during their construction), all inversions should be marked by ticks, above each observation point. Dashed lines should be used for stable layers that are not inversions. Such markings are entered in Fig. 8-19a.

Troposphere and Stratosphere

The isopleths drawn in Fig. 8-19a reveal two important regions, the troposphere and the stratosphere. The isotherms in the troposphere are fairly horizontal and comparatively densely spaced, and they form a rather regular "ladder." The troposphere can be identified up to about 400 mb above SSM and up to about 200 mb above HAT.

The stratosphere is characterized by sparse, scattered isotherms, much less orderly than in the troposphere. Some isotherms cross from the troposphere to the stratosphere. An example is the −48°C isotherm that is typically tropospheric above HAT, Washington, D.C. (IAD), and Pittsburgh, Pennsylvania (PIT). Northwest of PIT this isotherm rises to near 160 mb above Flint, Michigan (FNT) and SSM.

The curve on the −48°C isotherm between PIT and FNT indicates the transition between the troposphere and stratosphere. This curve should be changed to a kink as soon as the tropopause is placed, as shown on the finished chart in Fig. 8-19b. Other isotherms that cross the tropopause are similarly arranged. The change in the spacing between isotherms also shows the tropopause. Using this technique, the tropopause can be found, as shown in the final product in Fig. 8-19b.

Tropopause

The level of the tropopause changes significantly between PIT and FNT. This change occurs near the maximum wind speed in this example (above PIT). Previously studied models showed the break in the tropopause and the jet stream appearing together. This justifies the break in the tropopause, as illustrated in Fig. 8-19b.

The isotherms of −40 and −48°C are nearly symmetrical to each other about the level of 265 mb. This level is the level of maximum wind (Fig. 8-19a). An intermediate isotherm will be horizontal at that level and will show no thermal contrast at the level of maximum wind. A solid line with T's may be used for the tropopause. Tops and bottoms of inversions are shown by thicker lines.

By comparing this chart with the model in Fig. 8-11, it can be seen that the tropopause near 200 mb is the middle tropopause. The air mass under it is the

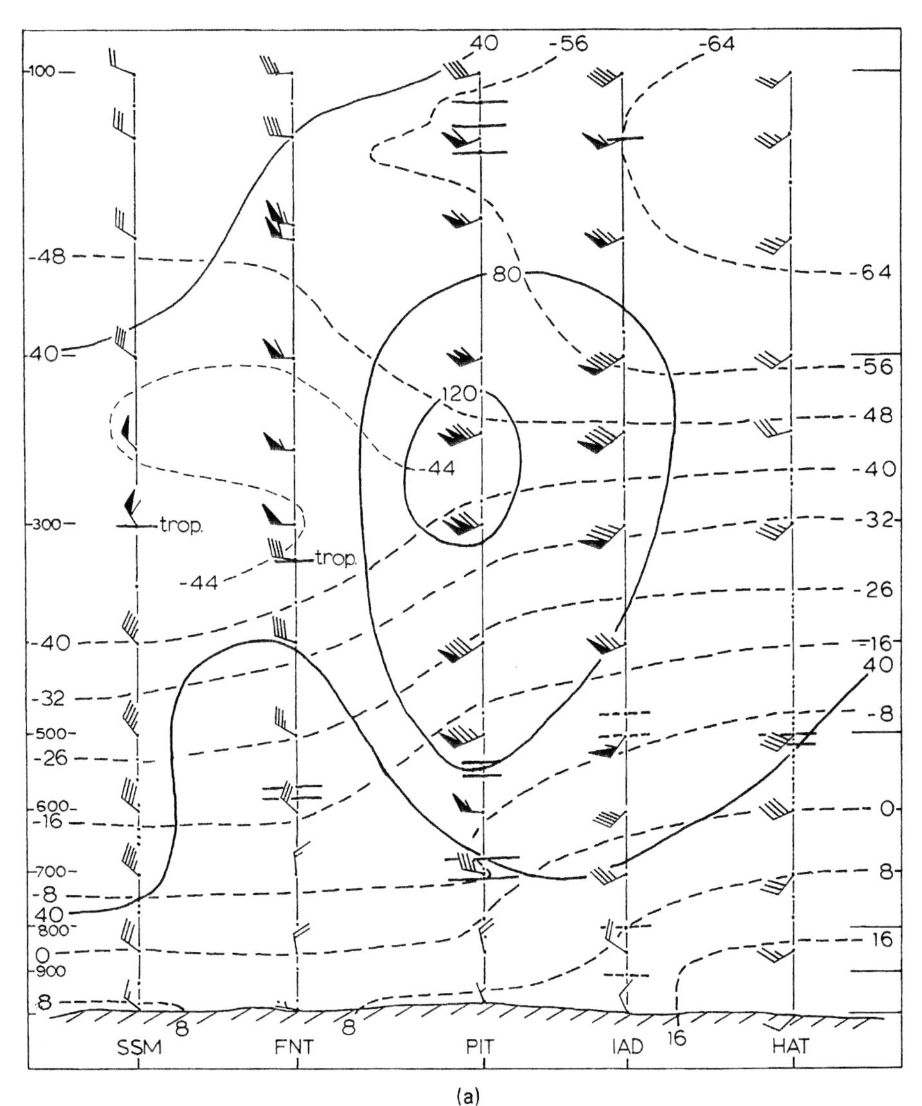

FIGURE 8-19. Vertical section through the atmosphere for 1200 UTC 28 August 1986 from Sault Ste. Marie, Michigan (SSM) to Hatteras, North Carolina (HAT) with isotherms (dashed, °C) and isotachs (solid, kt). (a) Sketch of isopleths and marked stable layers. (b) Analyzed chart with polar front and jet stream.

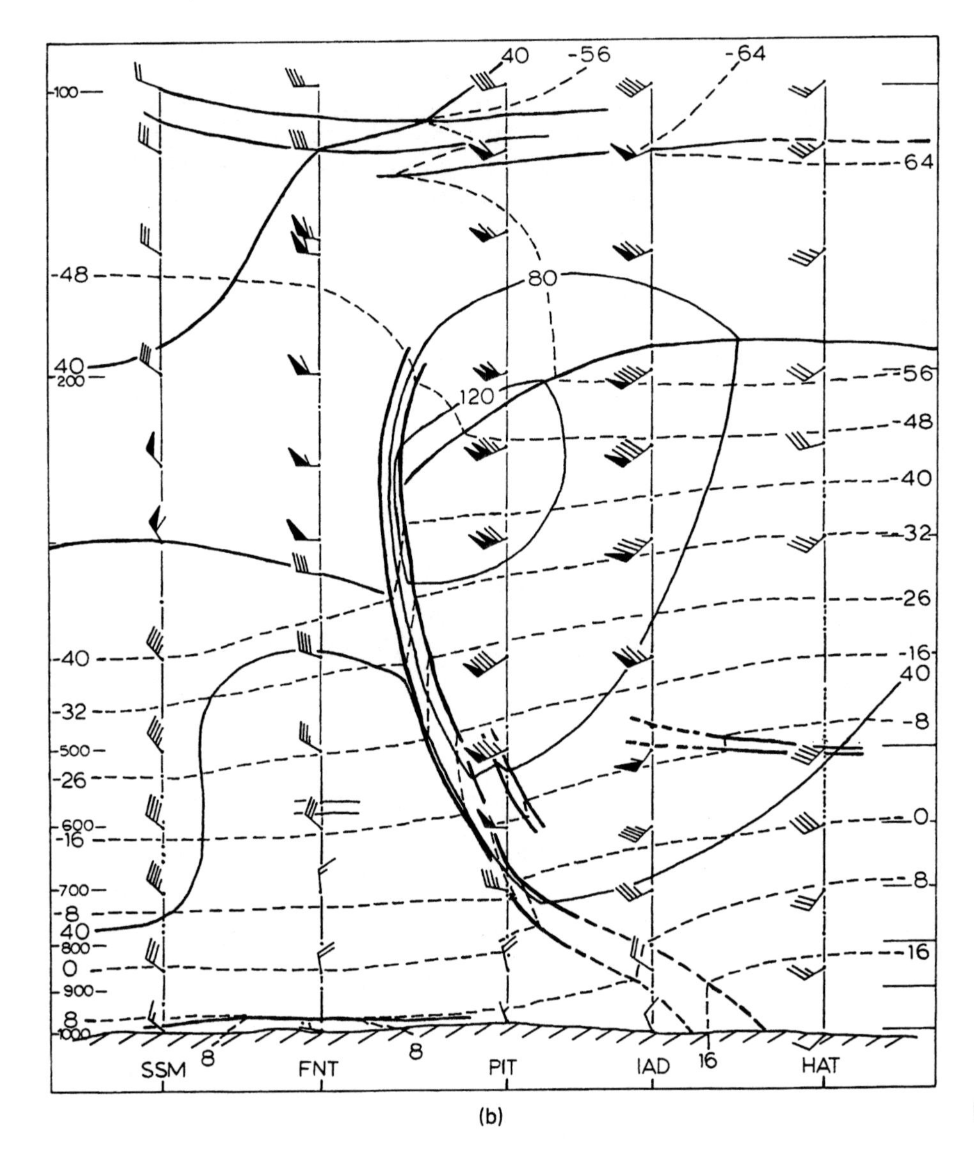

FIGURE 8-19. Cont.

middle-latitude air mass. The tropopause near 360 mb is the polar tropopause and the air mass under it is the polar air.

It is possible that the tropopause is at 350 mb over SSM, since the layer between 350 and 300 mb is comparatively stable. The isotherms are sparse. The level of 300 mb has been selected in this case because there is a prominent minimum of temperature there.

The stable layer between 300 and 250 mb over SSM may be interpreted as a tropopause "layer," as opposed to "level." The top of this tropopause layer is not indicated. This justifies the common practice to indicate only the bottom of the tropopause layer, and not the top. In this way, the tropopause is indicated by one line only, whether it is a level or a layer.

The middle tropopause in Fig. 8-19 can be found as the level where the tropospheric isotherms of −48 and −56°C bend up. This is the level where the tropospheric and stratospheric thermal regimes meet. The isotherms are nearly horizontal in the troposphere ("ladder"), but they are steep and sparse in the stratosphere. These two isotherms (−48 and −56°C) are kinked at the tropopause to show the abrupt transition between the troposphere and the stratosphere.

Frontal Layer

The isotherms in Fig. 8-19a are greatly slanted in two places in the troposphere: (a) between PIT and FNT from 300 to 600 mb and (b) between PIT and IAD from 600 mb to the surface. Since the slanted isotherms occur under the jet stream, we look for the polar front in these places. The inversion above PIT at about 610 mb is between the two regions (a) and (b) of slanted isotherms, suggesting that this inversion is the frontal inversion. We can also find the surface front between

IAD and HAT, using temperature differences and wind shear. By now we have three locations for the polar front: at the tropopause break, at 615 mb above PIT, and between IAD and HAT at the surface. Along this line we can also spot a stable layer above IAD between 900 and 790 mb. All these indicators make it possible to draw the front as shown in Fig. 8-19b. The thickness of the inversion above PIT and the stable layer above IAD reveal the thickness of the frontal layer.

Other Inversions

Several more inversions occur in this vertical section. If they appear above more than one station, we may assume that they are parts of the same inversion and we may connect them.

The inversions are most often horizontal, unless they are fronts. Such is the case with the inversion and stable layer above IAD and HAT near 500 mb. The arctic front, which appears regularly in winter, is usually a horizontal, low-lying inversion. The inversion above PIT at about 540 mb is near the frontal layer; for this reason it is, exceptionally, drawn slanted, parallel to the front.

When the fronts, tropopause, and inversions are drawn, the isopleths are adjusted to show their expected theoretical forms. The isotherms and isotachs are kinked on discontinuity surfaces of the first order, where the temperature gradient is discontinuous. Isotherms are also inverted in the inversion layers, showing an increase of temperature with height. The final form of isotherms is shown in Fig. 8-19b.

The bottom of the inversion in the stratosphere near 100 mb is possibly a far extension of the tropical tropopause that overlaps the subtropical jet stream. There are also several inversions in the troposphere that are not parts of the polar front. The thin surface inversion above SSM and FNT is a nocturnal, radiation inversion that is rather common on the morning sounding.

Jet Stream and Vertical Frontal Layer

The isotachs in Fig. 8-19 also show the model of the front and jet stream. The jet typically occurs near the tropopause break, within the middle-latitude air. Maximum wind is shown by the letter J, for jet. The strongest wind shear (shown by dense isotachs) is within the frontal layer. Once the frontal layer has been identified, the isotachs from Fig. 8-19a have been redrawn in Fig. 8-19b so to bring the strongest shear into the frontal layer. The wind shear in the frontal layer is cyclonic, as required by the Margules equation (Appendix J).

The frontal layer is vertical at the level of maximum wind, or jet level, near 270 mb. There is a discontinuity of wind at the tropopause and at the frontal layer. Therefore, the isotachs kink at the lines that show thermal discontinuity.

There is no thermal contrast across the front at the level of maximum wind; instead, a strong cyclonic wind shear characterizes the frontal layer. The isotherms kink differently across the front under the level of maximum wind than they kink above that level. This shows that the thermal contrast across the frontal layer is of opposite sign below the jet core than above it.

Finished Analysis

The finished analysis in Fig. 8-19b contains the recognized conceptual models: stratosphere, troposphere, tropopause, air masses, the polar front, and several inversions. The inversions in Fig. 8-19b that are not part of the polar front do not have names. Such inversions often appear in many other places. These "nameless" inversions are usually less prominent than the frontal inversion.

There are technical difficulties for the practical execution of analysis as shown in Fig. 8-19. Computer-made isopleths look similar to the sketch in Fig. 8-19a. Entering corrections on a cathode-ray tube is currently close to impossible. Meteorologists will have to use the smooth isopleths as in Fig. 8-19a and will have to infer about all models on the basis of these smooth lines. When the product is available on paper where drawing is possible, improvements should be entered.

8-9 SPLIT STRUCTURE OF THE FAST-MOVING COLD FRONT

Split Front Model

The middle- and upper-tropospheric part of the cold front in the westerlies sometimes advances faster than the surface front. Then the upper-level cold front may overtake the surface front. Such a front is shown in Fig. 8-20, after examples by Browning and Monk (1982). Figure 8-20 represents a zonal section through the troposphere. Arrows show the air currents, and circled X's represent arrows seen from the rear, giving some three-dimensional character to the picture. The symbols for fronts are the ones normally used on a monochromatic surface chart: The line with dark tips

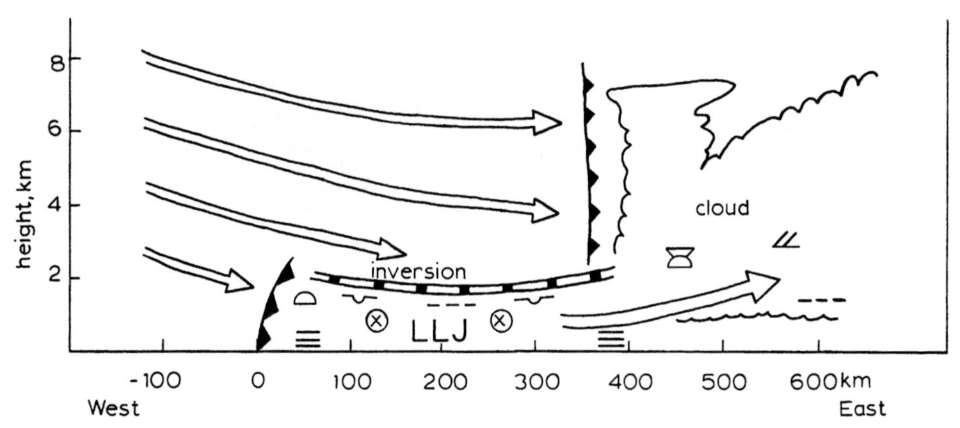

FIGURE 8-20. Vertical section through the split cold front. Principal air currents are shown by double arrows. Circled X's are arrows seen from the rear. Clouds are outlined, and typical weather phenomena are shown by conventional symbols.

(triangles) is the front that reaches the earth's surface, and the line with hollow tips is the upper-level cold front. The fronts move to the side with tips.

Several streamlines in the cold air are shown by double arrows. The flow of warm air is shown by the circled X's and by the double arrows on the bottom right. This is the *conveyor belt*. The warm air in the conveyor belt usually comes from the south, in form of the low-level jet (LLJ). The conveyor belt transports the warm air and water vapor through the extratropical cyclone. Therefore, the southerly flow shown by circled X's continues along the double arrow on the right. The three-dimensional structure of the air flow is depicted by a change of the symbols along the conveyor belt: The southerly flow progresses as shown by the X's and continues out of the figure along the double line.

The LLJ region in Fig. 8-20 is the warm tropical or middle-latitude air mass. Fog and low convection, sometimes with showers, are often in this air. An inversion usually occurs on the top of the southerly LLJ (double black–white line in Fig. 8-20). This inversion restricts the vertical development of clouds.

The cold air coming from the west is dry since it descended before it came to the vicinity of the cold front. This descent causes adiabatic warming, so this

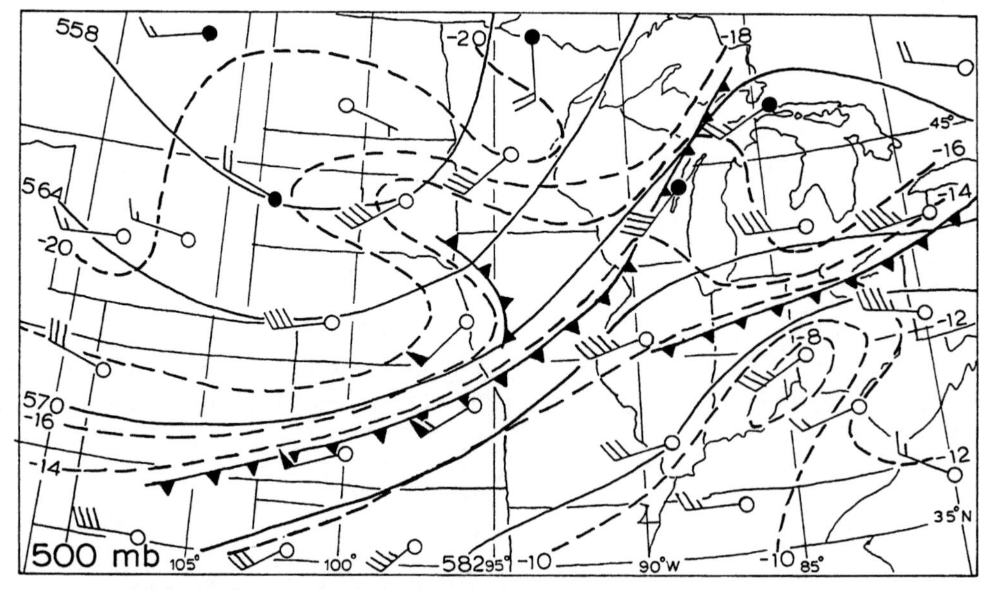

FIGURE 8-21. 500-mb geopotential (gp dam) and temperature (dashed, intervals of 2°C) for 0000 UTC 16 October 1983. Isotherms indicate a multiple front.

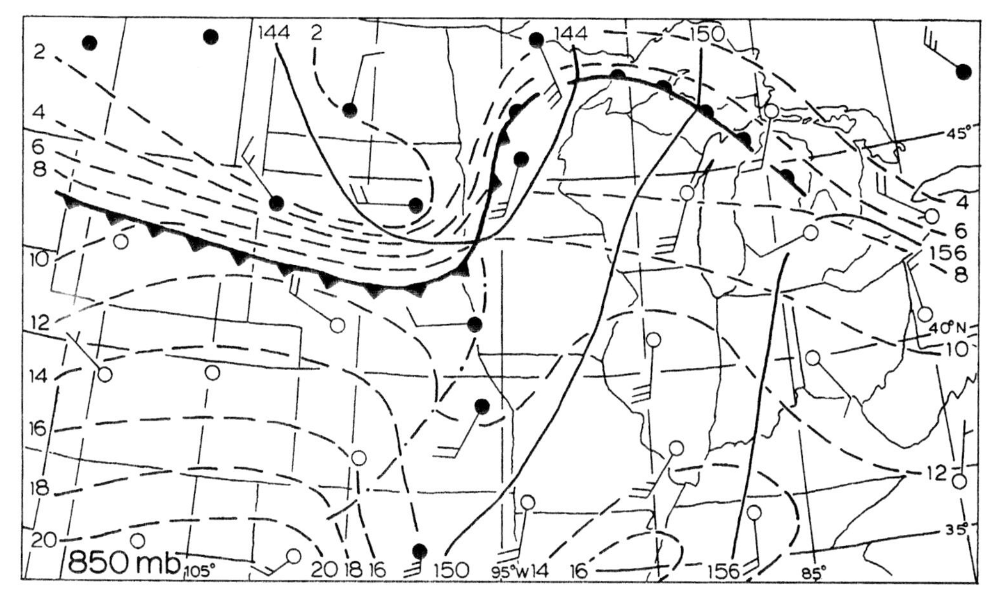

FIGURE 8-22. 850-mb geopotential (gp dam) and temperature (dashed, intervals of 2°C) for 0000 UTC 16 October 1983. The polar front is north of its 500-mb position, showing a split front structure.

"cold air" is warmer than the "warm air" under the inversion. In this way, a stable stratification is maintained at the inversion. The appearance of dry air above humid air below the inversion gives this inversion the characteristic subsidence type, as described in Fig. 5-4. The stable stratification at this inversion may be potentially unstable due to the dry air above the low-level humid layer. Thunderstorms may develop in such cases.

The combination of the upper-level and surface cold front must not be interpreted as an occluded front since this system did not develop as a combination of cold and warm fronts. The two fronts in Fig. 8-20 are both "cold" and we find only two air masses: "warm air" in the conveyor belt and "cold" (or dry) air of the polar westerlies.

The temperature difference across such a cold front is small, both at the surface and aloft, due to

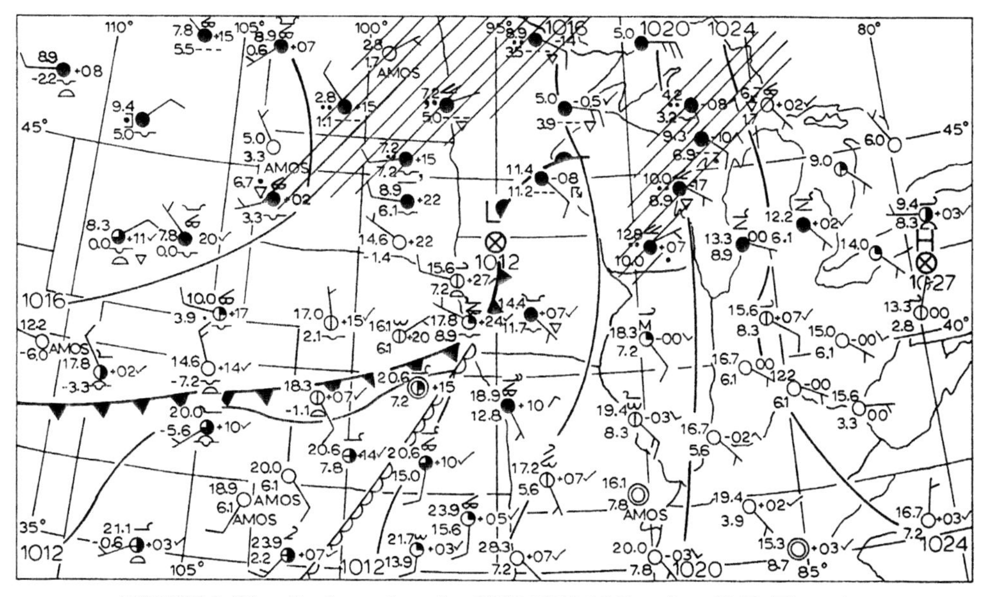

FIGURE 8-23. Surface chart for 0000 UTC 16 October 1983. The rain clouds over eastern Wisconsin are in the warm conveyor belt near the cold front at 500 mb.

subsidence in the cold air. Sometimes the temperature is higher at the surface of the surface front's cold side due to diurnal heating over the sunny surface in the cold air and due to diminished solar heating under the clouds in the conveyor belt. The designations "cold" and "warm" are used in a structural sense: The cold air is generally on the poleward side of the polar front, even if this air may be warmer locally than the adjacent warm air. Similarly, the front is designated a "cold front" when polar air advances against tropical air, even when the two air masses cannot be distinguished by temperature or when polar air is locally warmer than tropical air.

Wet-bulb potential temperature (5-19) is a practical tool for distinguishing air masses. It is individually conserved during the pseudo-adiabatic process. The fronts like the one in Fig. 8-20 are usually easier to detect using wet-bulb potential temperature than using (dry-bulb) temperature.

The Margules equation cannot be used for conclusions about the shape and slope of fronts described in this section because this situation is far from the stationary equilibrium that is normally considered in the theory of the Margules equation. When the westerlies are going in excess of 25 m s^{-1}, and sometimes with the front moving that fast, the accelerations may drastically change the stationary frontal slope. We often lack evidence concerning all the details on these fronts. Without better information, both segments of the front in Fig. 8-20 are drawn roughly vertically.

An Example of a Split Front

Figures 8-21 through 8-27 illustrate a split cold front. The 500-mb chart (Fig. 8-21) shows the lower parts of the polar jet stream, together with a baroclinic zone from Wyoming to the Great Lakes. The polar front is over Kansas, as seen by the accumulation of isotherms. The front extends northeast, but there are only few isotherms left with this front over Illinois and Wisconsin. The isotherms suggest that other discontinuities may exist east of Omaha, Nebraska (OMA) and northeast of Dayton, Ohio (DAY). The front itself is associated with a line of convective clouds that can be seen in the satellite photograph in Fig. 8-24. The 850 mb (Fig. 8-22) and surface (Fig. 8-23) charts show the humid southerly flow over Iowa well separated from the westerly flow over Nebraska and South Dakota. The cold front just passed OMA, moving from the west. The difference between this front and the one explained by the non-split model is apparent: The upper-level cold front is farther east than the front at 850 mb and at the surface. Overlapping over Iowa and Wisconsin is still more prominent. There can be little doubt that St. Cloud, Minnesota, (STC) is in the warm sector at 850 mb and at the surface, whereas it is in the cold sector at 500 mb.

The vertical section in Fig. 8-25 illustrates this situation from another angle. This section is located between Grand Island, Nebraska (LBF) and DAY.

FIGURE 8-24. Enhanced infrared satellite image for 0001 UTC 16 October 1983. Clouds in the conveyor belt from northwestern Texas to lake Michigan are along the 500-mb front.

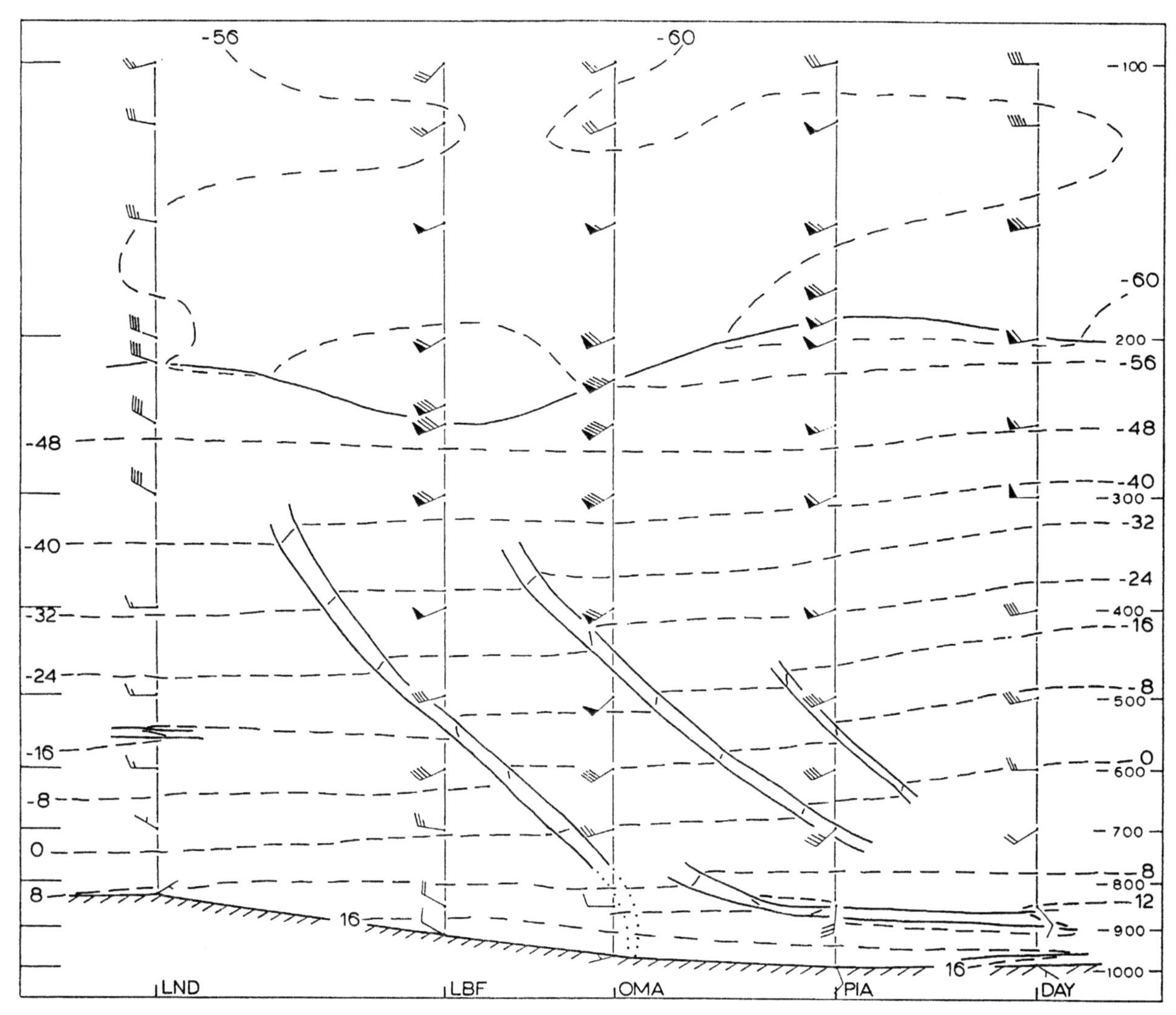

FIGURE 8-25. Zonal vertical section for 0000 UTC 16 October 1983. Several slanted inversions are candidates for the fast-moving upper-level cold front.

The section lies roughly parallel to the break in the tropopause at the polar jet stream; thus this break is not present there.

There are several presumably slanted inversions in this section. Each of these inversions can be potentially explained as the polar front. The largest thermal contrast occurs near the 540–560 mb inversion over Peoria, Illinois (PIA). Since the line of convective clouds (Fig. 8-24) is seen north of Peoria, along Lake Michigan, it is plausible to take the nearby inversion as the polar front in the middle troposphere. This is also supported by the thermal contrast at 500 mb. This justifies placing the polar front. The two other slanted inversions in Fig. 8-25 may be secondary discontinuities that developed in the westerlies within the polar air. The almost horizontal inversion near 900 mb above Peoria and Dayton is the subtropical trade-wind inversion that is normally advected far north in the warm sector of the extratropical cyclone. The space under this inversion contains the humid air of the warm conveyor belt. There is a cold front just east of Omaha. Its thermal contrast is vanishing, but there is significant contrast in wind and humidity near the surface. This is the surface position of the arctic front seen better in Fig. 8-26.

The meridional vertical section in Fig. 8-26 also shows the upper front overlapping the surface front confirming the overlapping in the Nebraska–Kansas region. There is also significant baroclinicity between Huron, South Dakota (HON) and Omaha, near the strong inversion at about 800 mb. This inversion can be interpreted as the arctic front.

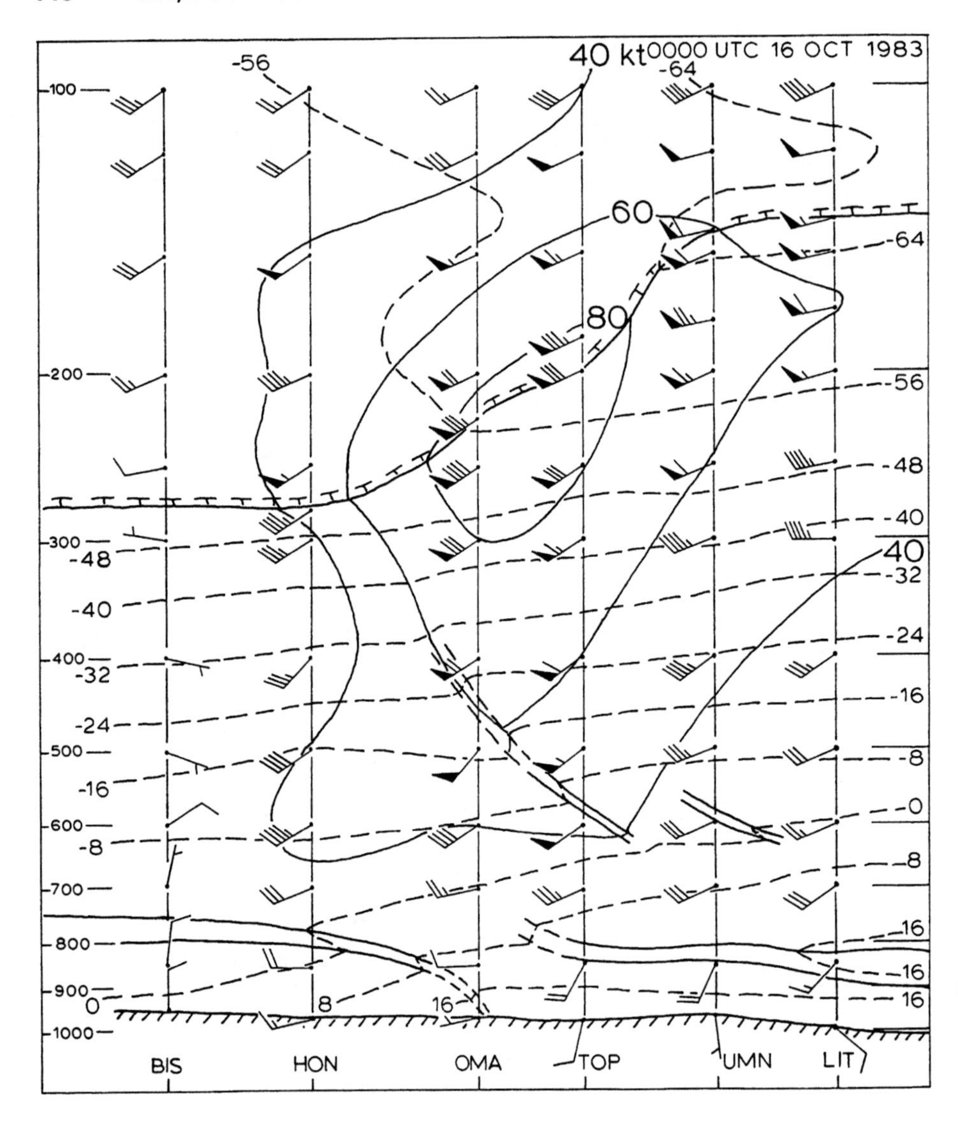

FIGURE 8-26. Meridional vertical section for 0000 UTC 16 October 1983. The polar front is developed only in the middle troposphere.

8-10 PASSIVE COLD FRONTS

Gravity Current

In absence of forcing due to the dynamics of the polar jet, polar air may spread over the earth's surface under its own weight. This process typically occurs south of the inactive region of the polar jet, where the surface front has already moved far from the jet. The position of the line A in Fig. 8-16 describes the typical large-scale situation where such a *passive front* occurs. The motion of this front is governed by the mechanism of the *gravity current* or *density current*. This current is influenced mainly by the hydrostatic pressure force ("due to gravity") resulting from the difference in density across the front. The dynamics of this situation are described in Appendix K. A short description of the passive front is given in the text with Fig. 8-15a.

Besides at the polar front, such a mechanism works at the arctic front, at the outflow boundary near thunderstorms, and in various cases in oceans and rivers. A flash flood is one of these cases.

Narrow Frontal Cloud Band

The vertical section through the passive front, as shown in Figs. 8-15a and K-1 (in Appendix K), implies concentrated convergence at the front line at the surface. This, in turn, implies a narrow zone of intense lifting at the front. Such a narrow zone of lifting often causes a formation of *narrow frontal cloud band* that is often observed on the front. A narrow cloud band can be observed in the field, especially from an elevated position, such as a tower. The clouds in this cloud band often acquire large rain drops and give a strong radar echo. Satellite images sometimes show the narrow frontal band as a *rope cloud.*

A narrow frontal cloud band does not appear on all fronts, even if convergence and lifting occur. However, Doppler radar is capable of detecting concentrated convergence on the front even in the absence of clouds. Also, an appearance of a narrow cloud band does not necessarily mean that the front is passive. Such a cloud band may also appear at other forms of a front. In any case, when it is observed, the narrow cloud band is a useful tool for locating the front.

Dynamic Pressure in Warm Air

An explanation of the movement of the passive front can be given using the normal velocity component in the warm air, as described in Appendix K. Two important cases of frontal movement should be considered.

(a) The passive front is stationary. This is due to the wind in the warm air that causes the right amount of dynamic pressure to keep the cold air from moving.

(b) The passive cold front moves against a stagnant warm air mass. A dynamic pressure

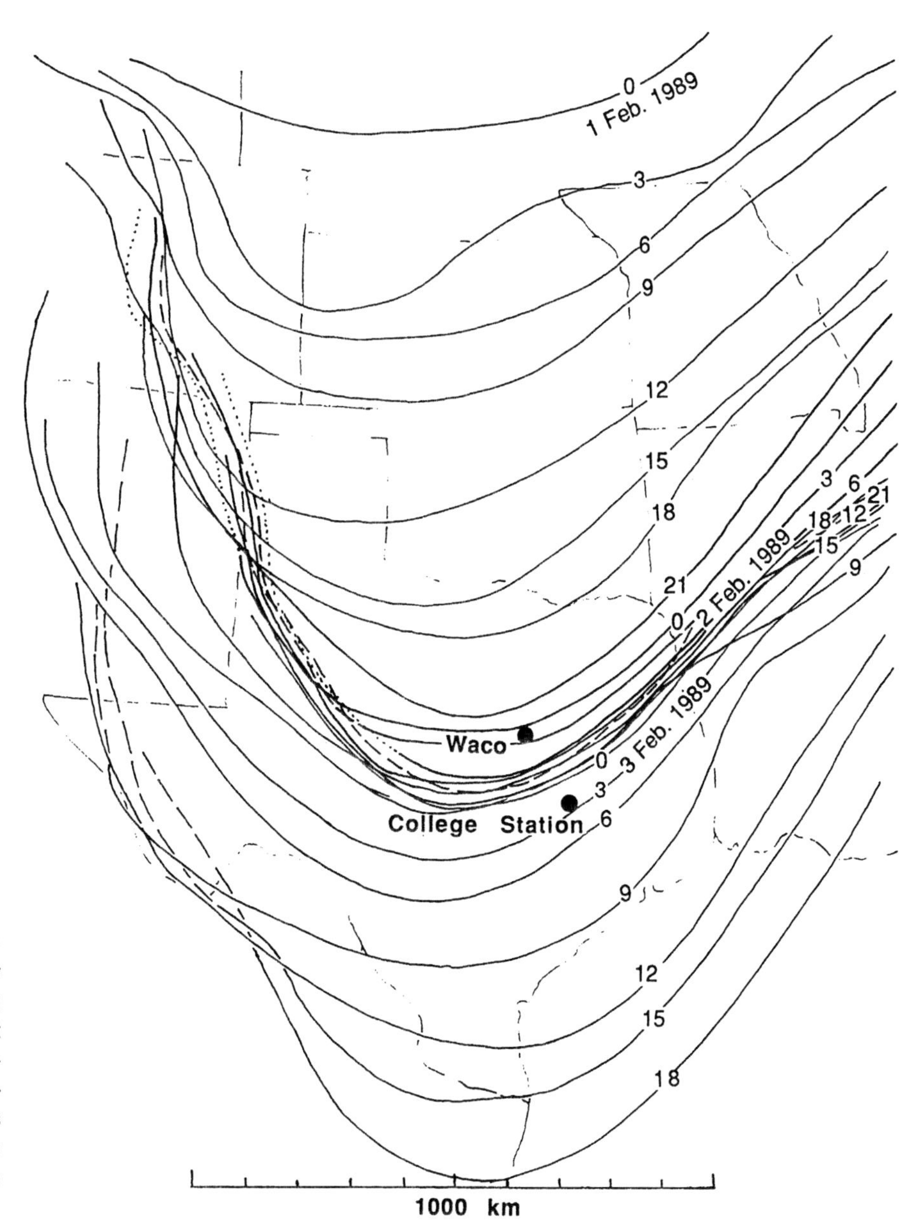

FIGURE 8-27. Positions every 3 h of the cold front of 1-3 February 1989. The hour is indicated with each frontal position. The front was stagnant in central Texas between 0000 and 2400 UTC 2 February. Parts of several isochrones are drawn dotted or dashed in an attempt to distinguish them from each other.

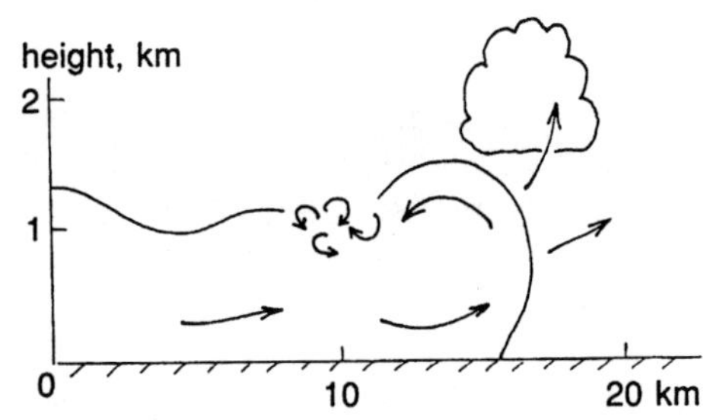

FIGURE 8-28. The shape of the moving passive front. Lifting occurs on both sides of the front. Lifting on the warm side is the cause for formation of the narrow frontal cloud band.

excess appears in the warm air, due to compression from the approaching cold front. The front may move with constant speed if a balance arises between the pressure increase in the warm air and hydrostatic pressure increase in the cold air.

Other possibilities of frontal movement are also possible. Cases of movement slower than described under (b) are often observed. In such cases, wind in the warm air blows against the front, but does not produce enough dynamic pressure to halt it. Another interesting case is when very fast warm air pushes the cold front back.

Stagnation of Fronts

Cases that cause problems in forecasting are when the speed of the passive front varies with time. This occurs when the wind speed in the warm air varies. Each time the warm air moves faster against the cold front, the front slows down. Processes in the high troposphere may account for variable speed of the warm air. Stagnations of the polar or arctic front typically last 6 to 24 h.

An example of frontal stagnation is shown in Fig. 8-27. The isochrones of a cold front show approxi-

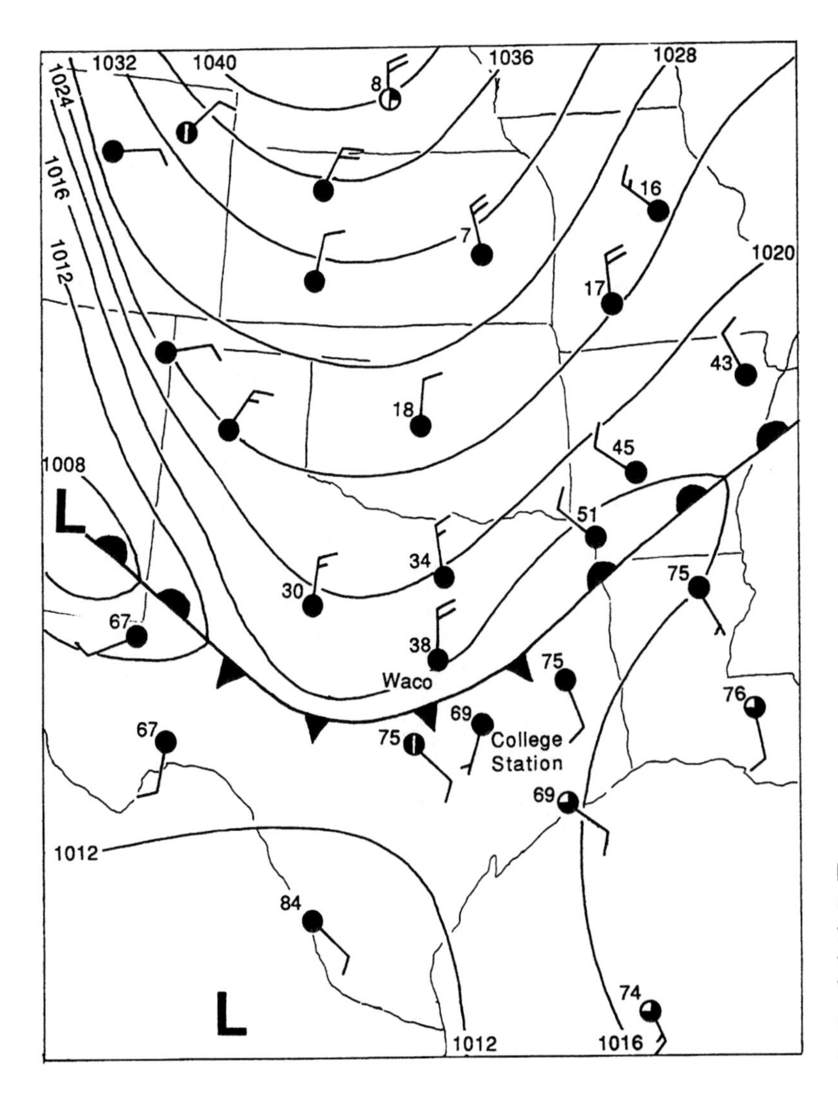

FIGURE 8-29. Sea-level isobars and selected observations of wind, temperature, and sky cover for 0000 UTC 3 February 1989. The arctic front moves approximately at the speed of cold air near the front.

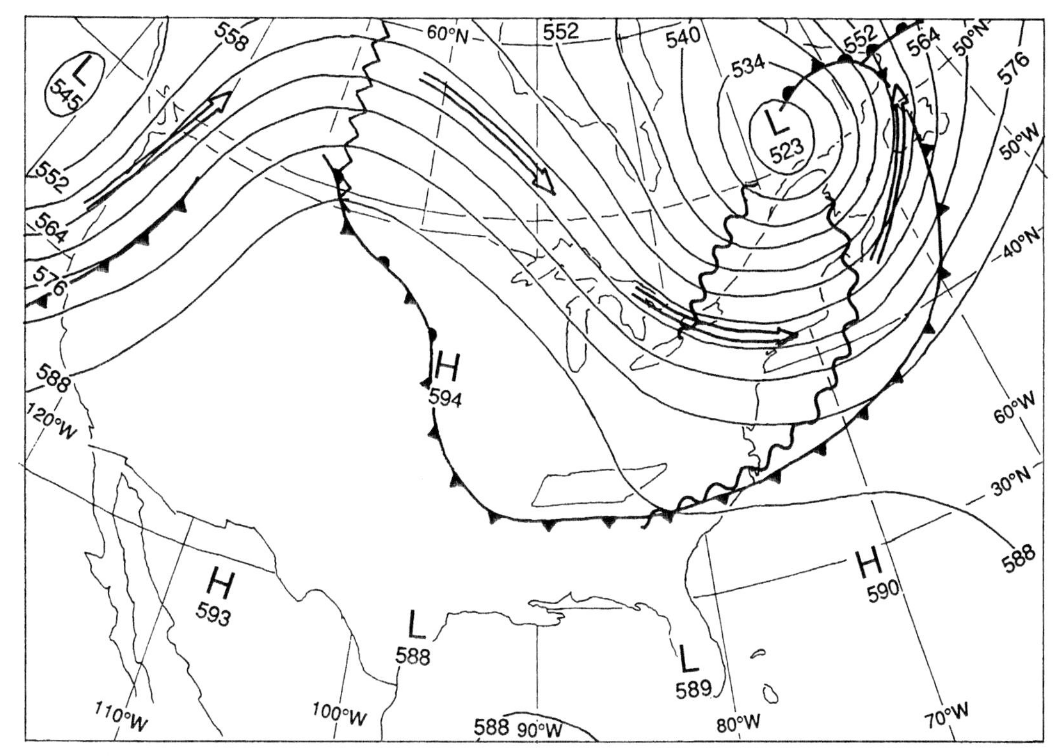

FIGURE 8-30. Contours of the 500-mb surface at 0000 UTC 18 June 1986, intervals of 60 gpm. The polar jet axis is shown by arrows; the troughs at 500 mb and the surface front are shown by conventional symbols. The front is passive in the inactive region of the westerlies upstream of the upper-level trough. The state of Tennessee is outlined, for orientation.

mately steady displacement of the front until it reached Waco, Texas, where it seemed to pause inexplicably for about 24 h. This case represented a puzzle to the forecasters, who could not anticipate comparatively small changes in the southerly wind on the warm side of the front.

Shape of the Front

The shape of the moving front is frequently different from the shape shown in Fig. K-1 (in Appendix K). Laboratory models with fluids, numerical models, and rare observations in the atmosphere show that the leading edge of the passive front shows lifting within the cold air. Then a form as in Fig. 8-28 appears. This provides for a stronger deflection of the warm air and enhances the formation of the narrow line of convergence and associated narrow frontal cloud band. The theory about the speed of the passive front in Appendix K is not detailed enough to make a distinction between the shapes of the front in Figs. K-1 and 8-28.

The shape of the front as in Fig. 8-28 can be observed when the air flow is made visible by the presence of dust or clouds. This most often occurs on the thunderstorm scale, with the cold outflow from convective cells.

Wind often blows at nearly a right angle to the isobars in the cold air near the passive cold front. This is illustrated on the sea-level chart for 0000 UTC 3 February 1989 (Fig. 8-29), which shows the front from Fig. 8-27. This is the time when the front resumed motion after it had been stagnant for 24 h. The flow in the cold (arctic) air goes nearly normal to the isobars. The friction at the ground is very strong compared with the Coriolis force. Strong friction (large C) in (7-4) implies a large cross-isobaric flow.

An Example of a Passive Front

An example of a passive front is shown in Figs. 8-30 to 8-32. These figures illustrate a weather situation that is typical for summer. The polar jet is farther north than in other seasons. The jet core is shown by arrows on the 500-mb chart in Fig. 8-30. The two high centers, one in eastern Kansas, the other over the Atlantic,

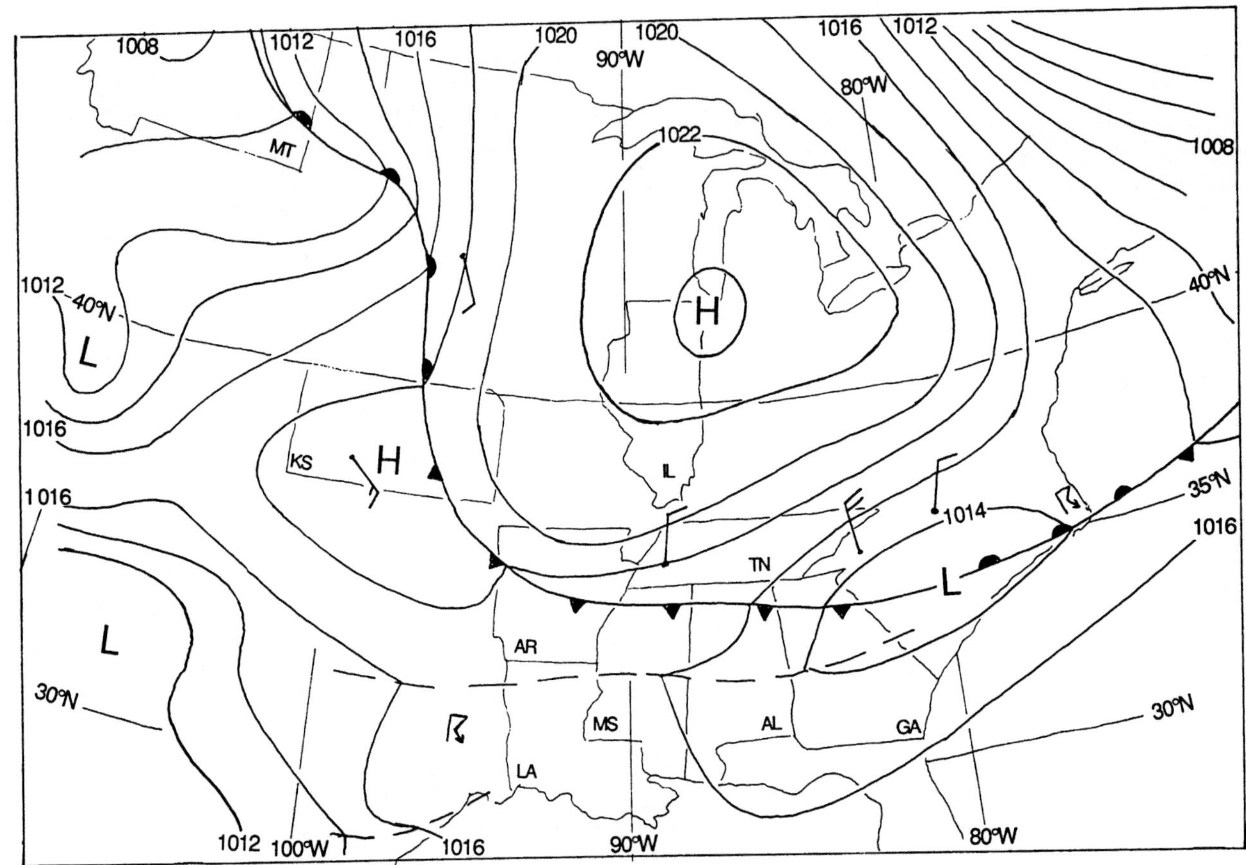

FIGURE 8-31. Sea-level isobars, intervals of 2 mb, the surface front, and troughs (dashed) at 0000 UTC 18 June 1986. Several wind and thunderstorm observations are shown. The passive part of the front extends from Montana to northern Georgia. The same front should be designated as active over the Atlantic since that part is in the active region of the westerlies. A warm-sector trough has developed at a line of thunderstorms in central Mississippi and Alabama. The states mentioned in the text are outlined.

belong to the warm (dynamic) anticyclones of the subtropical belt. Easterly flow that is typical for the tropics can be seen south of these highs.

The polar front at the surface is in its usual position in relation to the polar jet. The front at the surface is closer to the jet axis in the active region east of the upper-level trough. The front is farther away from the jet axis in the inactive region, upstream from the upper-level trough. The anticyclone at the surface, with the center in northeastern Illinois, is a cold high, consisting of polar air. This high does not extend through deep layers of the troposphere. This cold high cannot be found at 500 mb.

The far outreach of the polar air is delineated by the surface front from Montana to the Atlantic. There is not much difference in temperature across the front. The late afternoon reports show the temperature around 24–29°C on both sides of the front. Only the dew point is markedly different: The dew point is about 24°C in the warm air and about 14°C in the cold air. Convergence and associated vertical motion at this front were not large, but they proved sufficient to trigger the development of thunderstorms by release of potential instability. Tops of thunderstorms can be seen as bright spots on the satellite image in Fig. 8-32. The clouds that can be seen in eastern Tennessee, northern Georgia, northern Alabama, and northern Mississippi have developed on the passive front. The clouds in Alabama and Georgia cast a shade on their eastern and northern edges. More clouds can be seen along the same front in Arkansas, but these are not yet developed like the ones in Alabama and Tennessee.

Mesoscale Line of Thunderstorms

There are several more groups of thunderstorms in the same example. The most prominent are the ones in central Mississippi and Alabama and in Louisiana, extending to southern Alabama. The storms in central Mississippi and Alabama have developed along a

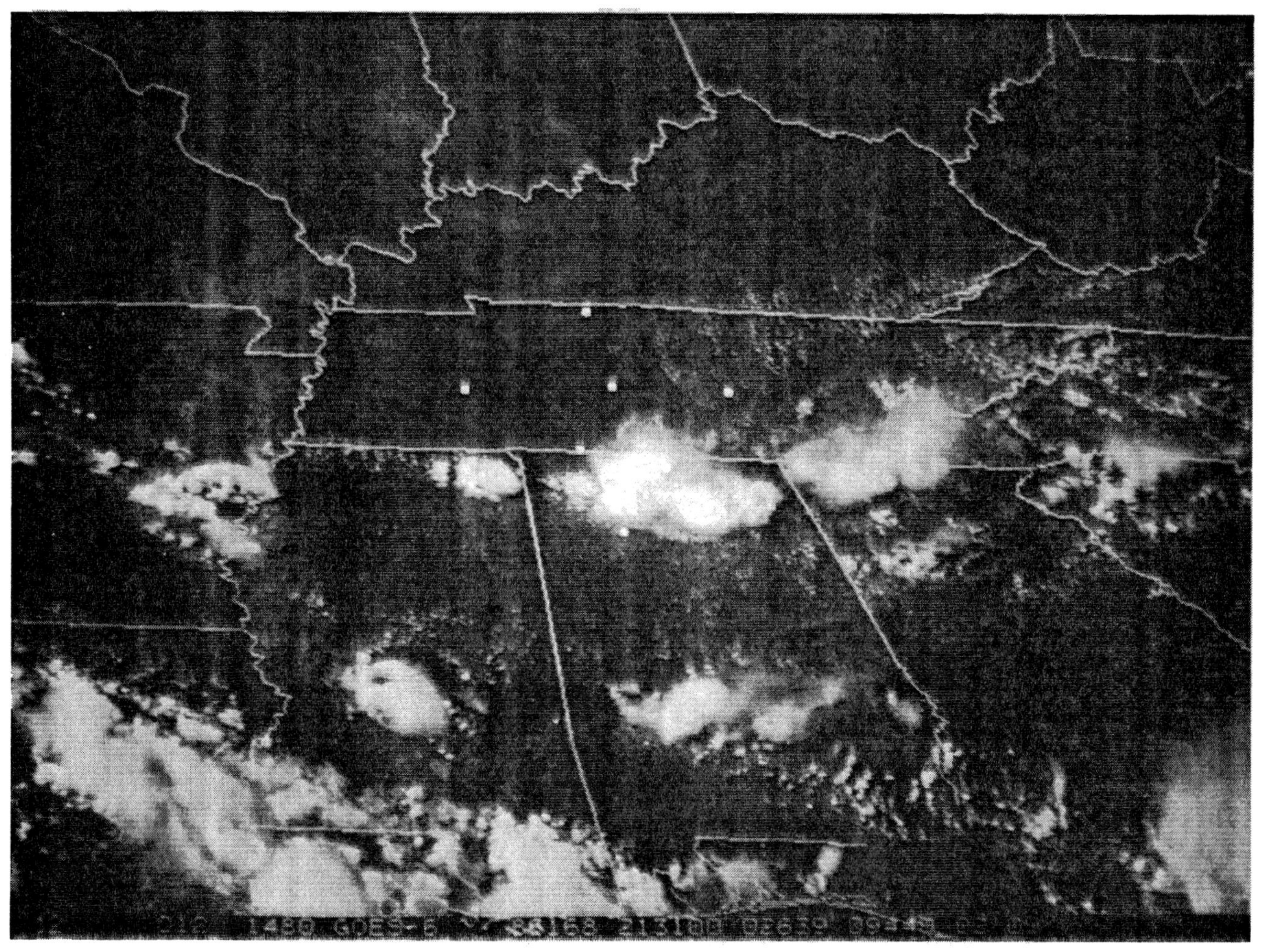

FIGURE 8-32. Satellite photograph (visible) from 2131 UTC 17 June 1986, resolution of 2 km. Thunderstorm tops are situated in three groups: (a) along the passive front from northern Mississippi to eastern Tennessee, (b) along a trough in central Mississippi and Alabama, and (c) in Louisiana and southern Alabama.

trough in the sea-level pressure field. This is a trough along an outflow boundary of a group of storms that preceded the present group of storms.

Breeze Front

The thunderstorms in Louisiana and along the Gulf coast are somewhat different from those in central Mississippi and Alabama. A connection of these storms with wind or pressure signatures cannot be concluded from the weather chart of this hour. These thunderstorms in Louisiana started their development at the breeze front along the coast earlier in the day. Toward evening, these storms moved 100–200 km deeper inland and spread over a large area where the potential instability was favorable. The breeze front is a gravity current just like the passive front or thunderstorm outflow; however, it often can be overseen on a surface chart of this scale.

The existence of several groups of storms (mesoscale convective systems) indicates that the atmosphere is potentially unstable. The high dew point and the outburst of storms show that the supply of water vapor was adequate for storm development. Besides potential instability and an abundance of water vapor, a lifting mechanism is needed to start storm development (see Chapter 12). If the lifting mechanism can be identified, the position of the passive front may point out the exact location of storm development. This shows the importance of careful analysis of the passive front. The narrow line of lifting and clouds on this front may be sufficient to stimulate development of thunderstorms.

9

Air Masses and Weather

The concept of air masses is very useful for describing and explaining the differences in weather over geographical regions and over time. Therefore, the practice of weather analysis is very much based on the distinction of physical properties of air. Traditional air-mass analysis has been based on synoptic weather charts since the first weather services started routine work in the middle of the nineteenth century. Due to technical restrictions, the synoptic chart of that time contained only surface data. The analysis of the surface chart was greatly enhanced in the second quarter of the twentieth century when the polar front became widely accepted in weather analysis. The surface chart still represents the main tool for weather depiction, since it normally contains most of the observations. Users of weather information outside the meteorological profession also benefit from the weather data on the surface chart. Therefore, this chapter is dedicated to describing the surface chart.

9-1 AIR MASSES

Discovery of Fronts

Until the early twentieth century, study of atmospheric processes was restricted to weather and climate elements and a few dynamical structures such as clouds, cyclones, and anticyclones. Considerable progress was made when fronts, air masses, and jet streams were identified. These newer structures could be used to systematize most of the tropospheric processes.

There was a widespread reluctance to accept the existence of fronts in the atmosphere, just as all radically new ideas encounter resistance. However, by the mid-1920s, the polar front was recognized in many positions around the globe and the concept gained acceptance.

The discovery of the polar front by Jacob Bjerknes and Halvor Solberg around 1920 was one of the greatest achievements in science, both for its practical merits and for the introduction of radically different ideas. After this discovery, it became easier to accept other discoveries and ideas that came along.

The polar front appeared in meteorology at the time in history when discontinuous (quantum) physics became first known. Thus the discontinuities in the atmosphere became a meteorological counterpart of newly discovered discontinuities in physics.

Polar and Tropical Air

Each segment of the polar front stretches over several thousand kilometers. Thus it is understandable that the polar front became the first well-described front in the atmosphere. The existence of fronts logically leads to the concept of air masses, since one role of fronts is to separate air masses.

The polar front makes it possible to distinguish between air masses by serving as a boundary between air masses of different properties. In this way, the first distinction was made between polar and tropical air masses. These names were intended to point to the polar and tropical *sides* of the polar front, and not to associate air masses with polar and tropical regions. Both these air masses are normally found in the middle latitudes.

Soon after the identification of the polar front, secondary discontinuities were found within polar air. So the separate arctic front and arctic air were recognized. The arctic front is a shallow, widespread, polar-night inversion. It does not extend vertically to the levels of the polar jet stream.

With widespread radiosonde observations in the 1940s and 1950s, the polar jet stream was discovered, together with its relation to the polar front.

Middle-Latitude Air

The tropical air mass was so often observed away from the tropics that there were attempts to change its name to the "air of middle latitudes." These attempts became practical after the polar jet, the subtropical jet, and the subtropical front were discovered as permanent features in the atmosphere (around 1950). Then it became possible to separate *middle-latitude air* from tropical air.

One difficulty with the subtropical front is that it does not extend to the earth's surface. Therefore, it is not possible to separate tropical air from middle-latitude air at the earth's surface precisely. This distinction can be made only on upper levels, where the subtropical front can be identified. It is still common to use the name *tropical air* for the air mass that normally occupies the warm sector of the extratropical cyclone.

Mesoscale Fronts

Although the attention of weather analysts is usually turned to the large-scale fronts, the natural tendency of the atmosphere to form discontinuities on all scales must be kept in mind. The fronts (discontinuities) that appear on the scales between 10 and 1000 km can often be recognized on weather charts. Bodies of air form with different physical properties and moving in different directions around these fronts. Both meso- and large-scale fronts often stimulate the formation of thunderstorms. Intersections of two fronts are particularly favorable for new thunderstorm development.

A difficulty with analyzing mesoscale processes is that they usually last only several hours, in great contrast to the polar front. A segment of the polar front can usually be followed for a week before it disappears. Thunderstorms, coastal breezes, and similar phenomena seldom last longer than 6 h and may not be present on two successive synoptic charts. This makes them difficult to recognize and study. Satellite images greatly alleviate this problem, since thunderstorms and other phenomena with clouds can be directly observed.

Transformation of Air Masses

Air masses form when the air resides in a geographic region for a long time, about 1 week. The air acquires the properties of the underlying surface: heat and humidity. The physical properties of air masses slowly change when the air moves to the regions normally occupied by other air masses. An air mass transforms by acquiring new properties from the underlying surface.

The exchange of air masses between various latitudes provides the transport of heat and water vapor from the tropical to the polar regions. Mixing between latitudes is an essential link in the transformation of heat into kinetic energy. Thus mixing of air masses is a part of the thermodynamic process that drives the general atmospheric circulation. If the exchange of air masses between polar and tropical regions were prevented, the general circulation would be greatly altered.

Mixing of air masses renders the distinction between air masses very difficult in some cases. For instance, it is not always feasible to determine when a polar air mass has been transformed enough to be called "tropical." The best advice is to use the designations of air masses only when there is some distinguishing sign. The best distinction between air masses can be made when a front develops between them. Then the air mass on the polar side is polar air. The other air mass may be middle-latitude or tropical air. If this other mass is hot (for the season) and humid, it may be tropical air. Generally, we should refrain from classifying air masses in the absence of a front.

The transition between air masses becomes gradual when the fronts dissipate, showing no clear signs where one air mass stops and the other begins.

When polar air spreads toward the low latitudes, it transforms into tropical air at the earth's surface. The transformation of tropical (or middle-latitude) air into polar air occurs in the cyclones of middle and higher latitudes where warm air is lifted to the polar jet level. This lifting mechanism, together with the precipitation patterns associated with fronts, is described in Chapter 10.

9-2 CLASSIFICATION OF AIR MASSES

Structural Classification

The earliest criteria for classifying air masses were structural: Polar and tropical air masses appeared on the corresponding sides of the polar front. Also, it was noticed that these air masses enter the cyclone from opposite sides. In this way, the structural classification was based on the existence of fronts and on circulation of air in extratropical cyclones.

Physical Classification

The roots of the physical description of air masses can be found in the nineteenth century when the adiabatic process was discovered. Then it was possible to distinguish stable and unstable air masses. Physical classification was greatly developed after the discovery of the polar front. Then the designations *warm*, *cold*, *stable*, *unstable*, *humid*, and *dry* were introduced. Although these designations are objective in the sense that they depend on direct weather observations, objectivity of this classification is difficult to defend when distinguishing between air masses. The physical properties may assume all intermediate values, thus rendering the classification very difficult.

The expressions "cold" and "warm" are used in two different meanings: (a) colder or warmer than the underlying surface or (b) on the cold or on the warm side of a front. There is no "cool" air mass in the classification. Of course, the weather may be "cool," if this is the subjective judgment of the weathercaster, but this is not related to the classification of air masses. When referring to colder or warmer than the underlying surface, cold very often means that the air mass is unstable. Similarly, warm means that the air mass is stable.

Geographical Classification

By working with weather charts, meteorologists became aware that air masses assume their physical characteristics from the underlying surface. Thus it became customary to name air masses by their geographical origin. This geographical classification yields generic names: *continental*, *maritime*, and specific names for various regions of the world such as: *Pacific*, *Canadian*, *Arctic*, and *Siberian*. The names of main air masses of the contemporary structural classification (polar, middle-latitude, and tropical) also have geographical origins.

The geographical classification is also sometimes ambiguous. For example, the categories of continental and maritime can be used (a) to designate the air mass as dry or humid, or (b) to refer to the origin of the air mass even if the humidity had already changed within the air mass. In the first sense, a maritime air mass is the same as a humid air mass. However, sometimes we keep the same designation for a day or two after the air mass has already lost its water vapor, as when it crossed a mountain range. In this second case, the designation maritime refers to the origin of the air mass more than to its humidity. For example, there is no agreement on how long designations such as Pacific should be kept with the air mass that travels over the Rocky Mountains. One choice is to keep the name as long as we still can attribute some property of weather to the Pacific origin of the air mass. That property may be a higher temperature than in arctic air that may be in the vicinity.

Contemporary Structural Classification

The structural classification became widespread during the 1950s when radiosonde observations became available for construction of weather charts and when we became aware of the permanent nature of the subtropical jet and subtropical front. The primary classification of air masses and fronts is outlined in Sec. 1-3 and is explained on the basis of Fig. 1-3. According to the structural classification, we have three main air masses: polar, middle-latitude, and tropical. These masses are on the appropriate sides of the polar and subtropical fronts. Since the arctic and equatorial fronts are frequent phenomena in their seasons, we recognize arctic and equatorial air masses, but we keep them formally as subdivisions in the superjacent polar and tropical air masses.

The structural classification does not obliterate the physical and geographical classifications. Each of the structurally classified main air masses can have physical and geographical properties that come from other classifications. This is a matter of terminology: We name air masses structurally and give them physical and geographical attributes. Also, it is common to drop the word *mass* from the names of air masses; for example, maritime polar air, Canadian arctic air, Mexican continental tropical air, stable maritime middle-latitude air, and so forth are used.

Classifying air masses is not always possible. When the front between air masses disappears, a clear distinction between air masses disappears as well. This is regularly the case with the subtropical front.

This front usually does not reach the lower half of the troposphere. Therefore, middle-latitude air cannot be easily distinguished from tropical air near the earth's surface. A distinction between tropical and middle-latitude air at the surface can normally be found only over great distances, and this with a gradual transition from one air mass to the other.

9-3 IDENTIFICATION OF LARGE-SCALE WEATHER PATTERNS ON A SURFACE CHART

The word *surface* in the name of a chart means that this chart contains observations made from the earth's surface, without airborne instruments. Since this chart depicts several phenomena in the free atmosphere, Tor Bergeron appropriately called it "the composite chart" (Godske and others, 1957, p. 620).

Among all the charts covering an area, a surface chart normally contains the most information. The number of stations that report "surface" observations is greater than the number of standard sounding stations. Surface observations also contain a greater number of elements than upper-level observations.

Weather Symbols

Weather systems on a surface chart are represented using the international symbols listed in Table 9-1. The tips on the monochromatic fronts point in the direction of motion. The stationary front (the front that does not move) has sharp tips on the warm side and rounded tips on the cold side. When an area is covered by symbols or shading, these symbols or shading apply to air masses or geographical areas where a physical process occurs. Therefore, the symbols should not be used only at the places where the corresponding weather phenomena are observed; instead, they should cover the region where the same weather type exists. During the analysis of weather charts, observations at neighboring stations should be compared to determine the extent of weather systems. Knowing the typical properties of atmospheric processes is of great help in this process.

In weather analysis, we are looking for areas with a weather type. We do not mechanically enhance the plotted reports. If the process of analysis consisted only of enhancing the plotted symbols, the Weather Service could save money by employing unskilled workers at minimum wage, giving them green and red pencils, and letting them enhance the charts. Unfortunately, some make-shift "meteorologists" use this quality of work.

Computers can be programmed to draw lines and to shade surfaces. However, the automation of weather analysis is a long way in the future, if it can ever be achieved. We should not confuse drawing lines and shading with weather analysis. Drawing lines and shading are mechanical actions that even the unintelligent computers can perform. Analysis, though, consists of identifying of physical processes in the atmosphere. Trained meteorologists are expected to be much better than computers in the process of analysis.

Satellite images and radar echo charts are very useful tools for supplementing the analysis of the surface chart and should always be used, when possible. Outlines of clouds and rain areas can be worked out with the help of satellite images and radar charts, time permitting.

The symbols used for plotting weather charts are shown in Appendix E. We should attempt to use the standard symbols whenever possible to facilitates communication between people.

Steps in Analysis

The following nine steps, largely based on the system introduced by Bergeron (Godske and others, 1957) and later elaborated by F. Defant and H. T. Mörth (1978), can be used to analyze a surface chart. Steps 3 and 8 here are different from the steps in previous books.

Step 1. Identifying Thermally Stable and Unstable Air Masses

Stable air masses are warmer than the underlying surface and unstable masses are colder. Instability is characterized by showers, by the clouds of Cu, Cu cong., and Cb, and by thunder and lightning. When an area of showers is detected, it should be uniformly covered with the shower symbols. Preferably the symbols should be drawn in the blank spaces between stations, so that the data can be easily read. It may be stressed again that we are looking for areas of showers, not for individual observations. We use station data to discover those areas. In this procedure, we should not forget that a report of a thunderstorm means that showers are in the area, since thunder occurs only in convective rain. Consequently, an area of thunderstorms should contain uniformly spaced symbols for showers in addition to the symbols for thunderstorms. A shower is sometimes reported as steady rain, since the observers may be unsure about the precipitation type. Clouds should be noticed when there is doubt whether a report means a shower or steady rain. If rain is reported with broken cloudiness, it must not be interpreted as continuous rain; it is a shower or

TABLE 9-1

Suggested symbols for use on weather charts

Feature	*Monochromatic*	*Polychromatic*
Cold front		Blue line
Cold front above the surface		Blue dashed line
Warm front		Red line
Warm front above the surface		Red dashed line
Occluded front		Purple line
Stationary front		Alternating red and blue line
Tropopause		Same pattern, blue
Shear line		Dash-dotted line
Line of instability		Same as monochromatic
Intertropical convergence zone		Same as monochromatic
Intertropical discontinuity		Dashed, alternating red and green
Trough line		Same as monochromatic
Ridge line		Same as monochromatic
Zone of continuous precipitation	Shading or cross hatching	Green shading or crosshatching
Zone of intermittent precipitation	Hatching	Green hatching
Larger snow or drizzle symbols should be distributed over the area in both preceding zones, if these forms of precipitation occur. Absence of these symbols in green-shaded regions show rain.		
Area of showers	Large shower symbols, uniformly distributed	Same as monochromatic, but green
Area of thunderstorms	Large thunderstorm symbols, uniformly distributed	Same as monochromatic, but red
Area of fog	Large fog symbols	Yellow shading
Area of dust storm, sandstorm, or dust haze	Appropriate large symbols, uniformly distributed	Brown shading with appropriate symbols distributed

intermittent rain. Showers in the vicinity of the station, plotted with the symbols)·(and (·), are the most reliable reports of showers. A rainbow is an excellent indicator of showers, but unfortunately this phenomenon is not provided in the meteorological code.

Lightning also indicates showers; however, at night lightning may be seen far away, up to 300 km, and does not necessarily represent showers in the vicinity. In daylight, lightning may be interpreted as showers in the vicinity of the station. Thunder normally cannot be heard more than about 15 km from the station, whereas even in daylight lightning can be spotted farther away. Reports of showers or rain in the past hour or the past 3 h usually mean that there are still showers in the area.

Precipitation symbols are green, if colors are available. Electrical phenomena are marked with red symbols.

Fog signifies a stable air mass, and yellow shading is standard for it. However, observers often report fog together with showers. This type of fog comes from evaporated rain and some condensation near showers or over wet ground. Such fog should not be indicated on a chart; shower symbols suffice. Hydrostatic instability is the dominant property of an air mass with showers.

Drizzle is also typical of stable air. Drizzle is a precipitation that starts in heavy fog, when fog drops become large. Therefore, the best indicator of drizzle is fog in the area. Observers often do not distinguish weak rain from drizzle. A weather analyst has a different opportunity to distinguish between these precipitation forms, since he or she can compare each observation with other stations in the vicinity. In a region with fog and drizzle, reports of light rain may be often interpreted as drizzle. On the other hand, a report of drizzle may mean weak rain and, in rare cases, even showers.

When two or more weather types are observed and reported, the one with the highest number in the *ww* part of the International Meteorological Code should be given the most credibility. The code numbers have been carefully arranged with this purpose in mind. For instance, when rain and fog are observed, rain is normally the more important element and should be indicated on the surface chart.

Step 2. Physical Properties of Air Masses

The origin of an air mass can be determined before the fronts are detected on the surface chart. Using the knowledge of climatology, we can get a good idea about the physical properties of air masses (warm–cold, humid–dry, stable–unstable) and about their geographical origin. The likely names of air masses (abbreviated) should be recorded on a chart only in regions where there is little doubt. Final naming of air masses will come later, when the fronts have been identified. Since upper-level charts should have been analyzed already, we have an approximate idea about air masses by the structure of the atmosphere. Jets and tropopauses are good indicators about air masses in the upper half of the troposphere.

Step 3. Outlining Areas with Continuous Precipitation

Areas with showers have been identified under step 1. Continuous rain (and snow) can be recognized by layered clouds and overcast sky. Green continuous shading, or cross hatching on black-and-white charts, is applied to this region. A continuous cloud cover likely indicates an area of stable upgliding; this area should be outlined with a solid green line. A continuous precipitation area should be shaded uniformly green. Shading, like all other symbols, should be clearly visible, but it should not be so dark that plotted data cannot be read.

Sometimes, reports from stations do not reveal continuous precipitation. If rain is light, the observers may report intermittent rain or drizzle. In all cases, we have to judge the reports by the characteristics of the air mass and by comparing reports from neighboring stations.

Bands of precipitation are frequent in cyclones, both tropical and extratropical. Therefore, the shape of a precipitation area is usually elongated along some front or along the conveyor belt. In doubtful cases, due to sparse observations, elongated areas of precipitation should be preferred. It is more likely that a precipitation area is elongated than compact.

Step 4. Detecting Pressure Variation

In higher latitudes, the movement of highs and lows can be followed conveniently with isallobars. Pressure tendency is reported in units of 0.1 mb $(3\ h)^{-1}$. Isallobars are drawn in units of 2 mb $(3\ h)^{-1}$. Intermediate values of every 1 mb $(3\ h)^{-1}$ are drawn in regions where more detail is needed. Isallobars should be drawn only in regions where values larger than 2 mb $(3\ h)^{-1}$ (or smaller than -2 these units) appear. A zero isallobar should be drawn between $+2$ and -2 isallobars, but this isallobar should not be extended in regions where all values are small. In regions of small values of pressure tendency, less than ± 2 mb $(3h)^{-1}$, are numerous processes of smaller scale that cannot be individually analyzed. Such processes (mostly showers) are marked on synoptic charts only as a group ("region of showers"). Since we cannot identify individual showers, we should also not draw corresponding isallobars.

The standard style for isallobars is dashed: red for pressure fall and blue for pressure rise. The zero isallobar is black. The centers of pressure fall are indicated with a red letter F and the number of millibars (rounded to the nearest millibar per 3 h) of the estimated tendency in the center, even if there is no observation exactly in this center. For example, F4 means "pressure fall of 4 mb $(3\ h)^{-1}$." Similarly, a blue R3 is for $+3$ mb $(3\ h)^{-1}$ for pressure rise.

Step 5. Locating Pressure Patterns

Most work on a surface chart is invested on drawing isobars. If the speed of drawing is important, a computer is used. A graphite pencil (not a black color-pencil) is normally used, because this pencil erases easily. The standard isobar intervals are 5, 4, or 2 mb, but in each case the 1000-mb isobar should be included in the set. Wind direction and speed should be observed in this step, since wind tends to be in geostrophic or frictional balance. The wind speed on the land surface may reach only 50% to 90% of the geostrophic speed due to friction. The direction will often show cross-isobaric flow coming from the high side of the isobars. Over the oceans, wind is usually closer to geostrophic wind than over land. There is often significant deviation from geostrophic wind near thunderstorms within the cold outflow.

Step 6. Identifying Fronts

Fronts on the synoptic scale appear between air masses when there is frontogenesis. When frontolysis (negative frontogenesis) appears, the fronts dissipate. Therefore, we should expect that often there is no front between air masses and that the transition between air masses may be gradual.

The first indication for the position of the polar front at the earth's surface comes from the thickness chart between some upper-level (often 500 mb) and the 1000-mb geopotential. We are searching for the front on the equatorial side of the frontal zone on a thickness chart. This point was discussed in more detail in Chapter 6.

Another important criterion for locating the polar front on a surface chart is the location of this front on the 850-mb chart, and possibly on higher-level charts. Smaller-scale discontinuities (squall lines, accumulation of cold air in mountain valleys) usually do not appear on upper-level charts. Therefore, if the polar front can be detected on the 850-mb chart, it should be detected on the surface as well.

Density differences indicating a frontal discontinuity are detected by differences in temperature. This difference, however, may be obscured by local differences in cloudiness and heating. If a density difference cannot be detected across the front, there may still be such a difference in the atmosphere just above the surface. In such cases, a pressure trough and cyclonic wind shear will reveal the position of the front. Discontinuous wind shear is always cyclonic on a front (anticyclonic discontinuity is very unstable and cannot persist in the atmosphere). Often a contrast in dew point will show the front better than the contrast in temperature.

Another good procedure that helps in detecting fronts is temporal continuity from the previous synoptic maps. It is a good idea to draw the front from the chart 6 h before (or 3 h before, if available) using a ballpoint pen. A thin, inky line does not obscure the data much on the current chart, but helps to locate the front. An inky line is acceptable since we seldom reanalyze old charts. There is little probability that we would like to change the position of the front on the previous chart.

At this stage it is good to draw the front in graphite pencil, before the type of the front is determined. A careful analyst will not confuse this line with an isobar.

Step 7. Determining the Front Type

Establishing the movement of a front when a previous chart is not available, or when a front was not properly placed on that previous chart, is not always easy. In such cases, we use the direction of the geostrophic wind at the front. If the isobars are coming from the cold side (along the geostrophic wind), this is a cold front. If the isobars are coming from the warm side, this is a warm front. Somewhere near the col in the isobars will be a point of transition of the cold front to the warm front. A front should be designated as stationary only in rare cases when the isobars do not cross the front and the front has not been moving for the last 6 h.

An occluded front consists of a surface front and an upper-level front: One is a warm front, the other cold. The upper-level front moves over dense (cold) air at the surface. It is very difficult to find such pairs of upper-level and surface fronts on weather charts. For this reason it is sufficient to draw only one occluded front in the position of the surface front.

The temperature difference across the front does not always determine the front type. Cold air, advancing through warm subtropical regions, may come with a local increase of temperature. For example, in the afternoon hours there may be bright sunshine behind the cold front and it may be cloudy and even rainy on the warm side of this front. Then the air may be warmer behind a cold front than ahead of this front. Other elements and upper-level charts still may show that this is a cold front.

By placing the fronts, air masses are identified. On a surface chart we can easily identify arctic and polar air, since these are normally separated by the arctic front. While middle-latitude air can be distinguished from polar air easily, this air cannot be easily distinguished from the tropical air because the subtropical front does not appear at the earth's surface.

Equatorial air can be identified on the equatorial side of the monsoon front (or the intertropical front). In the interior of each air mass, a label should be placed: A for arctic air, P for polar air, ML for middle-latitude air, T for tropical air, and E for equatorial air. A lowercase letter c or m is commonly used for continental or maritime air, if the distinction can be made. This letter is written as a prefix. Thus we have cP for continental polar air, mT for maritime tropical, and so forth.

Step 8. Outlining Main Air Currents

Principal air streams are important for the development of weather systems. Therefore, it is useful to introduce large arrows that point along the principal streamlines where air masses move the fastest. Polar outbreaks and low-level jet streams are the most frequent currents that should be identified on a surface chart. Colors are commonly used for the air currents: blue for cold air, red for warm. Coloring should stand out clearly, but without obscuring the plotted data.

Step 9. Finishing Touches

All elements of the analysis are now brought into agreement with each other. The fronts are colored in the final position, kinks of isobars are adjusted on fronts, spacing and direction of isobars are corrected, and arrows of the air currents are colored and enhanced. Last corrections of shading are now made.

Comments on the Steps in Analysis

The order of executing the preceding nine steps of analysis need not be strict, since it is not possible to recognize later whether the isobars or green shading were done first.

The purpose of analysis is to detect and explain the physical processes in the atmosphere. Therefore, it is not advisable to work only on one chart until it is "finished." Other levels and other available data (soundings, station reports, satellite images, and so forth) should be consulted throughout the analysis process.

The standards in symbols are needed so that the communication between meteorologists and between meteorologists and their clients can be effective.

An Example of Surface Analysis

A classical example of the surface chart is shown in Fig. 9-1. Originating from a famous 1924 paper by T. Bergeron and G. Swoboda that has been rewritten for the book by Godske and others (1957), this chart depicts a number of weather processes and illustrates standard symbols in weather analysis. Besides having had a great value for the development of our concepts of weather models, the original paper used many symbols that have since been accepted as international standards. As for the data plotted in Fig. 9-1, it is of interest to know that in 1923 weather reports did not contain anything other than pressure, temperature, and wind. It is hard to imagine today that meteorologists could be serious about weather without reports of precipitation, humidity, clouds, and so forth. Bergeron and Swoboda collected these data by writing letters to all weather stations. At that time there were no strict standards; therefore, different stations observed different weather elements. Also, not all stations made observations at the same time. Bergeron and Swoboda's paper played an important role as an argument to introduce a unified system of observation and reporting.

The chart in Fig. 9-1 is of 0070 and 0080 UTC 13 October 1923, covering parts of central and western Europe, with Great Britain on the western side, Italy in the south, and part of the Scandinavian Peninsula in the north. The isobars are drawn every 5 mb, an international standard. Dotted lines are used for mountain isobars and for troughs in the pressure field over Great Britain and Denmark. The isallobars are dashed lines, every 1 mb $(3\ h)^{-1}$. Rain areas are of two mutually excluding types: showers and continuous rain. In color, both types are represented in green: triangles for showers and shading for continuous rain. Black-and-white crosshatching is used in Fig. 9-1 instead of light green shading. Showers can be recognized in the observations by their essentially intermittent character. Some stations do not have rain, but show rain during the past 1 or 3 h. These reports are interpreted as "showers in the vicinity", thus these reports reveal the region of showers. The region of showers is extended over the seas where there are no direct observations. This is done by extrapolation of observations at coastal stations and by assuming that the cold air is heated from the sea, thereby becoming unstable.

Shower symbols (triangles) on this chart are drawn without the dot, following the original Bergeron's style. The region of fog is represented with dashed zonal lines, since large fog symbols in black and white obscure the data. Yellow symbols are preferable for fog. Yellow fog symbols do not obscure the other symbols much.

The weather in Fig. 9-1 shows a strong cyclone centered in the bay of Skagerrak, between Norway and Denmark. This storm developed into one of the strongest storms in several years in southern Sweden. The polar front can be recognized by a trough in pressure and contrast in temperature. The warm sector of the cyclone is separated from the cyclone center. The front through the cyclone center is the occluded front, which consists of a warm front at the earth's surface (dark tips on the front symbol) and an upper-level cold front (hollow tips). The upper-level cold front climbs over the cold air in southern Sweden. The cold air that progresses behind the upper-level cold front is warmer than the surface air.

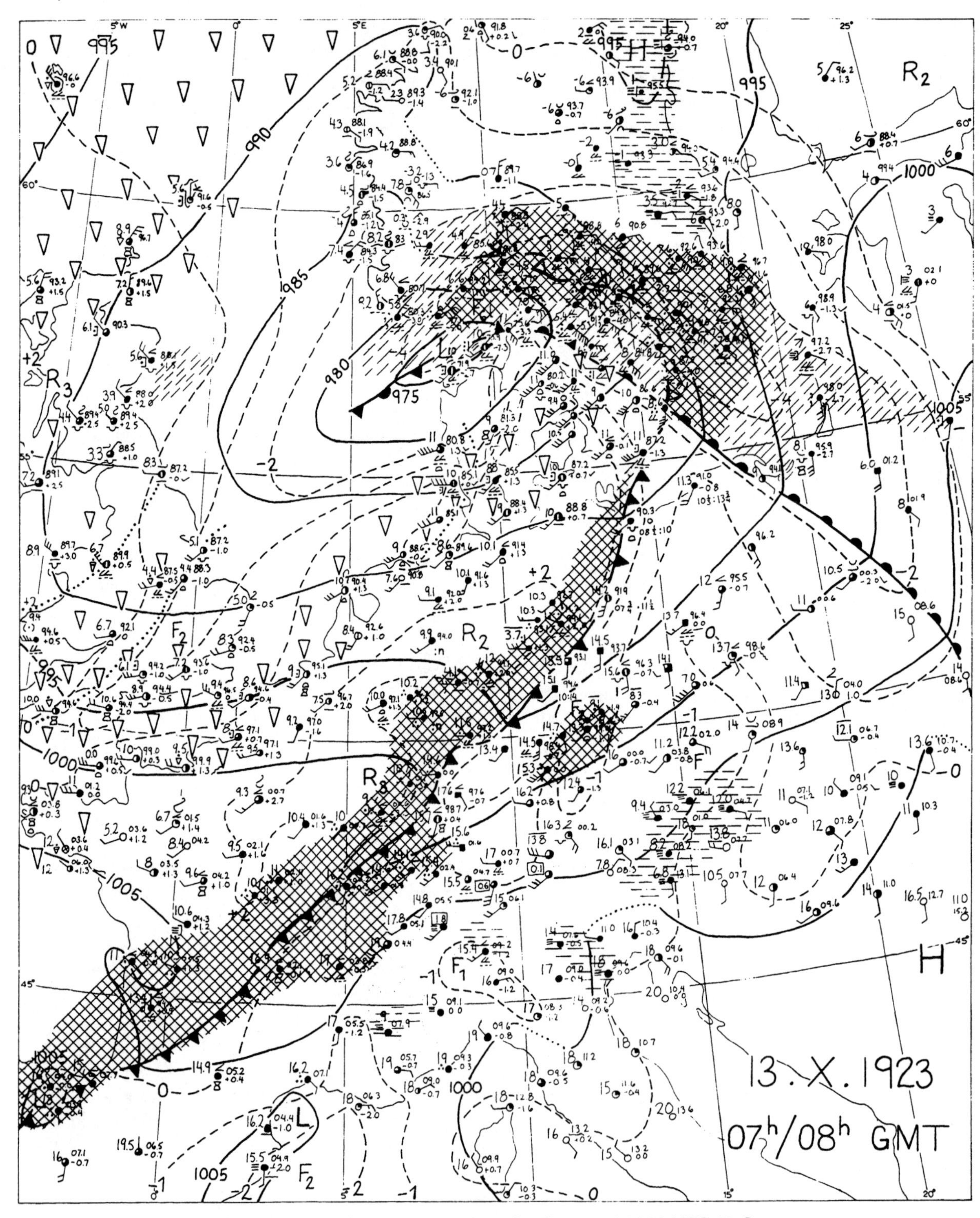

FIGURE 9-1. Surface weather chart for 0700 and 0800 UTC 13 October 1923. A cyclone over northern Europe is shown, with a cold front through Germany and France. Redrawn from Godske and others (1957).

9-4 MESOSCALE ANALYSIS

Section 9-3 contains a review of the analysis of large-scale processes. The mesoscale processes, whenever they are detected, should be added to the analysis of large-scale phenomena. The weather and weather changes are often dominated by mesoscale processes. The most severe storms are of mesoscale categories, and attention must paid to represent such phenomena on weather charts (see Figure 9-2).

In most instances, mesoscale processes are accompanied by fronts that separate various air currents. Therefore, the practical work on the mesoscale is concentrated on identifying and marking various discontinuities-fronts.

There are additional difficulties with mesoscale processes since they appear on limited space and time scales. Mesoscale processes may occur between reporting stations and between reporting times. Such difficulties should not be an excuse to avoid a thorough analysis. Special reports should be sought, especially if large-scale conditions are favorable for mesoscale phenomena. The prospects are that an improved radar network will greatly help to detect smaller storms.

Mesoscale analysis is facilitated when the two following categories are distinguished from each other:

(a) Processes governed by geographic features such as coastal fronts, local heating, deflection of air flow by mountains, and so forth.

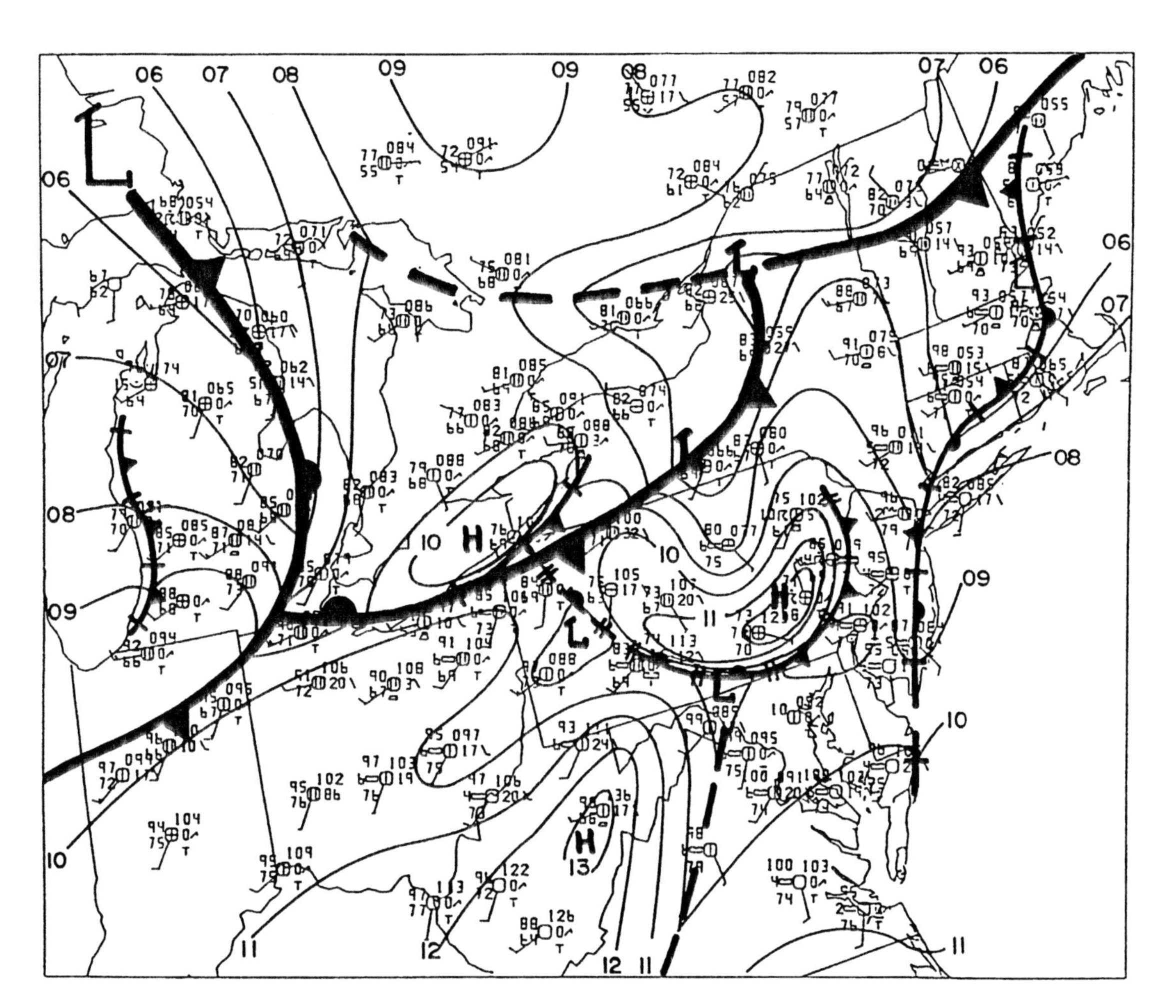

FIGURE 9-2. Isobars and fronts on the surface chart of 2100 UTC 16 July 1980. Isobars are at 1-mb intervals. The polar front is drawn with usual symbols. Mesoscale fronts have smaller triangular and semicircular ticks and ticks between tips. Single intermediate ticks indicate fronts that are geographically induced. Double ticks are on outflow boundaries of thunderstorms. From Young and Fritsch (1989).

(b) Processes caused by other atmospheric processes, without direct relation to geographical features. Thunderstorms and related phenomena are the most prominent features in this category. Another is the dry line that forms in the extratropical cyclone.

Geographically induced processes repeat under similar conditions; thus it is possible to recognize them repeatedly. Some processes of both kinds have a prominent diurnal variation since they depend on heating. Therefore, it is expected that some processes will be obscured or absent at some hours.

Excellent texts with description and explanation of mesoscale meteorology are the books by Doswell (1982; 1985) and Ray (1986).

Symbols for Mesoscale Fronts

To make a distinction between large-scale and mesoscale processes on weather charts, the following conventions are proposed by Young and Fritsch (1989):

(a) Mesoscale fronts in geographically induced mesoscale discontinuities should have single ticks (short crossbars) between the conventional triangular or semicircular tips.

(b) Mesoscale fronts that are induced by thunderstorms or other atmospheric processes should have double ticks between the usual triangular or semicircular tips.

Also, to keep track of the scale of various processes, the recommendation is to make the tips smaller on mesoscale fronts. In this way, the large-scale polar front stands out, as in the analysis without mesoscale fronts. Larger tips should be about twice as large (in length) than smaller tips

An Example of Mesoscale Analysis

A variety of weather processes are depicted in Fig. 9-2. The polar front stretches diagonally from northern Indiana to Maine and it is drawn with larger conventional symbols for cold or warm fronts.

A geographically induced phenomenon is the coastal front showing the extent of the sea breeze along the Atlantic coast. This is a stationary front; thus triangular and semicircular tips alternate from one side to the other. Between the tips are short ticks indicating the mesoscale character of type (a). A similar breeze front has been detected and marked along the east coast of Lake Michigan.

Thunderstorm outflow is noticeable from the two mesohighs: one over Lake Erie, the other over Pennsylvania. The intersection of the mesoscale front over western Pennsylvania and the polar front was significant in this case, since on this intersection new thunderstorms developed in next several hours.

Sharp troughs are drawn here as dashed lines, at variance from the recommended wavy line (Table 9-1). When speed in analysis is important, it is understandable that a dashed line is used. A standard wavy line for the trough takes more time to draw than a dashed line.

10

Cyclones and Anticyclones

Unstable fluid flow usually breaks up into roughly circular vortices. When such vortices appear in the atmosphere and are larger than several hundred kilometers in diameter, they are described as *cyclones* or *anticyclones*. In the middle latitudes, these vortices usually appear as a consequence of baroclinic instability of the westerlies. In the tropics, cyclones can originate because of convergence in regions of warm and humid air lifting. Weather development mostly depends on the behavior of these vortices. This chapter describes the most important properties of middle-latitude cyclones and anticyclones.

Cyclogenesis is treated in depth in books on atmospheric dynamics. Here, however, only a section is devoted to the descriptive aspects of cyclogenesis.

10-1 WARM, COLD, AND TRANSITIONAL CENTERS

Cyclones coincide with low sea-level pressure or with low geopotential height of isobaric surfaces. However, the pressure (or height) distribution is not sufficient to identify a cyclone. A cyclonic wind circulation is also expected in a cyclone. Therefore, whereas every depression in the pressure (height) field is a *low*, only a low with recognizable cyclonic circulation also may be called *cyclone*. Similarly, a region with pressure (or height) higher than the surrounding pressure is a *high*, and a high with anticyclonic circulation is also an *anticyclone*.

Symmetrical Centers

An elementary classification of highs and lows is based on thermal considerations. Each center (high or low) may have another center (again high or low) above it, in the upper troposphere. When a center has a high above it, it is a warm low or a warm high. This is explained by thicker layers in a region that is warmer than its surroundings. Similarly, cold centers have lows above them in the upper troposphere. Thus there are four elementary classes: *warm low*, *warm high*, *cold low*, and *cold high* (Fig. 10-1).

Surface centers (high or low) are also classified according to the classes in Fig. 10-1 if the upper-level features are not closed centers. A center is warm if there is a ridge above it, and, similarly, a center is cold if there is a trough above it. Because highs and lows appear on different scales the horizontal scale is omitted in Fig. 10-1.

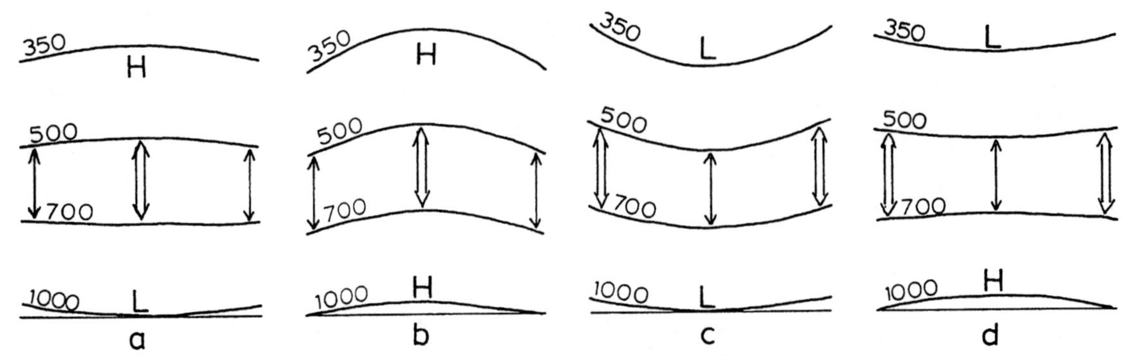

FIGURE 10-1. Elementary thermal structure of highs and lows in vertical section. (a) Warm low, (b) warm high, (c) cold high, and (d) cold high. The line with ticks is sea level. Isobars of 1000, 700, 500, and 350 mb are drawn with exaggerated slope. Double-headed arrows show the thickness between isobaric surfaces; double arrows show greater thickness than single arrows. Warm centers have H aloft; cold centers have L aloft.

Warm lows, such as tropical cyclones and polar lows, are usually rather small horizontally. Other lows of this type appear as local circulations over islands and peninsulas under daytime solar heating. A warm low over land usually does not develop into a cyclone. Its circulation normally has only vertical and radial components of significance. The largest warm lows appear over parts of continents in warm seasons, typically over India, Mexico, and northern Australia. Polar lows are about as large as tropical cyclones and, remarkably, are also warmer than their environment. Tropical cyclones and polar lows are described in Sections 12-8 and 12-9.

Dynamic Anticyclones

The largest warm highs are the subtropical anticyclones that constitute the belts of highs between the easterlies and westerlies. These highs are very tall, normally extending from the surface to the stratosphere, and are typically also very wide, usually several thousand kilometers. Such warm highs are also called *dynamic anticyclones* since their continuous existence cannot be explained thermally. Frictional and other ageostrophic cross-contour flow can rapidly destroy every warm high that does not have anticyclonic circulation. However, the subtropical highs can be explained by the quasi-geostrophically balanced pressure field between easterlies and westerlies. Large subtropical highs are also sometimes called *centers of action*, but this is hardly justified in view of the rather quiescent weather in them.

Cut-Off Centers

Another frequent warm high occurs when a jet stream makes a loop far north (south in the Southern Hemisphere), forming an anticyclonic cut-off loop. Another loop on the same jet may form in the south. This is a cyclonic eddy or a cold cut-off low. These cut-off vortices often constitute a block, or blocking situation, described in Chapter 3, that disrupts the "normal" zonal flow of the westerlies. The formation of cyclones and anticyclones in the loops of the jet stream occurs primarily at the expense of the kinetic energy of the zonal jet stream.

Asymmetrical Centers

Other classes of highs and lows occur as transitional cases between the four elementary categories of Fig. 10-1. When one side of the center (high or low) is cold and the other side is warm, the center is slanted. A smaller thickness on the cold side of a center explains the lower geopotential height aloft on that side. A larger thickness on the warm side of the center explains the higher geopotential height on the upper levels on that side. As a result, the lows tilt with elevation toward the cold side and the asymmetrical highs tilt toward the warm side. This is illustrated in Fig. 10-2. The thickness is greater on the warm side, as it is shown by separation of the isobars (isobaric surfaces). The high or low is shown on each isobar, so that the slant can be noticed.

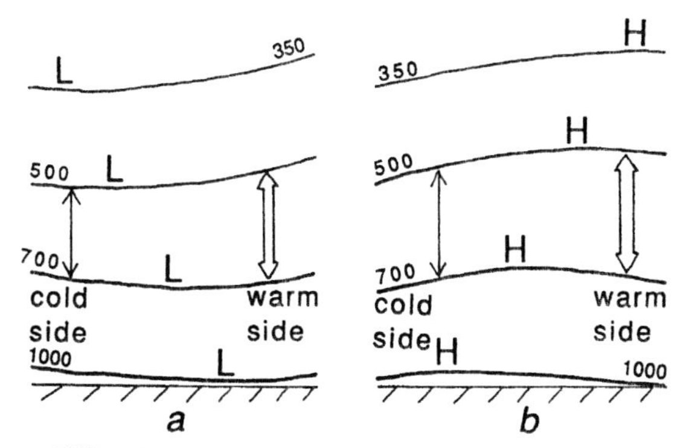

FIGURE 10-2. Thermal structure of asymmetrical centers. The notation is the same as in Fig. 10-1. (a) Low slants toward the cold side; (b) high slants toward the warm side.

As with symmetrical centers, upper-level highs or lows are not necessarily closed centers. They may be troughs or ridges in the environmental geopotential fields.

The transitional centers are actually much more important for physically interpreting weather development than symmetrical centers. The transitional centers usually move over the earth's surface and bring about weather changes. Therefore, several of the following sections describe transitional types of cyclones and anticyclones.

10-2 EXTRATROPICAL CYCLONES

An extratropical cyclone is typically 1000 km wide, lasts several days, and governs widespread weather changes of the middle latitudes. Precipitation, climate, and nature are modified by recurring extratropical cyclones.

The wind in an extratropical cyclone often causes damage, even if the wind seldom reaches tropical cyclone speed. Precipitation may be the cause for flooding. Thunderstorms develop in cyclones, and some of the thunderstorms may become violent. Snow may accumulate in dangerous amounts in some places. Thus it is understandable that a great part of meteorological activity is devoted to studying extratropical cyclones.

Structure of the Extratropical Cyclone

As mentioned, an extratropical cyclone develops as a consequence of baroclinic instability of the westerlies. Since the polar front is in the most baroclinic region, this explains the ubiquitous presence of the polar front in a cyclone.

The polar front at the earth's surface often assumes the characteristic position of the cyclone model proposed by the Norwegian School (Fig. 10-3.) Some comments on the Norwegian and Chicago schools are at the end of Chapter 1. The cold front is usually positioned in the western or southern quadrant of the cyclone. The warm front is in the eastern quadrant. The polar front divides the area of the cyclone into a *cold sector* and *warm sector*. The warm sector is smaller than the cold sector at the surface. However, the warm sector occupies an important part in the middle of the cyclone, and this sector is normally wider on upper levels.

Significant lifting of air occurs in the region of cyclone development. As a result, clouds and precipitation develop and fronts intensify. Lifting is most developed in the warm air, sometimes above the warm air at the surface.

A developing extratropical cyclone cannot be classified as warm or cold since the polar front is positioned about in the middle of the cyclone. An upper-level ridge is usually northeast of the low and a trough

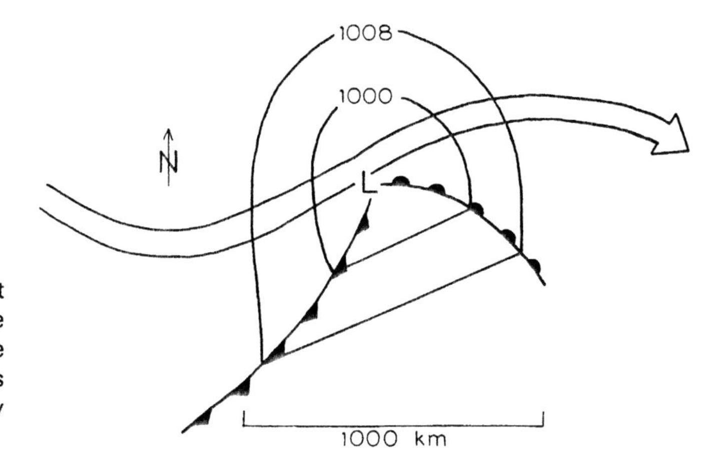

FIGURE 10-3. The polar front and isobars in the model cyclone of the Norwegian School. The core of the polar jet is shown in its typical position, as described by the Chicago School.

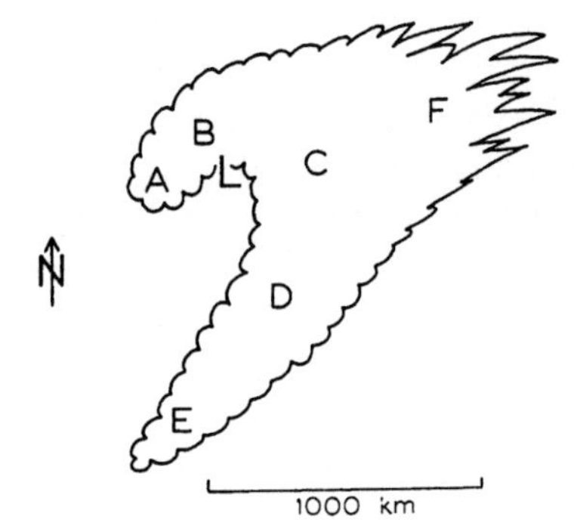

FIGURE 10-4. A comma cloud. A comma proper is ABCDE. Anticyclonically curved outflow occurs at F.

is southwest of it. This shows that the frontal cyclone, as the one in Fig. 10-3, is of transitional type, between warm and cold. As mentioned before, such a transitional type occurs most often and is of high interest in the study of weather.

The shape of the front, as shown in Fig. 10-3, is called a *frontal wave*. A cyclone that contains a frontal wave is sometimes called a *wave cyclone*.

Clouds in a Cyclone

When a cyclone develops in middle latitudes, it is commonly associated with a *comma cloud* that is visible in satellite images (Fig. 10-4). The comma is the curved shape along the letters ABCDE. When the *outflow from the cyclone* near F is prominent, the comma assumes the *hammerhead shape*, as described in Section 4-6. The cyclone center at the surface is at L.

Warm and humid air flows into the cyclone from the south. Cloud development starts at a point near E. This is the *burst point of convection*. Lines of convective clouds and squall lines develop near such points; those are the *wide cloud bands* in the warm sector. The clouds develop downstream along the EDC line.

Near C, the wind at the cloud level (in the middle troposphere) branches out in two directions: one around the low center (CBA) and the other exiting toward F. Clouds in the outflow from the cyclone near F have a diffuse and ragged shape since the air starts to sink and the clouds evaporate due to adiabatic warming.

Convective clouds develop vigorously at the burst point near E and move as a steady plume into the cyclone. The burst point moves slower than the wind. Clouds that form there move away with the wind, but new clouds form in almost the same point, giving the effect of a quasi-stationary burst point. As the whole cyclone moves in the direction of the upper-level flow, but slower than the flow, so moves the burst point. Multiple burst points are also frequent, as in the examples below. Rainbands and sometimes squall lines form downwind from burst points.

There are often multiple burst points of convection in the warm sector of the cyclone. Some burst points and associated cloud lines develop downstream from other burst points. These downstream burst points may be hidden under the cirrus canopy that originates from convective lines upstream. In such cases, the new burst point cannot be seen on the satellite image. Squall lines, even severe, may appear in the cloud bands that are obscured under the cirrus canopy.

The air current EDCF is the *conveyor belt*. Most of the humid air enters the cyclone along the conveyor belt. This current provides water vapor for clouds, precipitation, and release of latent heat of condensation. The conveyor belt rises near C, the area of the warm front. The outflow from the cyclone near F contains the conveyor belt that climbed near the tropopause. The part of the cloud AB that bends around the cyclone is the *comma head*.

The part of the conveyor belt in the warm sector (EDC) is concentrated in the *low-level jet stream*. This is a jet stream that has a vertical extent of about 2 km and wind speed typically up to 30 m s^{-1}. It is about 500 km wide and 1000 to 3000 km long and lasts 2 to 4 days, as long as the cyclone that contains the low-level jet.

The low-level jet part of the conveyor belt is usually confined under the subtropical trade-wind inversion in the cyclone's warm sector. The air is much less humid above this inversion, and this is the place where potential instability may be created. Due to large-scale lifting in the cyclone, the inversion on the top of the humid layer is frequently lifted above the lifting condensation level. Then vigorous convective clouds develop out of the humid low-level jet. As seen from the satellite, most of the clouds in the cyclone consist of cumulonimbus anvils. These cumulonimbus develop in the warm sector.

Banded shapes of clouds appear predominantly along the streamlines. It is a kinematic property of fluid flow that the dilatation axis turns along the flow, as discussed in Section 4-6. In this way, elongation of clouds in bands helps detect the direction of air currents from satellite images. Most bands in the warm sector of the cyclone emanate at burst points of convection.

Not all comma clouds are as large as the extratropical cyclones. A smaller comma cloud is often an indicator of a *polar low* or a *polar trough*. These phenomena are described in Section 12-9.

FIGURE 10-5. GOES-E infrared image for 2000 UTC 21 March 1981. From Carr and Millard (1985), by permission of the American Meteorological Society.

The main variations from the model comma cloud in Fig. 10-4 are due to varying intensity of the exiting branches at A and F. Depending on the configuration of the upper-level flow, the conveyor belt may pass through the cyclone along the line EDCF when there are only a few low clouds in the branch AB. On other occasions, the outflow at F may be insignificant, and the circulation goes along the comma proper: EDCBA. In most cases, both exit branches are present.

An Example of Clouds in a Cyclone

Figure 10-5 shows a typical case of cloud distribution in a cyclone over North America. The photograph of Fig. 10-5 is redrawn in Fig. 10-6, so that the geographical location and main features can be pointed out. The letters are introduced at the points that correspond to the model in Fig. 10-4. The two main branches of the conveyor belt are shown by two arrows, as this flow can be observed on the 500-mb chart of 0000 UTC 22 March 1981 (Fig. 10-7). Two main exits from the cloud system are suggested by these arrows.

Several other deviations from the comma cloud of Fig. 10-4 also are visible. The entrance to the cloud near E is not a single burst point of convection. There are several burst points west and south of E where the cloud plume generates. Numerous smaller groups of clouds are present on various sides, which may sometimes obscure the basic comma pattern.

For many cyclones, the northern part of the comma (BA in Figs. 10-4 and 10-6) is not a continuation of the cloud mass in the warm sector. Instead, this is a layer cloud (altocumulus and stratocumulus) that forms in the polar air under the warm front and under the main cloud outflow CF. This cold air current constitutes the *cold conveyor belt*, which is often concentrated in the low layers as another low-level jet in the cold sector of the cyclone. Cloud moisture in the cold conveyor belt (BA) originates mainly from evaporated rain that fell from the clouds of the warm conveyor belt at CF. There is lifting near the cyclone center, with clouds developing near points B and A.

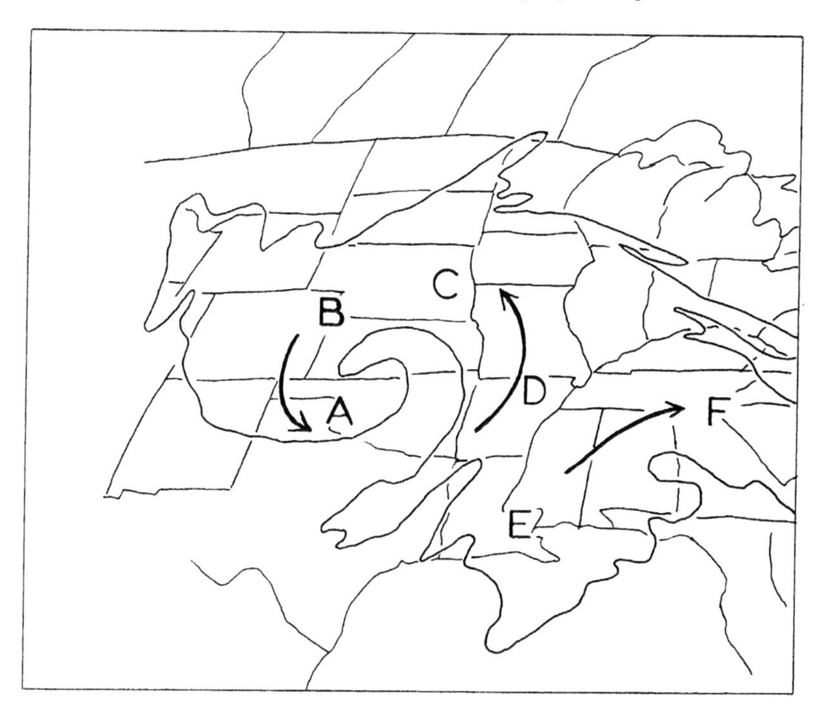

FIGURE 10-6. An outline of clouds from Fig. 10-5, with letters referred to in the text.

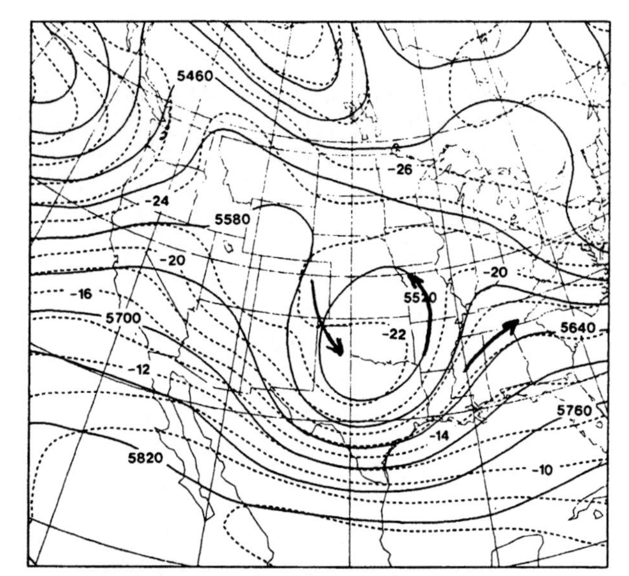

FIGURE 10-7. 500-mb contours (gpm) and temperature (°C) for 0000 UTC 22 March 1981. The arrows over Colorado–Texas, Arkansas–Missouri, and Mississippi–Alabama are in the same places as the arrows in Fig. 10-6. Adapted from Carr and Millard (1985), by permission of the American Meteorological Society.

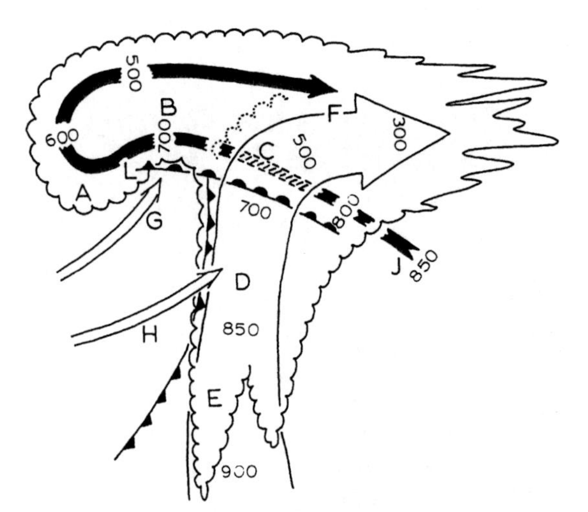

FIGURE 10-8. A conceptual model of the extratropical cyclone. The center of the sea-level low is at L. The broad arrow EDCF is the warm conveyor belt. Narrower arrows are in the polar air. The dark arrow JCBAF is the cold conveyor belt. The white arrows G and H are in the dry polar westerly flow. Surface fronts and cloud limits are outlined.

Conceptual Cyclone Model

A conceptual model of the cyclone is shown in Fig. 10-8, after similar models by Carlson (1980), Mason (1985), and Young and others (1987). The warm conveyor belt is shown by a broad arrow, with numbers (in millibars) denoting its top. A typical cloud is outlined with a cusped line, which changes to a ragged line on the northeastern end of the cloud. In this place, we find the main outflow along the conveyor belt. The sea-level low is at L, and surface fronts are shown by conventional symbols.

The form of the cloud mass on the south side is depicted with two burst points of convection, from which the rest of the clouds emanate. These points are very typical in extratropical cyclones. Often they represent the starting points of squall lines (back-building type, Section 12-6) or wide cloud bands of the warm sector.

The core of the cold conveyor belt in Fig. 10-8 is represented by a dark arrow (JCBAF). This belt is about as wide as the warm conveyor belt. In Fig. 10-8, however, the cold conveyor belt is arbitrarily drawn narrower, so it could be fitted easier in the figure. Most of the moisture of the cloud near B and A originates from rain that falls ahead of the warm front, under the warm conveyor belt. As at the warm conveyor belt, the numbers along the cold conveyor belt indicate the level in millibars.

Often a distinction can be made between levels of cloud tops in the warm and cold conveyor belts. The boundary of the higher cloud of the warm conveyor belt is drawn by a dotted cusped line above the cold conveyor belt, between B and C in Fig. 10-8.

Most cold air enters the cyclone from the west, as represented by arrows G and H in Fig. 10-8. Cold air enters the cyclone as a broad current, occupying the space between and around arrows G and H. This current is also very tall, occupying the whole troposphere. The cold air coming from the west is often not colder than the warm air above the warm conveyor belt at D. The cold air had subsided significantly before entering the cyclone. Therefore, it is often difficult to find the cold front at 500 mb if only temperature is considered. The cold front near D is often easier to locate by the contrast in a humidity variable as wet-bulb potential temperature or dew-point depression. The form of the front between H and D may exhibit the split structure, shown earlier in Section 8-9. For this reason, the arrow at H is extended over the surface cold front. An upper-level cold front is shown traveling above the warm conveyor belt.

The descent of cold air at G and H is associated with lowering and folding of the tropopause as de-

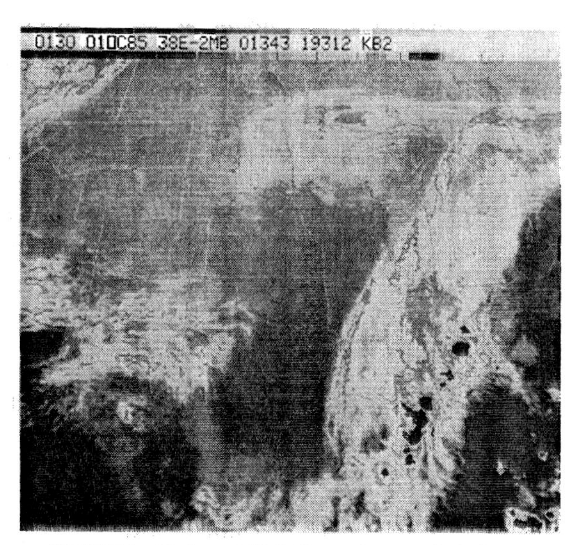

FIGURE 10-9. GOES-E enhanced infrared image for 0130 UTC 1 October 1985.

scribed in Section 8-4. A tropopause at 500 mb is not uncommon in the upper-level low, southwest of the surface cyclone. Bands of very dry stratospheric air can be found in this region.

The dry air in the westerly current near G and H may lift in the vicinity of the cyclone center, and clouds may not form in it. When the air is very dry, it may sustain lifting of 1–2 km before saturation is reached.

An Example of a Conveyor Belt

An example of a cyclone where the cold conveyor belt is clearly separated from the warm conveyor belt is illustrated in Fig. 10-9. The main bright cloud band is within the warm conveyor belt. The smaller cloud mass northwest of the warm conveyor belt is the part of the comma cloud that formed in the cold conveyor belt. Some details of Fig. 10-9 are highlighted in Fig. 10-10. The characteristic parts of the comma cloud are marked with letters A through F in positions corresponding to the letters in Figs. 10-4 and 10-6. The clouds near E, over the Gulf of Mexico, show several burst points of convection, from which the cloud lines develop toward the north. The most prominent burst point of convection is the one near the coast at the Texas–Louisiana border. Several other burst points can be discerned east of this point.

The center of the low at various levels is indicated by an L with a number, as explained in the figure caption. The center of the low tilts toward the cold (polar) air, as is typical for such cyclones. The tilt of the low implies that the wind drastically veers with height in the region between various L's in Fig. 10-10. Wind variation with height is shown by the wind observations at International Falls, Minnesota. The wind is plotted at three levels: 850, 500, and 200 mb. The 850-mb wind is from the north, since the low is east of the station at that level. The wind at 500 mb is east southeast, and at 200 mb the wind is from the south. The elongation of the cloud band takes place along the easterly wind near and under the 500-mb level. The

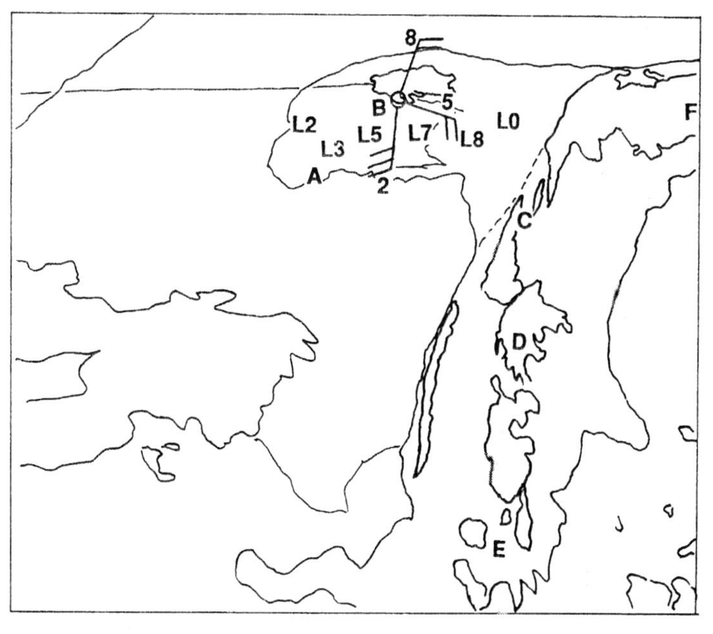

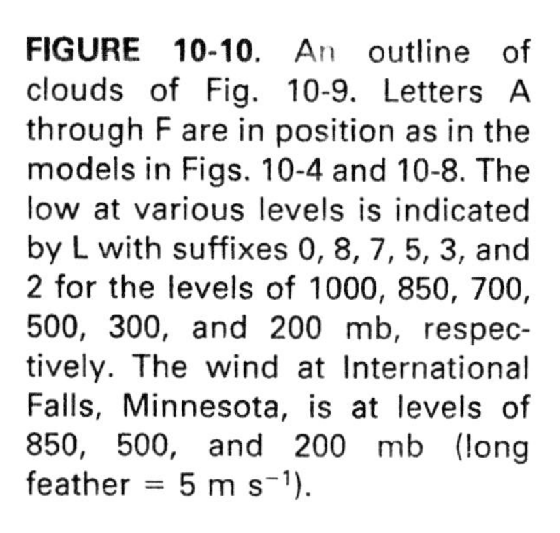
FIGURE 10-10. An outline of clouds of Fig. 10-9. Letters A through F are in position as in the models in Figs. 10-4 and 10-8. The low at various levels is indicated by L with suffixes 0, 8, 7, 5, 3, and 2 for the levels of 1000, 850, 700, 500, 300, and 200 mb, respectively. The wind at International Falls, Minnesota, is at levels of 850, 500, and 200 mb (long feather = 5 m s^{-1}).

bright northern rim of this cloud mass extends zonally north of International Falls. In that region, the air current that was southerly over International Falls becomes westerly in the jet stream north of the cyclone.

The two cloud masses in the cyclone in Fig. 10-9 are not completely separated. The region between C and L0 (as shown in Fig. 10-10) is covered with low clouds of the cold conveyor belt.

Rainbands in the Cyclone

Radar observations show that rain clouds usually form *rainbands*. Cloud bands that correspond to rainbands can be seen, although less clearly, on satellite images. Rainbands are the dominant patterns in most rain events. A categorization of rainbands is shown in Fig. 10-11 (from Hobbs, 1981). This figure shows the relation between the most prominent weather patterns, mostly fronts, and rainbands. The shown *cold front aloft* and the *prefrontal cold surge* may each be of the split front type, as described in Chapter 8.

Rainbands are most often arranged parallel with fronts. The bands are named primarily after the nearby front to which they are parallel. Wide bands (numbers 1, 2, 3, and 5) are about 50 km wide and may stretch several hundred kilometers. The narrow cold frontal band is a thin line of cumulonimbus, about 5 km wide. The model rainbands do not appear in every cyclone. There is a great variety of cyclones, with more or less rain. If it rains, it likely occurs in bands of some of the types from Fig. 10-11.

The structure of fronts in cyclones is also not uniform. The models in Figs. 10-8 and 10-11 are usually varied in cyclones. The model in Fig. 10-8 is more suitable for a younger cyclone where the warm sector is large and contains a strong low-level jet. In a later stage, when the cyclone deepens and moves further poleward, it assumes more the type of Fig. 10-11. Then the cloud mass spirals around the center, and the exit F from Figs. 10-4 and 10-8 is hardly noticeable. The warm sector at the surface is reduced to a small segment of this cyclone.

Cells of Convection

Figure 10-11 shows another prominent feature on the west side of extratropical cyclones: the development of *open cells* of convective clouds in the cold sector. When cold air is heated from below, especially over oceans, cumulus and cumulonimbus grow in narrow lines. Those are the *convection cells*. They are *open*, so their protrusions reach to the neighboring cells. There are clear, polygonal spaces between the cloud lines. These cumulonimbus often do not reach the tropopause; their tops are normally at 3–6 km elevation. Convection in polar air of this type was noticed as an important part of a cyclone already at the time of the Norwegian School. At that time the shape of the open cells was not known. The cellular pattern was discovered only after satellite images became available.

A somewhat similar pattern of convection, although of opposite contrast between clouds and clear areas, is the case of *closed cells*. These appear in less active regions, when a layer of stratus is cooled from above. This is opposite of the heating from below that occurs with open cells of convection.

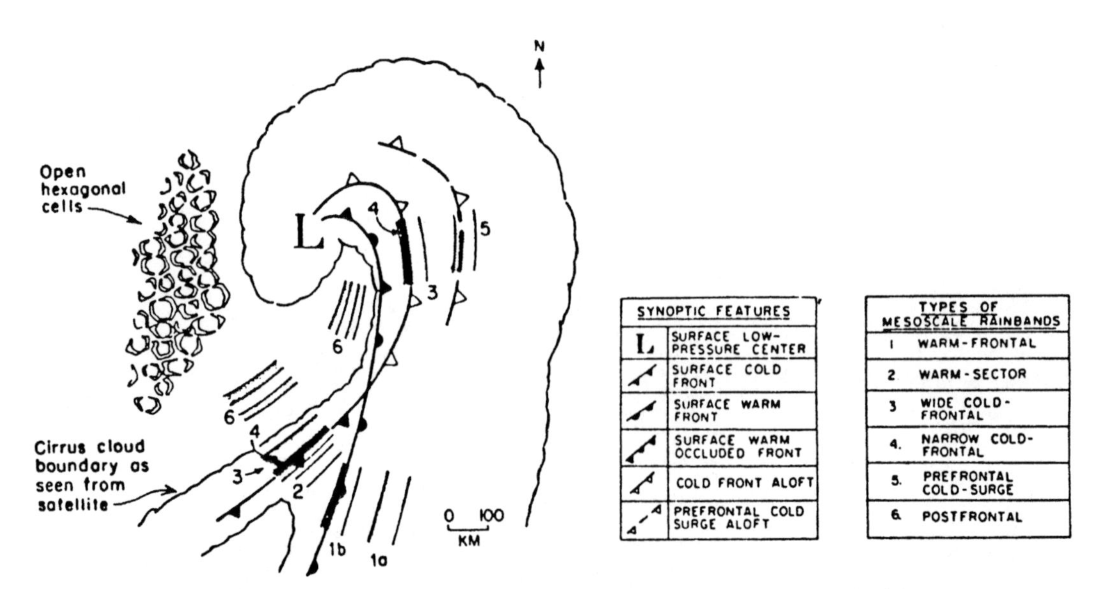

FIGURE 10-11. Rainbands in the extratropical cyclone. From Hobbs (1981).

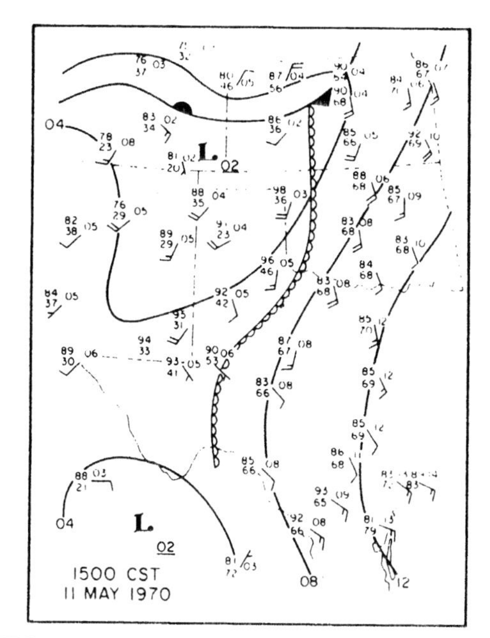

FIGURE 10-12. Sea-level isobars and the dryline for 1500 CST 11 May 1970. From Schaeffer (1986), by permission of the American Meteorological Society.

The Dryline

A contrast in humidity often develops in cyclogenetic conditions between the southerly humid flow and westerly current of continental origin. This is the *dryline*, which is important in mesoscale weather analysis since its passage will change the character of weather and since thunderstorms often develop on it.

An example of a dryline is shown in Fig. 10-12. The trough between the lows in southeastern Colorado and in Mexico is of the orographic type. It formed due to adiabatic warming of the westerly flow descending from the mountains. A southerly humid current formed east of the trough and southeast of the Colorado low. The western extent of the humid flow is the dryline. In this example, the dew point over 65°F is typical for the humid air. The dew point in the dry air in the west is under 46°F.

The dryline sometimes coincides with a *lee trough*, which appears east of the mountain chains when westerlies blow over the mountains. The dryline can be found easiest when it coincides with the lee trough.

The humid air flow is normally shallow. Its top is near 1 km. This is the low part of the conveyor belt, often in form of a low-level jet.

Temperature contrast across a dryline is not consistent. Sometimes one side is warmer, sometimes the other. Commonly, the dry side is warmer in daytime and the humid side is warmer at night. This is due to an enhanced diurnal temperature variation in the dry air. Solar and, especially, terrestrial radiation is stronger with low humidity.

Lows of the orographic type often do not have a cold front in their southwestern quadrant. This is the case in Fig. 10-12. There the dryline resembles the cold front of the Norwegian model. Analysts sometimes confuse the dryline with a front.

On other occasions, thunderstorms develop along the dryline. Then the dryline is included in the squall line and cannot be identified separately anymore.

Drylines are most prominent over the High Plains of North America (North Dakota to northwestern Texas). This is the region in the lee of the Rocky Mountains where orographic cyclones frequently form. The dryline develops in their warm sectors.

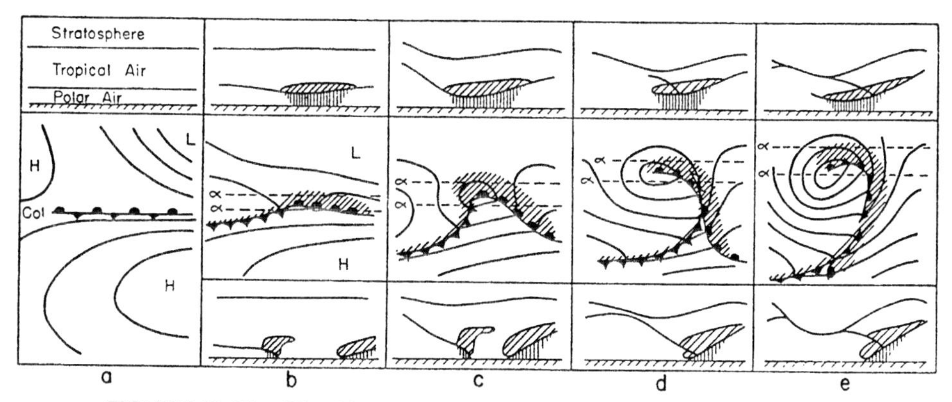

FIGURE 10-13. The life cycle of a frontal cyclone over about 2 days. The horizontal scale of each frame is about 2000 km. The vertical sections (top and bottom rows) are at lines α of the horizontal charts (middle row of drawings). From Godske and others (1957), by permission of the American Meteorological Society.

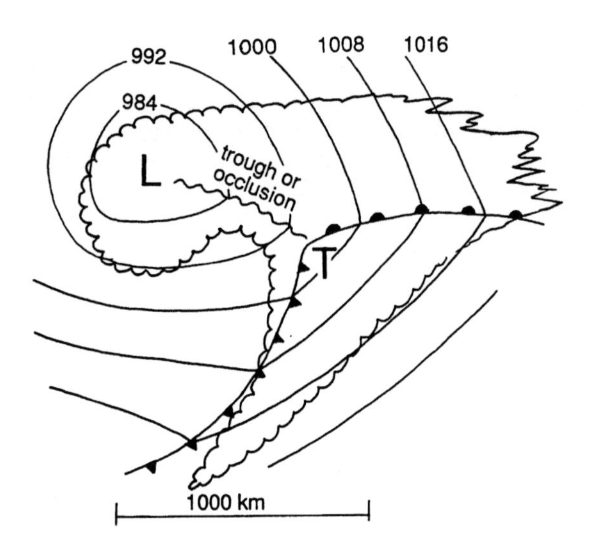

FIGURE 10-14. Isobars (mb) and the polar front in an occluded cyclone. A satellite view of the cloud is outlined. The trough between the cyclone center (L) and the front can be interpreted as the occluded front. The polar front and occluded front meet in the *triple point* T.

The dryline, however, forms in cyclones in other locations as well. Those other cyclones usually move faster than orographic cyclones. Drylines do not last long in moving cyclones. They are soon overtaken by the cold front. Orographic cyclones also move several days after formation, and they lose their dryline in the process.

Occlusion

In later stages of cyclone development, cold air occupies most of the area of the cyclone. Such a cyclone is a cold cyclone. It is also called an *occluded cyclone*, since the warm air is separated (*occluded*) from the cyclone center at the earth's surface. The words *occluded* and *occlusion* are related to *seclusion*, or separation. Bergeron suggested the expression *seclusion* for cases when warm air can be identified at the surface near the cyclone center but when the cold front has reached the warm front away from the center. Cases of seclusion are extremely rare, but cases when the warm air is completely eliminated from the cyclone center are frequent. For these the word occlusion has been coined. This word is used for three concepts: an occluded front, an occluded cyclone, and the process when the occluded front is formed.

Several mechanisms have been suggested to describe the formation of the occluded front (occlusion). The oldest explanation (attributed to Bergeron) is based on the observation that the cold front moves faster than the warm front. When the cold front reaches the warm front, a combined front is formed. That is the occlusion. This folding theory of the occlusion process is described in Fig. 10-13. The top and bottom rows of the sketches represent vertical sections through the cyclone, with the tropopause and frontal surfaces. The middle row gives a succession of surface charts with a developing cyclone and occluding front. The appearance of an occluded front is strongly linked to the development of a cyclone. Therefore, large extratropical cyclones are *occluded cyclones*. They contain an occluded front, and warm air is eliminated from the cyclone center at the surface.

There is some doubt that the cold front catches up with the warm front during the life of a cyclone. It is possible that the frontal wave travels to the northeast, without folding of the front. However, it is undisputed that the cyclone center in strong cyclones often separates from the cold and warm fronts. Then a pressure trough can be found between the cyclone center and the wave on the front, and this trough may be inter-

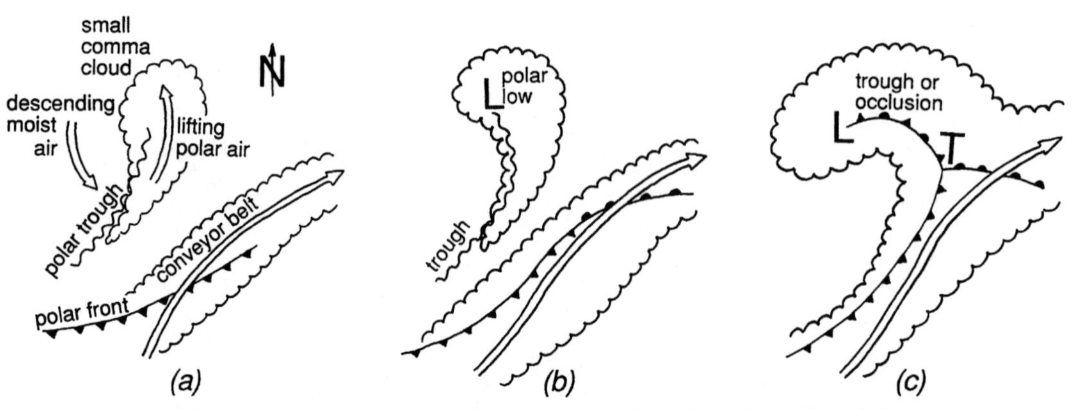

FIGURE 10-15. Instant occlusion. An occluded cyclone forms by the joining of a comma cloud in the polar air with the cloud mass along the polar front.

preted as the occluded front. In this case, the cyclone model looks as in Fig. 10-14.

The point T where the front and the occluded front meet is the *triple point*. The three fronts that meet in this point are: cold front, warm front, and occluded front. Often a cyclone regenerates in the triple point.

Satellite imagery has revealed that the occluded cyclone can form without folding of the front and without separation of the low center from the front. This occurs in cases of *instant occlusion*. It has been observed that a pressure through (or even a small low) in polar air forms a comma cloud. This trough–comma-cloud system sometimes approaches the elongated cloud mass along the polar front. When the two clouds join (the comma cloud and the frontal cloud), the pattern of an occluded cyclone emerges that has the shape as in Fig. 10-14. The formation of the instant occlusion is described by the sequence of drawings in Fig. 10-15.

10-3 EXTRATROPICAL CYCLOGENESIS

The vigorous development of synoptic-scale cyclones in middle latitudes is a consequence of *baroclinic instability* in the westerlies. This is the process of wave development. Large unstable waves result in vortices.

Baroclinically unstable westerlies are characterized by large horizontal thermal contrast between northern and southern air masses. The thermal contrast implies a large vertical wind shear, or thermal wind, and is strongest in frontal layers. Therefore, the cyclone is usually associated with fronts. Further, vertical wind shear is largest with the jet stream. Thus, cyclone development is also associated with the jet stream.

The development of an existing cyclone (lowering of central pressure) is often called *deepening*.

Environment of Cyclogenesis

Ambient conditions for cyclogenesis involve lifting and the formation of the conveyor belt. Lifting is usually provided by the general configuration of the flow in the active region of the westerlies (Section 6-5). It was shown that the active region of the jet stream (polar or subtropical jet) provides for lifting in the area between the trough and the downstream ridge. This region also is the location where the conveyor belt develops as a low-level jet before the cyclone forms. The clouds grow from the low-level jet when lifting starts, aided by water vapor brought by this jet. The appearance of a dense, high cloud top, formed by the merger of individual clouds in the active region, is a sign of cyclogenesis. This joined cloud top stretches along the conveyor belt and takes the shape of a leaf; thus it obtained the name *baroclinic leaf cloud*. Typical dimensions of a baroclinic leaf cloud are 500×200 km^2, within a factor of 2. When a baroclinic leaf deforms due to the circulation around the low, it becomes a comma cloud. The expression *cloud head* is also used for the cloud form of comma cloud.

An illustration of a baroclinic leaf cloud can be seen in Figs. F-1, F-3 and F-4 in Appendix F, where different enhancement of infrared imagery is illustrated. The cloud mass that stretches from west central Oklahoma to southwestern Nebraska is a baroclinic leaf cloud (Appendix F). In its later development, this cloud bends around a new cyclone and forms a comma cloud. In the case when the atmosphere is sufficiently unstable, a baroclinic leaf cloud may be a squall line of the type described in Section 12-6.

Types of Cyclogenesis

Several indicators of the ambient conditions where cyclogenesis is expected allow us to classify cyclogenesis into three main types (Radinović, 1986): (a) an amplifying frontal wave, (b) a disturbance in the upper troposphere, and (c) an orographic disturbance. This classification must not be taken rigorously, since many cyclones have properties of more than one type. However, for a physical interpretation of weather processes, it is instructive to consider the properties of cyclones of types A, B, and C separately. Types A and B are well described by Petterssen and Smebye (1971).

Type A cyclone

A type A cyclone appears under an almost straight polar jet, without a significant active region. The initial development of a type A cyclone is characterized by warm advection in the region of pressure fall. Cold advection appears several hundred kilometers upwind from the region of warm advection. A cyclone develops into a strong storm only when an upper-level trough develops over the region of cold advection. At that time, a ridge forms over the region of warm advection, forming an active region conducive to further development. All three features (ridge, surface low, and trough) move down the jet stream, keeping this order. The movement of this cyclone is slower than the wind in the jet stream by about half. Movement of the system slows down with the deepening of the low. The arrangement of isobars and fronts in the developed cyclone, together with the core of the polar jet, becomes similar to the case in Fig. 10-3.

Differential heating of air masses intensifies baroclinicity and in this way also stimulates the develop-

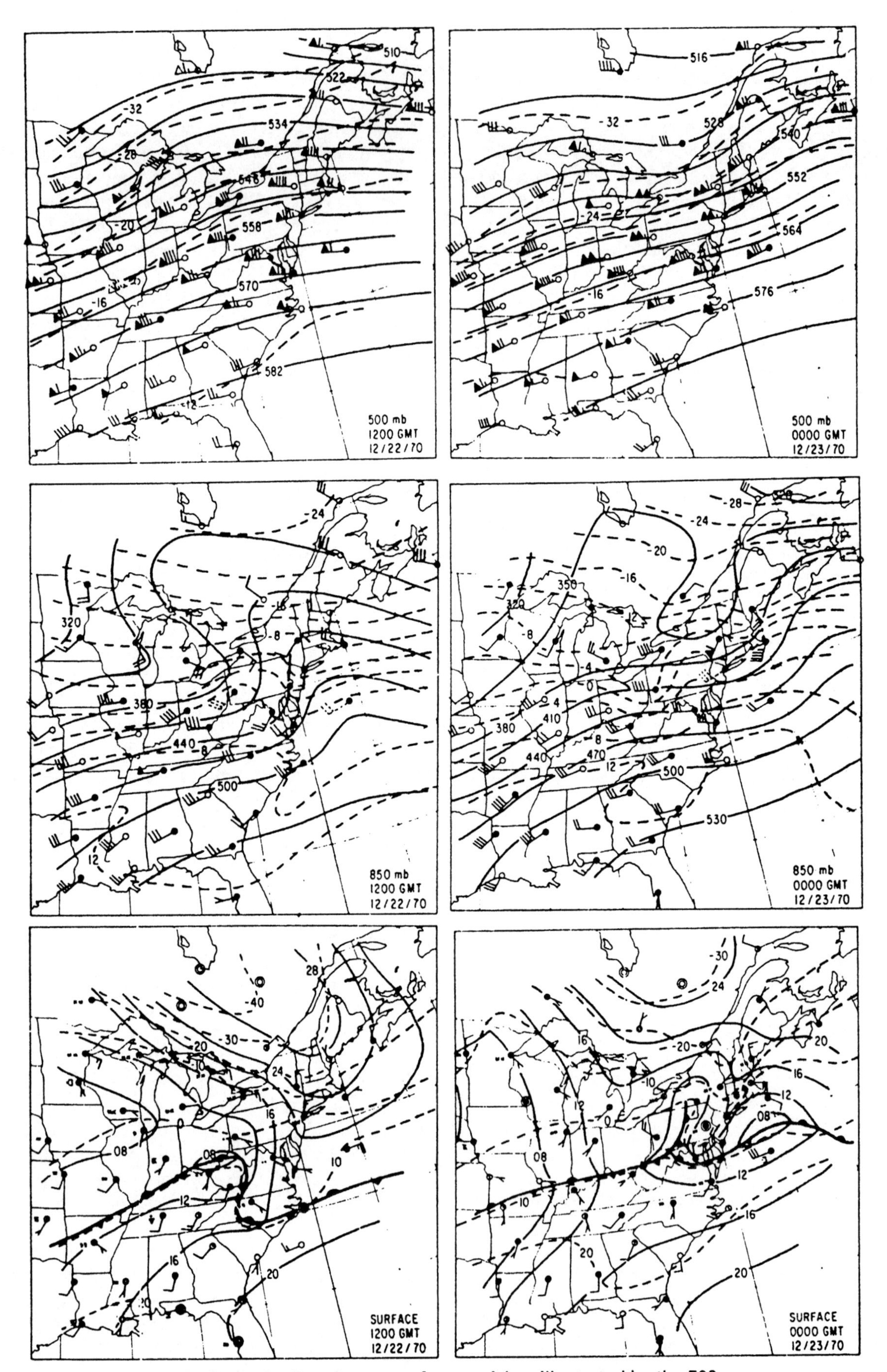

FIGURE 10-16. Development of a type A low illustrated by the 500-mb, 850-mb, and surface charts for 1200 UTC 12 December and 0000 UTC 23 December 1970. From Pagnotti and Bosart (1984).

ment of the amplifying frontal wave. Therefore, thermal contrasts found in the oceans near major currents provides favorable areas for development of type A cyclones. When atmospheric conditions are favorable and the thermal contrast on the ocean surface is large, the amplifying frontal wave may develop extremely fast, with pressure falling up to 50 mb in 24 h. Such an extreme cyclone resembles a hurricane by the wind speed and pressure gradient and, because of its violence, has acquired the name *bomb*. The development of a type A cyclone is not as violent over continents.

Figure 10-16 shows the early development of a type A cyclone, from a paper by Pagnotti and Bosart (1984). The 500-mb chart shows an almost straight upper-level flow. A baroclinic region is evident by the presence of numerous isotherms at 500 and 850 mb. The sea-level anticyclone over Quebec is cold and shallow, as is shown by the isotherms. This cold high weakens with height and the 500-mb flow shows only a weak ridge. Warm advection in the lower troposphere can be seen at 850 mb over the northeastern United States, where the wind barbs point across isotherms.

Two lows appear at the surface during the 12 h illustrated in Fig. 10-16. The first low, characterized by one closed isobar (at 1200 UTC 22 December) over southern Ohio, is not a pure type A low; it is partly orographic since the frontal wave had been amplified by mechanical retardation of warm air on the Appalachian Mountains. Warm air coming from the southwest entered Kentucky and West Virginia, but has not yet entered North Carolina. The mountains provide a barrier that retards the progression of warm middle-latitude air. The frontal wave from southern Ohio moved to Pennsylvania in 12 h.

At the end of the considered period (0000 UTC 23 December), another low formed over the ocean east of New Jersey, in the region of maximum warm advection. This other low is a more typical baroclinic wave of type A. A cold trough is now developing over Lake Erie, where cold advection had been predominant. A ridge formed at 500 mb over Nova Scotia, in the region of warm advection. The low over the Atlantic is now a cyclone of type A, since the cyclonic circulation is established. This cyclone is now similar to the model of the Norwegian School in Fig. 10-2. This low developed into a major Atlantic storm in the few days after the period shown in Fig. 10-16.

Type B cyclone

A cyclone of type B normally develops faster than type A. In type B, an upper-level wave is already present, with vorticity advection. Strong upper-level vorticity advection is a sign that large-scale lifting and cyclogenesis are in progress. The development of a cyclone due to a wave in the upper troposphere typically occurs when strong upper-level vorticity advection overtakes a strongly baroclinic region in the lower troposphere (*Petterssen's rule*). The most baroclinic region is normally recognized as the polar front. The point of maximum vorticity advection aloft is downstream from an upper-level trough, in the active region of the westerlies.

Particularly strong development occurs when the upper-level trough is diffluent, that is, when the contours separate downstream from the trough. Diffluence implies that there is a net inflow of kinetic energy in the region of cyclogenesis. The diffluent upper-level flow implies the *delta* or exit from the jet streak and is a region of enhanced indirect thermal circulation.

Figure 10-17 shows strong development of a type B low. The existing surface low in northeastern Wyoming developed earlier; its origin could be traced to California. Further development of type B occurred under the active region of the polar jet, as the low moved to southern Manitoba while remaining downstream of the trough along the Rocky Mountains. Vigorous development occurred under conditions of strong vorticity advection aloft. At the same time, the cold front over the Great Plains intensified. The pool of cold air in the low moved eastward into Wyoming and Minnesota. The ridge at 500 mb along the Mississippi Valley on 17 December was warm and its western part was still warming up due to warm advection. On the other hand, in the lowest 1-3 km of the troposphere, polar air was covering approximately the eastern half of the contiguous United States. Therefore, the cold front on 18 December had the appearance of an occluded front with polar air on both sides. It is typical for strong cyclones, as in Fig. 10-17, to contain an occluded front.

Cyclones of types A and B have many similarities. They are especially similar in their later stages when a type A cyclone normally shows a trough and a ridge on the jet stream, as in type B. The most apparent difference between types A and B is in the initial shape of the polar jet. Type A has an almost straight jet, where vorticity advection is weak. Type B has a well-developed wave in the polar jet, with significant vorticity advection in the region of cyclogenesis.

Type C cyclone

A cyclone of type C also has many similarities with types A and B. However, the physical connection with orographic forcing justifies a separate classification of the orographic cyclone as type C. The most prominent characteristic of this cyclone is that the low may form

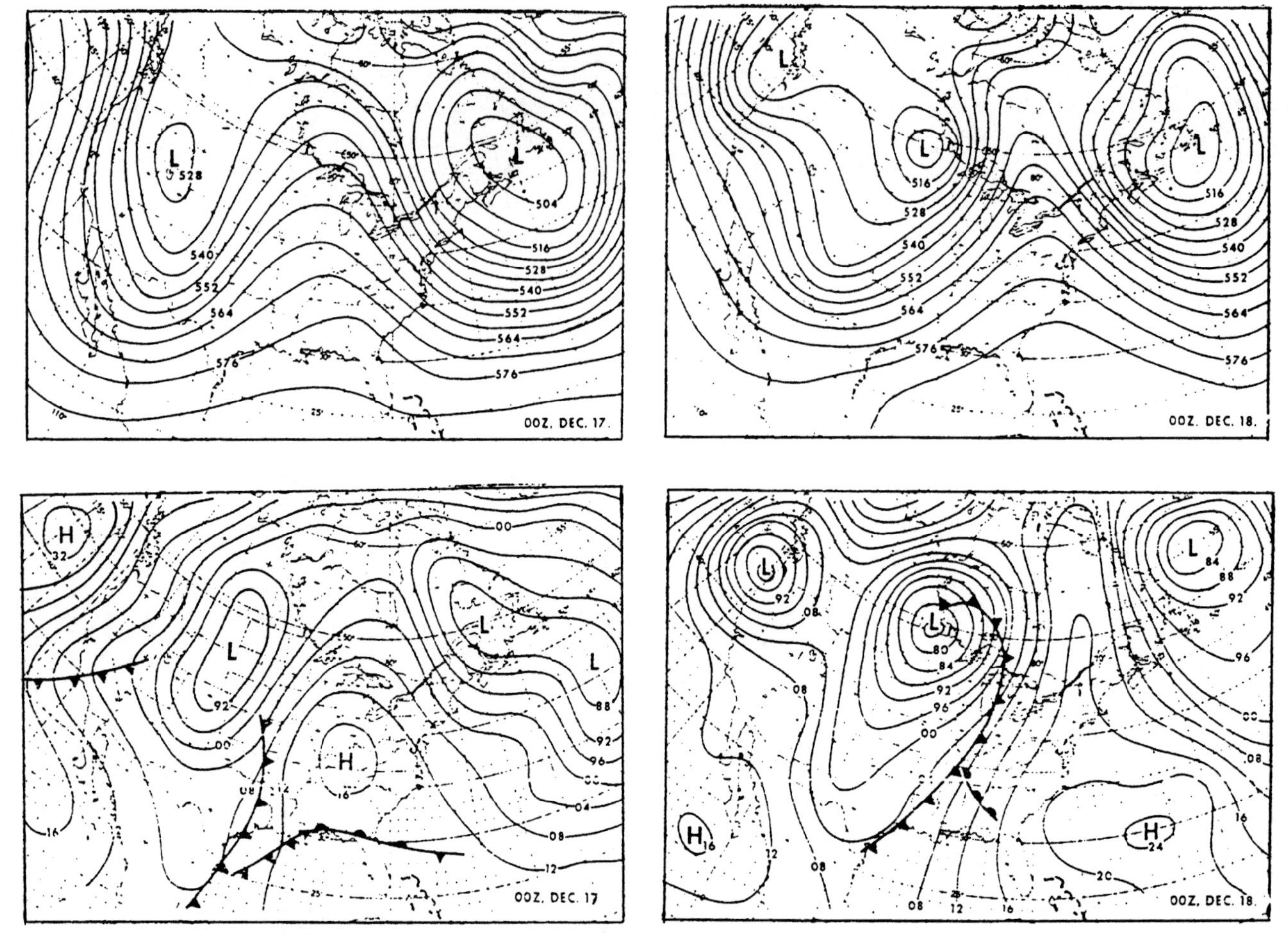

FIGURE 10-17. Development of a type B cyclone. The charts of 500 mb (top frames) and surface (bottom frames) are for 0000 UTC 17 December and 0000 UTC 18 December 1967. From Petterssen and Smebye (1971), by permission of the Royal Meteorological Society.

before the clouds form. This points to the mechanical influence of the mountains. Heating due to latent heat release is not a significant factor.

Numerous cyclones of type C generate in the lee of large mountain chains. In the Northern Hemisphere, orographic cyclogenesis is most often found south of the Alps (*Genoa cyclone*) and east of the highest mountain ranges in North America: the Rocky Mountains of Colorado (*Colorado cyclone*), and the Coastal Range of British Columbia (*Alberta cyclone*).

The Genoa cyclone can be explained by the barrier effect, which prevents cold air in the lower troposphere from readily entering the region behind the mountains. The upper-level trough above the cold outbreak is not hindered by the mountain range and forms an active region in the lee of the mountains. Since cold air is held back by the mountains, lifting in the active region is much more active than in the presence of cold air in the lower troposphere. For this reason, cyclogenesis preferably occurs in a comparatively small region behind a mountain range.

An example of the Genoa cyclone is shown in Fig. 10-18. The thickness chart for 1200 UTC 17 February 1958 shows a well-defined zonally oriented frontal zone from Great Britain through central Europe and farther to the east. The surface front also occupies a zonal position, at the southern edge of the frontal zone. Both the frontal zone and the front travel south. In 24 h, the frontal zone can cover about 10° of latitude, and it does so over western Europe and Spain (until 1200 UTC 18 February 1958). The steep mountain range of the Alps from southeastern France to Austria keeps the frontal zone back. This results in deformation of thickness contours. The low formed over the Gulf of Genoa (hence the cyclone name) before the front crossed the mountains. Soon (about 6 h) after initial cyclogenesis, the front entered the low. From then on the cyclone resembles other frontal extratropical cyclones.

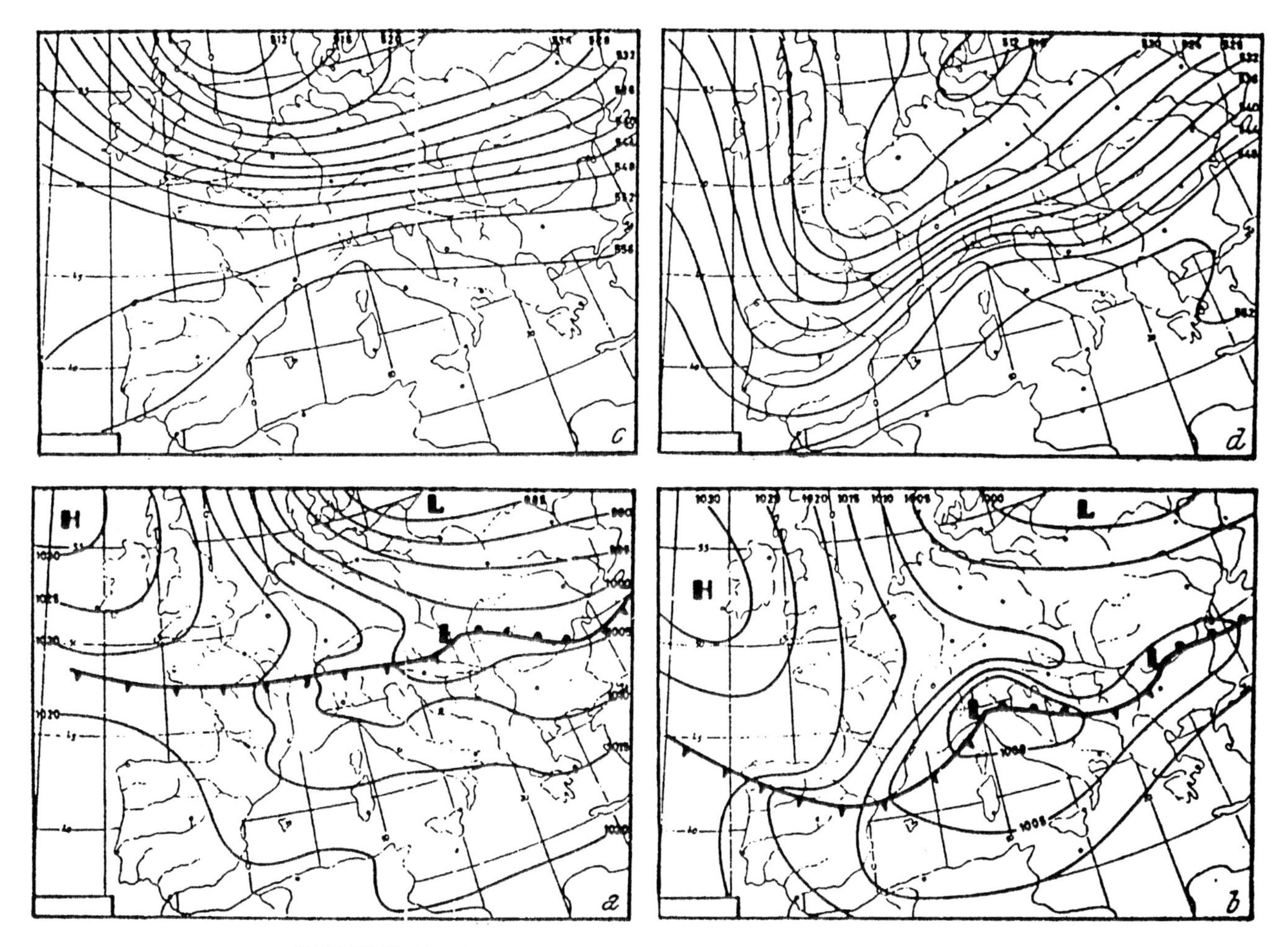

FIGURE 10-18. Development of the Genoa cyclone. Thickness 500/1000 (intervals of 40 gp dam) and sea-level isobars (5 mb) are for 12 UTC on 17 and 18 February 1958. From Radinović (1965), by permission of Springer Verlag.

Cyclogenesis over eastern Colorado is similar to cyclogenesis over the Gulf of Genoa due to mountains. The initial cyclogenesis in Colorado is not characterized by a local deformation of the thickness contours. Instead, the pressure fall occurs under a zonal westerly flow, when the air warms in the lee of the Rocky Mountains due to adiabatic descent. The initial pressure fall is not very strong. It is sufficient to form a depression with one or two closed isobars. The polar front arrives from northwest only after the depression has been formed. At that time, the baroclinicity (horizontal thermal contrast) is intensified. Baroclinicity also increases on the eastern side of the incipient low, where a low-level jet develops and brings warm and, later, humid air. The baroclinicity increases during 1 or 2 days after the formation of the low. Also during this time an active region forms in the jet stream. Then the cyclone assumes the familiar look of a frontal cyclone.

The development of the Colorado cyclone is illustrated in Fig. 10-19. The frames of this figure are about 12 h apart. This figure, a composite of many winter cases, contains the sea-level isobars, the surface position of the polar front, the core of the low-level jet (typically about 1 km above the surface), and the core of the polar jet at about 9 km elevation. The average wind speed in the low-level jet is marked on the core of this jet.

Figure 10-19a shows the synoptic hour before the closed low was noticed. When the first low appears, shown in part (b), this region is still on the upwind side of the upper-level trough. This is the inactive

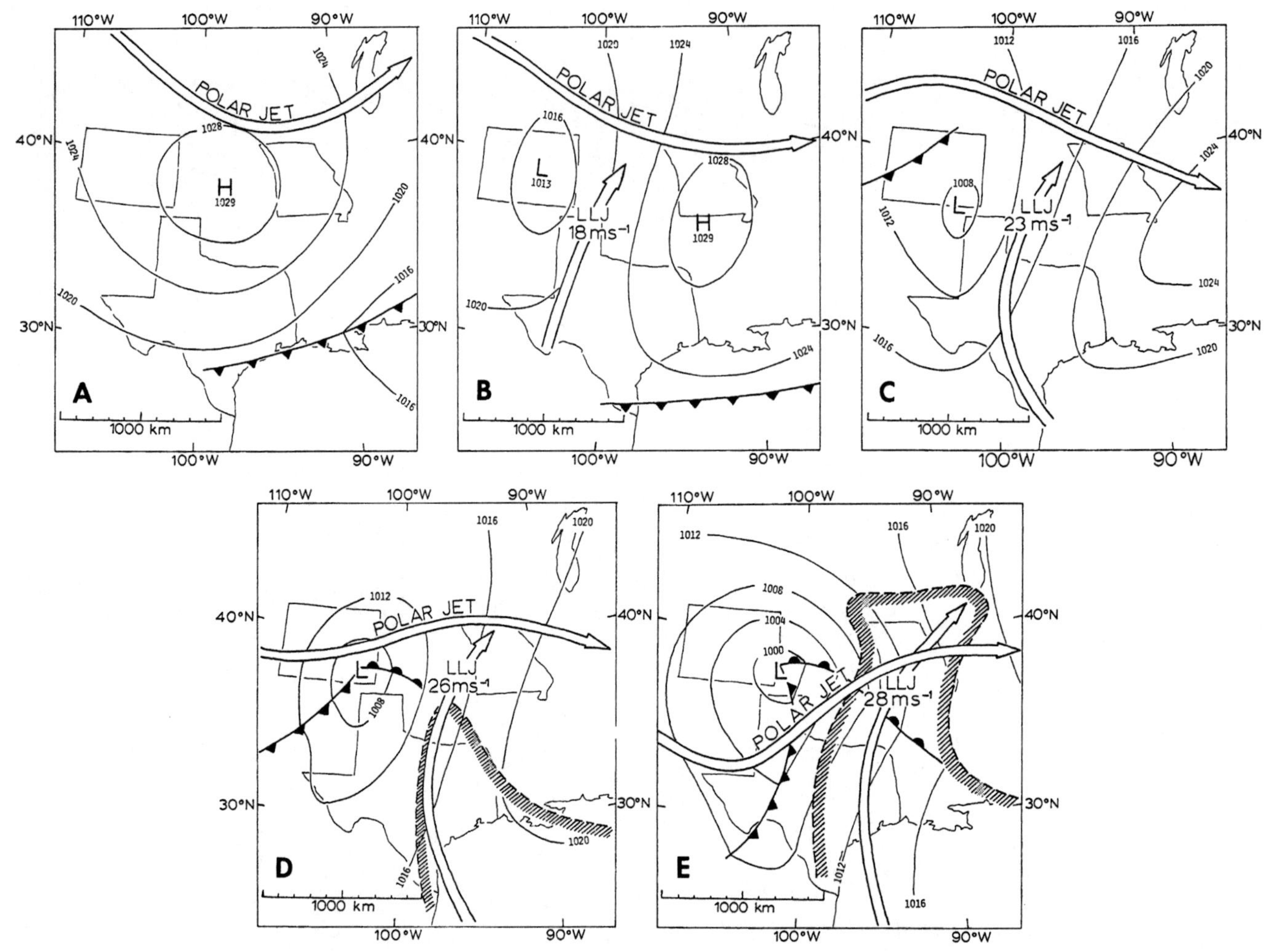

FIGURE 10-19. Development of the Colorado cyclone, illustrated by a sequence of charts approximately 12 h apart. This is a composite model, based on numerous observed cases. The cores of the polar and low-level jets are shown by double arrows. The polar front is shown at the earth's surface. Humidity over 70% at the low-level jet is enclosed by a dashed line with shading. Typical maximum wind in the low-level jet is entered at the jet core. From Djurić and Ladwig (1983), by permission of the American Meteorological Society.

region, where negative vorticity advection indicates subsidence. The pressure falls because of adiabatic warming in the westerlies as they descend the mountains. Southerly flow develops in the lowest kilometer of the atmosphere, near the low. This is the beginning of the low-level jet that later serves as a vehicle for transporting warm air and water vapor into the cyclone. In part (c), the low-level jet expands so that its entrance reaches the Gulf of Mexico. Starting at that time, significant amounts of water vapor are drawn into the circulation of the cyclone. The polar front has not yet entered the cyclone. In part (d), the polar front arrives from the west, ahead of the trough in the upper troposphere. A warm front is now noticeable in the low, so that the cyclone assumes the familiar shape of a frontal cyclone. In both parts (d) and (e), we also find the trough and ridge at the polar jet level, again similar to the cyclone of the Norwegian School (Fig. 10-2).

Low-Level Jet

Besides cyclogenesis, Figure 10-19 also shows the low-level jet stream (LLJ). This is an important part of the extratropical cyclone since it provides the mechanism for the supply of sensible and latent heat. The heat, in turn, is used for the development of middle-latitude storms: cyclonic and convective.

The LLJ develops in all significant extratropical cyclones. The core of the LLJ is drawn by a double line in Fig. 10-19. Probably due to difficulties with collection of upper-level wind data, the attention of researchers in the past was seldom drawn to the LLJ. This may explain why the LLJ is absent in earlier models of the extratropical cyclone. Whereas the LLJ had been known for decades, it was first introduced in the model of the extratropical cyclone by Newton (1967). Figure 10-18e is very similar to Newton's model, but the core of the LLJ is placed further away from the cold front than in Newton's model, as is justified by more recent evidence.

The LLJ is usually between 500 and 1000 km wide, 1000 to 2000 km long, and about 2 km high. Maximum speed in it is about 35 m s^{-1} (65 kt). Maximum wind appears about 1 km above ground. Maximum wind speed seldom appears at the standard level of 850 mb. There is often the subtropical trade wind inversion about at the level of maximum wind.

The LLJ is best developed before sunrise. In the afternoon, the southerly flow may not have a prominent maximum in the lowest 1–2 km. Then the LLJ turns into a general southerly flow in the lower half of the troposphere.

10-4 EXAMPLES OF WARM AND TRANSITIONAL ANTICYCLONES

Several anticyclones are illustrated in Figs. 10-20 and 10-21, where we see the westerlies in the northern part of the charts. The polar front is maintained in several troughs, mainly north of the belt of anticyclones.

The large anticyclone over the Atlantic (Fig. 10-21) is a warm anticyclone since there is a high center above it (at 300 mb, Fig. 10-20), although the chosen contours do not close around the center at 300 mb. The ridge (serrated line, Fig. 10-20) from the high over the Atlantic extends towards the southwest and assumes a zonal position about 15°N over the Caribbean Sea. Therefore, the Atlantic anticyclone and ridge can be described as parts of the belt of subtropical highs.

The sea-level high centers near Florida and over the Adriatic Sea also belong to the belt of subtropical highs (Fig. 10-21). This belt is normally farther north at the surface than in the higher atmosphere. This example (Figs. 10-20 and 10-21) also illustrates the tilt of the subtropical belt: Each sea-level high has a corresponding upper-level high or a ridge farther south. All these highs are warm, or dynamical, of the type that appears between the westerlies and easterlies.

The anticyclone centered south of the Great Lakes (Fig. 10-21) is a transitional anticyclone. Its eastern part, over the East Coast of North America, is under a trough in the westerlies (Fig. 10-20). Its western part (Great Lakes and Mississippi Valley) is under a 300-mb high from which several ridges branch out. Therefore, the eastern part can be described as a cold anticyclone and the western part as a warm anticyclone. Such anticyclones normally form in polar air at the surface. The polar front is still drawn around this anticyclone, from the Atlantic Ocean into the southern United States. The eastern part of this front, near Newfoundland, is easily identified. There is a sharp trough in the surface pressure field and a 5–10°C difference in temperature across this front. This part of the front is under the active region of the polar jet stream, downstream from the trough, with lifting and frontogenesis. The other part of this front, over the eastern United States, almost vanished. This is an inactive region of the polar jet stream, upstream from the trough, with subsidence and frontolysis. On the polar side of this front we find the dew point of 20°C in Knoxville, Tennessee, and 19°C in Greensboro, North Carolina. On the equatorial side, the dew point is 21°C in Montgomery, Alabama, and 19°C in Macon, Georgia. Since the difference in dew point is so insignificant, this front is not easily found. The front can be recognized by a weak wind shear and by following its position from previous charts. There is a great similarity between this example and the illustration of the passive front over the central United States in Figs. 7-24 through 7-26.

The anticyclone over the eastern United States can be characterized as transitional not only in space, but also in time. The air mass in this anticyclone was clearly polar a few days before the time of the chart in Fig. 10-21. Polar air had been traveling over the warm and partly wet surface of the continent. Consequently, polar air heated and acquired a higher dew point. The above example of dew point in several stations shows how this air became humid. The dew point of 20°C is too high for genuine polar air. In addition to heating and absorbing water vapor, subsidence in the central and western part of the anticyclone contributed significantly to the warming of the anticyclone.

Another process contributes to the anticyclone's transformation in the region of the warm front on the transitional anticyclone's western end. In that region, there is warm advection in middle and lower layers of the troposphere. This is followed by the progression of the warm front on a surface chart. Sometimes, especially in winter when air in upper layers has been substituted, the surface layer of air may stay cold for a day or two. Then the surface layer may warm up rather

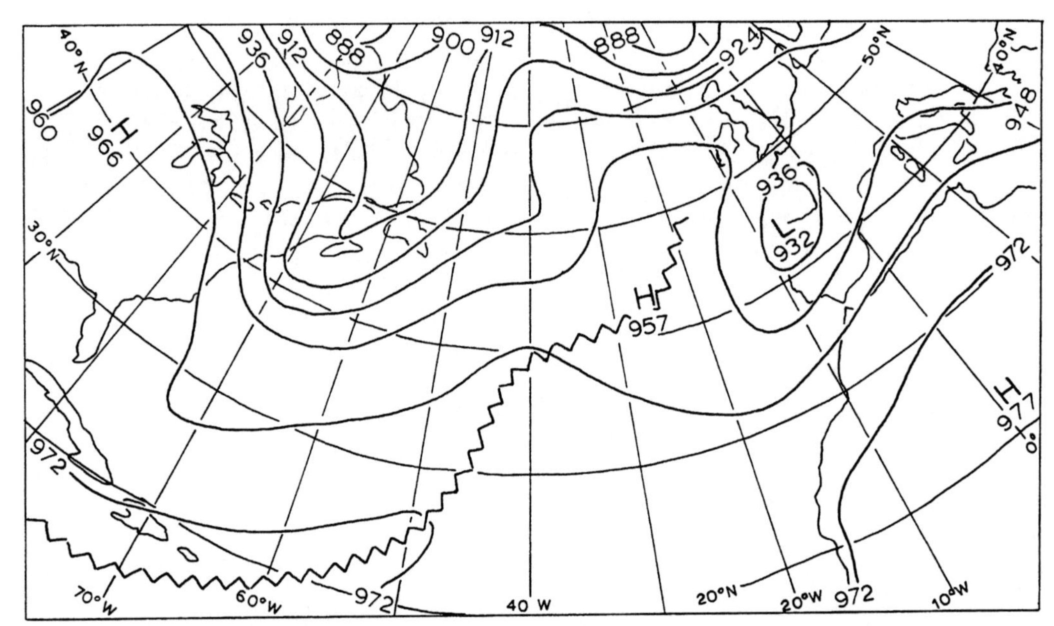

FIGURE 10-20. 300-mb contours (intervals 12 gp dam) for 0000 UTC 26 June 1983. A ridge over the Atlantic is shown by a serrated line.

quickly and the warm front on the chart needs to be moved downwind (usually toward the northeast) discontinuously. On other occasions, it is very difficult to detect the warm front at all and it is omitted from the chart. For example, on the western side of the large warm anticyclone over the Atlantic in Fig. 10-20 there is no warm front. Instead, the next cold front slowly approaches. There are two troughs in the southerly flow in that anticyclone. Each may be considered as a candidate for the warm front, but in absence of firm

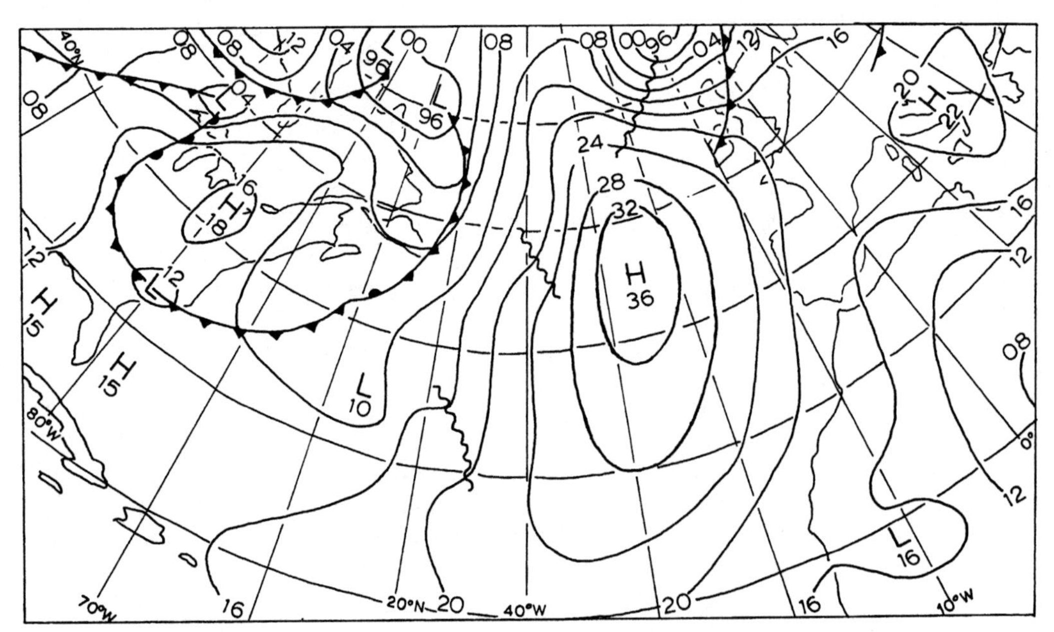

FIGURE 10-21. Sea-level isobars (mb, last two digits) and surface fronts for 0000 UTC 26 June 1983. The anticyclone over the Atlantic is warm since it is under a ridge at 300 mb (Fig. 10-20). The anticyclone at the Great Lakes is transitional since it is located between a ridge and a trough at 300 mb.

evidence, it is advisable to omit placing the warm front.

The anticyclone over the Adriatic Sea (Fig. 10-21) still has some transitional character, but it has evolved more toward the warm type than the anticyclone over the eastern United States. The 300-mb flow shows an inactive region above this anticyclone (the region between the trough and the upstream ridge). The cold front around the anticyclone has already dissipated. The center of the anticyclone now tilts towards the south, so that its upper-level part is over Africa. We conclude that this is now a warm anticyclone in the belt of subtropical highs.

11

The Upper Troposphere and Jet Streams

Weather processes crucially depend on the structure and dynamics of jet streams. For a proper interpretation of weather processes, weather analysts must be familiar with the structure and dynamics of jet streams. The structure of major tropospheric jet streams is described in this chapter. The dynamics are postponed to the next course.

The structure of jet streams is hydrostatically and geostrophically related to the distribution of temperature. Therefore, the major thermal structures in the atmosphere are described in this chapter, in relation to jet streams.

11-1 UTILIZATION OF SOUNDINGS AND VERTICAL SECTIONS

Model Thermal Profiles

Soundings make the comparison of the temperature of air masses in deep layers possible. The model for thermal comparison of air masses is shown in Fig. 11-1, which presents three schematic soundings for the three main air masses: polar, middle-latitude, and tropical.

In the troposphere, polar air is the coldest and tropical air is the warmest. The situation is reversed in the stratosphere: The polar stratosphere is normally warmer than the middle-latitude stratosphere, and the tropical stratosphere is the coldest. The level of the tropopause is also characteristic for air masses. We expect a low tropopause above polar air (*polar tropopause,* not too far from 350 mb), a *middle tropopause* near 200–250 mb above middle-latitude air, and a high *tropical tropopause* near 100 mb above tropical air. A secondary tropopause near 200 mb often appears in tropical air. This is the *secondary tropical tropopause*.

Arctic air can be found under a low inversion within polar air, up to 1–3 km above the surface. Arctic air generally does not extend to the tropopause and does not occupy much space in the atmosphere since it is shallow. However, this air mass covers wide areas of the earth each winter.

Air masses can be identified by comparing the observed soundings with the model. If a sounding penetrates a front, its graph will follow one model and jump to another model at the front. The jump between air masses will appear as a frontal inversion on the graph. Unfortunately, analysis of soundings is not so simple. In routine work, many intermediate cases and

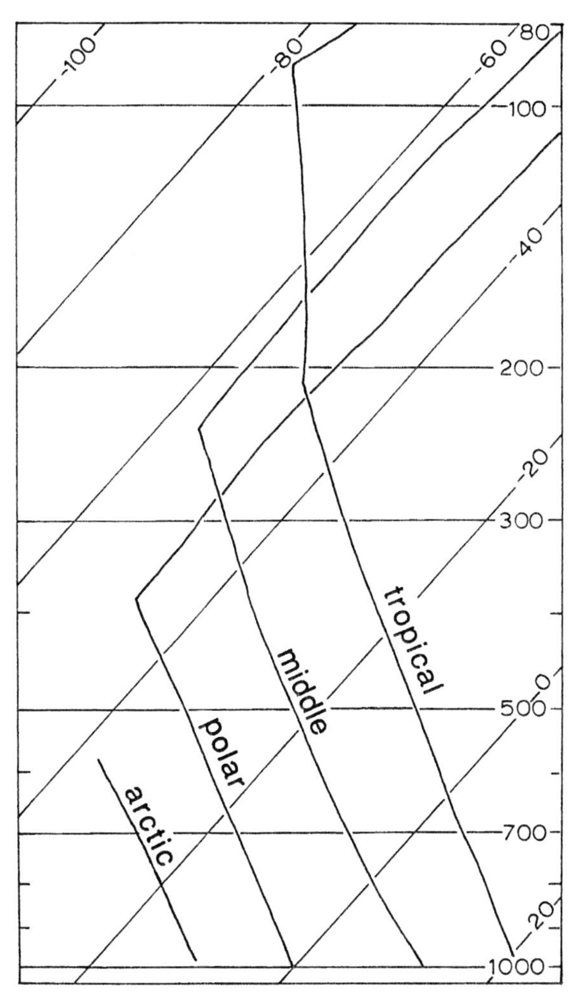

FIGURE 11-1. A schematic variation of temperature with pressure for main air masses, as represented on the skew *T*–log *p* diagram.

exceptions complicate the work. The examples that follow illustrate how soundings can be used to identify air masses.

Standard Atmosphere

It is useful to compare soundings with the standard atmosphere. The U.S. standard atmosphere has been selected to be similar to a typical sounding in the middle-latitude air mass. This air mass is most frequently found over the 48 contiguous states. The standard atmosphere is printed on the skew *T*–log *p* graph, so it is readily available. The tropopause in the standard atmosphere is at 226 mb. In the absence of measurements, and when no great accuracy is needed, the standard atmosphere provides a reasonable approximation of the thermal state of the atmosphere.

The sounding of the standard atmosphere is not an average sounding, as can be recognized by the kink at the tropopause. In an average sounding, this kink is smoothed out. The standard atmosphere thus represents a typical sounding rather than an average sounding. The standard atmosphere is drawn as a smooth line with one kink in each of the following examples.

11-2 EXAMPLES OF SOUNDINGS AND VERTICAL SECTIONS

Air Masses

The soundings in Fig. 11-2 are examples of a graphical, comparative identification of air masses. The temperature and dew point are plotted on the standard skew *T*–log *p* chart.

Figure 11-2a shows that the atmosphere is warmer than the standard atmosphere almost everywhere below 165 mb. The tropopause is at 107 mb. Both these characteristics indicate that the troposphere is occupied by tropical air. However, the layer from the surface to about 860 mb is colder than expected in tropical air. A comparison with other charts reveals that the inversion between 840 and 860 mb is the polar front. This front passed the station 2 days before.

The inversion near 605 mb is a typical subsidence inversion, very similar to the model in Fig. 5-3a. Such inversions develop in anticyclones with subsidence.

The layer between 615 and 650 mb is adiabatic and the dew point follows the w_s = const. isopleth. This means that the inversion between 610 and 615 mb is reinforced by strong turbulence in the adiabatic layer under it. Turbulent mixing affected equalization of the potential temperature and mixing ratio in the adiabatic layer between 615 and 650 mb. This caused a lowering of temperature at 615 mb and a reinforcing of the subsidence inversion between 610 and 615 mb.

The sounding in Fig. 11-2b shows that the temperature above 700 mb is much closer to the standard atmosphere than in parts (a) and (c). This part of the sounding illustrates middle-latitude air. The middle tropopause is at 261 mb, which is typical for this air mass.

Figure 11-2c shows the polar tropopause at 430 mb. The troposphere is colder and the stratosphere is warmer than the standard atmosphere. These are characteristics of polar air. Arctic air can be identified under 800 mb in this sounding, since the temperature is another "step" lower than the continuation of the polar air curve in the layer between 430 and 600 mb.

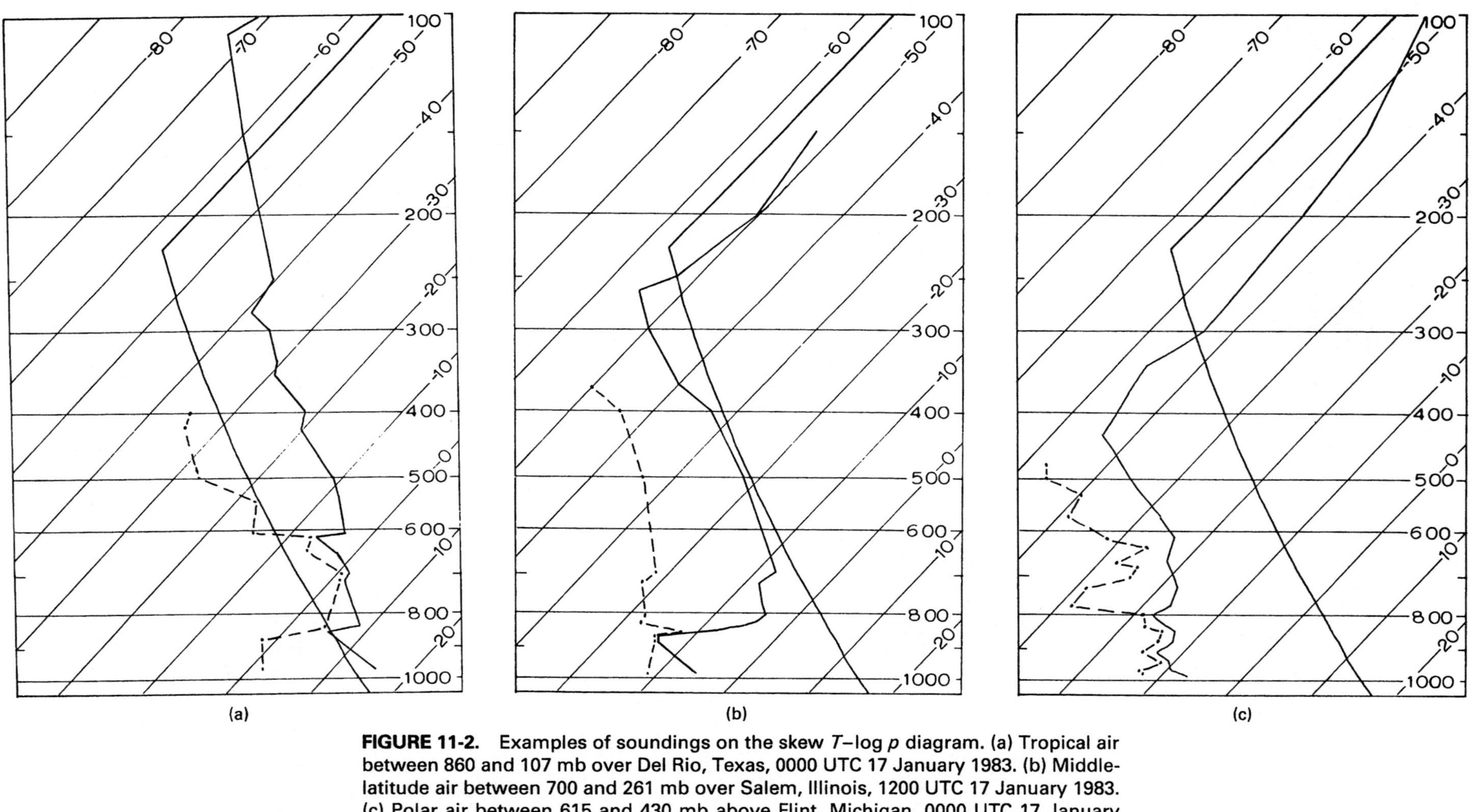

FIGURE 11-2. Examples of soundings on the skew T–log p diagram. (a) Tropical air between 860 and 107 mb over Del Rio, Texas, 0000 UTC 17 January 1983. (b) Middle-latitude air between 700 and 261 mb over Salem, Illinois, 1200 UTC 17 January 1983. (c) Polar air between 615 and 430 mb above Flint, Michigan, 0000 UTC 17 January 1983.

Fronts

The soundings often penetrate through several air masses, one above the other. Then the frontal inversions or frontal stable layers can be observed between air masses. The soundings in Fig. 11-2 all show more than one air mass and the fronts between them.

As mentioned above, the inversion between 840 and 860 mb is expected to be the polar front. Both the temperature and dew point increase with height through this inversion, assuming the frontal character of Fig. 5-3b. However, the temperature in the presumed polar air near the earth's surface is close to the standard atmosphere, differently than required by the model in Fig. 11-1. Also, the wind near the earth's surface at Del Rio is already from the south. Such are the characteristics of polar air that is already returning from the south, where it had arrived 1 or 2 days earlier. This returning polar air has been heated from the underlying surface, which explains why its temperature is more typical for middle-latitude air than for polar air.

The Salem sounding (Fig. 11-2b) shows two fronts: a weak polar front between 700 and 725 mb and a prominent arctic front between 810 and 870 mb. The arctic front is atypical, since polar air above is of low humidity. Low humidity in this area can be explained by subsidence since this region is situated upstream from an upper-level trough. This is an inactive region in the westerlies.

The Flint sounding (Fig. 11-2c) shows a multiple arctic front. The stable layer of 615–670 mb and the inversions of 780–800 mb and 880–915 mb can each be interpreted as the arctic front. Such multiple inversions on top of the arctic air mass are common and may create an uncertainty as to which inversion to designate as the arctic front. One practical way is to take the highest inversion. This leaves the lower inversions to be designated as radiation inversions or elevated radiation inversions within arctic air.

In rare cases, some soundings do not show a tropopause. This may occur in soundings that ascend through the vertical part of the polar frontal layer, in the break between the polar and middle tropopauses. Another case of a vanishing tropopause may occur in the cold weather of the polar night, as mentioned at the end of Chapter 1. Then the whole troposphere may disappear and a stratospheric temperature increase with height starts at the earth's surface.

Vertical Sections

The technique for constructing vertical sections is similar to the technique used with other charts. First, isopleths are drawn of relevant scalar variables. (This first step can be done by a computer). Next, the models are recognized and sketched. Finally, the isopleths and models are adjusted to each other. The final analysis satisfies both models and data. This technique is also shown in Section 8-8.

In vertical sections, the common choice is to draw either isotherms and isotachs or isentropes and isotachs. For some purposes it may be useful to construct isotachs of only one wind component, preferably the component normal to the section.

An Example of a Vertical Section

The vertical distribution of temperature and wind drawn in Fig. 11-3 contains a finished set of isotherms, isotachs, fronts, and tropopauses. Several instructive steps in the analysis process occurred. As the isotherms were drawn, the analyst had the opportunity to spot several inversions.

When the inversions appear over neighboring stations at about the same level, they can often be connected. This was done with the inversions between 800 and 950 mb in the three northern stations in Fig. 11-3. This inversion is interpreted as the arctic front. The polar front can be found by the abrupt change in tropopause level between JFK (New York, New York) and DCA (Washington, D.C.) and by the thermal contrast between DCA and GSO (Goldsboro, North Carolina) between 350 and 650 mb. Also, inversions at DCA (312–323 mb) and GSO (726–758 mb) are connected in such a way that the model of the front is formed. The level of maximum wind near the front is at 250 mb over DCA, implying a vertical frontal layer at this level.

Several inversions and stable layers at other locations do not form known models. These layers should always be indicated by heavier lines exactly as the fronts, even if we cannot give them a more specific name. These layers may play an important role in later weather development.

The tropopause over DCA, GSO, and CHS (Charleston, South Carolina) is near 200 mb, which is typical for the middle tropopause. The inversion at 103 mb over GSO and the stable layer at 102 mb over CHS may represent a distant protrusion of the tropical tropopause. The region of fast wind, with speed over 100 kt, is located in middle-latitude air, as is typical for the polar jet stream. The level of maximum wind is near 250 mb, which is higher than average, but not unusual for the polar jet. This region is located in a ridge in the westerlies, and in ridges the polar jet tends to be at higher elevation than in the troughs. Otherwise, a more common level for the polar jet stream core is near 300 mb.

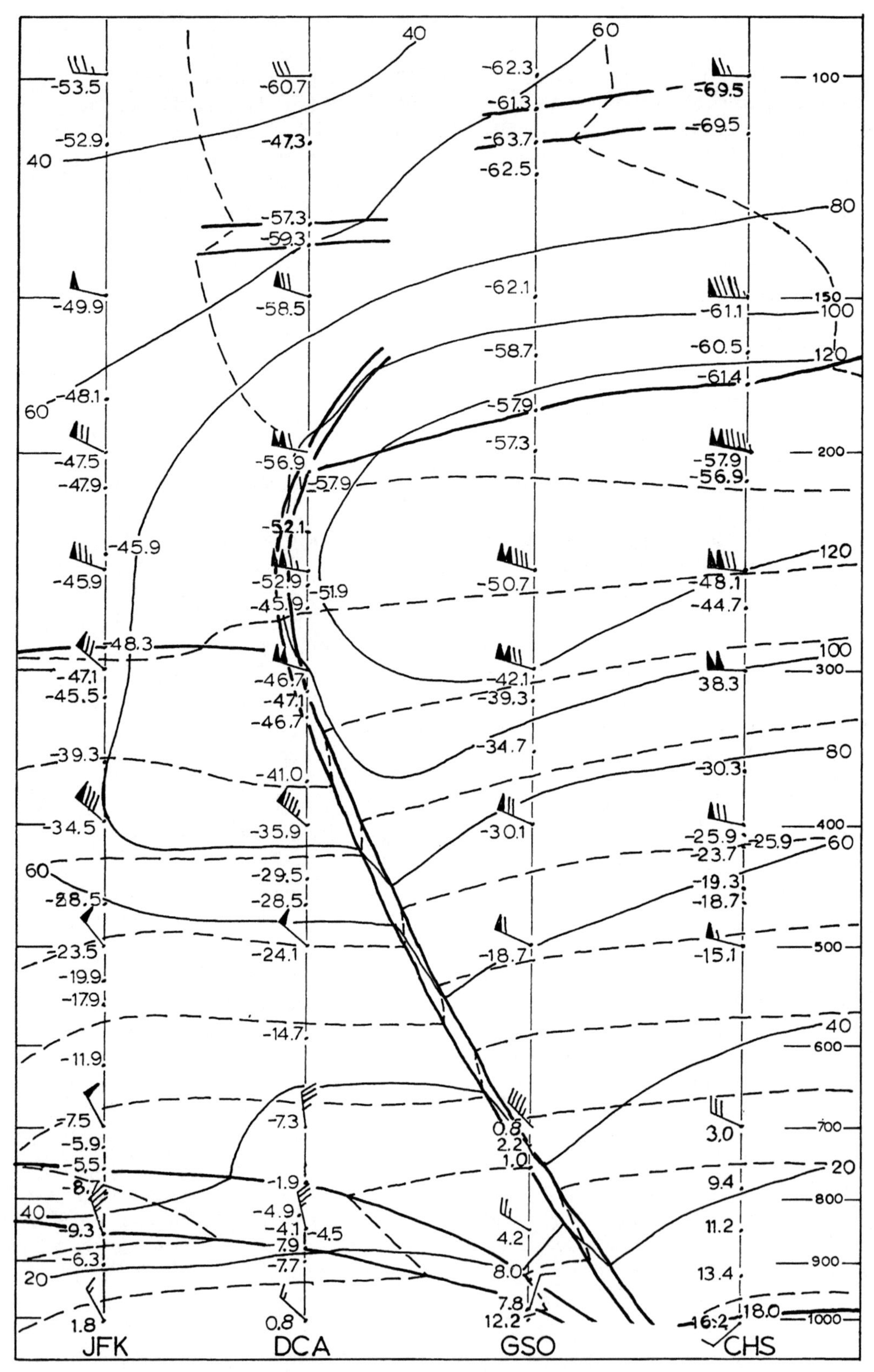

FIGURE 11-3. Vertical section through the atmosphere for 0000 UTC 21 January 1973 from JFK (New York, City) to CHS (Charleston, South Carolina). The temperature is plotted in °C at observation points. Wind is in knots. Isotherms are dashed, unlabeled, in intervals of 8°C; isotachs are solid lines, in intervals of 20 kt.

Sections with Isentropes

Figure 11-4 shows a vertical section with isentropes. The isotachs are drawn for the geostrophic wind component normal to the section. Both sets of isopleths have been drawn automatically using a method that interpolates polynomials between the observation points. The routine used also pays attention to the preservation of stability in the spaces between stations. This is a very successful technique, providing a high level of continuity of layers in the spaces between stations. The analysis (placement of models) can be done easily in a section where the isopleths of such quality are drawn.

The basic model in the field of isentropes is the appearance of dense isopleths in stable layers. The stratosphere in the upper part of Fig. 11-4 is an area of dense isentropes. Inversions are also layers with dense isentropes. Some of the inversions represent atmospheric fronts. The western half of Fig. 11-4 shows the polar front as a belt of dense isentropes between the surface at OAK (Oakland, California) and about 400 mb at LND (Lander, Wyoming). This western part of the polar front is typically less steep than the eastern part between LBF (North Platte, Nebraska) at 360 mb and OMA (Omaha, Nebraska) near 900 mb. The difference in the frontal slope is related to the general lifting and sinking in the respective active and inactive regions of the jet stream.

The steep eastern frontal layer in Fig. 11-4 shows signs of splitting. In this region are two layers of enhanced stability: one with the lowest four isentropes immediately west of Omaha and the other around the isentrope of 300 K between OMA at 500 mb and LBF at 300 mb. Cold fronts in the westerlies often assume such a shape. The two layers of the polar front are of the type shown schematically in Fig. 8-15c. The upper part of the baroclinic zone (inclined isentropes) moves eastward faster than the part near the earth's surface, giving rise to a split structure. The upper cold front in such situations may even move ahead of the surface cold front (see Section 8-9).

The isotachs in Fig. 11-4 are negative in the western part of the section. This shows a northerly component of geostrophic wind. The northerly jet stream between SLC (Salt Lake City, Utah) and LND makes a loop south of the section and returns as a southerly wind between LBF and OMA. The trough in the westerlies is in the region between the two wind maxima, where the tropopause is low between LND and RAP (Rapid City, South Dakota). The polar tropopause in the loop of the jet stream is near 360 mb. A higher middle tropopause is seen outside the loop of the polar jet stream.

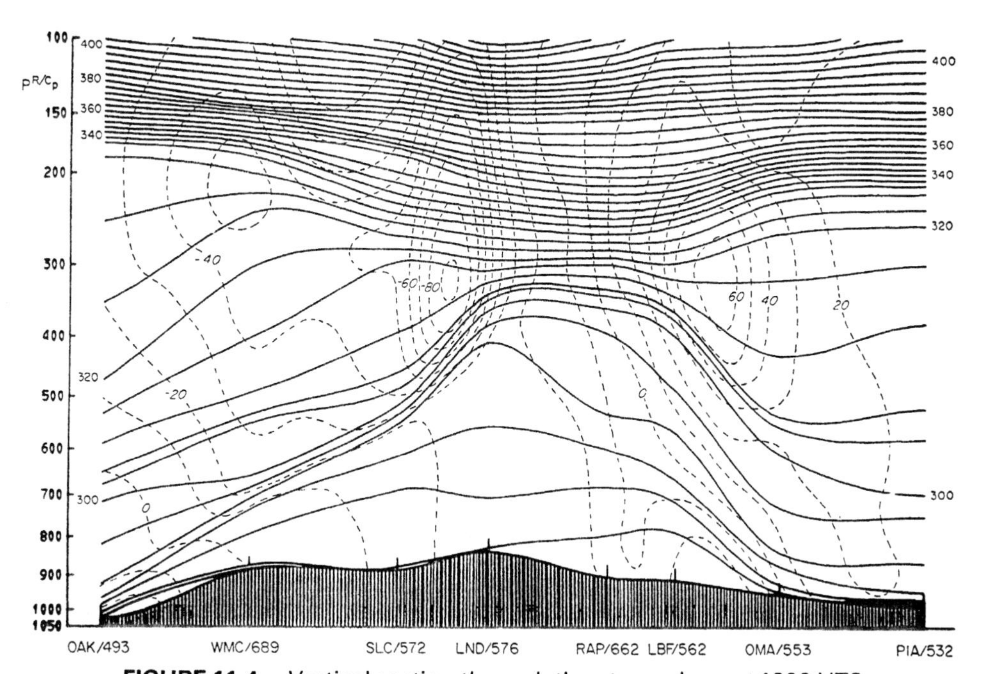

FIGURE 11-4. Vertical section through the atmosphere at 1200 UTC 7 December 1963 from OAK (Oakland, California) to PIA (Peoria, Illinois). Isentropes are in intervals of 4 K. Isotachs are for the normal component of geostrophic wind (m s^{-1}), with negative numbers showing air coming out of the plane of the section. From Shapiro and Hastings (1973).

11-3 JET STREAMS AND FRONTS

This section describes the polar and subtropical jet streams primarily as they appear on isobaric charts. The relationship of the jets to the fronts and tropopauses is emphasized, and the models of jets and fronts are shown. The regular occurrence of these models in the atmosphere suggests that the atmosphere possesses stable states that the air flow preferably assumes.

Dimensions of Jet Streams

The polar jet stream usually has a large vertical extent, allowing it to be identified on several standard pressure levels. The maximum wind in the polar jet is usually near 300 mb, but the location of this jet can also be seen on the 500- and 200-mb levels. The subtropical jet, however, tends to be rather shallow, sometimes thinner than 3 km. Therefore, it may not show up on either 200- or 300-mb standard levels. If the subtropical jet stream appears on standard levels, the 250- or 200-mb charts are preferable for its location.

Jets, both polar and subtropical, extend around the globe, but they have spots with lower wind speed. Normally the slower spots in the jet streams are at intervals between 1000 and 5000 km. A segment of the jet between the slower spots is the *jet streak.*

There are no natural signs determining the width or the vertical extent of the jet stream. For convenience we often determine the width of the jet stream between points where the wind speed decreases to one-half its maximum value in the core. This rule is also practical in the vertical. When two jet streams are close to each other, such a determination of width may fail. The wind speed may not decrease to one-half before it increases again in the other jet. In such cases, the line of minimum wind may formally play the role of the boundary between the two jet streams.

Polar Jet at 300 mb

The model of the polar jet stream, as it may appear at the 300-mb chart, is shown in Fig. 11-5. This figure covers a jet streak of the polar jet, the segment between the points A and E. These points lie in adjacent minima of the wind speed along the jet core. The upstream part of a jet streak, near point B, is the *entrance*, and the downstream part is the *exit*, near point D. The maximum wind speed is at point C. Care must be taken when interpreting this as the maximum wind in the whole jet streak. The wind speed may be higher at some other level, between standard levels. The contours are not shown in Figs. 11-5 and 11-6, so the picture is not cluttered. The contours are approximately aligned along the jet core. Some common deviations of the contour direction from the jet core are shown in several examples below.

Intervals of every 20 m s^{-1} (or 40 kt) in isotachs are usually sufficient to show the jet stream. Isotachs are typically oriented along the jet core. The greatest curvature of the isotachs can be found in the core of the jet stream. The isotachs and other isopleths kink on the front, since the front is a discontinuity.

The density of the isotachs can be large on the polar side of the jet stream, since there is no dynamical restriction to the magnitude of the cyclonic shear. The strongest (and always cyclonic!) wind shear is within the nearly vertical frontal layer. This region may be accentuated on weather charts by drawing the frontal layer enclosed with two lines (blue, if color), similar to Fig. 8-18. Strong cyclonic wind shear on the polar side of the jet core is the best cue for positioning of the

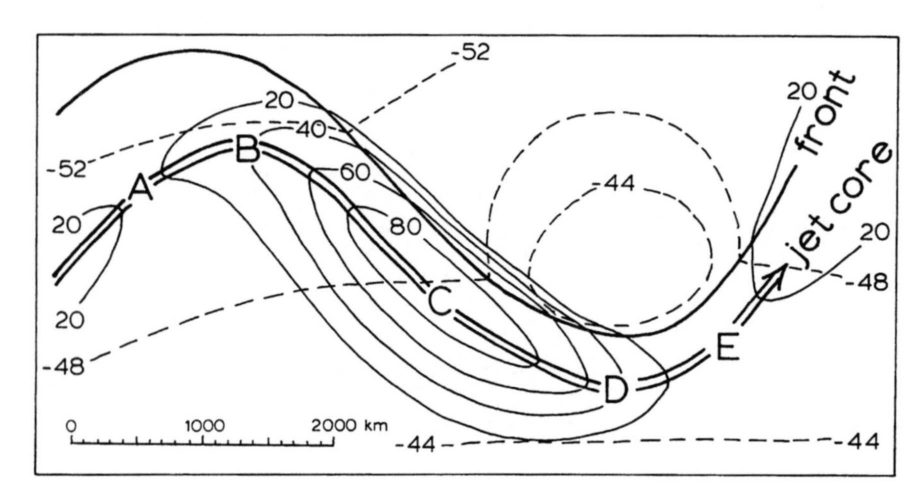

FIGURE 11-5. A model of the polar jet stream at 300 mb. The core of the jet is shown by a double arrow. Isotachs (m s^{-1}) are solid, isotherms (°C) are dashed, and the polar front is the heavier solid line.

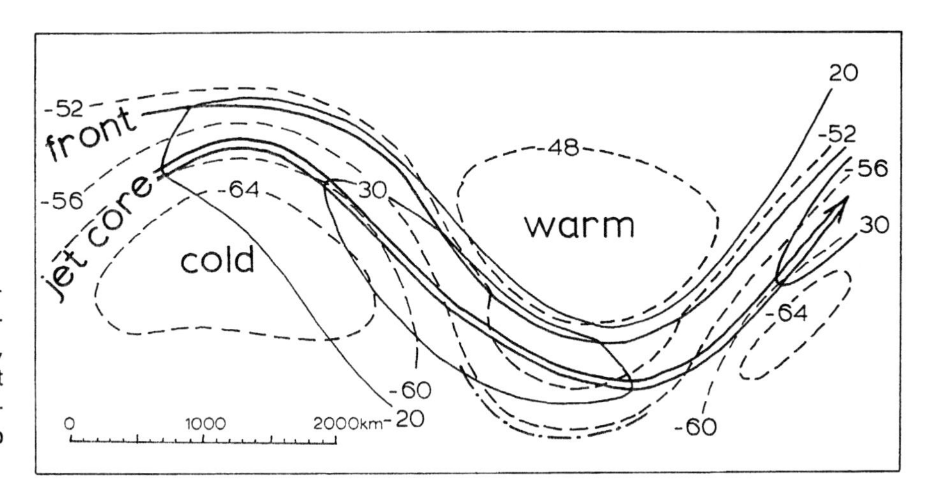

FIGURE 11-6. A model of the polar jet stream at 200 mb. The symbols are the same as in Fig. 11-5, except the dash-dotted line that shows the intersection of the middle tropopause with the 200-mb level.

polar front. Wind shear takes the place of thermal contrast when the front is vertical (Section 8-3).

The anticyclonic shear on the equatorial side of the jet stream is restricted to absolute values less than or equal to the Coriolis parameter (Section 4-3). Observations confirm that the strongest observed anticyclonic wind shear does not exceed the Coriolis parameter.

The isotherm of −48°C in Fig. 11-5 illustrates a weak thermal contrast at the front at 300 mb. This isotherm crosses the front, showing the same temperature on both sides. Weak thermal gradient, shown by a scarcity of isotherms, is also typical near an upright polar front. However, the isotherms kink at the front, showing discontinuity in the thermal gradient (∇T). The region on the front's polar side is the polar stratosphere. The warm center of over −44°C is due to subsidence of the air from higher stratospheric layers, which warms adiabatically. This explains the warm centers on the polar side of the jet stream. The polar stratosphere reaches its lowest levels in cyclonic loops of the jet streams. On occasion, the tropopause can be observed between 500 and 600 mb.

Thermal Structure of the Polar Front at 200 mb

The polar front assumes a very steep, possibly vertical, position on the polar side of the jet. The thermal contrast across the front is usually very weak, making it difficult to find the front on the basis of isotherms. If a temperature contrast appears, it is often of the reverse type, with the polar side warmer than the equatorial side. This is often the case in the troughs, where the tropopause descends deepest.

The isotherms in Fig. 11-6 illustrate the reversal of the thermal gradient at 200 mb between polar and middle-latitude air masses and in the nearby layers of the stratosphere. Caution should be exercised when talking about gradients near fronts. Gradients, like all derivatives, are not defined on the discontinuities. However, the same scalar field can be formally smoothed, whereby the discontinuity is eliminated. In this way, the gradient becomes determinate.

Polar Jet and Front at 200 mb

The main features around the polar jet at 200 mb are illustrated in Fig. 11-6. Similar to Fig. 11-5, we have a long jet core showing a jet streak in the middle. The wind speed is lower at 200 mb than at 300 mb.

Wind shear is still the most prominent signature of the front. Therefore, we can identify the front as a zone of strong cyclonic shear. The polar front can be recognized by the isotherms a little easier at 200 mb than at 300 mb, although the temperature is systematically higher on the polar side of the front. This is attributed to the usual absence of temperature lapsing with height in the polar stratosphere. The 200-mb level is deep within the polar stratosphere. On the other hand, this same 200-mb level is near the bottom of the middle stratosphere. Since the temperature lapses with height in the troposphere, the higher the tropopause, the lower its temperature. Therefore, warmer air is normally found on the polar side of the polar front at 200 mb.

Within the stratosphere, the equatorial side of the polar front is characterized by strong baroclinicity, that is, by large $|\nabla T|$ or dense isotherms. Judging by dense isotherms, it may appear that the front is on the other side of the jet core, as illustrated by the heavy dash–dot line south of the jet core in Fig. 11-6. When such a discontinuity line appears, it may represent the interception of the middle tropopause with the isobaric surface (level of the chart). South of this line is the troposphere of middle-latitude air mass; north of the

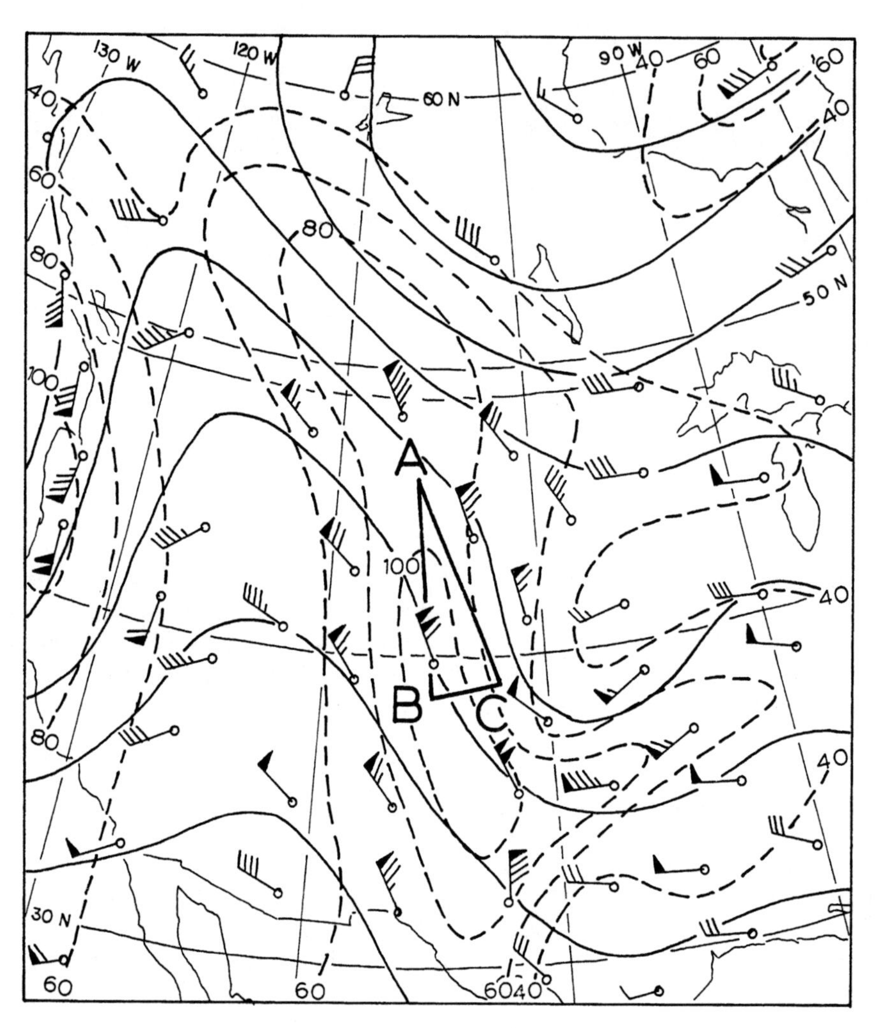

FIGURE 11-7. Wind barbs, isotachs (kt, dashed), and contours (solid, intervals of 60 dam) at 300 mb, 1200 UTC 24 January 1983. The core of the jet stream stretches along the segment *AB* at an angle across the contours, showing a moving wave in the westerlies.

line is the stratosphere. This interception looks like the polar front, with concentrated isotherms. If detected, this line of discontinuity should be drawn as a blue line on weather charts, like the front. Another blue line for the front proper should be used on the polar side of the jet, so that the agreement with the model in Fig. 6-9 is kept.

The slope of the front near 200 mb is usually opposite from the slope at 500 mb. This is explained by the reversal of the temperature difference across the front. Therefore, the front at 200 mb normally appears closer to the equator than the front at 300 mb. Figures 11-13 and 11-14 (later in this chapter) represent the core of the polar jet at exactly the same location, but the front is closer to the jet core at 200 mb.

Wave Motion

An appreciable angle appears between the jet core and contours in some situations. One such situation where the core of the jet and the contours intersect appreciably is near fast-moving waves in the jet stream. An example of an intersection of contours and jet core is shown in Fig. 11-7. The direction of the polar jet stream at 300 mb (line *AB*) cuts the contours (line *AC*) at about 20°.

The relationship of the velocities of the wind and the wave pattern is interesting. The wave pattern in this example includes the jet streak, the ridge upstream, and the trough downstream from the jet streak. In this case, the wave pattern moves toward the east with a speed (S) of about 30 kt. This speed was evaluated from the positions of the ridge (presently at 115°W) and trough (at 100°W) on the charts 12 h before and 12 h after the chart time in Fig. 11-7 (those other charts are not shown here). In contrast to the wave speed, the characteristic wind speed in triangle *ABC* is about 100 kt. The longest and shortest sides of triangle *ABC* have about the same ratio as the speeds of the system S and wind V:

$$\frac{AB}{BC} \approx \frac{V}{S} \approx 3$$

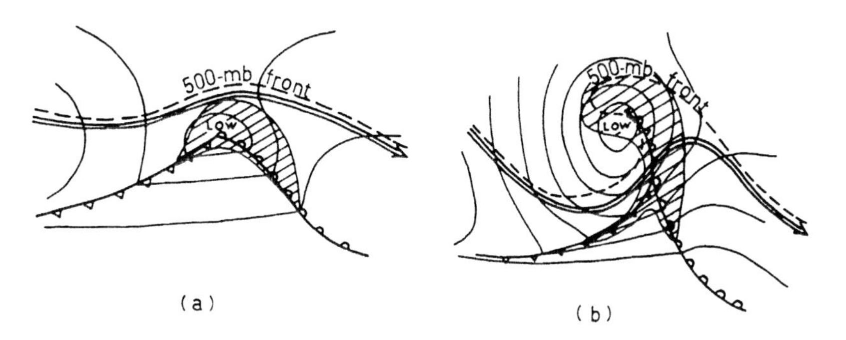

FIGURE 11-8. Relation between the polar front and the polar jet stream in the wave cyclone (a) and in the occluded cyclone (b). Precipitation areas are hatched. The dashed line is the 500-mb front. The double arrow is the jet stream. From Palmén and Newton (1969, p. 357), by permission of the Academic Press.

This illustrates that the pattern ridge–jet streak–trough moves so that the air parcels in the jet core stay in the core, even if it looks as if they are crossing the core at about 20°. The core moves to catch up with the parcels.

Warm Cut-Off

The polar jet and the polar front usually follow each other closely. The most noticeable systematic separation of the two occurs in occluded cyclones, where the jet stream may cut across the upper-level warm sector of the cyclone and the front may loop far north (Fig. 11-8). In Fig. 11-8a we have a wave cyclone, that is, a surface low on the polar front. The frontal position at 500 mb is close to the position of the jet core.

In Fig. 11-8b, the same cyclone is shown about 1 day later when the amplitude of the frontal wave increased greatly. Now the cyclone is occluded.

The surface cold front has reached the surface warm front, and the warm sector is eliminated from the cyclone center at the surface. However, the warm sector may be large at the upper levels, say between 500 and 300 mb. The front makes a loop around this warm sector, mainly in the northeastern half of the occluded cyclone. In this stage, the jet stream commonly cuts across the warm sector, cutting off an anticyclonic loop northeast of the occluded cyclone.

Figure 11-8 also explains the transformation of warm air. Warm air travels north in the warm sector of the cyclone. Northern protrusion of the old warm sector coincides with a ridge or a warm high. When the jet stream is established south of the warm sector, a new polar front generates near the jet stream.

The development of a warm cut-off high is similar to the development of the cut-off low in Fig. 5-6. However, the sides are reversed. If we flip Fig. 5-6 around a west–east line, we get a fair model of the formation of a warm cut-off high. The anticyclonic cut-off high slowly cools due to radiation, and in about a week the air in it becomes new polar air.

Subtropical Jet at 200 mb

The most prominent characteristics of the subtropical jet stream are represented in Fig. 11-9. Streaks are present along the jet. Branching occurs more often in the subtropical than the polar jet. The curved isotachs in the northeastern part of Fig. 11-9 show a case of branching. Branching need not be in the location shown relative to the jet streak. It may appear on the other side of the core or near the entrance of the jet streak.

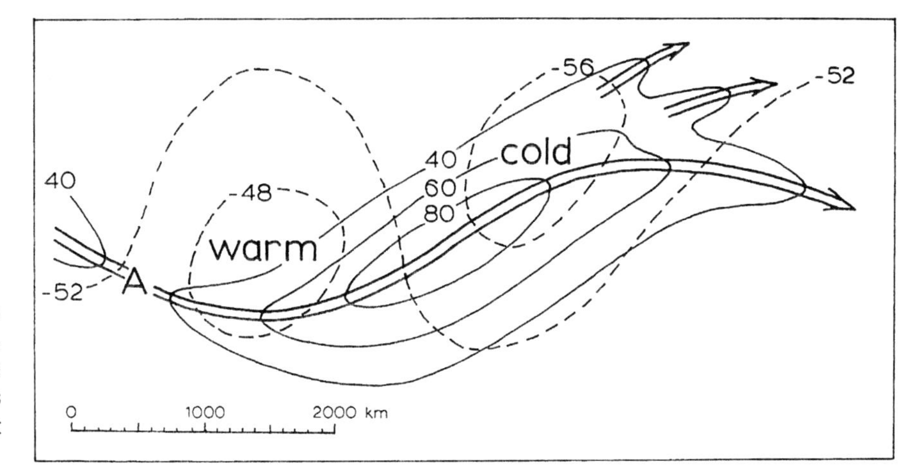

FIGURE 11-9. A model of the subtropical jet stream at 200 mb. The symbols are the same as in Fig. 11-5. Branching of the jet is shown on the exit end of the jet streak.

The minimum wind along the core (A in Fig. 11-9) is not necessarily the minimum speed in the jet stream. The jet stream may have sunk under the level of this chart. The subtropical jet stream is shallow and may appear and disappear on standard levels, resembling a dolphin on the surface of the ocean.

The cyclonic wind shear on the polar side can reach large values, say two or three times the Coriolis parameter. To the contrary, the absolute shear on the anticyclonic (equatorial) side does not exceed the Coriolis parameter.

The temperature distribution near the subtropical jet stream does not exhibit prominent patterns. Generally, it is colder near the ridges and warmer near the troughs around this jet stream. The temperature contrast around the subtropical front is usually small at 200 mb, and at this level the front is very steep or vertical. Since the isotherms are sparse, the jet in Fig. 11-9 is drawn without the accompanying front. The front may be entered on the charts if it is recognized, but it is a common practice not to draw the subtropical front in routine weather analysis. The examples that follow, however, give several empirical rules for recognizing the subtropical front on constant-pressure charts.

11-4 AN EXAMPLE OF JET STREAMS

Chart Construction

Figures 11-10 through 11-13 show the two upper-tropospheric jet streams over North America. The isotachs (Figs. 11-10 and 11-12) show comparatively narrow and elongated regions of high wind speed. The ridges in the isotachs, indicated by double lines, are the cores of the jet streams. The contours of the isobaric surfaces are not drawn since they lie along the wind vectors (barbs) and, therefore, do not give much additional information. To avoid cluttering, the isotherms are drawn in separate charts, Figs. 11-11 and 11-13.

During routine work in weather service, most of the lines from one level are drawn on the same chart,

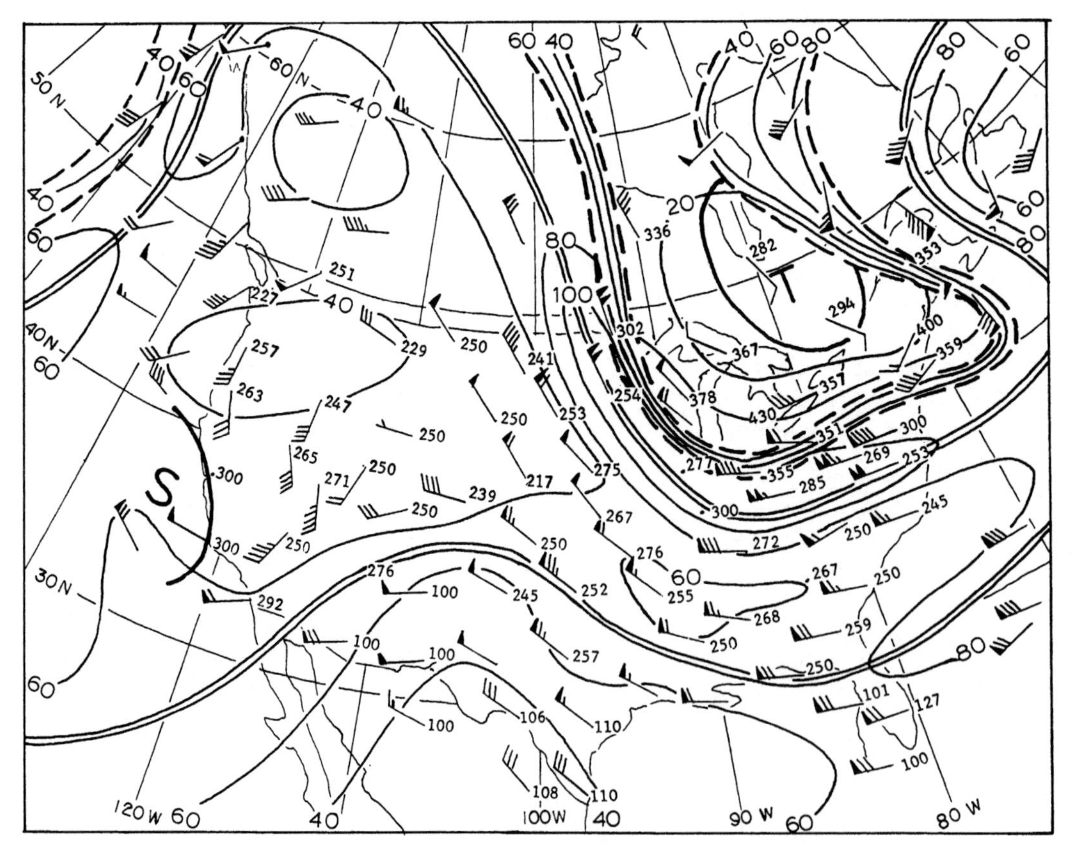

FIGURE 11-10. Wind barbs and isotachs (intervals of 20 kt) at 300 mb and the reported tropopause level (mb) for 0000 UTC 17 January 1983. The jet core is the double line, the polar frontal layer is bounded by thick dashed lines, and the tropopause is the thick line.

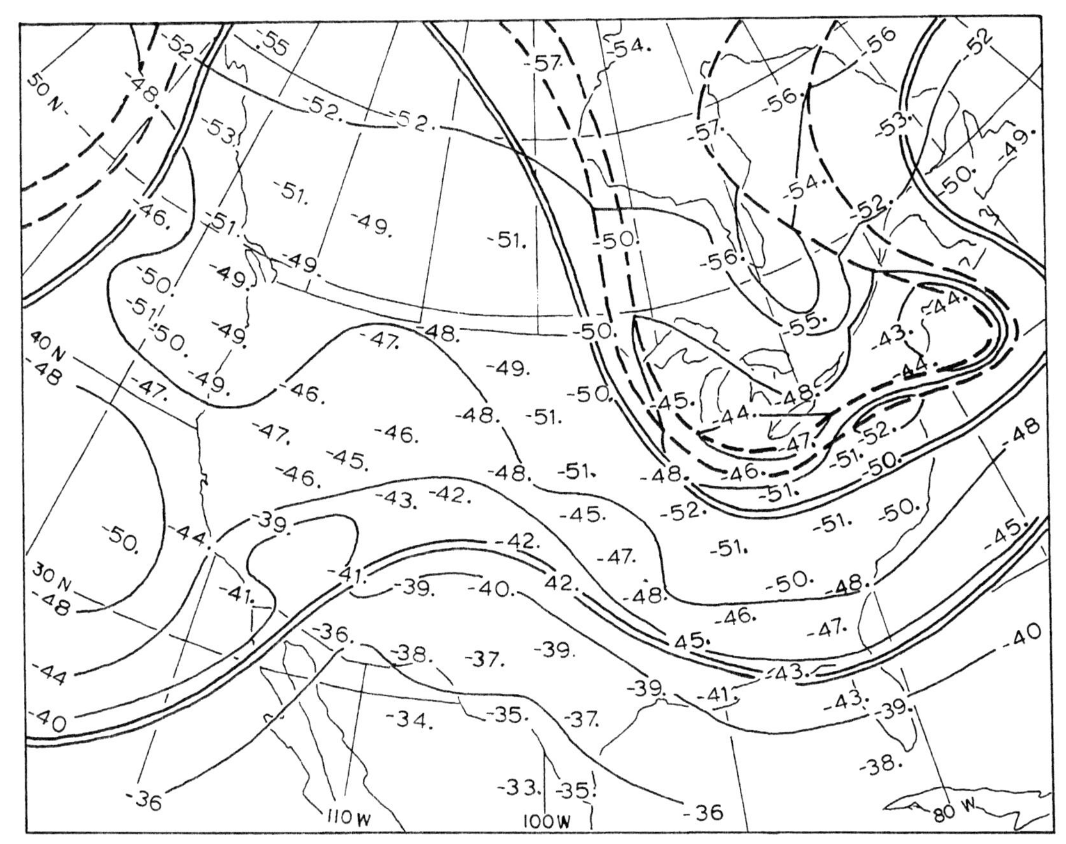

FIGURE 11-11. Temperature and isotherms (intervals of 4°C) at 300 mb for 0000 UTC 17 January 1983. The jet core and the front are from Fig. 11-10. Stations are in places of decimal points. Numbers without decimal points are isotherm labels.

preferably in different colors or at least in different styles: contours as solid black lines, isotherms as solid red or dashed black, isotachs as solid green or dotted black, and the jet core brown or black, with arrows. The tropopause level is plotted in Fig. 11-10. The values of the tropopause pressure are entered here as they were reported from the observing stations. Some may be disputed, as shown below. The polar front is indicated in these figures by thicker dashed lines, using the style with two lines: one on each side of the frontal layer.

Polar Front

The polar front in Figs. 11-10 through 11-13 was found using the wind shear, position of the polar jet stream, and level of the tropopause. The temperature did not reveal much contrast across the front at 300 mb, as seen by several of isotherms that cross the front (Fig. 11-11). When an isotherm crosses the front, this formally shows that the temperature is equal on both sides. The isotherms are kinked at the front, on both sides of the frontal layer, in agreement with the Margules equation (Appendix J). The observations confirm, or at least do not contradict, the existence of kinks on the isotherms. The kinks are strongest near the warm centers within the cyclonic bends of the jet stream.

The warmest places (warmer than −44°C) on the polar side of the polar front are in the troughs, where the polar jet is farthest south. In the same troughs of the polar front, the tropopause is lowest, that is, its pressure is highest. Here the stable stratospheric air subsided and its temperature increased by adiabatic warming. We find several isotherms within the frontal layer in the loops of the front around the warmest stratosphere, reminiscent of the situation at lower levels; however, here the temperature difference across the front is reversed: The polar side is warmer than the equatorial side. The reversal of thermal difference is associated with the reversal in the frontal slope, as required by the Margules equation (Appendix J). Consequently, the front slopes up toward the equator between the 300- and 200-mb levels. This rule has been of help in determining the front's position in Figs. 11-10 through 11-13.

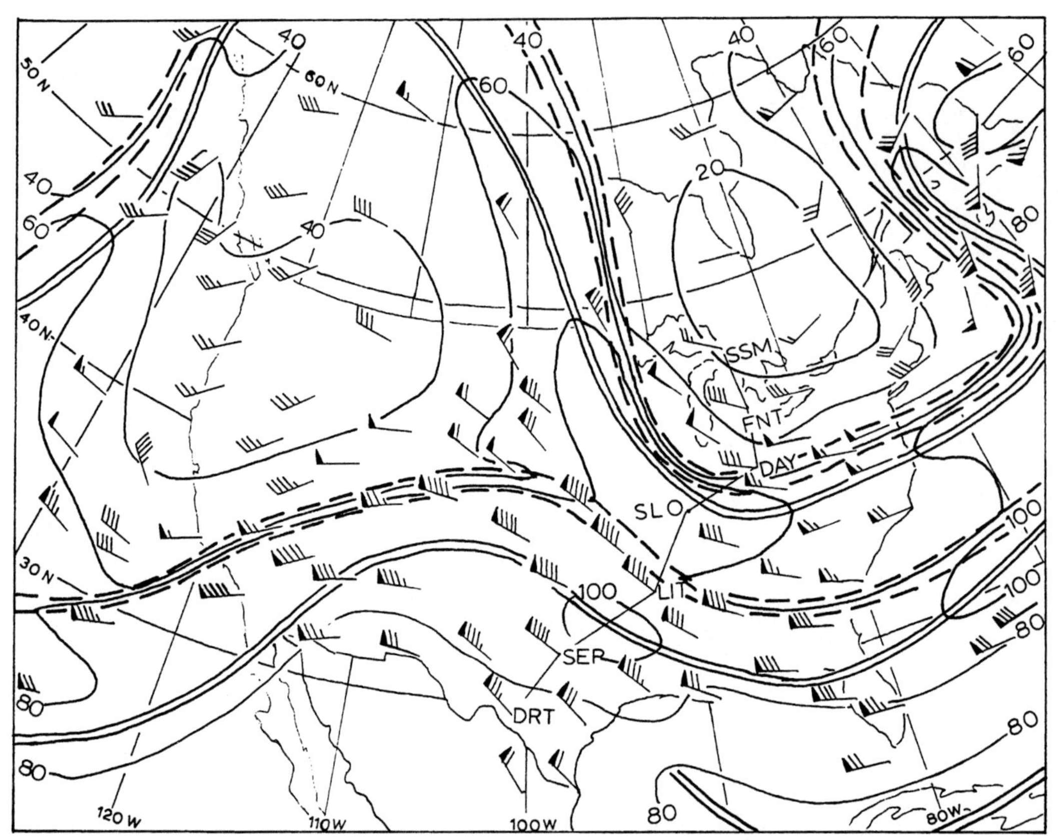

FIGURE 11-12. Wind barbs and isotachs (intervals of 20 kt) at 200 mb for 0000 UTC 17 January 1983. The jet core is the double line, and the polar frontal layer is bounded by thick dashed lines. Stations used in the vertical section in Fig. 11-14 are shown by three-letter abbreviations.

Troposphere and Stratosphere

Most of the area on the polar side of the polar front at 300 mb (Fig. 11-10) is occupied by stratospheric air. However, in a region near the Ontario–Quebec border in Canada, the tropopause is very high, with reported pressure under 300 mb. This region of tropospheric air at the 300-mb level is outlined by a thicker line and marked by the letter T. As is typical for the top of a high troposphere, it is colder than the superjacent stratosphere. This part of the high troposphere has two stations with temperatures of −55 and −56°C (Fig. 11-11), when other stations at this level show temperatures between −43 and −48°C in the stratosphere on the polar side of the polar front.

On the equatorial side of the polar front at 300 mb we find high tropospheric air of the middle-latitude air mass (Fig. 11-11). This air mass is characterized by the middle tropopause being mostly in the range of 220–260 mb. However, a portion of the stratosphere can be located between the two jets: Two stations in California report a tropopause level of 300 mb. Consequently, west of these two stations there is a lower tropopause (higher tropopause pressure), which normally appears above cyclonic vortices in the troposphere. In this area near the California coast there is a tropospheric cyclone, as shown by the wind observations in Fig. 11-10. Within this loop, the temperature at 300 mb is lower than outside the loop (observations of −44 and −50°C in Fig. 11-11). This is a remnant of a cut-off low, in which polar air travels south. At the time of the chart there is only a low tropopause left, without a front, at 300 mb; the polar front sank under the 300-mb level. The vortex in the wind field can also be described as a remnant of a loop on the polar jet stream. The cut-off loop of this type is also described in Fig. 6-6.

Subtropical Jet

The subtropical jet stream in Fig. 11-10 is discernible between two isotachs of 60 kt. It is possible to redraw these 60-kt isotachs in the southwestern United States

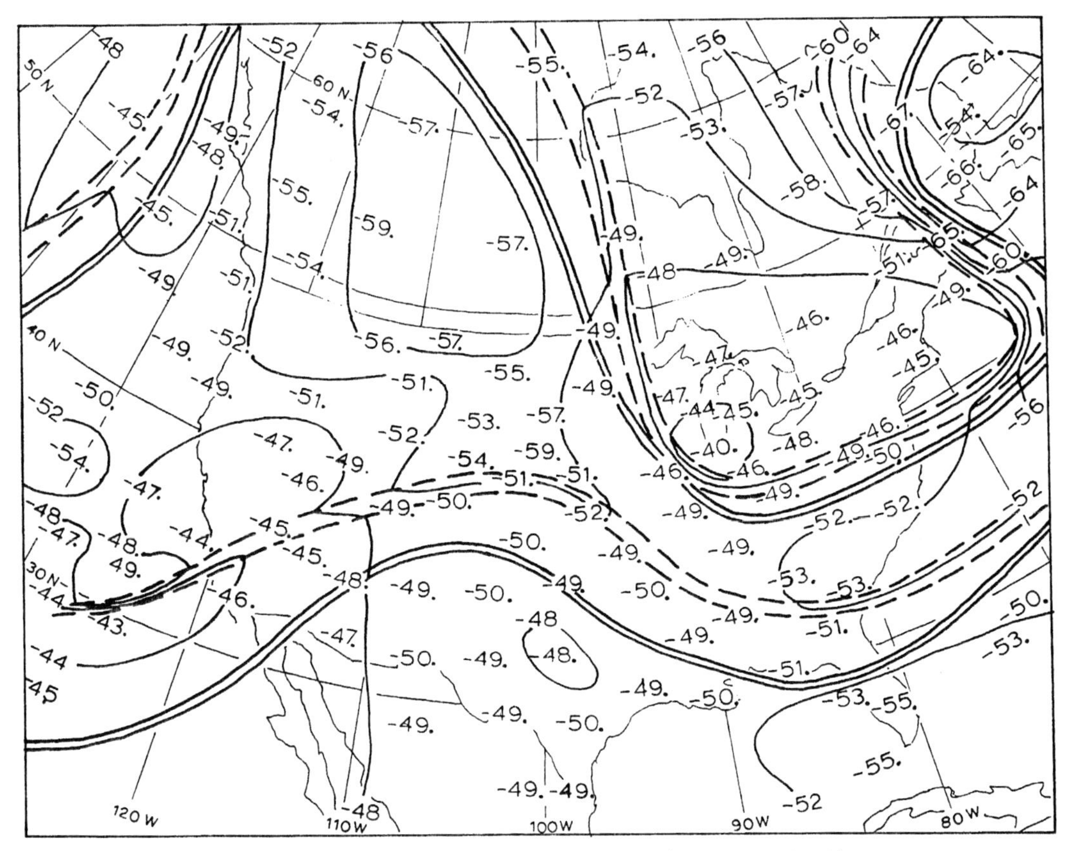

FIGURE 11-13. Temperature and isotherms (intervals of 4°C) at 200 mb for 0000 UTC 17 January 1983. The jet core and the front are from Fig. 11-12. Stations are in places of decimal points. Numbers without decimal points are isotherm labels.

to show a speed along the core of less than 60 kt. However, a continuous band of over 60 kt is chosen here, promoting the idea that the subtropical jet stream should be continuous around the world at about 30° latitude.

The continuous form of the subtropical jet is confirmed by the isotachs at 200 mb. A strong and continuous subtropical jet stream, with wind speed over 80 kt, can be noticed in Fig. 11-12. Typically, wind in the subtropical jet at 200 mb is faster than the top part of polar jet at this level. The example in Figs. 11-10 and 11-12 also shows that wind in the polar jet is considerably slower than in the subtropical jet at 200 mb. Such observations have led to the model of the atmosphere in Fig. 1-3.

Another wind maximum in Fig. 11-12 lies over the Gulf of Mexico and Cuba. This is a separate branch of the subtropical jet that deviates from the main jet core near Stephenville, Texas (SEP). Such common branching of the subtropical jet was also shown schematically in Fig. 11-9. The wind maximum over the Gulf of Mexico does not appear at 300 mb in Fig. 11-10 because it is too shallow to reach that level.

Subtropical Front at 200 mb

Isotherms may reveal the location of the subtropical front at 200 mb if this is not exactly the level of maximum wind with the associated vanishing thermal contrast. The subtropical front is located in the region immediately north of the subtropical jet stream, where the cyclonic wind shear is strongest. Some isotherms on the polar side of the subtropical jet (Fig. 11-11) lie in the frontal layer of the subtropical front. The isotherm of −44°C is one of them. However, there is not much thermal contrast here for a safe determination of the front. Therefore, the subtropical front is not drawn in Figs. 11-10 and 11-11.

Unlike Figs. 11-10 and 11-11, Fig. 11-13 offers a better-than-average example of the subtropical front. Therefore, this front is drawn on this chart and on the companion chart in Fig. 11-12.

The steps to find the subtropical front are: (a) determining the cyclonic shear near the subtropical jet, (b) identifying isotherms that may show the front, (c) locating a discontinuity in the tropopause, and (d) consulting vertical sections and other charts. Wind shear by itself is an insufficient indicator of the subtropical front. Cyclonic wind shear is present on the polar flank of every jet, due to the geometry of the isotachs, and the problem is to find maximum shear that may indicate the front. However, observing stations may not allow us to find such detail in the wind field. Therefore, in addition to wind shear, we need more indicators.

Several isotherms are lying parallel to the subtropical jet, on the polar side (Fig. 11-13). This is roughly in the expected position of a steep subtropical front. Therefore, these isotherms are drawn in the frontal layer. Isotherms alone are also not sufficient to establish the existence of the front, but a combination of thermal contrast and wind shear gives a sufficient indication of a frontal presence. If, at the same time, we find the typical break in the tropopause, with different levels on the two sides, then we have enough evidence to justify the placement of the front.

In some places with wind shear, isotherms fail to show the subtropical front. Such a region with no significant difference in temperature across the front can be found in the central United States in Fig. 11-13. Several stations in that region have readings of −49°C. This is a region where the front is vertical and the thermal difference vanishes, in agreement with the Margules equation (Appendix J).

Comparison of 200- and 300-mb Charts

The jet stream cores at 200 mb in Figs. 11-12 and 11-13 are in the same positions as at the 300-mb level (Fig. 11-10). There is no indication that the jet core is slanted between these levels. This is not a general rule. The maximum wind may be at different locations on the 300- and 200-mb charts. However, without other evidence, it is best to assume that the geographical location of the jet core is the same on these two charts.

The isotachs at 200 mb (Fig. 11-12) further demonstrate that wind in the polar jet is so slow in some places that the jet cannot be identified without comparison with other charts. The branch of the polar jet over the Pacific, off the coast of Canada, has been entered in the chart only after a comparison with the 300-mb chart. Contrary to the polar jet, the subtropical jet is easier to find at 200 mb than at 300 mb. Also in Fig. 11-11, the subtropical jet stream is well defined, with wind speed exceeding 80 kt.

Reversal of Temperature Gradient

The polar front at 200 mb from central Canada to the Atlantic is well marked with isotherms in Fig. 11-13. At this level, the polar side of the front is warmer than the equatorial side. This is in agreement with the model in Figs. 8-6 and 8-7, showing the polar stratosphere warmer than the middle stratosphere. As a result of the reversed temperature difference across the polar front, the front slopes up toward the equator between the 300- and 200-mb levels.

The isotherms at 200 mb often show the polar front well: The isotherms are concentrated in the frontal layer. However, the difference in temperature across the front is reversed, if compared with the 500-mb level. Above the jet-core level, the polar side of the polar front is warm and the equatorial side is cold. This is also related to the different tropopause levels, as shown in Sections 11-1 and 11-2. The tropopause is lower above polar air. The reversal-of-temperature difference across the polar front is illustrated over the northeastern United States in Fig. 11-13.

Vertical Section

Fronts and other structures are often found more easily in vertical sections than on isobaric charts. The vertical section from DRT (Del Rio, Texas) to SSM (Sault Ste. Marie, Michigan) is drawn in Fig. 11-14. The geographical location of this vertical section is shown in Fig. 11-12 by a kinked line that connects the stations in the section. In this section, the subtropical front is at 400 mb near SEP (Stephenville, Texas). This section is selected to show both upper tropospheric jet streams. Three soundings from this section are the examples in Fig. 11-2.

Isotherms in Vertical Sections

The shape of isotherms in Fig. 11-14 is typical for vertical sections. In the troposphere, we find a more or less uniform decrease of temperature with height, characterized by the "ladder" of isotherms. This "ladder" is interrupted on slanted inversions and stable layers. Usually, the most prominent inversions are the fronts. The polar and subtropical fronts characteristically bend around the corresponding jet streams. These two fronts separate the three principal air masses: polar, middle-latitude, and tropical. The two inversions within polar air may be interpreted as two layers of the arctic front. The inversion near 610 mb above DRT may be an elevated subtropical trade-wind inversion. This inversion typically stretches north and

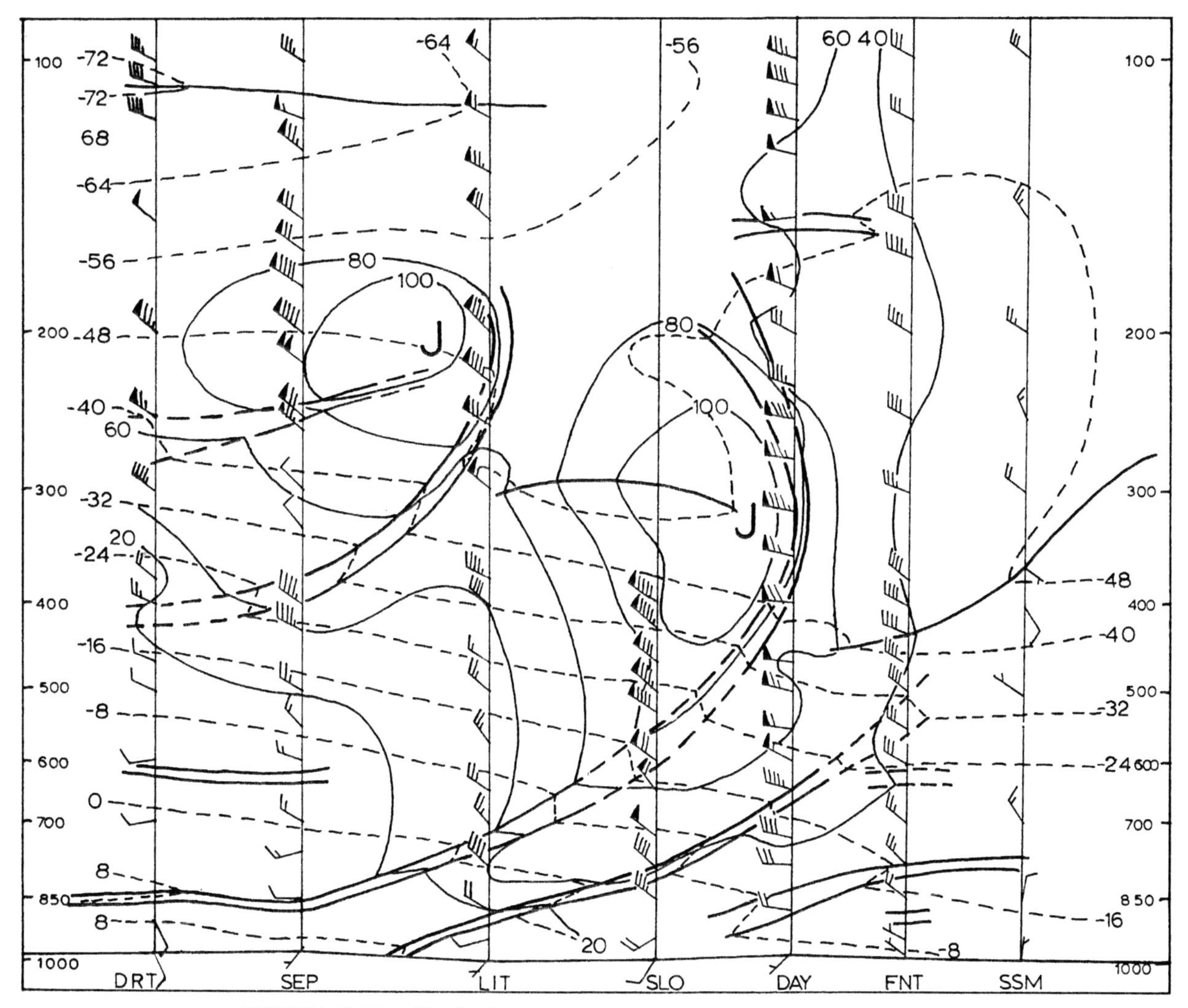

FIGURE 11-14. Vertical section through the atmosphere from DRT (Del Rio, Texas) to SSM (Sault Ste. Marie, Michigan) for 0000 UTC 17 January 1983. Frontal layers, inversions, and tropopauses are drawn with thicker lines. Stable layers that are not inversions are drawn with thicker dashed lines. Isotherms are dashed thinner lines, in intervals of 8°C; isotachs are solid thin lines, in intervals of 20 kt.

rises in the south wind on the west side of the Atlantic subtropical anticyclone (Bermuda High). The other inversion above DRT (at about 830 mb, also shown in Fig. 11-2a) is a far protrusion of the polar front. This is also a region of subsidence and frontolysis. The nearly horizontal polar front over SEP and DRT will dissipate or gradually take the place of a new subtropical tradewind inversion. This is an inactive region, upstream from the trough in the subtropical jet; therefore, frontolysis is expected. A new polar front will form in the north when the polar jet stream again forms an active region.

Figure 11-14 offers another example of the familiar difference in the shape of isotherms in the troposphere and stratosphere. Contrary to tropospheric isotherms, which are predominantly horizontal, isotherms in the stratosphere are often vertical, indicating isothermal layers. Sparse isotherms in the stratosphere are characteristic of a stable atmosphere. Isotherms that cross the tropopause kink sharply and thereby exhibit different stability regimes near the tropopause: very stable above and less stable or near-neutral below the tropopause.

The Tropopause in Vertical Sections

All three tropopauses, with distinct breaks between them can be seen in Fig. 11-14, and all have similarities with the conceptual models in Chapter 1 and in Fig. 11-1. The tropical tropopause is around 110 mb in the three southern stations. The middle tropopause is lower than usual, about 300 mb above SLO (Salem, Illinois). The middle tropopause is also low in places with cyclonic curvature. The polar tropopause is at 430 mb above FNT (Flint, Michigan), as in the sounding in Fig. 11-2c. The tropopause is considerably higher on the northern end of the section. It reaches 294 mb in northeastern Canada and is so drawn in Fig. 11-10.

South of the subtropical jet, in the area where values near 100 mb are expected, we find a few reports of the tropopause near 250 mb. By recalling the model in Fig. 11-1, this can be explained as the secondary tropical tropopause. A prominent tropical tropopause above SEP (Stephenville) at 110 mb is shown by strong kinks of the −64 and −72°C isotherms. The other inversion above SEP at 257 mb can be interpreted as the secondary tropical tropopause. We can ascertain that the troposphere extends above 257 mb at SEP, since the temperature uniformly decreases with height above that level, as is typical for the troposphere.

Jet Streams in Vertical Sections

Both upper-tropospheric jet streams in this section are prominent in vertical section in Fig. 11-14. The subtropical jet stream is centered between SEP and LIT (Little Rock, Arkansas) at about 200 mb. The core of the jet, J in Fig. 11-14, is placed at a level where the temperature difference across the front is smallest. We know from the Margules equation (Appendix J) that these three elements appear on the same level: vertical front, maximum wind speed, and vanishing temperature difference across the front. For this reason, the isotach of 100 kt is stretched around the jet core at 200 mb. After examining isotherms and isotachs in the vertical section in Fig. 11-14, the vertical subtropical front can be verified at 200 mb near LIT.

A similar pattern of isopleths occurs around the polar jet. Again we find the vertical front at the maximum wind level and, at the same place, the least temperature difference across the front.

The levels of the two jets in Fig. 11-14 are typical. The subtropical jet has a smaller vertical extent than the polar jet. This explains why sometimes the subtropical jet does not appear at either 200 or 300 mb. It may be hidden between these levels.

The presence of the jet and meridional variation of the tropopause level are strong indications that a front is present. A vertical section is usually the best chart to identify the subtropical front, although we hesitate to use the section because it is technically cumbersome to construct. The task of constructing vertical sections should be much easier with the coming new generation of computers.

12

Mesoscale Storms

This chapter contains a description of storms that are normally accompanied by electrical discharges, thus the name *thunderstorms*. Only on rare occasions does a tropical cyclone appear without electrical discharges. Thunderstorms are normally associated with convective currents; for this reason they are also called *convective storms*. The mesoscale phenomena are included in this book because they are routinely identified on weather charts. Smaller storms may not be adequately represented by routine weather observations, but analysts must still know the basic properties of such storms. These storms influence the weather and thereby alter the appearance of larger-scale phenomena.

12-1 DIMENSION OF THUNDERSTORMS

Mesoscale Categories

By their size, thunderstorms are usually described as being of mesoscale. This scale is smaller than the waves in the westerlies but larger than the cumulus scale. By their length, mesoscale storms are commonly categorized as meso-alpha: 1000 km; meso-beta: 100 km; meso-gamma: 10 km.

The order of magnitude of these categories should be understood within a factor of 3. The meso-alpha scale is the *synoptic scale* or *cyclone scale*. Extratropical cyclones, larger hurricanes, and large squall lines are normally of this category. The use of the word *synoptic* for the scale means that such phenomena are adequately represented on routine synoptic charts.

The meso-beta scale is the most common thunderstorm scale. The meso-gamma scale is the *cloud scale*, since larger individual convection cells appear on this scale.

Thunderstorms appear on all three horizontal scales. They emerge on the smallest, meso-gamma scale, but often grow to meso-beta size. Some squall lines attain the meso-alpha scale.

Categories of Storms

Due to the typical size of mesoscale storms, their detailed structure can seldom be discerned on routine weather charts. However, the density of observing stations is usually sufficient for these storms to show on charts. A weather analyst does not aim to discover all the details of storms during routine work. Instead, an analyst locates the regions with groups of storms on weather charts. Analysts are expected to predict further motion and development of groups of storms. The study of known models of storms is of great help, since with this knowledge an analyst can identify the thunderstorm patterns on weather charts easier. Elementary information on mesoscale storms is indispensable for successful weather chart analysis and is exposed in this chapter.

A word of caution is needed again: Storms may appear in forms and with properties that are not described in books. Therefore, attention must be paid to each storm separately, as if the current case is different from previous experience and different from the book examples.

Research over the last few decades established several categories of thunderstorms. Scientific studies have shown that the categories of storms differ not only by size, but also by the dynamics of the air flow in them. The subject of storm classification is not yet settled. We still lack unambiguous indicators that show the type of storm. Besides, storms often evolve from one type to another, or consist of several types. Still, for the sake of systematic exposition, a classification is helpful. The following outline of storm categories will be used in this chapter:

(a) single-cell local thunderstorm
(b) supercell storm
(c) thunderstorm cluster (also called *multicell storm*)
(d) mesoscale convective complex
(e) squall line
(f) tropical cyclone
(g) polar low

The mesoscale convective complex, the squall line, and sometimes the thunderstorm cluster may be classified together as a *mesoscale convective system.*

Mesoscale Convective Systems

Due to its size, a mesoscale convective system (MCS) is the most interesting category for synoptic-scale weather analysis. An MCS is an "organized" region of thunderstorms that have a common origin or that cause each other. Therefore, an MCS is not an accidental aggregate of randomly occurring storms. Normally, an MCS is easily detected on synoptic weather charts since it occupies a large area and is observed from several stations.

Weather charts that cover large geographic areas are not appropriate for a detailed study of local storms. These storms may even be lost on charts if they appear between observing stations and if the observers do not spot them. We need radar and satellite observations to detect smaller storms. Only larger MCSs can be adequately followed on the basis of observations from weather stations. However, identifying regions with thunderstorms and tracing of single storms (mainly from radar reports) is an essential part of weather analysis.

Some Common Storm Properties

Thunderstorms of all types can produce severe weather: wind in excess of 25 m s^{-1}, flash flood, hail larger than 2 cm in diameter, or tornadoes. Since they also can cause severe weather, storms that are smaller than MCSs must not be underestimated in weather analysis. Also, smaller storms may develop into MCSs.

Studying small-scale local storms is important in weather analysis, even if the details of these storms cannot be detected on weather charts of usual scales. Also, we can properly understand larger storms only if we know the physics of smaller storms. MCSs consist of smaller storms, such as single-cell, supercell, and smaller multicell thunderstorms.

Tropical cyclones are easily distinguished from other storms. They show a prominent cyclonic circulation and originate only over warm oceans (water temperature above 27°C).

Other kinds of storms also appear in the tropics. Single-cell local storms and squall lines in the tropics are similar to such storms in the middle latitudes. Only supercells have not been observed in the tropics.

An isolated report of a thunderstorm on a weather chart does not reveal which of the storm categories is present. To determine the significance of thunderstorm reports, a weather analyst must look for additional data from other stations, from radar, or from satellite images. To facilitate weather analysis in presence of thunderstorms, it is necessary to know the basic properties of various storms.

The wind near thunderstorms may show intricate flow patterns, due to various currents in and around storms. One common property of storms is that the wind in them is usually greatly different from geostrophic wind. A typical thunderstorm lasts 1 to 6 h, which is too short to establish geostrophic balance. This short duration is also observed in convection cells constituting larger and longer-lasting thunderstorms. This explains why the geostrophic balance is not established even in squall lines lasting several days: Every cell in an MCS lasts too short for geostrophic balance to develop.

Thunderstorms exhibit a prominent diurnal variation. In most places on the continents, late afternoon and evening hours are favorite times for thunderstorms. This points to the importance of heating from the earth's surface. Convection is most stimulated at the time of the warmest surface, but it develops even more later, when latent heat becomes the primary source of energy.

Offshore, convection and thunderstorms develop at night. This is when the water is warmer than the nearby shore.

Severe storms often appear after midnight in Kansas, Nebraska, and neighboring states. This is not completely explained. It is likely that the nocturnal intensification of the southerly low-level jet contributes to the development of storms.

12-2 THUNDERSTORM ENVIRONMENT

Stability

Potential instability in the atmosphere is a necessary condition for development of thunderstorms. Ordinary conditional instability, based on the lapse rate of temperature with height, is insufficient. If the atmosphere becomes unstable in terms of the lapse rate, as in Fig. 5-3c and d, convection develops on a scale too small to produce a cumulonimbus. The situation is different in presence of an inversion that caps a humid layer near the earth's surface. In this case the energy of instability is not released until higher values of potential instability are reached. Then the convection develops more vigorously and may form thunderstorms. An inversion in about the lowest 1 km of the atmosphere is common in the warm sector of the extratropical cyclone; therefore, this is the location of most thunderstorms. These circumstances are described in Section 5-8 on integrated indicators of stability.

Availability of Water Vapor

Water vapor is necessary for thunderstorm development since it supplies latent heat for cloud growth. An increased amount of water vapor in the lowest layer of the atmosphere results in increased potential instability. The needed amount of vapor varies widely in comparable storms in different geographical locations. For example, it is commonly regarded that at least 25 mm (1 in.) of precipitable water should be available in eastern Texas for thunderstorms to form. This is in contrast to a shallow humid layer, containing about 5 or 10 mm of precipitable water, that will support development of severe storms in the plains of eastern Colorado.

Wind Shear

An important quantity that determines the development of thunderstorms is the environmental vertical wind shear. In weak wind shear (say under 2×10^{-3} s^{-1} in the lower half of the troposphere), only weak cells of local thunderstorms develop. Stronger vertical shear, of the order of 10^{-2} s^{-1}, favors development of the strongest storms: supercells. When the shear is still higher, thunderstorms are not very strong again. They may develop squall lines, but they will be less severe since they will not contain supercells. Too strong a shear tears apart the balanced, quasi-stationary circulation of a supercell.

Environmental Lifting

To overcome the initial stability in the lowest layer (high CIN, Section 5-8) of the atmosphere, some stimulus is needed that will lift the air to the level of free convection. This stimulus is often a larger-scale process that can be identified on weather charts or satellite images. Common stimuli for initiation of convection are: the approach of a front or a dryline, a prominent identifiable pattern of vertical motion, or motion of one or more outflow boundaries of previous thunderstorms. The stimulus may be of a geographical nature, such as orographic obstacles in the low-level jet. For example, it has been noticed that the thunderstorms on the high plains of Colorado and Wyoming often form on the rather small ridges of Trinidad, Palmer Lake, and Cheyenne.

The formation of rain-producing thunderstorms is sometimes triggered by weak lifting on gravity waves in the stable layer (or inversion) in the lowest 1–2 km of the atmosphere. This stable layer near the surface is often humid, while most of the troposphere is dry. Gravity waves in the stable layer normally cannot be observed before clouds develop. Their wave length is often shorter than the distance between observing stations. Therefore, the pressure or wind readings cannot show these waves. These readings can show a front or an outflow boundary, but hardly a sinusoidal wave. Once clouds start forming, the wavelike structure can be seen on satellite photographs. Humid air in the stable surface layer provides water vapor for clouds.

An outflow boundary generated by thunderstorms is often a stimulus for the generation of new storms. Such a boundary moves like a solitary wave. It can be observed by Doppler radar even in a cloudless atmosphere.

A quantity related to lifting is wind divergence of the flow in the upper troposphere above storms. Divergence is regularly observed above larger storms, such as a mesoscale convective complex or a tropical cyclone. This divergence can be spotted on the basis of radiosonde observations at 200 mb. Also, looping of satellite images reveals divergence at the cirrus level above developing and mature storms.

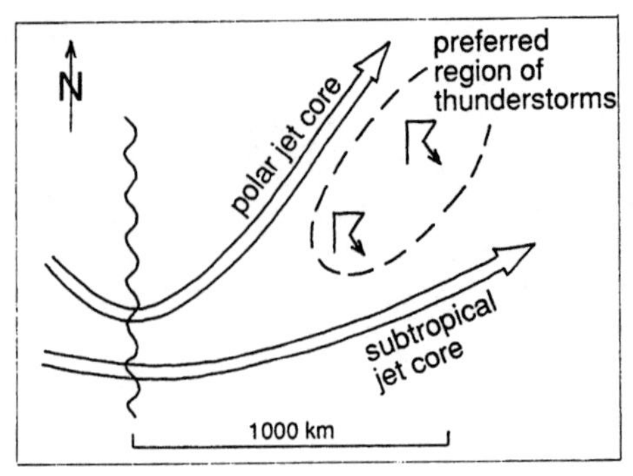

FIGURE 12-1. The active region between two separating jet streams that is favorable for development of severe thunderstorms.

Thermal Advection

Warm advection in the lower troposphere contributes to the formation of thunderstorms in three ways: (a) It enhances the thermal instability of the atmosphere; (b) it provides an environment with lifting so that the potential instability can be released; and (c) it provides directional wind shear that is favorable for the development of supercells. The relation of warm advection and environmental lifting is handled in Section 7-6 about the **Q** vector with differential thermal advection.

Warm advection can be recognized in the atmosphere by the patterns of contours and isotherms or by the vertical directional wind shear. Warm advection can also be detected by the helicity or relative helicity of the flow (Appendix I).

Active Region

A majority of middle-latitude thunderstorms develop in active regions of the polar jet, downstream of upper-level troughs. Large-scale lifting in these regions triggers the release of potential instability.

The formation of severe storms occurs primarily in middle-latitude air, where the tropopause is near 250 mb or 10 to 11 km. The region between the two major upper-tropospheric jet streams is characteristically less thermally stable than other regions. Storms are not as severe in tropical air, where the troposphere is much higher. Thunderstorms are also not as severe under a low, polar tropopause.

A particularly favorable situation for severe storms is an active region where the polar and subtropical jets flow away from each other. Vorticity advection is enhanced in such an area because the vorticity in a diffluent trough is stronger than in other troughs. This pattern is illustrated in Fig. 12-1.

The active region with storms is not always in a southwest flow. Other directions of upper-level wind are also conducive for thunderstorms. Often a severe storm appears in a northwesterly flow, only if a trough in it develops to form an active region.

Another common pattern that favors thunderstorms is the presence of a southerly low-level jet that brings warm and humid air, increasing the energy of instability. The cloud mass in a warm conveyor belt often grows to become a group of cumulonimbus. These cumulonimbus grow out of a humid low-level jet, and their tops are very prominent on satellite images. As described in Chapter 10, a low-level jet develops regularly in the active region of the polar jet stream.

The middle tropopause commonly elevates in the region of severe storms of the middle latitudes. The tropopause may reach 140 mb or 15 km above supercells or squall lines.

Positioning of Thunderstorms

The net of observing stations, as usually found over populated continents, is sufficient to determine the regions of thunderstorms. However, thunderstorms may stay undetected over regions with sparse observing networks, such as oceans and deserts.

The location of individual thunderstorm cells can best be determined with radar. Thus a weather analyst should regularly consult the radar screen and radar reports. Doppler radar gives more information than ordinary radar, particularly for storms characterized by vortices or downdrafts too small to be detected with standard observations.

Satellite images are also very useful in detecting thunderstorms, especially over regions where other observations are missing. A difficulty may arise when the anvils of cumulonimbus cells obscure some neighboring thunderstorms. On the other hand, the flat anvils of cumulonimbus make a clear background for detecting strong updrafts characteristic of severe weather in squall lines. These updrafts penetrate the flat cloud top and form turrets 2–3 km high and 2–10 km wide.

Regions of showers can be recognized from reports on a standard surface chart. It may be not immediately clear which storm category is present. A typical area of showers is of the order of magnitude of 1000 km in diameter. The conventional symbol for showers is a green triangle (∇), and such symbols should uniformly cover the area with showers. Thunderstorm reports must also be interpreted as showers, since lightning originates in rain clouds. When thunderstorms are reported, both symbols (green triangles for showers and red thunderbolts for thunderstorms) should be used, as described in Section 8-3. A squall line is indicated by the dash–dot–dot line, as suggested in Table 9-1.

12-3 LOCAL THUNDERSTORM

Formation of a Single-Cell Thunderstorm

A single-cell local thunderstorm is a rare event. Usually, when thunder is heard, several convective cells are already in progress. However, the conceptual model of the single-cell thunderstorm forms the basis for all other types of thunderstorms. Therefore, this section gives information on cloud dynamics and physics that is necessary for understanding storms. Elementary properties of a local thunderstorm are important for describing other thunderstorms that consist of a conglomerate of cells similar to those of a local thunderstorm.

Like all other categories of thunderstorms, general weather conditions under which single-cell thunderstorms may appear are characterized by instability. Rather small vertical wind shear (of the order of 10^{-3} s^{-1}) is also present, implying that the ambient wind is not too fast at any level. As seen later, an environment with strong vertical wind shear promotes other categories of storms.

The Updraft

The updraft in a cumulus tower may penetrate several kilometers up into the atmosphere in about 20 min. The whole event of a local thunderstorm typically lasts about 1 h and extends to about 10 km in diameter. Vertical development of isolated thunderstorms does not always reach the tropopause. Instead, the tops of thunderstorms may reach only 4–8 km above the earth's surface. Some thunderstorms develop in polar air, and there the tropopause is only at about 8 km.

If the updraft in the cumulonimbus is faster than about 5 m s^{-1}, all raindrops, including the largest ones, are carried upward. The speed of 5 m s^{-1} is the stationary fall speed of the largest raindrops (5 mm diameter). If a raindrop grows larger than 5 mm, it soon breaks up into several smaller drops. Therefore, rain cannot fall through a fast updraft, and all raindrops are removed from the updraft. In such cases, radar reveals a weak echo area in the updraft within the cumulonimbus. The precipitation in a single-cell storm does not occur far from the updraft, perhaps 3–5 km away. As mentioned above, small wind shear implies slow ambient flow, and the precipitation is not carried far from the updraft.

The updraft in a local thunderstorm consists of air that was heated near the earth's surface in the vicinity of the cloud. This differs from other storms, where air for the updraft comes from a great distance with a conveyor belt.

Cloud Physics

A developed thunderstorm cell extends from the lower atmosphere where the temperature is usually above freezing to the higher troposphere where the temperature is below −40°C. These two values of temperature are of basic significance in the physics of clouds and precipitation. Ice elements (snow, hail) melt at temperature above 0°C. All cloud elements freeze below −40°C.

The region where the temperature is between 0 and −40°C is of great significance for cloud development. In this region, the water elements (drops of cloud and rain) usually do not freeze, unless they are mechanically disturbed or come in contact with a freezing nucleus. Therefore, this is the region where ice and water may coexist.

The simultaneous existence of ice and water is strongly influenced by the lower saturation vapor pressure over ice than over liquid water. Therefore, if there is a mixture of ice and water elements, the water evaporates and the vapor condenses on the ice elements, which makes the ice elements larger and fall through the air faster.

When glaciation occurs, a cumulus becomes a cumulonimbus. Glaciation can be recognized by the appearance of a fibrous or fuzzy structure on or near the top of the cloud. The fibrous appearance is due to snowfall. As mentioned above, ice elements (crystals, snowflakes) grow in the presence of water drops and fall faster than cloud drops. Falling ice crystals thin the cloud top. This produces fibrous stripes that are typical for cumulonimbus. Before the frozen top of a cumulonimbus spreads horizontally, falling stripes are seen only on the sides of the cloud; this is the *cumulonimbus calvus* (bald cumulonimbus). In this stage, a cumulonimbus still has rounded protuberances on the top of the cauliflower shape. When the top of the cloud consists only of snowy falling stripes that spread horizontally, this is *cumulonimbus capillatus* (capped cumulonimbus).

Formation of the Downdraft

A local thunderstorm develops in the absence of fast wind. This implies that there is little vertical wind shear. In such an atmosphere, the cumulonimbus development is primarily vertical. An important consequence of vertical development is that the precipitation enters the upper portion of the updraft.

Air with precipitation is colder and denser than the rest of a cloud for several reasons. Precipitation originates in higher and colder layers; therefore, its presence cools the air. A large part of rain originates from the melting of ice, creating cold drops, and these drops cool the air. Air with precipitation is also cooled due to the spending of sensible heat for the melting of ice. Water evaporates from the drops and cools the air, since the updraft may be unsaturated.

In addition to the increase of density due to cooling, the density of air also increases by the density of falling precipitation. The density of precipitation is the mass of liquid water per 1 m^3 of (air) volume.

When colder and denser air appears in a cloud updraft, it obstructs the updraft and eventually forms a cold downdraft. The formation of a cold downdraft is the main mechanism that hinders the cumulonimbus's growth and leads to its dissipation.

Cooling of air due to the above-mentioned processes is usually stronger than the adiabatic warming that occurs in the downdraft. This enables the cold downdraft to reach the earth's surface. The cold downdraft is observed as a squall of cool wind under a shower cloud.

As seen below, longer-lasting thunderstorms develop downdrafts that flow on the side of the main updraft. In that way, the main updraft proceeds unhindered by the cold downdraft.

The Downburst

A downdraft with precipitation under the cumulonimbus deserves special attention, since it frequently causes damage to aircraft and to objects on the ground. Such a downdraft is a *downburst*. The downburst often appears suddenly, with wind speed of 15 to 30 m s^{-1}. Such wind may down aircraft, capsize boats, damage houses, and uproot trees. Its typical lifetime is 1 to 5 min; therefore, it cannot be traced and forecast with a comfortable lead time. The area covered by one downburst is 1–10 km in diameter. When there is a downburst on the runway of an airport, flight controllers in the nearby tower may not be aware of it. Fortunately, only a few downbursts are of hazardous intensity. Weaker downbursts occur regularly in cumulonimbus. Some appear even without clouds, as a result of convection in a dry atmosphere.

Great progress in detecting downbursts has been made recently. Doppler radar is a device that gives additional observations on the scale of the downburst. An example of a downburst is shown in Section 2-6.

Gust Front

The edge of a downburst is an *outflow boundary* or *gust front*. The downburst is observed as a wind gust or a squall. The name "squall line" is not used for the outflow boundary of individual convective clouds instead, it is used for a group of thunderstorms whose outflow boundaries merge. This is described in Section 12-5.

Passage of a gust front is accompanied with an abrupt pressure rise, noticed as a "nose" on the barogram. A sudden increase of pressure is due to the hydrostatic weight of cold air in the downburst. A local increase of pressure is common to all types of thunderstorms. If a thunderstorm is large, the increase of pressure may form a small-scale high (*mesohigh*). This is described in Section 12-7.

A Model of Thunderstorm Development

The life cycle of a local thunderstorm is illustrated in Fig. 12-2, drawn after the classical model of Byers and Braham (1948). In this life cycle we can recognize three stages: the cumulus, mature, and dissipating stages.

In the cumulus stage (Fig. 12-2a), all cloud drops are fairly uniform. The drops are so small that they can be lifted by the updraft. There is no precipitation, and no significant separation of electricity occurs. The cloud is of species cumulus congestus. Individual convective cells are about 3–5 km in diameter. Vertical velocity may reach 10 m s^{-1}.

In the mature stage of a thunderstorm (Fig. 12-2b), the cloud has reached the levels in the atmosphere where the temperature ranges from −20 to −40°C. In this region, some droplets freeze. The resulting ice particles grow much faster than the water drops. The ice elements (crystals) often grow while the drops evaporate, due to the different saturation vapor pressure over ice and water. This process is described in the literature as the Bergeron–Findeisen process of precipitation formation. Air with precipitation is colder and denser than the adjacent air. Therefore, the precipitation acts as a significant agent in the formation of the downdraft. In this stage, the precipitation

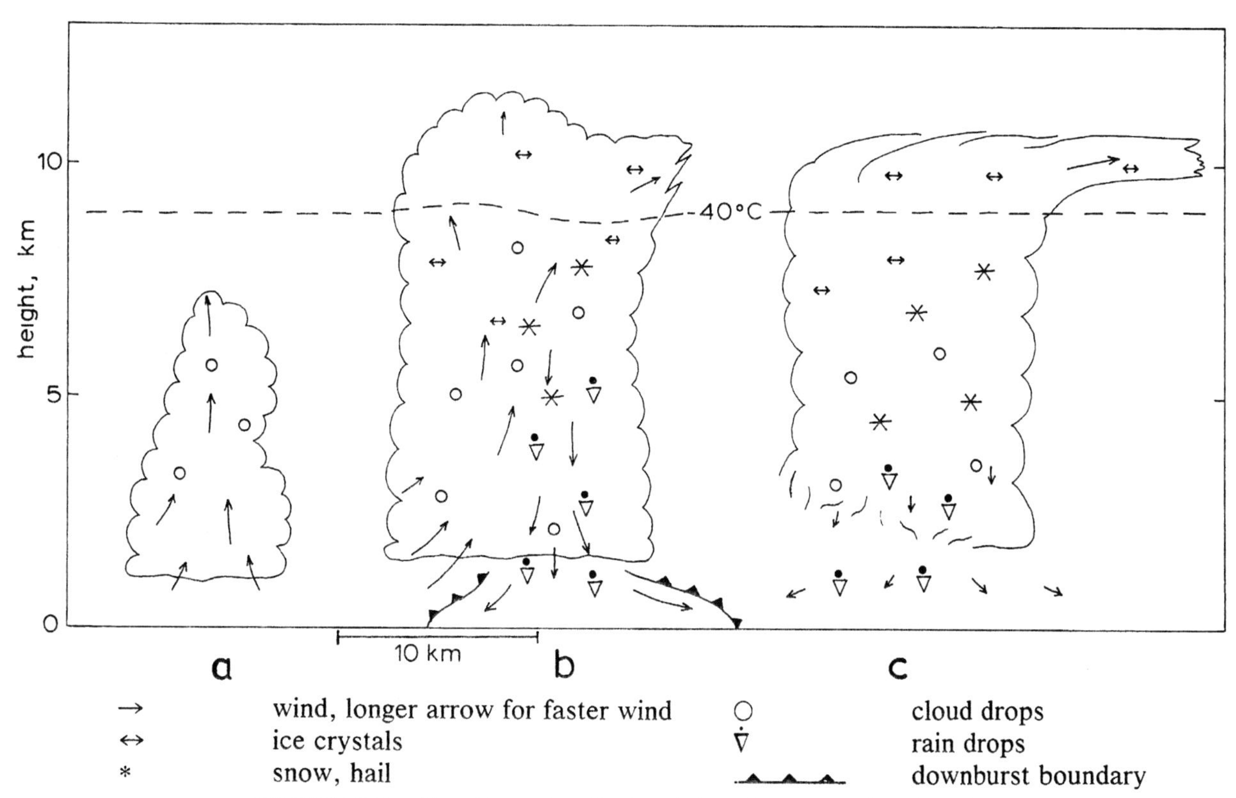

FIGURE 12-2. The life cycle of a local thunderstorm, showing the three stages of development of a thunderstorm cell. (a) Cumulus stage, (b) mature stage, (c) dissipating stage.

reaches the ground through the downdraft. An anvil-like top can be recognized on the top of the cloud, which is now a cumulonimbus capillatus. As estimated by the falling velocity of hailstones that are suspended in the air during hail formation, vertical velocity may exceed 50 m s^{-1}. The diameter of the convective cell is about 10 km.

In the dissipation stage (Fig. 12-2c), most of the cloud is occupied by air with precipitation. There are no significant buoyant air parcels. The rain is weak, and the cloud dissipates. After the shower rains out, the remnants of cumulonimbus spread out as stratocumulus in the lower layers (1–3 km above the ground) and as cirrus densus cumulonimbogenitus near the tropopause.

This life cycle of a thunderstorm has been observed many times. However, many departures have also been observed. For example, showers often develop in warm clouds (with temperatures above freezing) where there is no glaciation. This is a frequent occurrence in the tropics. In such cases, rain forms due to the different saturation vapor pressure over drops of different sizes and due to the growth of bigger drops by coalescence with smaller drops. Also, not every cumulus congestus develops into a cumulonimbus; other variations may also occur. Knowledge of the typical life cycle helps us to organize our thinking about clouds and gives us clues to successful physical analysis and forecasting of weather.

An Example of a Local Thunderstorm

It is difficult to recognize a genuine single-cell thunderstorm, since multiple updrafts may be hidden in the same cloud. If a thunderstorm is small, say 2–10 km in horizontal extent, and if there are no obvious multiple columns of falling rain, we may assume that the storm is single cellular. Satellite images and radar echoes give the best information on local thunderstorms.

Examples of single-cell thunderstorms can be seen in Fig. 12-3. Several single cells can be recognized near Galveston, Texas, over the Gulf of Mexico. The horizontal extent of these storms is about 10 km. The anvils, which are seen in the picture, can cover a wider area. The same picture shows two much larger storms in a line from northwestern Arkansas to Pennsylvania.

The small storm near Galveston was a precursor of a large mesoscale convective system that developed 6 h after this image was taken. In the afternoon of the same day, this other storm became about as large as

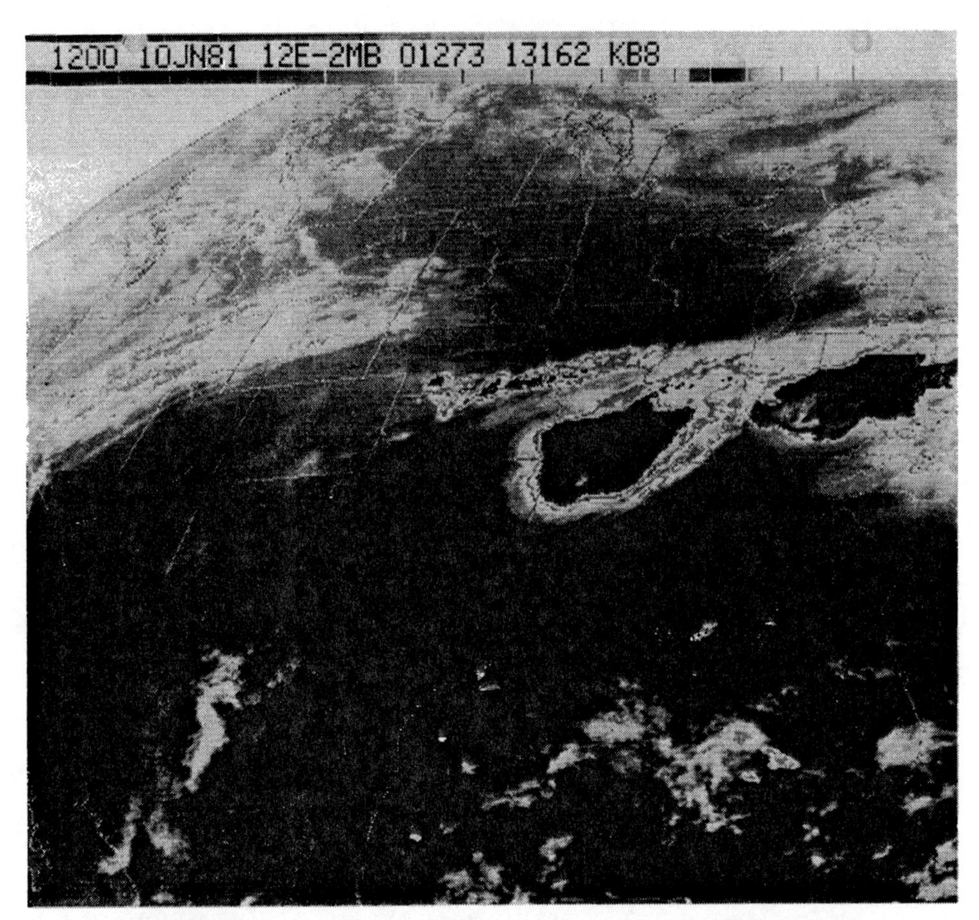

FIGURE 12-3. An enhanced infrared image at 1200 UTC 10 June 1981. Large MCSs are along a line from northwestern Arkansas to Pennsylvania. Several single-cell thunderstorms, besides other clouds, are over the Gulf of Mexico. From Read and Maddox (1983), by permission of the American Meteorological Society.

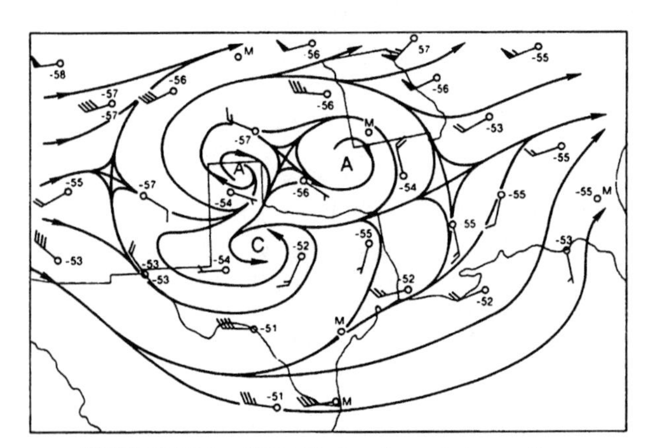

FIGURE 12-4. Wind in knots and streamlines at 200 mb, 1200 UTC 10 June 1981. Divergence can be spotted near points A and B. From Read and Maddox (1983), by permission of the American Meteorological Society.

the storm over Missouri in Fig. 12-3. There was a similarity of the 200-mb flow over those two storms. Wind analysis (by streamlines, Fig. 12-4) for this case shows divergence in two places: over the large thunderstorm over Missouri and over the Texas coast near Galveston. The storm on the coast developed only later, but the environment was already conducive to strong convection.

Electrical Phenomena

Separation of electrical charges occurs in clouds with strong updrafts and downdrafts and with falling rain and hail and other frozen precipitation. The electrical discharge (lightning and thunder) occurs when a high electrical potential is reached.

A thunderstorm occurs when thunder can be heard. Thunder can be heard only if lightning occurs not more than 10–15 km from the observer. The report of thunder is a reliable sign that the precipitation is in the form of showers (not steady rain) at the station or its vicinity.

Lightning can be spotted from much farther away than thunder, up to 50 km in daytime and up to 300 km at night. For this reason, reports of lightning should be considered with caution when thunderstorms are being located on the weather chart.

Lightning makes it possible to position the thunderstorm accurately using the lightning detector (Section 2-9). Networks of lightning detectors have been active since the early 1980s.

12-4 SUPERCELL THUNDERSTORM

Description

Some thunderstorms develop into long-lasting, rotating forms that typically last about 2–4 h. Those are supercells in which the configuration of the flow assumes a quasi-steady pattern of a nearly cyclostrophic balance. A supercell is characterized by two major vertical currents: one updraft and one downdraft. However, inflow to and outflow from the supercell acquire a convoluted pattern, as described below.

The relatively long-lasting supercell is much larger than a usual upright local thunderstorm. The diameter of the supercell is normally between 20 and 50 km. The time and space dimensions of the convective cell, and the severity of the storm, explain the name *supercell.*

The supercell is the most common source of damaging wind, hail, tornadoes, and flash floods. A supercell, or even a group of supercells, is often found within various MCSs.

The main features of the supercell can be observed easiest by radar; therefore, the radar is the best tool for detecting supercells. Observations from aircraft and from the earth's surface have also contributed to our knowledge of the supercell, but radar observations offer the least ambiguity in detecting supercells.

The supercell has been successfully simulated by numerical modeling, greatly increasing our understanding of its dynamics. The evidence from numerical models confirms the observations that supercell circulation is of a quasi-steady, approximately cyclostrophic type and therefore can persist for several hours.

Environment

The supercell develops in a potentially unstable atmosphere. Typical CAPE is of the order of 2000 $m^2\ s^{-1}$ and CIN 20 $m^2\ s^{-1}$. Moderately strong vertical wind shear, about $5 \times 10^{-3}\ s^{-1}$ between 1 and 3 km elevation favors supercells. Shear stronger than $10^{-2}\ s^{-1}$ seems to break up the balanced rotating equilibrium. Such stronger shear is favorable for the formation of squall lines.

A rather fast jet stream aloft, of over 50 m s^{-1}, is often present with supercells. The upper jet stream helps the outflow at the cloud top. On the other hand, the atmosphere is often less unstable before the supercell than before other storms.

The main low-level inflow into the supercell is a few degrees colder than the available air in nearby regions. From this it appears that a lower environmental instability may be sufficient for the initial development of a supercell than for the development of a squall line. Stronger instability enhances formation of multiple cells of various MCSs. Once a single rotating updraft develops, without extreme instability, it may develop further until it reaches the severe proportions of a supercell.

Vertical directional wind shear also provides an environment where an updraft and downdraft develop side by side. This is favorable for rotation development in the supercell. A likely mechanism of nascent storm circulation in a directionally sheared wind field is illustrated in Fig. 12-5a. In this drawing, flat arrows show the three-dimensional wind distribution. The vertical distribution of wind shows large thermal advection and helicity. By this time, no middle-level inflow has yet developed. The only inflow is in the low troposphere, from one side of the cloud. The downdraft develops sideways to the inflow, so it does not fall into the inflow. The downdraft is mechanically displaced sideways by the middle-level environmental flow. When the storm grows in this way, it starts rotating, acquires the middle-level inflow, and forms a genuine supercell.

The upper outflow is near the tropopause. The tubular arrow near the middle of the troposphere in Fig. 12-5a represents the three-dimensional vorticity vector $\nabla_3 \times \mathbf{V}_3$. This vector is mainly horizontal. The sense of "rolling" is shown by a small circular arrow. The vorticity vector in this example has a significant component along the mean wind, which shows large helicity and temperature advection.

For the sake of comparison, a different case is illustrated in Fig. 12-5b. This is a similar cloud in a sheared environment, but without directional shear. The vorticity vector is again horizontal, and its rolling is shown by the small circular arrow. The wind speed near the bottom and near the top is great, but the average wind is small. The direction of the average wind is either in the direction of the tropopause wind or of the

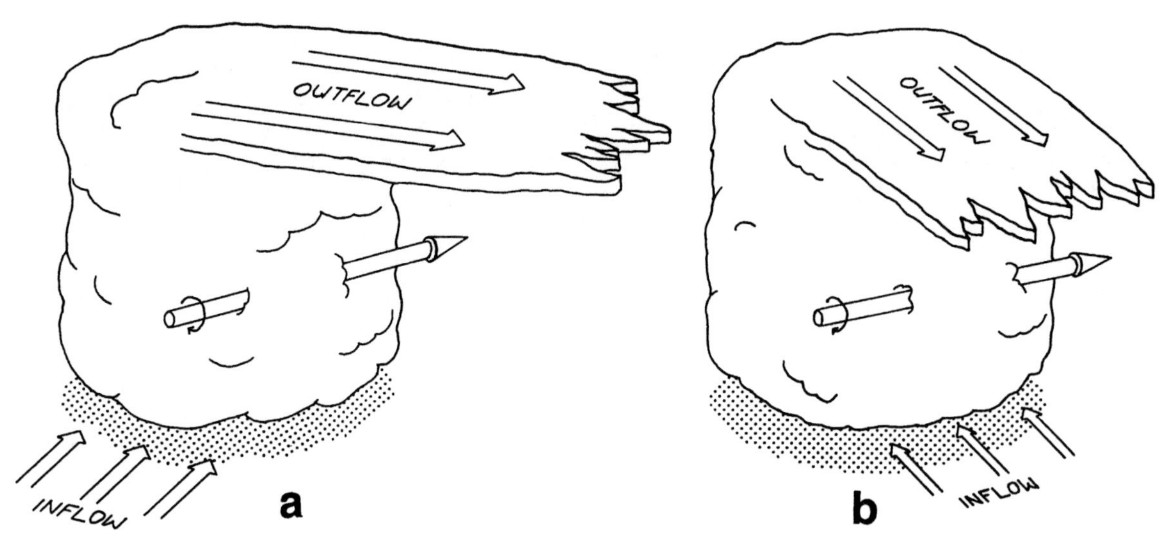

FIGURE 12-5. Cumulonimbus in a sheared wind field. The main air currents are shown by flat arrows. The vorticity vector is represented by the round arrow. (a) The case of large relative helicity: The wind at all levels has a significant component along the vorticity vector. (b) The case of small relative helicity: The wind at lower and upper levels does not have a component along the vorticity vector.

low-level wind, depending on which prevails in the averaging. In either case, the average wind is normal to the vorticity vector and therefore the helicity is zero. In such cases, the development of supercells is suppressed.

A Conceptual Model

Updraft and downdraft in a supercell constitute two major currents that coexist in a quasi-stationary state for several hours. The energy of instability is steadily transformed into the rotation of the storm. The conservation of momentum also contributes to the maintenance of the rotation of the storm. This explains the intensification of a supercell into the strongest form of thunderstorms.

A schematic view of a supercell is shown in Figs. 12-6 and 12-7. Warm and humid air enters the cloud from the layer under the subtropical trade-wind inversion, as in a single-cell, local thunderstorm. This current continues into the main updraft in the cloud. Since it is caught in the rotating cloud, the updraft rotates, and vertical stretching of air columns intensifies rotation. The updraft tilts downwind in the higher troposphere due to the environmental vertical wind shear. The tilted updraft rises up to the tropopause and then exits the cloud through the anvil. The outflow joins the environmental upper-level flow. Condensation of water vapor and the formation of precipitation take place in the updraft, but the products (rain, snow, hail) fall out of the updraft into the downdraft. When precipitation ends up in the downdraft, it can fall to the earth's surface.

The air current of the downdraft largely originates in the dry environment of the middle troposphere. This air enters into the cumulonimbus at an angle to the right of the low-level inflow, corresponding to the environmental vertical wind shear. The mid-level inflow bends under the tilted updraft and receives precipitation from the updraft. The cooling of the future downdraft occurs from the cold precipitation elements, evaporation of raindrops, and melting of snow. Cooling of the downdraft is stronger than adiabatic warming during the descent. In addition, downdraft gains density due to the load of falling precipitation. Cooling and precipitation load increase the negative buoyancy of the downdraft, thereby causing downward acceleration and a cold squall under the cloud.

Formation of Precipitation

Water vapor condenses mostly in the main updraft of the supercell and is carried upward in the form of cloud drops. The updraft gives a weak radar echo, since cloud drops are small. This *vault of weak echo*, also called *weak echo region* (W in Fig. 12-7) is typi-

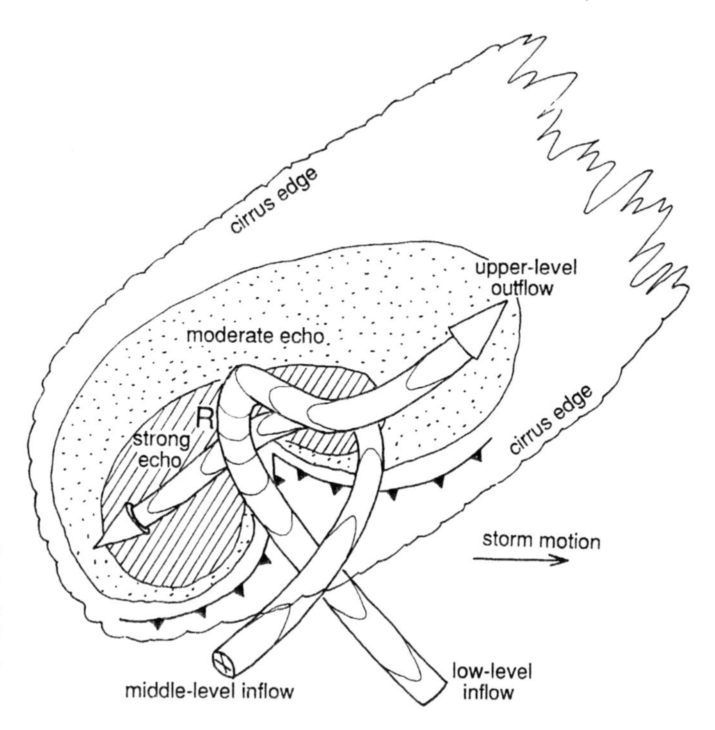

FIGURE 12-6. A model supercell, plane view. The main currents are shown by tubular arrows. Horizontal contours are drawn on the arrows to give the impression of space. The outflow boundary (front symbol) is at the surface.

cally 5–10 km wide and 3–6 km high. Updrafts of about 10 m s^{-1} under the cloud base have been observed from aircraft. The updraft is much stronger within the cloud and can be judged by the size of hailstones and from conversion of CAPE to the kinetic energy of vertical motion [Section 5-6, equation (5-34)]. A 5-cm hailstone falls with terminal speed of about 30 m s^{-1}, and there must be updrafts of this speed or faster to suspend such hailstones for about half an hour. The formation of large hailstones takes that much time. Larger hailstones have also been observed. Aircraft cannot risk observing such updrafts directly. Such a high updraft clears all raindrops and ice elements out of the vault. For this reason, the echo remains weak. Only cloud drops can be found in the vault, since they form in it. The bottom of the vault can be observed from the ground as a region with either a flat cloud base with a smooth-looking *wall cloud* around it or with fast-moving *scud* clouds. The wall cloud is also called a *pedestal cloud* since the base of the updraft is often at a lower level than the rest of the cloud.

Raindrops, hail, and other forms of precipitation in the supercell grow outside the weak echo vault since lifting is weaker in those regions. Raindrops scatter radar beams much more than cloud drops. Therefore, the rain region around the weak echo region gives a strong echo. Rain falls into the downdraft in the places

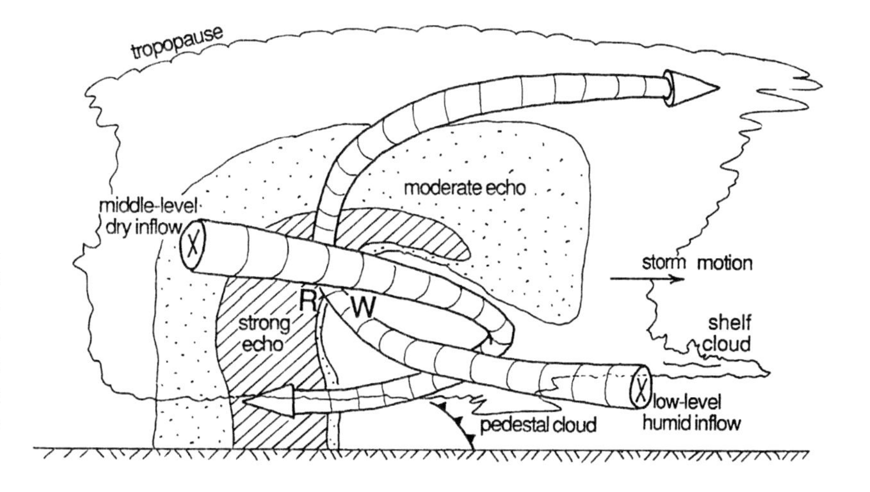

FIGURE 12-7. A model supercell, side view. Main currents are shown by tubular arrows. The arrows have "belts" and perspective, in order to enhance the space character. The top of the cloud is at the tropopause, possibly elevated up to 15 km.

where the updraft circulates above the downdraft (point R in Figs. 12-6 and 12-7). Hail may fall into the vigorous lifting of the main updraft. Then the hail can be recirculated into the top parts of the cloud.

The supercell is more suited to large hail formation than other clouds since some of the updraft is sufficiently strong to suspend and lift hailstones. When the stones are thrown out of the top of the updraft, they may fall in the lower portion of the same updraft. This enables the recirculation of hailstones needed for their growth. Larger hailstones have a better chance of penetrating the edge of the vault than smaller stones or other precipitation elements. In this way, the large stones get the best "harvest" of supercooled cloud drops. Only the remainder of supercooled drops is left for other precipitation elements farther away from the vault. Recycling of hailstones forms alternating layers of clear and opaque ice, as the stones travel through various parts of the cloud.

Mesolow

Precipitation in the supercell is most abundant in an arch of strong echo around the weak echo region of the vault. This is the downdraft where rain can reach the ground. A strong echo forms a curved pattern on the radar screen, known as a *hook echo*. The hook echo is represented by hatching and marked "strong echo" in Fig. 12-6. A low in the pressure field forms as a consequence of cyclostrophic balance of the rotating flow.

A cloud with a hook echo rotates cyclonically (with exceptions; see the comments below on anticyclonically rotating storms) and forms a *thunderstorm cyclone* or *mesolow*. The rotation of the thunderstorm cyclone is intensified by the convergence in the lower part of the updraft. Tornadoes (if any) develop from the borders of the weak echo region that shows a hook echo.

The gust front under a supercell often assumes a shape somewhat similar to the polar front in an extratropical cyclone, but very different in scale. The gust front is sensed by recording instruments on weather stations and is easily observed by people standing in an open field. However, a gust front of one supercell can seldom be detected on standard weather charts with the usual density of observations. The thunderstorm outflow that can be detected on large-scale synoptic charts is usually from multiple-cell storms.

Motion of the Supercell

A thunderstorm is carried by environmental wind (*steering*). The steering of a storm, however, is strongly influenced by the storm growth in the preferred direction from which the warm and humid low-level inflow is coming. Therefore, the updraft in a cumulonimbus usually displaces in the direction of the inflow. This causes a storm's movement to deviate from the environmental flow. The deviation of a storm track from the mean air flow can be as high as 60°.

The direction of supercell motion is usually such that the vault is on the front right side. If a supercell travels toward the east, the vault is on its southeastern side. Then the upper-level flow (polar jet-stream level) is from the southwest. Such a storm moves to the right of the upper-level flow; therefore, this is a *right-moving thunderstorm*. Other directions of motion and the corresponding different locations of the vault have also been observed, although less frequently.

Some supercells rotate anticyclonically. These supercells have the vault on the front left side and they

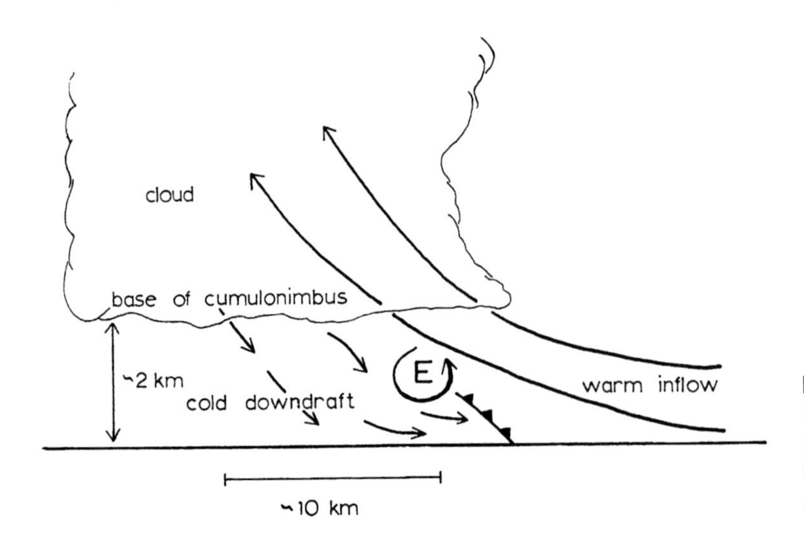

FIGURE 12-8. Air currents at the bottom of a cumulonimbus. The roll cloud is in the eddy E, between the inflow and the downdraft.

move to the left (*left-moving thunderstorm*). Such opposite-rotating supercells appear comparatively seldom, and when they appear they are weaker than cyclonically rotating storms.

Cloud Forms with a Supercell

A large part of the originally warm air that passed through the updraft exits the cloud on top, through the cirrus anvil. The anvil spreads far downwind as a uniform sheet. The top of the anvil normally coincides with the middle tropopause. This tropopause is significantly elevated over a supercell, up to 15 km. The length of the anvil can exceed 100 km. When a supercell is seen from the upwind side (as from "southwest," if north goes "up" the page in Fig. 12-6), the anvil is hidden and the cloud is reported as cumulonimbus calvus. The anvil is very prominent on the other side; the observers there report cumulonimbus capillatus. In a satellite image, an anvil is the most prominent feature of a thunderstorm. The updraft and gust front occupy a much smaller region than the anvil.

A horizontally rotating vortex, accompanied by a *roll cloud* or *arch cloud*, is often observed under a thunderstorm. This vortex appears on the top of the cold outflow under the cumulonimbus. Since condensation occurs in it, the vortex is visible, and it signifies a strong thunderstorm. The horizontal vortex is schematically represented at E in Fig. 12-8.

12-5 MULTICELL STORM

This section describes common features of various mesoscale convective systems (MCS) and the comparatively weak *cluster of thunderstorms*. Multicell storms of mesoscale convective complexes (MCC) and squall line types are described in separate sections below.

Formation of Multicell Thunderstorms

Convective thunderstorm cells seldom appear independently of each other. Most often several thunderstorm cells are physically connected and form a *multicell storm* or an MCS. Thunderstorms in a group may develop due to a common stimulus (a front, a dry line), or they may be caused by the cold outflow from neighboring cumulonimbus. The spreading outflow from a cell in its mature stage usually provides a sufficient stimulus for other cells to form. The outflow near the ground triggers new updrafts by wedging under the surrounding humid air. Therefore, when the first cell starts from some presumably small triggering impulse, other cells are likely to start in the immediate vicinity.

Larger multicell storms usually appear as mesoscale convective complexes (MCC) or squall lines. They often have supercells among them. Such MCSs appear on meso-beta to meso-alpha scale. Larger MCS's compete in size with extratropical cyclones.

The most apparent difference between a squall line and an MCC is the shape. A squall line is elongated; an MCC is more compact. This difference is a consequence of prevalent dynamics in a storm environment. The elongated squall line often forms above a low-level jet stream. Significant upper-level wind (as the polar jet) also contributes to the formation of a squall line. The upper-level jet moves the downdraft sideways so it does not interfere with the updraft.

The Role of Wind Shear

Wind shear in the lower half of the troposphere determines whether the storms in a cluster will arrange themselves into an MCC or a squall line or whether they will stay in the cluster form. Wind shear works by steering the updrafts and downdrafts away from each other. If this separation is successful, the updraft in convective cells can last longer than in a single-cell storm. If the wind difference between elevations of 1 and 6 km shows a wind shear of over 3×10^{-3} s^{-1}, supercells and larger multicell storms (MCCs or squall lines) may develop. These storms usually extend horizontally 30–300 km (meso-beta scale).

New cells prefer to develop on the side of low-level inflow of warm and humid air. Such storms may become severe, causing large hail or damaging wind.

Cluster of Thunderstorms

When the environmental vertical wind shear is smaller than about 2×10^{-3} s^{-1}, thunderstorms with multiple cells will neither develop severe weather nor organize into an MCC or a squall line. Such thunderstorms may stay in the form of a cluster. One such thunderstorm cluster is represented in Fig. 12-9. Several updraft cells are shown in various stages of development (A, B, and C), giving a multiple-cell character to this storm. Updraft air originates in the atmosphere outside the storm. Each cloud cell progresses through the stages of development described for the single-cell storm in the Section 12-4. Cell B is in the mature stage and has a cold downdraft that reaches the earth's surface. Cell A is in the developing stage. There is an embryo of a downdraft at D, which will develop further and cause cell A to grow into the mature stage.

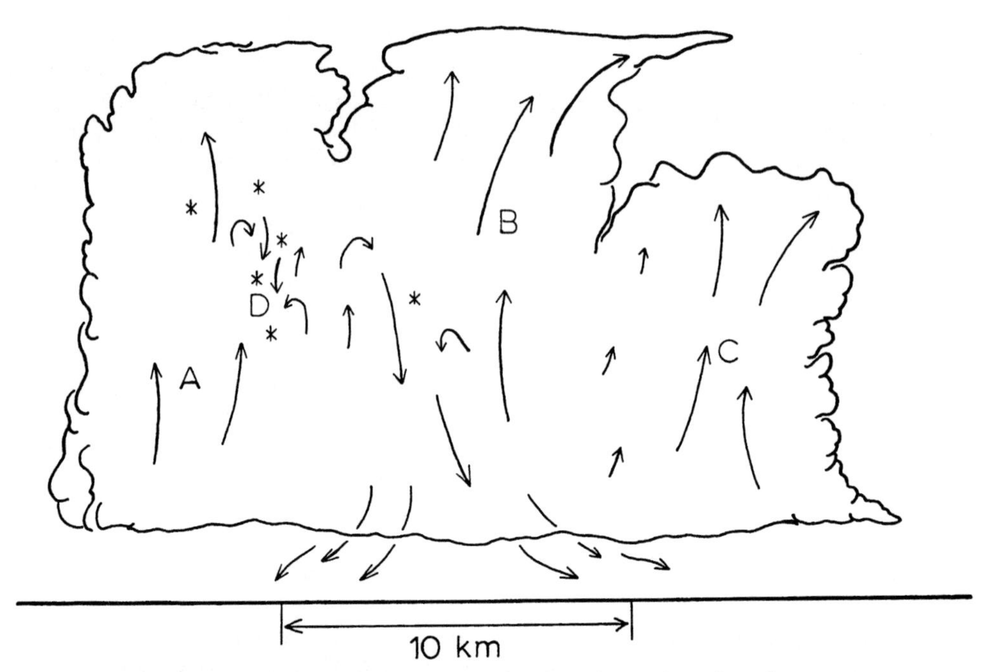

FIGURE 12-9. Schematic vertical section through a thunderstorm cluster in an environment with weak vertical shear. Three updraft cells are shown at A, B, and C. The middle cell (B) is in its mature stage with a downdraft; the two flank cells are younger.

Splitting of Supercells

Multicell storms sometimes form by the splitting of supercells into pairs of cells. One of the resulting cells moves to the right of the movement of the original single cell; the other cell moves to the left. Splitting occurs in response to lifting and tilting of the strong horizontal vortex tube in the middle atmosphere in the domain of the convection cell. Figure 12-10 illustrates the deformation of the horizontal vortex. The vortex is shown as the smooth tube AB between the lower-level inflow and upper-level outflow. The low-level inflow is shown by two stream bands at C and D. The two vertical vortices E and F form due to lifting of the formerly horizontal vortex, one on each side of the lifted part. Since rotation of the cells causes the inflow in each cell to deflect in opposite directions, the two new vertical cells continue to move in diverging directions. Therefore, the updraft in each cell displaces in a different direction, toward its respective inflow. This results in different directions of motion for each cell. Later, the new cells may split again. A line of storms may form in this way. An observed history of splitting storms is illustrated in Fig. 12-11. This case was also very well simulated by the numerical model of Wilhelmson and Klemp (1981).

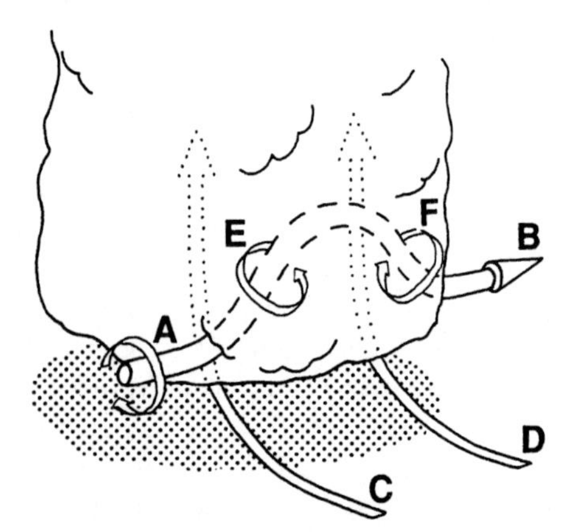

FIGURE 12-10. Formation of two vertical vortices by deformation of the horizontal vorticity vector AB. New rotating thunderstorm cells form from the vortices C and D.

Spearhead Echo

Some storms may rapidly develop multiple cells downwind from the main body of a cloud. Fast-propagating radar echo may assume the form of a *spearhead echo* (after Fujita and Byers, 1977). A case of spearhead echo is depicted in Fig. 12-12. The fast movement of the echo is explained by the emergence of fast-moving upper-tropospheric air in the downdraft. The downdraft in the spearhead echo originates in the high troposphere where wind is fast. Upper-level air enters the

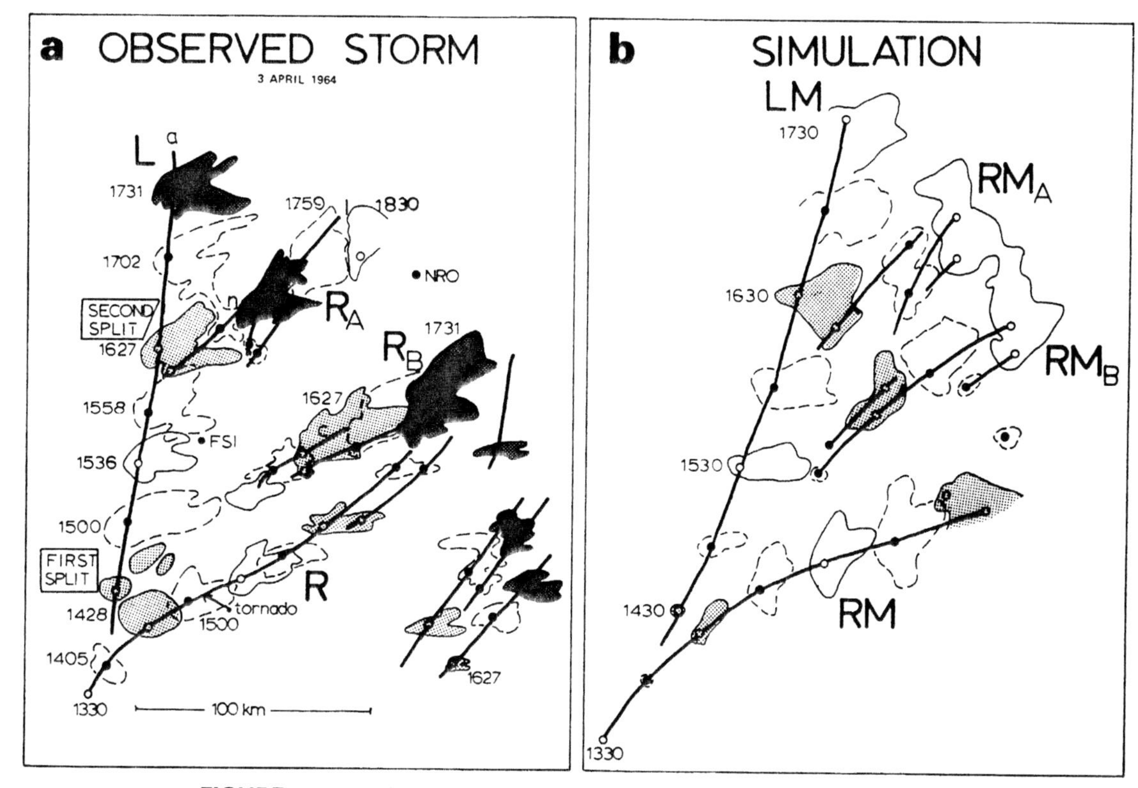

FIGURE 12-11. Observed (a) and modeled (b) storm development on 3 April 1964. Observed reflectivity > 12 dBZ at 0° and modeled rainwater contents of > 0.5 g kg^{-1} at the elevation of 0.4 km are enclosed by alternating solid and dashed contours about every 30 min. Maxima in the fields are connected by solid lines. The scale shown in (a) also applies in (b). From Wilhelmson and Klemp (1981), by permission of the American Meteorological Society.

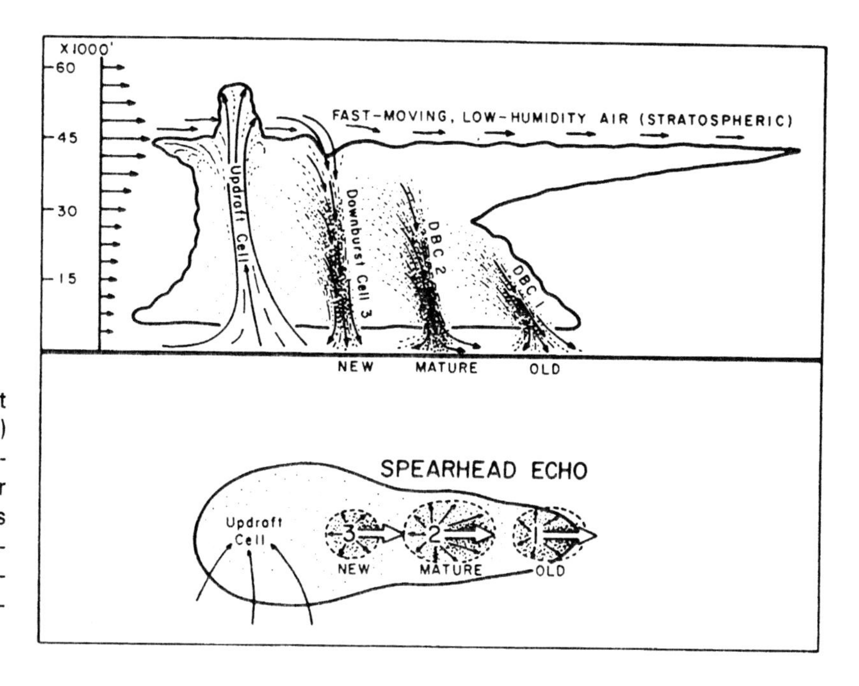

FIGURE 12-12. The storm that generated a spearhead echo (top) and the spearhead echo on the radar screen (bottom). The dry air that forms the downdraft enters the cloud near the top. From Fujita and Byers (1977), by permission of the American Meteorological Society.

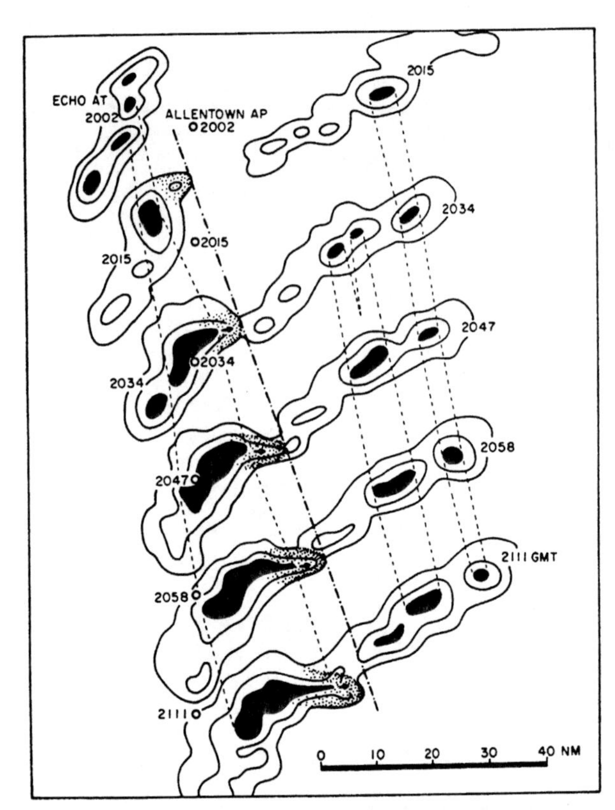

FIGURE 12-13. The observed spearhead echo of 24 June 1975 near Allentown, Pennsylvania. Individual patterns developed and moved predominantly to the east, not to the south as drawn. The patterns are drawn "south" of each other so that they are not obscured. From Fujita and Byers (1977), by permission of the American Meteorological Society.

cloud circulation near the tropopause, as shown in Fig. 12-12. The fast propagation of the spearhead echo signifies wind of damaging strength.

An observed case of a spearhead echo is shown in Fig. 12-13. As these examples show, the "spear" develops from the large echo mass. It moves faster than the spearhead system; the old cells have already moved away to the east.

12-6 SQUALL LINE

The arrangement of convective cells in a line is a frequent form of multicell thunderstorms. Most often the term *squall line* is used when the outflow boundaries of neighboring convective cells merge into a longer line. However, sometimes a line of isolated storms will be called a squall line if there are comparatively small spaces between the cells.

Squall Line Environment

A squall line normally forms in the warm sector of an extratropical cyclone, in the vicinity of the cold front (usually closer than 70 km). Once a squall line is formed, it may travel faster than the cold front behind it. In this way, a squall line can travel far from the front that apparently stimulated its formation. On other occasions, the front travels faster than the squall line, so that the two merge.

Squall lines in middle latitudes travel from the west, as do cyclones and active regions of jet streams. Tropical squall lines usually propagate toward the west, downwind in the tropical easterlies. Such tropical squall lines can be observed as far north as 35° in North America in the summer.

A squall line is often located under the elongated southwestern end of a comma cloud of an extratropical cyclone. A burst point of convection is usually the southern end of a squall line; however, not all burst points lead to squall lines. In satellite images, ordinary lines of convective clouds or rainbands look similar to squall lines.

A squall line does not occupy the whole elongated cloud of the conveyor belt. Strong radar echo of a squall line is usually confined to the southern end of the comma cloud, near the burst point of convection. The rest of the comma cloud may consist of merged spreading anvils of cumulonimbus cells in a squall line.

As described in Chapter 10, there may be multiple burst points of convection in a large comma cloud of a cyclone. Some burst points and even squall lines may be hidden under a cirrus canopy of other lines of convection. Also, not all lines of convection develop into squall lines.

Environmental conditions for squall line formation include potential instability, CAPE near 2000 $m^2\ s^{-1}$, strong vertical wind shear of $1–2 \times 10^{-2}\ s^{-1}$ in the lowest 2 km of the troposphere, and significant precipitable water (over 2.5 cm). These conditions are not too different from the conditions for the formation of other types of storms. The Coriolis force seems to be unimportant for the formation and propagation of squall lines, since squall lines also form in the tropics.

Formation of Squall Lines

Several ways of squall line formation have been observed. Bluestein and Jain (1985) categorize the main cloud shapes that characterize squall line formation as follows:

1. *broken line*, where individual cells start developing separately along a line and later join into a more continuous configuration;

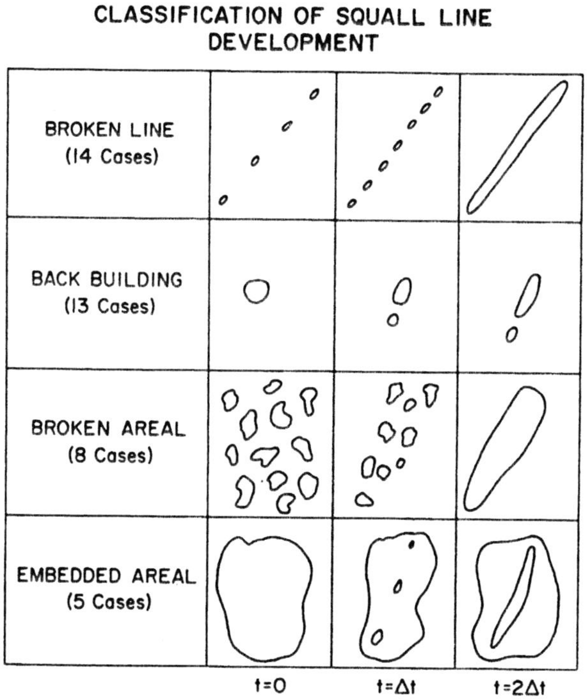

CLASSIFICATION OF SQUALL LINE DEVELOPMENT

BROKEN LINE (14 Cases)

BACK BUILDING (13 Cases)

BROKEN AREAL (8 Cases)

EMBEDDED AREAL (5 Cases)

t=0 t=Δt t=2Δt

FIGURE 12-14. Four main types of squall line formation. The number of cases refers to the period of study. From Bluestein and Jain (1985), by permission of the American Meteorological Society.

2. *back building*, which is similar to the development of a cloud plume in the burst point of convection, as described in Fig. 10-4;
3. *broken areal*, when seemingly randomly distributed convective cells enter a line formation; and
4. *embedded areal*, when stratiform precipitation gradually changes shape into a squall line.

These forms are sketched in Fig. 12-14. Different cloud formations in these cases imply that the mechanism is different in those cases as well. One property is common to all these types: the natural tendency to form lines of convection, as observed on various scales in the atmosphere. Lines of convection are also observed as precipitation bands in cyclones.

Individual convective cells in squall lines usually do not develop rotation in the presence of large vertical shear (about 2×10^{-2} s^{-1}). Strong shear is more conducive to the development of deep convection without rotation. The shear influences the downdraft to avoid interfering with updraft. Therefore, convection cells in such squall lines live considerably longer (perhaps 2–4 h) than in usual multicell storms.

Directional wind shear and associated thermal advection and helicity do not seem to be necessary for the development of deep, long-lasting convective cells (Rotunno and others, 1988). Directional shear helps the formation of supercells.

Detection of a Squall Line

Radar is the principal tool to recognize a squall line. Visual observations seldom distinguish a squall line from other convective storms. However, larger squall lines can be detected on weather charts since most squall lines are associated with a discontinuous wind shear line and pressure trough.

Satellite images are useful to detect the overshooting tops of cumulonimbus that are typical of a squall line. An updraft in the cells of a squall line usually penetrates through an otherwise flat anvil and can be seen from aircraft flying above the clouds. Satellite images with adequate resolution (1- or 2-km pixels) show the turrets of the protruding updrafts. These protuberances are most apparent in visible light near sunset, when they cast long shadows on the anvil. The turrets reveal the location of possible severe weather under the cloud.

Squall lines are similar to rainbands that appear in cyclones. A convenient distinction between squall lines and rainbands is the appearance of thunderstorms: Rainbands do not have prominent electrical phenomena and squall lines do. The intensity of radar echo is generally under 30 dBZ for rainbands and over 40 dBZ for thunderstorms in a squall line.

An Example of a Squall Line

An example of radar echo from a squall line is shown in Fig. 12-15. The outline of Oklahoma (except the far western part) is shown. An organized line of thunderstorms is noticeable at 1940 CST in Fig. 12-15a, with some echoes already exceeding 54 dBZ. This squall line is oriented north–south, and it moves toward the east. The closed centers in the north are still separated from the continuous line in the south. By 2040 CST a continuous radar echo spreads as a band across the domain (Fig. 12-15b). The squall line itself is the area with precipitation, as shown by the radar echo.

Typical Properties of a Squall Line

Numerous observations have shown that a squall line has several properties that repeat from case to case. Some of these features are present in Fig. 12-15. The most common properties of squall lines, with typical dimensions that should be understood within a factor of about 3, are as follows.

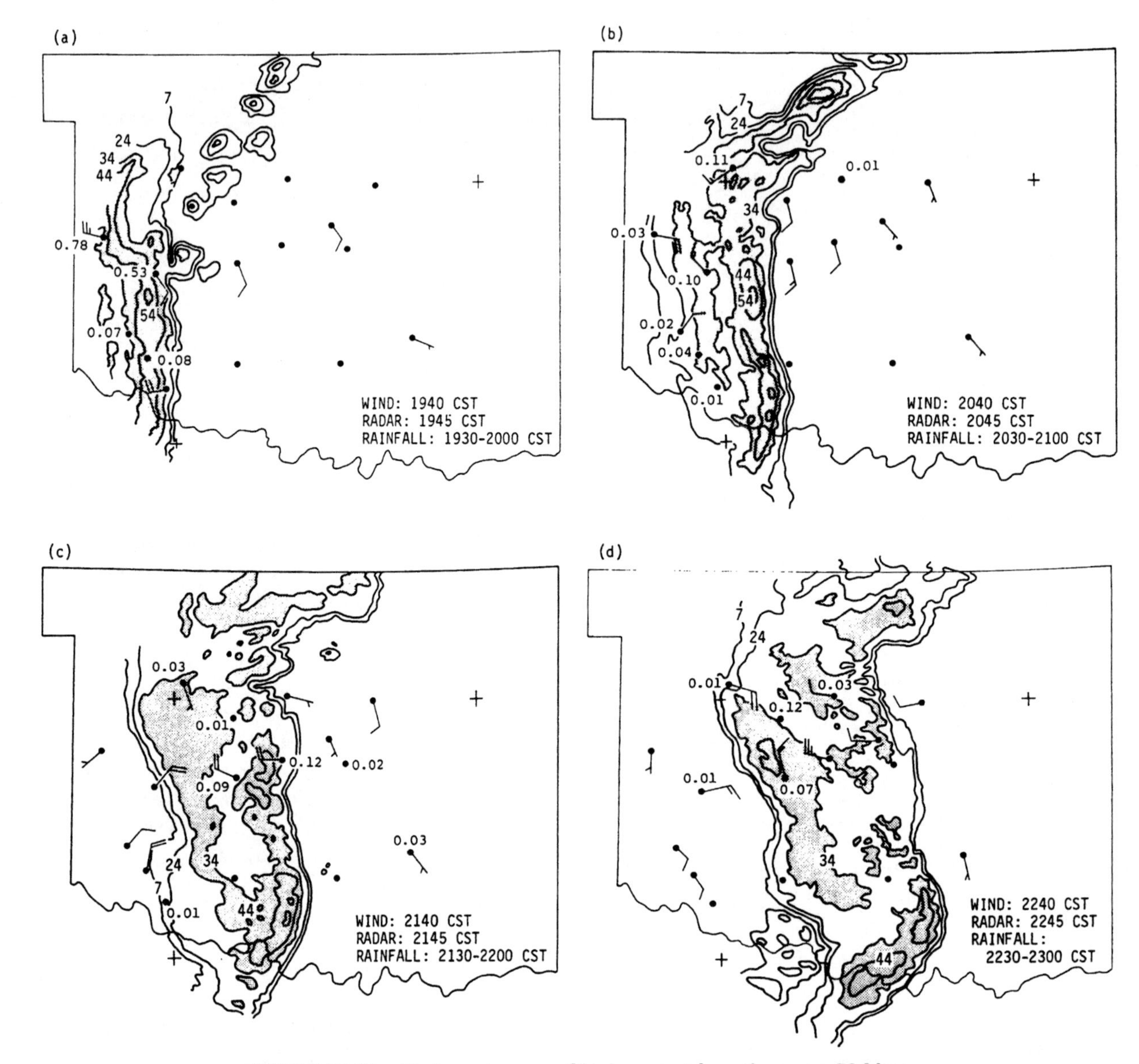

FIGURE 12-15. Radar echo over Oklahoma at four times on 22 May 1976. The contours are for 7, 24, 34, 44, and 54 dBZ, with shaded echo in excess of 34 dBZ. Rainfall rate [inches (30 min)$^{-1}$] and wind (knots) are shown for several observing stations. From Ogura and Liou, 1980, by permission of the American Meteorological Society.

1. A squall line is much longer than it is wide. A typical length is about 500 km. The line in Fig. 12-15 extends beyond the shown domain.
2. The width of a squall line is typically about 100 km.
3. There is a narrow line of strong echo, about 5 km wide, along the front side of the squall line (the eastern edge of the squall line in Fig. 12-15). Severe weather, if any, appears with these strong echoes.
4. There is a wide area of continuous, stratiform precipitation in the central and western parts of a squall line.
5. There is a merged outflow boundary along the front side of a squall line. The closely spaced isopleths of 7 and 24 dBZ at the eastern edge of the echo in Fig. 12-15

show a likely position of the outflow boundary.

6. The life time of a squall line is of the order of 10 h. Some of them formed and dissipated within 2 h, but some have lasted through 4 days.

A surface weather chart reveals that the merged outflow boundary of several convective cells in a squall line is very similar to the cold front. The most apparent phenomena that accompany squall lines and cold fronts are a discontinuity in temperature, a trough in the pressure field, a sudden wind shift, and vigorous cloud formation. However, on upper levels, a squall line is different from a front. For example, the squall line is not structurally connected with the jet stream.

Structure of a Squall Line

Many properties of squall lines are illustrated in the vertical section in Fig. 12-16. The cloud system of the squall line consists mainly of cumulonimbus on its front part and of altostratus and nimbostratus on its rear side. The cumulonimbus part is similar to other convective storms, with the updraft and downdraft. The downdraft forms the squall at the outflow boundary. Much of the downdraft originates in the middle troposphere at the rear of the squall line. The downdraft does not interfere with the updraft, thereby the convection cells can last longer. The updraft lifts slantwise over the downdraft, but becomes vertical in mature cells. Precipitation forms in the updraft, but falls through the downdraft to the ground. The updraft often produces overshooting turrets of cloud on the top of the squall line. The tropopause is often elevated over the squall line, similar to the elevated tropopause above the supercell. Satellite images sometimes show that the cloud top at the tropopause may be highest over the stratiform part of the squall line, even higher than the tops of the turrets over the cumulonimbus.

Middle-level inflow in the squall line may be responsible for the forward momentum of the outflow boundary and the whole cloud system. This forward momentum is similar to the development of a spearhead echo, as described at the end of Section 12-4, but the significant difference from the spearhead echo is that the youngest cell appears farthest south, at the burst point.

A characteristic of a squall line is a rather wide stratiform precipitation area on the rear side of a squall line. This can be noticed in Fig. 12-15 where the wide area of over 34 dBZ occupies the western part of a squall line.

Movement of a Back-Building Squall Line

The formation of a back-building type of squall line can be explained by the stimulation of cell growth near the older cells, somewhat similar to the growth of a multicell storm described in Fig. 12-9. However, in the presence of a low-level jet and with strong vertical wind shear, there is a tendency to form new cells on the side of low-level inflow. Therefore, the youngest cell in the group is the most successful in stimulating the growth of new cells since this youngest cell is most exposed to the inflow of warm and humid air. This mechanism supports the development of the burst point of convection in about the same place. The burst point moves much slower than the wind in the environment.

The formation of a line of thunderstorms is illustrated in Fig. 12-17. The southern point of the back-building squall line is the burst point of convection. Clouds emanate from this point. The burst point of convection typically travels slowly to the east at a significant angle to the upper- and lower-level flow.

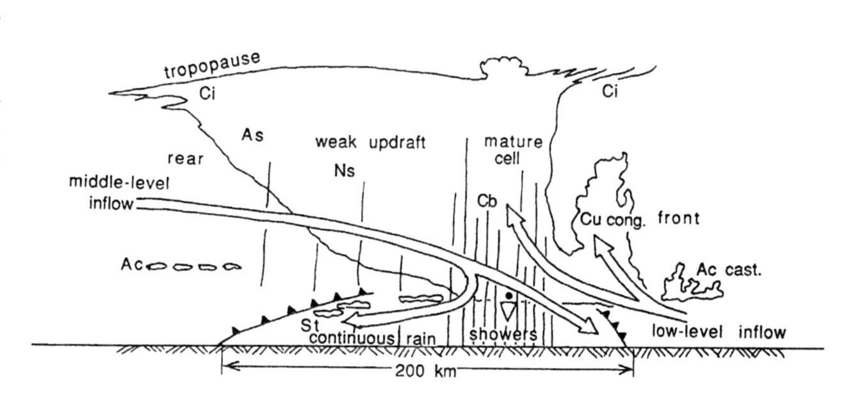

FIGURE 12-16. Vertical section across a squall line. Vertical lines indicate precipitation; denser lines correspond to stronger rainfall rate and stronger radar echo. An inflow of air occurs on both sides: on the front side in lower levels and on the rear side in the middle troposphere. Cold outflow occurs under the cloud (shown with the cold front symbol). Updraft exits on top, in the anvil. The updraft often forms protrusions on the top. Prevalent cloud genera are indicated by abbreviations.

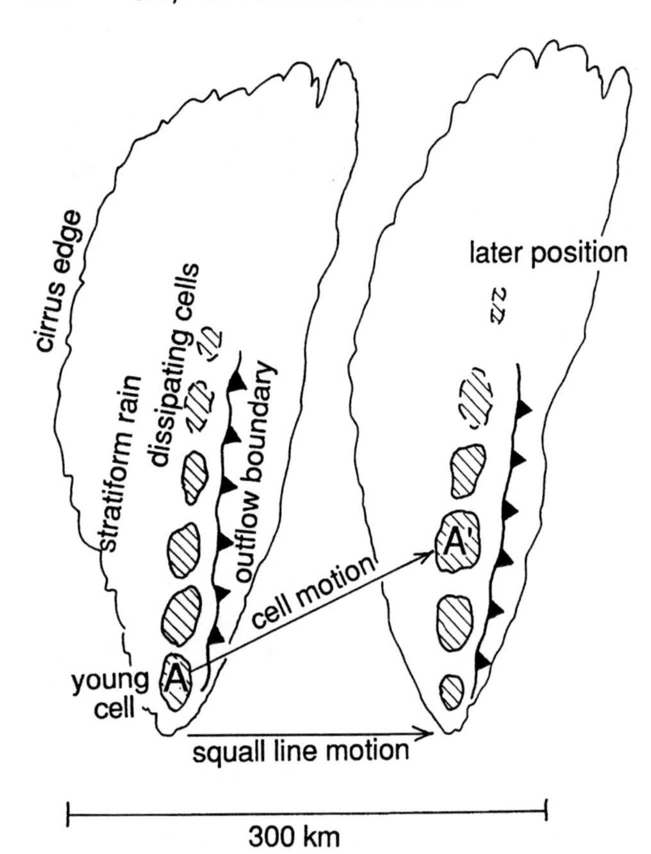

FIGURE 12-17. A squall line and its motion. The motion of the squall line is illustrated by its two positions. The cell A moved to position A'. New cells have formed on the "southern" end of the squall line, thus causing the system of cells to move differently from individual cells.

The whole squall line follows the burst point like the tail of a comet follows the head.

Comparison with the Supercell

To an observer on the ground, there are not many differences between a squall line and a supercell. However, on a radar display the structural differences are large and visible. A supercell rotates and therefore assumes a stable circulation state similar to cyclostrophic balance. The rotating pattern of a supercell produces a characteristic hook echo. The convection cells (cumulonimbus) in the squall line do not rotate. Instead, they produce a long line of echoes on the radar display. The cells in a squall line develop discontinuously, one after the other, near the burst point of convection. New cells, when they attain maturity, produce new protuberances on the top of the cloud at the tropopause.

The convective cells in squall lines normally do not rotate; however, some member cells in a squall line may acquire rotation about the vertical axis and thus become supercells. The squall line produces severe weather very similar to supercells (large hail, damaging wind). Tornadoes are less frequent with squall lines than with supercells. However, as mentioned, supercells may develop within squall lines.

12-7 MESOSCALE CONVECTIVE COMPLEX

Description

Multicell storms sometimes develop into compact oval shapes in satellite images. A common cirrus cover occupies an area of more than 300×300 km^2. If these areas are nearly circular, they are *mesoscale convective complexes* (MCC). Formally, the oval cirrus cover will be recognized as an MCC if it satisfies the following criteria:

1. a continuous area with a temperature of ≤ -32°C that occupies more than 10^5 km^2;
2. an interior cold cloud top of ≤ -52°C that occupies more than 5×10^4 km^2; and
3. a nearly circular cloud shape such that the ratio of the shortest axis to the longest is $\geq$ 0.7.

This description of an MCC is different from the description of a squall line: The description of an MCC is based on satellite images, while the description of a squall line is based on the radar echo.

As mentioned before, formal definitions cannot be taken rigidly in meteorology. So it is with an MCC: The above points are useful when the cases of MCC are counted. However, if one event is studied, it is possible that an MCC develops with an oval satellite image that does not satisfy the criteria (for example, it may be somewhat smaller or more elongated at some time during its lifetime). Such a storm may still be called an MCC for practical purposes. Definitions such as the ones above for an MCC should not always be taken with mathematical rigor; instead, they should be used only as guidelines.

The above description shows that an MCC is more oval and a squall line is elongated. These two kinds of MCSs have different physical characteristics. However, as with other categories of mesoscale storms, the distinction between an MCC and a squall line is not always clear. Sometimes both coincide: A squall line develops within an oval form of an MCC.

The distinction of an MCC from other mesoscale storms is also not clear-cut. There is often a supercell within an MCC; therefore, severe weather is common in an MCC. A great part of an MCC is covered with a region of stratiform rain, again similar to a squall line. Examples are shown in Fig. 12-18.

Environment

The large-scale environment that favors an MCC normally shows potential instability and lifting, as is the case with other convective storms. The active region of the polar jet is a favorite place, with its lifting and enhancement of instability. Lifting of 0.5–1 Pa s^{-1} (5–10 $\mu b\ s^{-1}$) is typical for formation of an MCC.

Warm advection in the lower troposphere also is advantageous for MCC formation. This is the process that most enhances instability and requires lifting in the balanced flow. Together with positive thermal advection, there is normally a significant advection of water vapor. Since both these advections occur primarily in about the lowest 2 km of the troposphere, these processes contribute to potential instability.

A remarkable property of an MCC is that in each case a strong divergence of horizontal flow is observed at 200 mb, about at the level of the cirrus anvil and top outflow from the storm. The upper-level divergence renders an MCC similar to a tropical cyclone. There is evidence that an MCC may rotate, somewhat like a tropical cyclone (Menard and Fritsch, 1989). Rotation is regularly weak over the continents, possibly due to friction at the ground. However, some MCCs over oceans do transform into tropical cyclones.

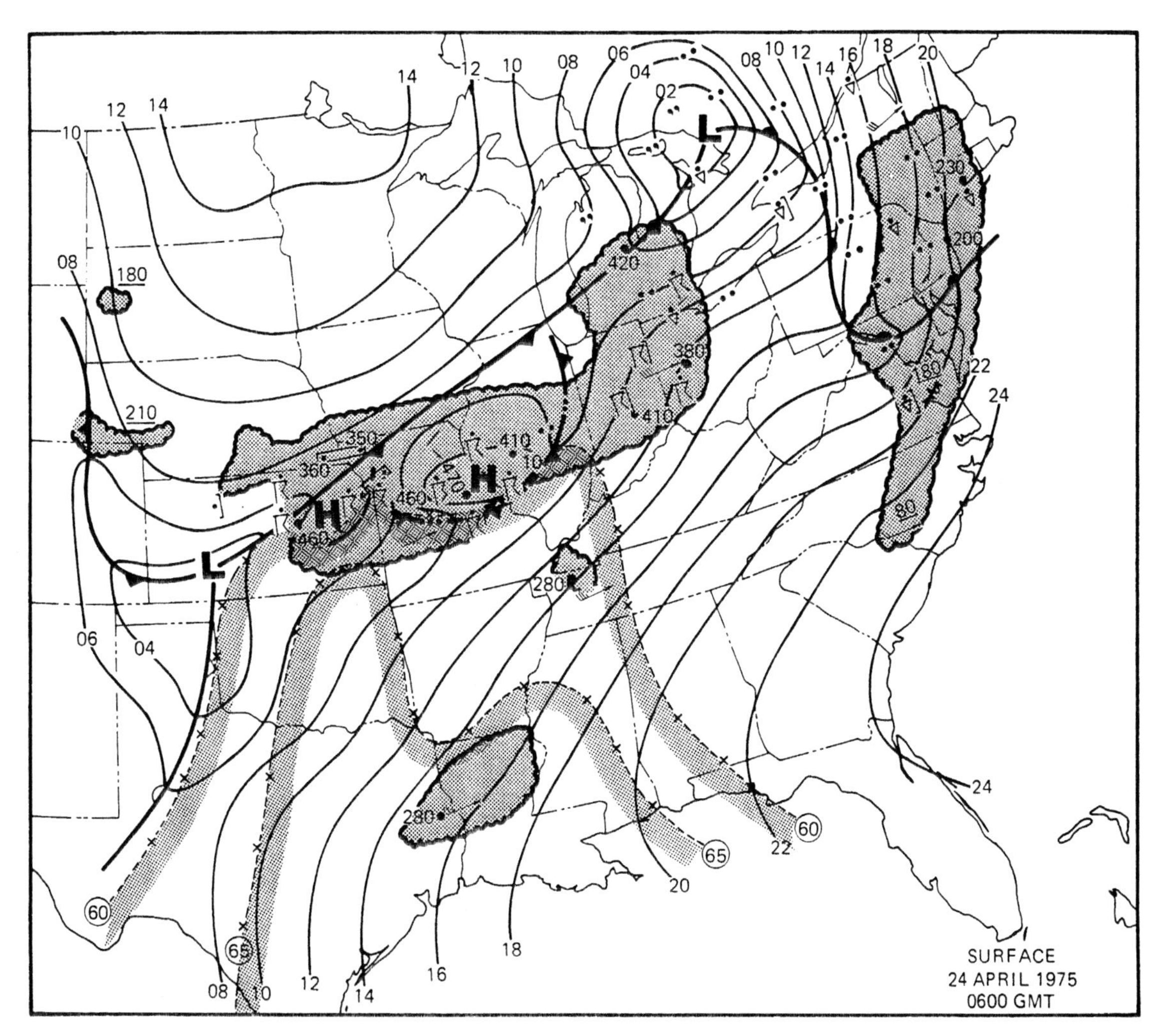

FIGURE 12-18. Sea-level isobars, fronts, and precipitation area (shaded) in a case of two MCCs over the central United States. Two mesohighs are in Kansas and Missouri. The isobar of 1008 mb is especially typical where it loops around the high in Missouri. Analysis by R. A. Maddox.

Radiosonde observations show that the lifting air in an MCC is warmer than its environment in the lower half of the troposphere. Also in this respect, an MCC resembles a tropical cyclone. The upper tropospheric part of an MCC is usually somewhat colder than the environment. This distribution of temperature shows that the hydrostatic stability is lower in the storm than in surrounding atmosphere.

Mesohigh

A joint downburst from the cells in an MCC produces a pool of cold air near the earth's surface. The weight of the cold air forms a high in the pressure field. The pressure increase spreads to the gust front around an MCC. The longest dimension of this high is 100–500 km, and it can be detected on weather charts. This mesohigh accompanies practically all thunderstorms. Due to the size of these storms, a mesohigh can be easily detected with a squall line and MCC.

An example of a mesohigh is shown in Fig. 12-18. This is a case of a group of MCCs over central United States. The sea-level isobars show several typical weather patterns, including the polar front, the dry line (over Texas), regions of showers (shaded, with appropriate weather symbols), and isobars. The high centers over Kansas and Missouri have a rather strong curvature of isobars around them. Such isobars are typical for thunderstorms.

A mesohigh is not an anticyclone since there is no corresponding anticyclonic circulation. The relation between wind and pressure is very unbalanced. The accelerations are large in this situation. The wind usually blows away from the center of a mesohigh.

Similar mesohighs are common with squall lines. A mesohigh of a squall line is accordingly elongated. Smaller storms also develop mesohighs, as is observed by a barograph at the time of gust front passage. Since it is usually located between stations, a small storm seldom influences the position of the isobars on the weather chart. Occasionally an observation appears in discrepancy with neighboring observations. This may be due to a small mesohigh.

Geographical Distribution

A favorite location of MCCs is the Great Plains of North America. MCCs are very rare over other regions near the west and east coasts of North America. MCCs appear selectively in several other areas of the world, and many MCCs have been observed over South America. Therefore, we may expect that these storms are likely to form over other continents as well. A great variability of frequency is expected too. Comparably few MCCs have been documented over the oceans. The polar lows, described below, have some similarities with MCCs. As mentioned before, some MCCs over tropical oceans have developed into tropical cyclones.

The continental MCCs prefer to form in the westerlies or easterlies *behind* large mountain chains (Velasco and Fritsch, 1987). The high plains of North America and the plains of Argentina are both in the westerlies and in the lee of high mountains. These plains are favorable places for MCCs. Another favorite place is in the lee of the mountains in Colombia, where the tropical easterlies descend from the Andes. An American example is shown in Figs. 12-19, 12-20, and 12-21.

Life Cycle

The duration of an MCC is of the order of 12 h. An MCC often develops from a group of local storms that form over the Rocky Mountains or the plains close to them. Those local storms develop during the afternoon. In the evening, when the local storms decay, an MCC may start developing. An MCC attains its maximum size around midnight local time and dissipates around sunrise. In rare cases, the remnants of an MCC survive until the next evening and then form another nocturnal MCC. The nocturnal development of the MCC is due to the intensification of temperature and

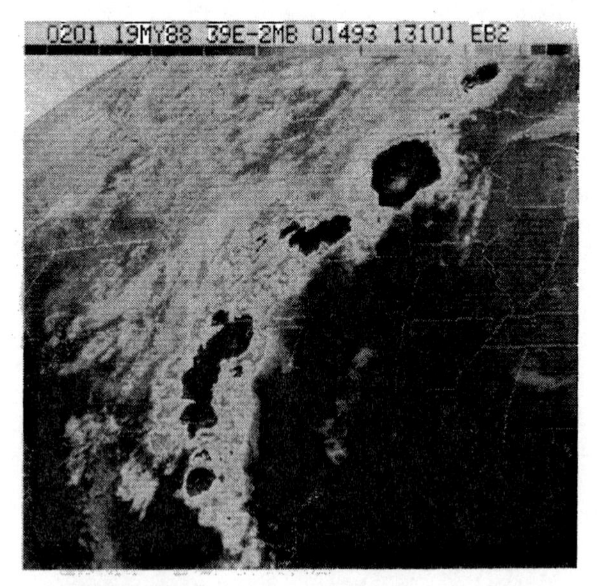

FIGURE 12-19. An enhanced IR satellite image for 0201 UTC 19 May 1988. An MCC is shown over South Dakota. The sharply delineated black surface is the high cloud top with temperature below −60°C.

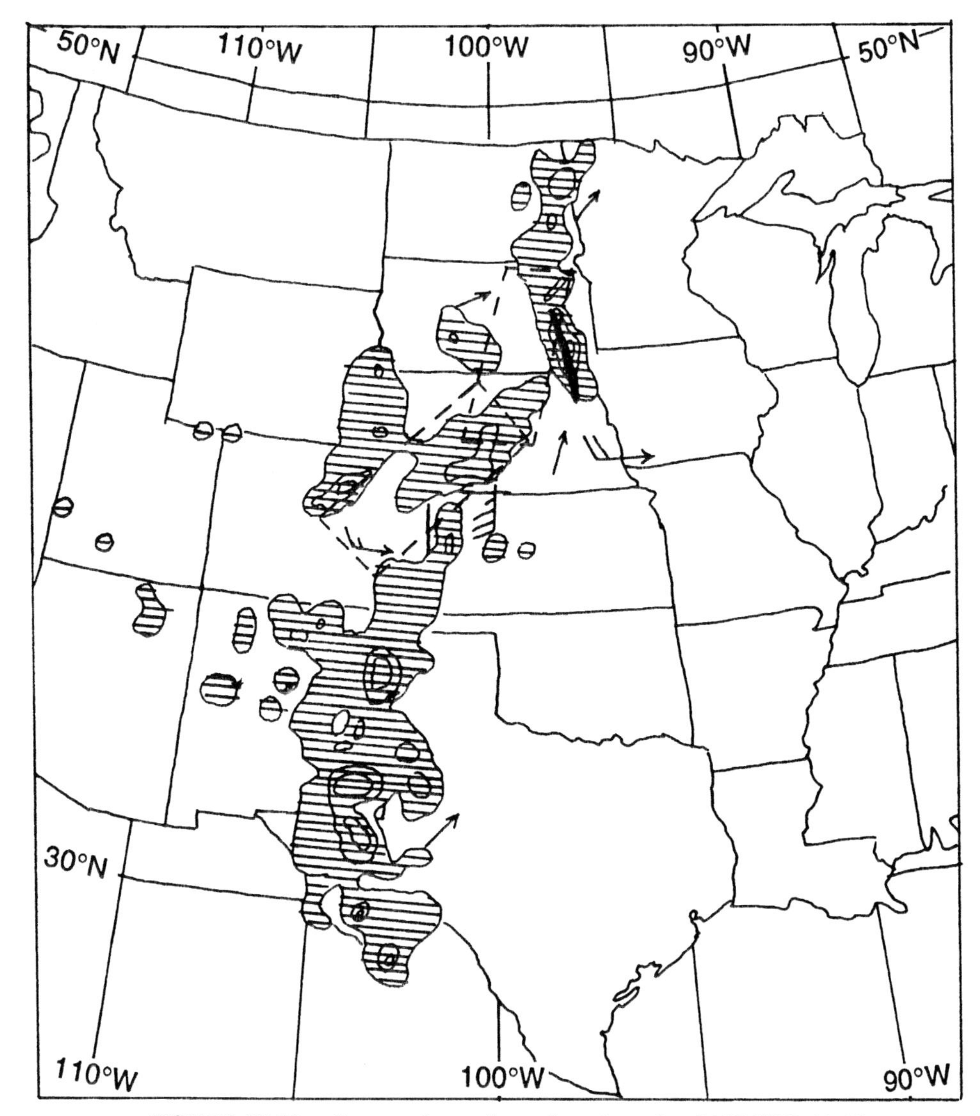

FIGURE 12-20. Composite radar echo chart for 0135 UTC 19 May 1988. Numerous echoes are along the conveyor belt from west Texas to North Dakota. There is a squall line in South Dakota that corresponds to the MCC in Fig. 12-19.

humidity advections in the low troposphere in the vicinity of the storm. The wind in the low levels (often the low-level jet) tends to intensify at night and thereby to enhance the advection. Increased thermal advection in lower layers enhances the thermal (also potential) instability, which enhances further development of convection. Nocturnal development of an MCC explains the climatological maximum of thunderstorms after midnight over the Great Plains of North America.

Severe weather associated with an MCC typically occurs before the MCC is formed. A study of 12 MCCs by Watson (quoted by Maddox an others, 1986) shows that most large hail and strong wind occurred before the MCC was formed. All tornadoes and funnel clouds also occurred before the MCC were formed. Only a few incidents of strong wind, hail, and heavy rain occurred within MCCs proper. This indicates that the smaller local storms that later formed MCCs, were more violent than MCCs.

Flooding

The potential for flooding stays high throughout the life of an MCC. Moving storms present less flooding hazard than an MCC since rain is distributed over a larger area. Major flood events are produced by slow-moving or stationary MCCs.

When an MCC moves more slowly than average, about at 5 m s^{-1}, then the danger of flash flood is more acute. This is understandable since in such a case rain is all delivered over one place. The outflow from some watersheds can exceed the capacity of riverbeds; in such cases flood or flash flood may occur. A moving MCC distributes rain over a long stretch, without excessive rainfall in any location. Flash floods are described briefly in Appendix K.

Movement

An MCC moves partly with the environmental wind, but deviates from it since the new convective cells develop on the side with the most low-level warm inflow. In this way, an MCC behaves like a back-building squall line that moves sideways to the environmental wind.

An Example

An example of an MCC is shown in Fig. 12-19. The enhanced IR image shows a circular "dark" cloud with temperature of less than -60°C. This interpretation is made by comparison with the MB curve shown in Fig. F-2 in Appendix F. Bright spots can be noticed inside the cold (black) area. This corresponds to the temperature below −63°C. The gray irregular ring surrounded by bright areas, and going around the highest cloud, is the region between −32 and −43°C. The surface within this ring satisfies the first definition for an MCC. The transition over the value of −52°C, for the second criterion is within the bright area between the ring and the black central region. The ellipticity from the third definition is also satisfied for the ring, since the cloud is almost circular.

The clouds that stretch from southwestern Texas to South Dakota and farther north are all in a long conveyor belt.

Radar echoes at about the time of the satellite image are shown in Fig. 12-20. There is only approximate similarity between the radar echoes and satellite image. There is a squall line within the MCC in South Dakota. This case shows that it depends on the data sources whether an event will be interpreted as a squall line or an MCC.

Weather charts for this case are shown in Fig. 12-21. They illustrate several typical circumstances in which MCSs appear.

The 300-mb chart in Fig. 12-20b shows that the nearest part of the polar jet stream is in Wyoming. As is typical, the MCC in South Dakota is located in the active region of the polar jet stream. Both lower-level

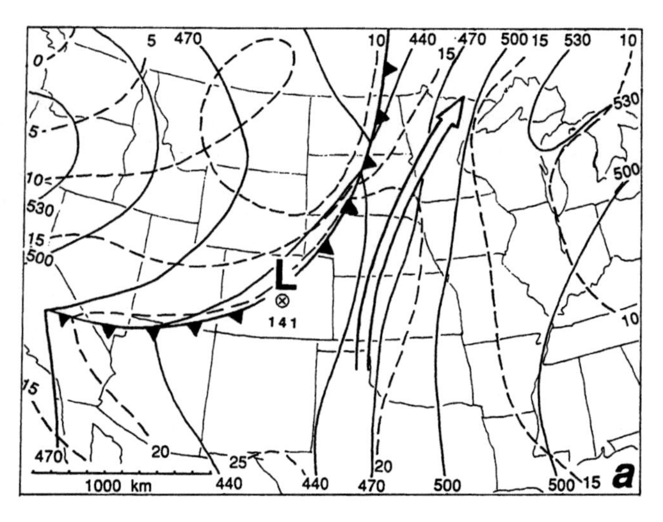

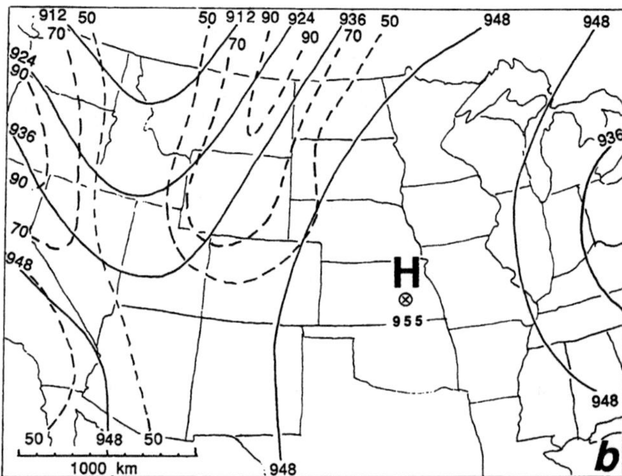

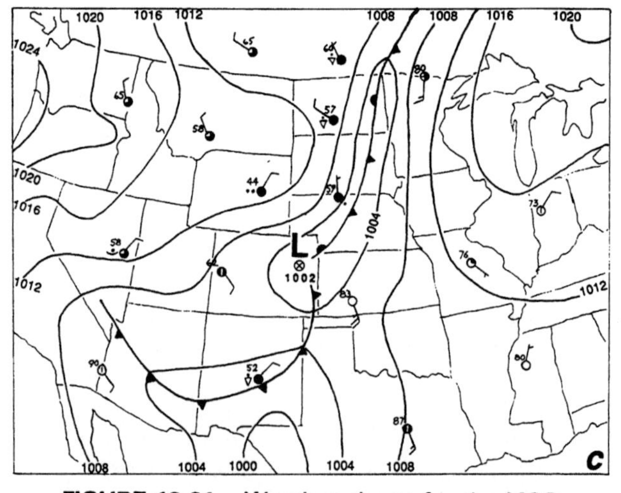

FIGURE 12-21. Weather charts for the MCC event in Fig. 12-19. (a) 850-mb contours at 30 gpm intervals, isotherms (dashed) at 5°C. (b) 300 mb, contours at 60 gpm, isotachs (short dashed) at 20 kt. (c) Sea-level isobars at 4 mb and fronts. The tallest clouds from Fig. 12-19 are in the conveyor belt, east of the trough.

charts (Fig. 12-20b and c) show a developed conveyor belt along the main stretch of high cloud tops.

12-8 TROPICAL CYCLONES

Thunderstorms in the tropics are primarily of the three categories: local thunderstorms, squall lines, and tropical cyclones. Local thunderstorms and squall lines are similar to extratropical storms and therefore will not be described further in this section. Tropical cyclones, however, are greatly different and are described next.

Description

A *tropical cyclone* (TC) forms in tropical easterlies. The isobars in a TC have an almost circular shape. A TC can be recognized by a low center in the sea-level pressure, cyclonic circulation of air around the center, and copious rain.

Originally all TC's were called *cyclones*. It is now common to classify TCs by wind speed as:

1. a tropical depression, with sustained wind slower than 18 m s^{-1} (35 kt);
2. a tropical storm, with sustained wind speed between 18 and 33 m s^{-1} (37 and 64 kt); or
3. a hurricane, with wind speed of 33 m s^{-1} (65 kt) and higher.

"Sustained" wind in these descriptions formally means the wind outside turbulent gusts. For this purpose, it is taken that a gust is an occurrence of higher wind speed that lasts less than 10 min.

The hurricane is called *typhoon* in Asia and Australia. In the Indian Ocean, it is also called *cyclone*, irrespectively of intensity. The weather services commonly give personal names to tropical storms and hurricanes; this facilitates the verbal description of weather charts in the cases when there are two or more cyclones in a region.

The duration of a TC is of the order of 1 week. A few cyclones dissipate within 12 h, but most last several days. The longest TC on record lasted 1 month.

Many Asiatic typhoons attain size and wind speed much in excess of the smaller typhoons and "ordinary" American hurricanes. The cirrus cloud top of a large hurricane may show a circular pattern as wide as 1000 km, and surface wind speed may attain 80 m s^{-1}. There is an indication that such extraordinary large and strong typhoons develop on the intertropical front, while smaller typhoons originate within the easterlies. Only a few hurricanes in the Gulf of Mexico have attained the size and intensity of large Asiatic typhoons.

Tropical cyclones attract much public attention because they cause severe danger to life and property. The most damage is caused by sea surges and the resulting waves. Away from the coast, damage is by strong wind and flooding. Supercells, squall lines, and tornadoes often develop in a TC, and these phenomena may cause additional damage. When speaking about damage from TCs, it is of interest to know that the economic benefit of TC's outweighs the damage, if we exclude the loss of life that cannot be measured in material units. The benefit of TCs is due to increased rainfall. Since the benefit of hurricanes is more evenly distributed than the damage, it makes much less news in the media.

Environment

Tropical cyclones form exclusively over warm tropical oceans, those with sea-surface temperature over 26°C. For this reason, South America south of the equator has never experienced a TC. Both oceans in this region are too cold for TCs to form. A warm sea surface may explain the hurricane season; in North America it is between June and November, when the sea surface is sufficiently warm. In addition, warm air is needed for a TC. This explains why there are two typhoon seasons in Asia: one before and one after the monsoon. The lower troposphere is colder during the peak of the monsoon (June–August) since it is cooled by frequent showers. During that time there are no TCs in India.

The Coriolis force is important in the formation of a TC. No TC has been observed to form within 3° latitude from the equator where the horizontal component of the Coriolis force is negligible, and the appearance of TCs closer than 8–10° to the equator is very rare.

The Coriolis force influences the circulation in a TC to be cyclonic, according to the gradient wind balance in each hemisphere (Section 6-4). This is always with the Coriolis force pointing out of the vortex.

No TC has ever been observed to cross the equator. Dynamic meteorology shows that if a cyclone somehow reaches the other hemisphere, it should attain devastatingly fast wind. The circulation of such a cyclone would become anticyclonic on the arriving hemisphere, with anomalously high wind speed (opposite rotating cyclone from the gradient wind theory, Section 7-4).

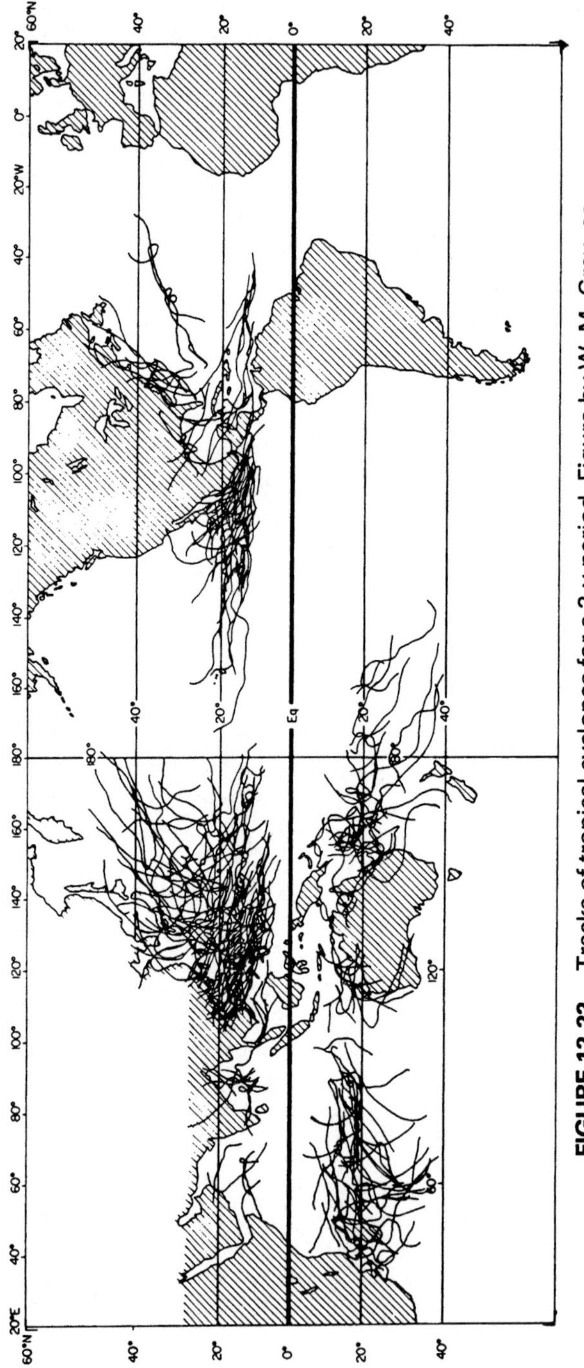

FIGURE 12-22. Tracks of tropical cyclones for a 3-y period. Figure by W. M. Gray, as reproduced by Elsberry (1987).

Geographical Distribution

The trajectories of cyclone centers observed in a 3-y period are drawn in Fig. 12-22. The trajectories begin on the end close to the equator, where the cyclones form. When a TC is only a few days old, it generally moves to the west, carried (*steered*) by the easterlies. After a few days, it turns away from the equator. Longer-living TCs sometimes move toward northeast (or southeast in the Southern Hemisphere), in the belt of the westerlies. When a TC arrives at a continent or at a cold ocean surface, it dissipates. This usually occurs between 20 and 40° latitude.

The absence of TCs in South Atlantic and Southeast Pacific is conspicuous. This is due to the cold water in the oceans around South America. Similarly, a TC weakens when it arrives on land or at a colder ocean surface. A TC cannot develop or persist when the flux of heat and latent heat from below is small.

There are no TC tracks in a narrow belt along the equator where the horizontal component of the Coriolis force is small or vanishes.

As can be seen in Fig. 12-22, the trajectories of TCs are very variable. Some trajectories have loops showing that a TC passed over the same place twice. It is not possible to forecast cyclone motion solely on the basis of previous TC tracks reliably. However, it is possible to evaluate the probabilities of TC motion, useful information several days before the cyclone may arrive.

Some tropical cyclones redevelop as extratropical cyclones in the middle latitudes. This occurs in active regions of the polar jet stream when tropical air with an imbedded TC possesses sufficient latent and sensible heats needed for cyclone redevelopment. The supply of heats is usually insufficient to feed a TC, but it may be adequate for redevelopment of extratropical cyclones in a baroclinically unstable frontal zone. Those cyclones are not considered "tropical" when they redevelop in the middle latitudes. However, the name given to the TC is sometimes carried as long as the descendant extratropical cyclone can be recognized on weather charts.

Wind

Wind distribution also reveals the circular shape of a TC. The streamlines are usually circular, similar to the isobars. However, there is considerable asymmetry in the distribution of wind speed. The highest wind speed occurs on the right front of the storm, relative to its motion. The example in Fig. 12-23 shows the wind speed around Hurricane Frederic. Wind speed of over 20 m s^{-1} was present in a circular area of about 300 km in diameter. The fastest wind of over 40 m s^{-1} occurred NNW of the cyclone center, which agrees with earlier experience. This storm moved toward 300° with speed of 5 m s^{-1}. It should be noticed that the isotach of 30 m s^{-1} occurs twice in the cyclone. The inner isotach of 30 m s^{-1} shows slower wind in the eye of

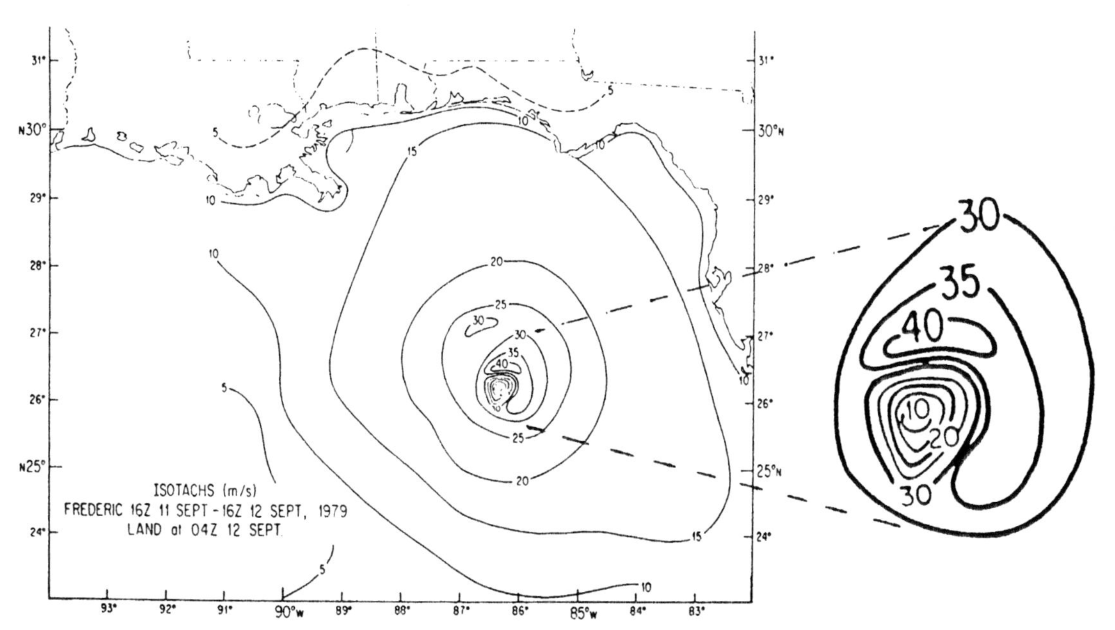

FIGURE 12-23. Wind speed (m s^{-1}) in Hurricane Frederic in the Gulf of Mexico. Storm motion was toward 330° at 5 m s^{-1}. The central portion is enlarged on the side so that the low speed (less than 10 m s^{-1}) in the eye can be noticed more easily. From Powell (1982).

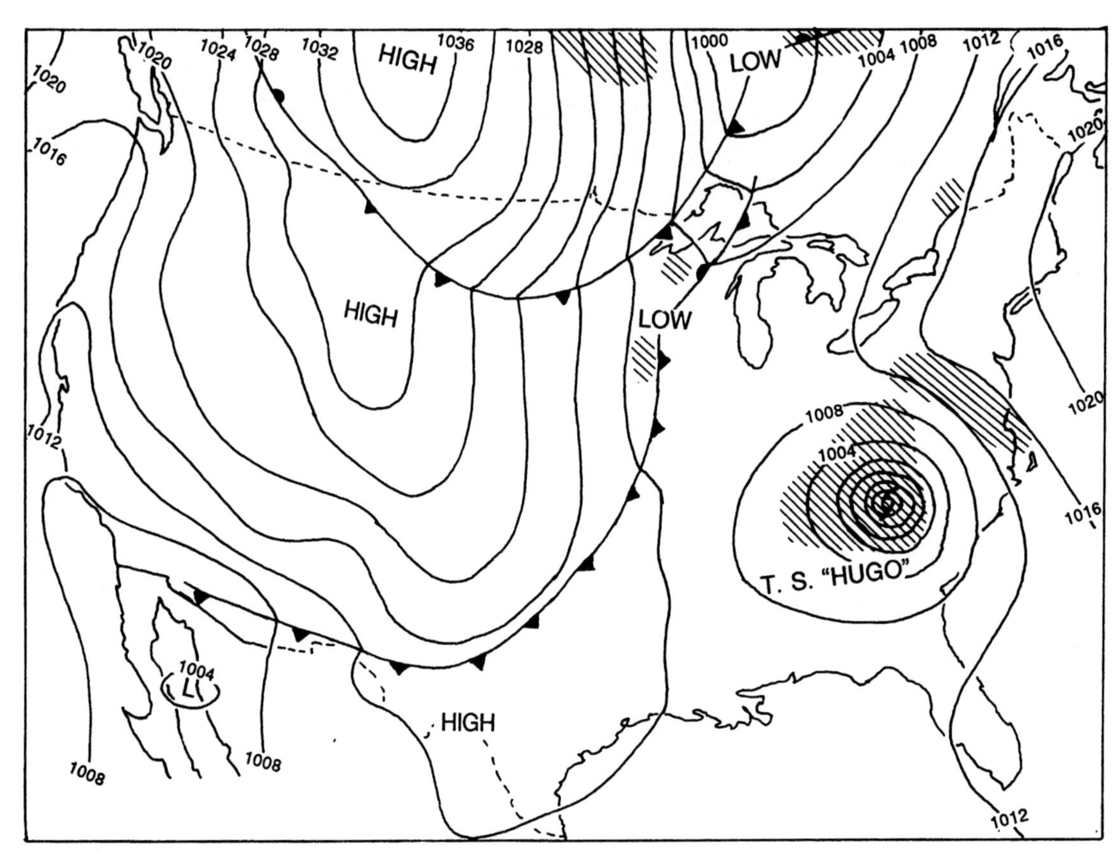

FIGURE 12-24. Surface weather chart for 12 UTC 22 September 1989 showing circular isobars in tropical storm Hugo. The innermost closed isobar in the tropical storm is for 988 mb. Rain areas are hatched. Redrawn from Daily Weather Maps of the National Meteorological Center.

the cyclone. The last closed isotach in the center is of 10 m s^{-1}. The low wind speed is in the eye of the hurricane.

Sea-Level Pressure

The isobars of tropical storms or hurricanes are almost circular. The example of tropical storm Hugo in Fig. 12-24 shows several closed isobars around the center. The outer closed isobar (1008 mb) has a diameter in excess of 1000 km. This storm was a strong hurricane until several hours before the time of Fig. 12-24. The wind speed decreased to tropical storm intensity because the cyclone arrived at land.

Upper-Level Circulation

Upper-level charts do not show a TC. By 500 mb there is only a trough, without a closed contour. The trough is much smaller than troughs in the easterlies. The example in Fig. 12-25 shows the 500-mb chart for the time of Fig. 12-24, for tropical storm Hugo. The low in Louisiana is not in the circulation system of Hugo; this low had been in the region for several days. The trough from northern Alabama to Virginia is the upper part of tropical storm Hugo. If we did not know that there was a tropical cyclone, we could not have recognized it from this 500-mb chart.

Circulation at cloud tops (about 17 km, at the tropical tropopause) is anticyclonic in tropical cyclones. The anticyclonic circulation is in sharp contrast with the cyclonic circulation of lower clouds. Both circulations can often be observed in time-lapsed sequences of satellite images. The lower clouds can be observed on a satellite image through the breaks in the upper-level cirrus shield. The satellite image is similar to Fig. 12-26, with smaller cumulonimbus bands having approximately the shape of radar echoes. These lower clouds flow cyclonically around a TC and enter it. At the same time, high clouds (cirrus from tall cumulonimbus) flow anticyclonically around a TC and exit the main cloud mass. The divergence of the flow at

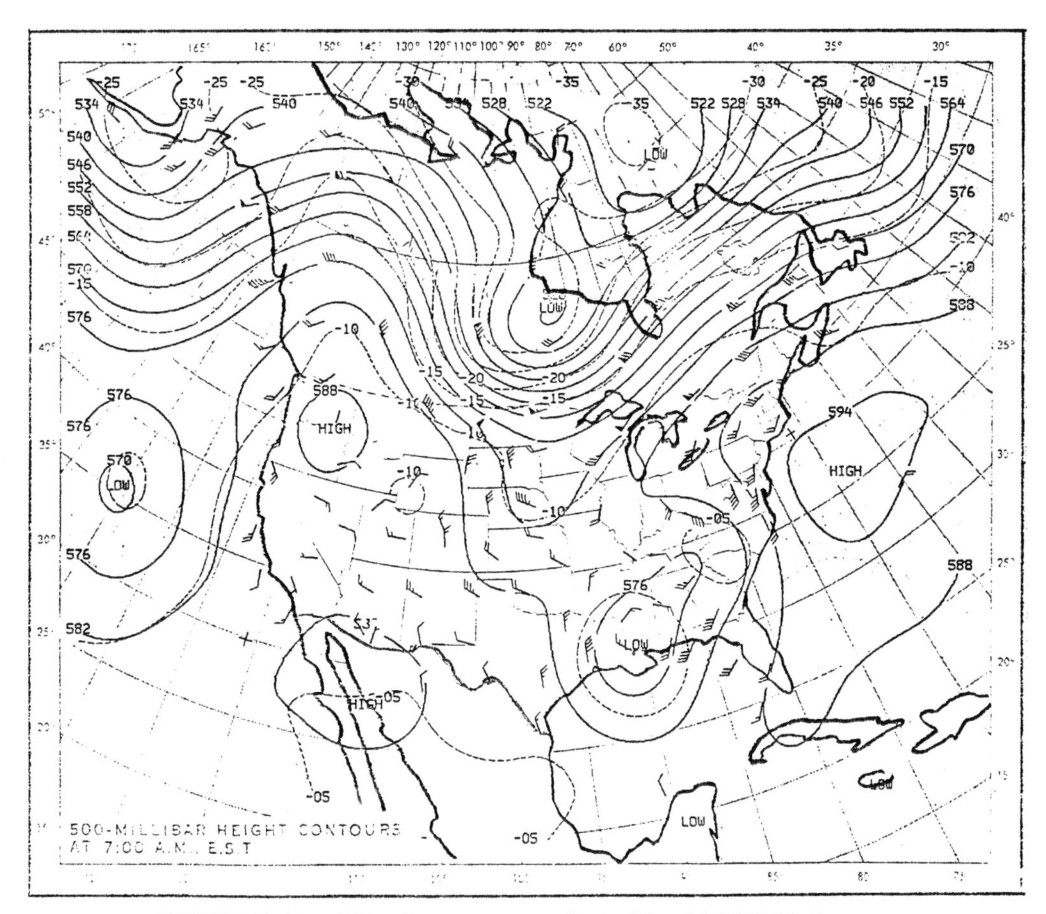

FIGURE 12-25. 500-mb contours and wind for 12 UTC 22 September 1989. This is the time of the chart in Fig. 12-24 for tropical storm Hugo.

the cirrus level can be visually observed on time-lapse sequences of satellite images. The upper-level divergence over a TC is similar to the divergence above an extratropical MCS.

The motion of the two cloud layers is illustrated in Fig. 12-26. Short, bold arrows indicate the motion of clouds in the lower troposphere. Long, thin arrows show the motion of the cirrus cover. The speed of both cloud layers is similar; the arrow length in Fig. 12-26 should not be taken as an indication of cloud speed.

Rainbands

Precipitation pattern (rain), as observed by radar, shows a roughly circular distribution, predominantly in *rainbands*. These bands are curved around the cyclone center, as illustrated by lumpy patterns outlined by heavier lines, in Fig. 12-26. The interiors of inner heavier lines are hatched, representing heavier rain. Circular rainbands are *rings*. Rainbands are also called *convective bands*, since they primarily consist of cumulonimbus. The longest rainband usually enters a TC from the equatorial side. This is the *principal band* that brings most heat and latent heat into a TC. The principal band lies in an air current similar to the conveyor belt of an extratropical cyclone. The principal band enters a TC, bending cyclonically. This band does not traverse a cyclone, as the conveyor belt does. The rain-free region in the center is the *eye*. Sometimes the clouds break in the eye and allow sunshine at the surface. The innermost rainband, or the inner side of this band, is the *eyewall*. Spaces between rainbands, and occasional clearing, are *moats*.

Rain intensity and the shape of rainbands are very variable from cyclone to cyclone, and also within the same cyclone at various times. Around the cyclone, the bands may be greatly asymmetric, observed as a grouping of rainbands on one side of the TC. Denser radar echoes, thus more rain, are often in the segment of a TC with highest wind speed. This is normally the forward right quarter, as mentioned above in the description of wind distribution.

Converging rainbands are too wide to be visually observed from the surface. They were first observed

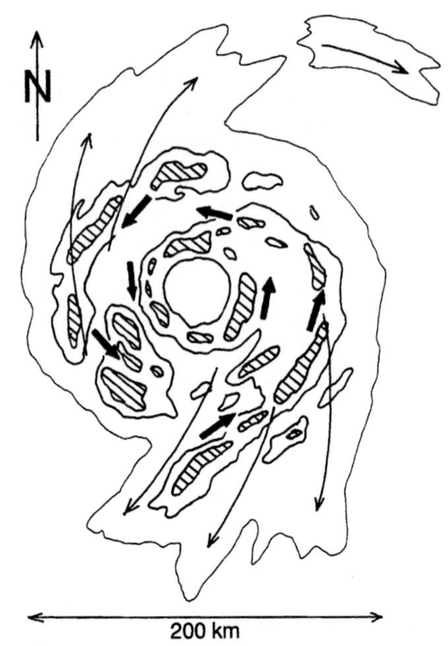

FIGURE 12-26. Rainbands and cloud cover in a hurricane as they may appear on an enhanced radar image. Rain clouds are outlined with heavy lines. The interiors of inner lines are hatched. The motion of rainbands is along thicker, short arrows. The cloud cover (cirrus) is shown by a thin envelope, as seen from the satellite. The air flow at the cirrus level is shown by long, thin arrows. Upper- and lower-level flows circulate in opposing directions.

with radar in 1940s. This explains the choice of the international symbol for the tropical storm and the hurricane (Fig. 12-27). The symbol is different for the Northern and Southern Hemispheres, in agreement with the shape of cyclonic circulation in each hemisphere. When the outflow from a TC was first observed (from satellites, around 1960), it was discovered that the shape of outflow bands was similar to the shape of inflowing low-level bands, only the sense of circulation was different. In this way, it happens that the TC symbols imitate the upper-level outflow as well.

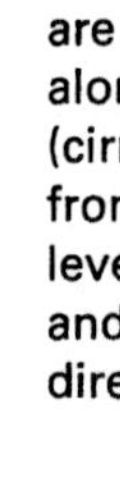

Tropical storm, Northern Hemisphere

Hurricane (or typhoon), Northern Hemisphere

Tropical storm, Southern Hemisphere

Hurricane (or typhoon), Southern Hemisphere

FIGURE 12-27. Symbols for tropical cyclones. The symbols are stylized pictures of rainbands, as observed by radar, or of cirrus outflow, as seen on satellite images.

Positioning Tropical Cyclones

Satellite images are very useful to locate a tropical cyclone. Clouds usually make a knotlike pattern with circular or spiral edges, around the cyclone center. Some tropical cyclones have cirrus cloud cover of an almost circular shape. Other cyclones have wide bands of cirrostratus that extend out of the center, exporting air from the central cloud mass. There is evidence that fast-deepening TCs have more and wider outflow cirrus bands.

Ship reports are certainly very useful in locating TCs. However, ships cannot be blamed for escaping out of a cyclone and thereby ceasing to observe it. Therefore, other observing means, such as radar, satellite, and reconnaissance aircraft, are of increased significance.

Reconnaissance aircraft have been deployed in the storms of North America since it became known (in the 1950s) that the turbulence in most parts of these storms was not too strong. Specially equipped airplanes report valuable data using on-board observations by radar and other instruments and by releasing dropsondes. A general warning must still be given to other aircraft: Strong updrafts and downdrafts do develop within tropical storms and aircraft without special equipment and trained crews are strongly advised to avoid flying in storms. Some convective cells in hurricanes develop into supercells. Numerous tornadoes have also been spotted in hurricanes.

12-9 POLAR LOWS

Description

Comma clouds of smaller sizes than in extratropical cyclones (≈500 km in length) reveal the existence of troughs and cyclonic vortices that often cannot be detected on standard weather charts. A favorite location of these smaller comma clouds is in the polar air, at a distance from the polar front. These are *polar lows* and *polar troughs*. They are frequent over the oceans, and they can hardly be observed by any means except satellites. There are too few observing stations or ships that can detect them. Sometimes the comma cloud in these vortices develops into a spiral or a circular cloud, similar to other cyclones, extratropical and tropical. Then the designation *cold air vortex* is used. A polar trough is illustrated in Fig. 10-13, with instant occlusion.

A polar trough may develop into a polar low, and vice versa. There are also cases when it cannot be determined if there is a low on the chart or if it is only a trough. Therefore, the term "polar low" may be used for both phenomena. The situation is somewhat similar to the waves in the easterlies when we cannot tell whether a closed low is already present due to sparse observations (Section 3-3, about the wave).

Clouds in the polar low (or polar trough) are stratiform, with only occasional showers. Wind speed seldom exceeds 20 m s^{-1}. However, despite their often innocuous look, it is advisable to pay attention to their development. Sea traffic may encounter hazards in stronger vortices of polar type. A few lows of this kind have developed into cyclones of tropical storm intensity.

Environment

Environmental conditions for the polar low include the existence of cyclonic shear, small static stability, and significant heating from the underlying sea (Reed, 1988). The cyclonic shear and vorticity in the lower troposphere usually indicate that lifting in the air mass has already taken place.

Low stability facilitates the formation of thermals and lifting of air masses. Heating from below indicates that a polar low is of the warm cyclone type. As mentioned in Section 9-1, the warm low is usually smaller than an extratropical cyclone. So is the case with a polar low: It is typically of a smaller size, like the tropical cyclone.

The polar low's similarity with a tropical cyclone is not only limited in size. The central region of a polar low is warmer than its immediate environment, as in a tropical cyclone. Also, a comma cloud may spiral all the way around the low center, to resemble the cloud bands of a tropical cyclone. A clear interior loop in the center may appear as a hurricane eye. This may be the basis for hardly believable stories by ship captains about encountering hurricanes in the polar air.

Arctic Frontal Waves

Horizontal thermal contrast (baroclinicity) seems to play a significant role in a polar low. Theoretical studies have shown that baroclinic waves of small extent, like polar lows, may develop under conditions of weak thermal stability. Indeed, most studied polar lows occurred in baroclinic air currents, although the baroclinicity was confined to the lowest 2–3 km of the troposphere.

A preference for a polar low to develop near the edge of the ice surfaces has been observed as well. This is the region where the arctic front forms due to contrasting underlying surfaces. An especially favorable condition for a polar low is when arctic air spreads over the sea. The seas in winter are relatively warm, while the ice regions (frozen sea or continent) may be much colder.

If the baroclinic zone is narrow, the development of a polar low may be interpreted as a wave on the arctic front. Otherwise, a wide baroclinic zone is also often present with a polar low. This situation is similar to other fronts: We often cannot tell if there is a discontinuous frontal surface (or layer) between observing stations or if there is a baroclinic zone without a discontinuity. The difference is probably not important for the development of the low. Nevertheless, in most such cases a front is drawn on the chart to show that the thermal contrast (baroclinic zone) has been detected.

Model of a Polar Low

An example of a polar low is shown in Fig. 12-28, as it may appear in relation to the polar jet and polar front. The contours at the jet level show a small trough west of the extratropical cyclone. An active region has developed, with lifting and comma cloud downstream of the upper-level trough. The surface trough (or a low) develops under the comma cloud.

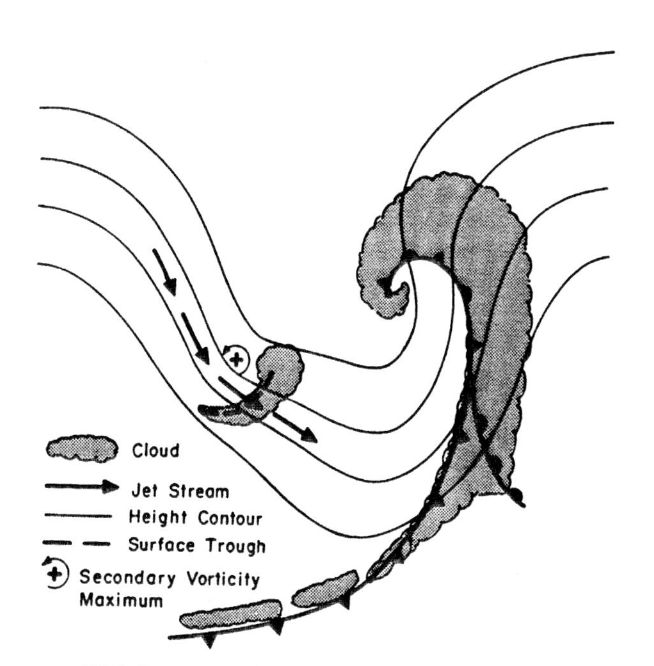

FIGURE 12-28. Relationship between the polar low (or a polar trough) with larger-scale structures. The polar low forms at the comma cloud near the secondary vorticity maximum. From Businger and Reed (1989).

Appendix A

List of Symbols

When the same symbol is used for more than one meaning, the different meanings are listed in the table below under (a), (b), and so forth. Values for coordinates are not listed. However, the dimensions of coordinates are listed under "a typical value." Values for vectors are given only for the magnitude, and this is indicated with each vector. Indexed variables indicate particular values of the letters before them. For example, p_{850} means pressure p at 850 mb, $\mathbf{V}_1$ means wind $\mathbf{V}$ in point 1, β_B means β in point B. Such particular values are not listed in the table below.

Partial derivatives are often abbreviated with indices. In this way, the derivatives are sometimes written as

$$\frac{\partial u}{\partial t} \equiv u_t \qquad \frac{\partial T}{\partial x} \equiv T_x$$

and so forth. Such derivatives can be recognized by the index that is otherwise a coordinate. In the above examples, t and x are coordinates. The quantities with indices that are not coordinates, u_1, T_A, and so forth, are not derivatives.

SI units are used, with a few traditional exceptions. Units are reviewed in Appendix B. General function values or distances, usually denoted by a, A, b, and so on, are not listed in the following table.

Symbol	*Meaning*	*A Typical Value, with Units*
A	(a) a material surface	any area
	(b) a part of deformation	10^{-5} s^{-1}
a	mean radius of the earth	6.366×10^6 m
B	a part of deformation	10^{-5} s^{-1}
C	surface friction coefficient	0.5×10^{-4} s^{-1}
c	speed of light	3×10^8 m s^{-1}
CAPE	convective available potential energy	500 m^2 s^{-2}
CIN	convective inhibition	30 m^2 s^{-2}
c_p	specific heat at constant pressure	1004 J kg^{-1} K^{-1}
c_v	specific heat at constant volume	717 J kg^{-1} K^{-1}
E	energy of instability	120 J kg^{-1}

Symbol	*Meaning*	*A Typical Value, with Units*
e	water vapor pressure	10^3 Pa
E_{kin}	kinetic energy per unit mass	200 $m^2\ s^{-2}$
e_s	saturation vapor pressure	10^3 Pa
e_{si}	saturation vapor pressure over ice	100 Pa
$\mathbf{F}$	friction force per unit mass	magnitude: 2×10^{-4} m s^{-2}
f	Coriolis parameter	$10^{-4}\ s^{-1}$
F_g	geostrophic frontogenesis	5°C $m^{-2}\ s^{-1}$
g	(a) acceleration of gravity	9.806. . . m s^{-2}
	(b) nondimensional constant	9.8
H	(a) relative helicity	0.8
	(b) height of the cold air	10^3 m
h	(a) a function of pressure, defined in Appendix M	5×10^4 $kg^{-1}\ K^{-1}$
	(b) thickness	5000 m
	(c) vertical distance between two material surfaces	3000 m
K	(a) kinematic eddy exchange coefficient	5 $m^2\ s^{-1}$
	(b) index of instability	30
k	streamline curvature	$10^{-6}\ m^{-1}$
L	latent heat of condensation	2.5E6 J kg^{-1}
l	(a) coordinate along the streamline	any length
	(b) wavelength of radar radiation	0.03 m
LCL	lifted condensation level	700 mb
LFC	level of free convection	820 mb
M	Montgomery stream function	$3 \times 10^4\ m^2\ s^{-2}$
m	map magnification factor	0.9
n	(a) normal coordinate	any length
	(b) constant of the cone	0.7
	(c) frequency of electromagnetic radiation	10 GHz
P	potential vorticity	10^{-9} kg $m^{-1}\ s^{-3}$ K
p	(a) air pressure	5×10^5 Pa
	(b) vertical coordinate	any pressure
p_0	(a) standard surface pressure	10^6 Pa = 1000 hPa
	(b) limit of integration in pressure	970 mb
p_e	equilibrium level	1.8×10^5 Pa = 180 mb
$\mathbf{Q}$	$\mathbf{Q}$ vector	magnitude: $10^{-9}\ m^{-1}\ s^{-1}$ K
q	(a) specific humidity	0.01
	(b) distance from the equator	4×10^6 m
q_s	saturation specific humidity	0.02
R	(a) distance from the axis of rotation	4×10^6 m
	(b) specific gas constant for dry air	287.04 J $kg^{-1}\ K^{-1}$
r	(a) radius of streamline curvature	10^6 m
	(b) relative humidity	50%
	(c) radius of the geographical parallel	5×10^6 m
Ri	Richardson number	0.2
R_v	specific gas constant for water vapor	461.55 J $kg^{-1}\ K^{-1}$
S	(a) lifted index of stability	−2 K (units usually omitted)
	(b) image scale of a geographical map	1.2
s	(a) a general space coordinate	any length
	(b) slope of the front in p system	10^{-1} Pa m^{-1}
	(c) map scale	10^{-7}
T	temperature	260 K
t	(a) time coordinate	any time
	(b) temperature in °C	−10°C
T_a	ambient temperature	260 K
T_d	dew point	−12°C
T_l	lifting temperature	277 K
tt	total index of instability	20
T_v	virtual temperature	262 K
u	x-component of wind	10 m s^{-1}
u_{ag}	x-component of ageostrophic wind	5 m s^{-1}
u_g	x-component of geostrophic wind	10 m s^{-1}

Symbol	*Meaning*	*A Typical Value, with Units*
u_{gr}	x-component of gradient wind	$5\ m\ s^{-1}$
V	wind speed	$20\ m\ s^{-1}$
$\mathbf{V}$	wind velocity vector	magnitude: $20\ m\ s^{-1}$
v	y-component of wind	$10\ m\ s^{-1}$
$\mathbf{V}_3$	three-dimensional velocity vector	magnitude: $20\ m\ s^{-1}$
$\mathbf{V}_a$	ageostrophic part of wind	magnitude: $5\ m\ s^{-1}$
v_{ag}	y-component of $\mathbf{V}_a$	$5\ m\ s^{-1}$
$\mathbf{V}_g$	geostrophic wind vector	magnitude: $20\ m\ s^{-1}$
v_g	y-component of geostrophic wind	$10\ m\ s^{-1}$
$\mathbf{V}_{gr}$	gradient wind vector	magnitude: $20\ m\ s^{-1}$
v_{gr}	y-component of gradient wind	$10\ m\ s^{-1}$
V_n	normal component of wind	$10\ m\ s^{-1}$
$\mathbf{V}_T$	thermal wind vector:	
	(a) difference type	magnitude: $20\ m\ s^{-1}$
	(b) derivative type, z-coordinates	magnitude: $10^{-3}\ s^{-1}$
V_t	tangential component of wind	$10\ m\ s^{-1}$
w	(a) vertical component of wind	$0.1\ m\ s^{-1}$
	(b) mixing ratio	10^{-2}
w_s	saturation mixing ratio	10^{-2}
x, y	horizontal coordinates (or isobaric, or isentropic)	any length
$\mathbf{x}_3$	three-dimensional position vector	magnitude: any length
x_s, y_s	horizontal spherical coordinates	any length
z	(a) vertical coordinate (height)	any length
	(b) height of the isobaric surface	5000 m
	(c) height in geopotential meters	5000 gpm
α	(a) azimuth	any angle
	(b) a general angle	any angle
	(c) specific volume	$1.6\ m^3\ kg^{-1}$
β, β'	(a) direction of the dilatation axis	any angle, up to 2π
	(b) direction of air flow	any angle, up to π
Γ	lapse rate of temperature	$-0.007\ K\ m^{-1}$
Γ_a	adiabatic lapse rate of temperature	$-0.01\ K\ m^{-1}$
δ	(a) polar angle	any angle
	(b) a small vertical distance	200 m
ζ	vorticity	$10^{-4}\ s^{-1}$
η	absolute vorticity	$10^{-4}\ s^{-1}$
θ	potential temperature	300 K
θ_w	wet-bulb potential temperature	300 K
κ	ratio R/c_p	0.28
λ	longitude	any angle, up to 2π
ρ	air density	$0.6\ kg\ m^{-3}$
ρ_d	density of "dry" air (without water vapor)	$0.6\ kg\ m^{-3}$
ρ_v	density of water vapor	$10^{-3}\ kg\ m^{-3}$
σ	nondimensional vertical coordinate	any number between 0 and 1
φ or ϕ	geographical latitude	any angle, between $\pm\pi/2$
χ	velocity potential	$10^8\ m^2\ s^{-1}$
ψ	(a) stream function	$10^9\ m^2\ s^{-1}$
	(b) slope of the front in z-system	10^{-2}
	(c) colatitude	any angle up to $\pi/2$
Ω	angular speed of rotation of the earth	$7.29 \times 10^{-5}\ s^{-1}$
ω	p-component of wind (vertical motion in p-coordinates)	$1\ Pa\ s^{-1}$

Numerous quantities are commonly expressed with Greek letters. Therefore, it is useful to know the Greek alphabet:

Α	α	alpha	Β	β	beta	Γ	γ	gamma	Δ	δ	delta
Ε	ε	epsilon	Ζ	ζ	zeta	Η	η	eta	Θ	θ	theta
Ι	ι	iota	Κ	κ	kappa	Λ	λ	lambda	Μ	μ	mu
Ν	ν	nu	Ξ	ξ	xi	Ο	o	omicron	Π	π	pi

P	ρ	rho	Σ	σ	sigma	T	τ	tau	Υ	υ	upsilon
Φ	ϕ	phi	X	χ	chi	Ψ	ψ	psi	Ω	ω	omega

There is practically no difference between script and block letters in the Greek alphabet. There are variations in symbols for a few letters:

φ is the same as ϕ
ϑ is the same as θ

Appendix B

Units Used in Meteorology

Measures are generally expressed in SI units (SI = Système Internationale, the French name for the internationally used system of units). The basic units are kilogram (kg) for mass, meter (m) for length, second (s) for time, and kelvin (K) for temperature. The nondimensional measures for the angle (radian, rad) and solid angle (steradian, sr) are frequently omitted in the list of units since their values can be written as numbers without units. The meter was first defined as 10^{-7}th part of the distance between the pole and equator. It was later redefined when better measurements became available. In view of imperfect accuracy of meteorological measurements, that first definition is still very useful. The kilogram was originally defined as the mass of pure water in 1 liter (10^{-3} m^3).

Units of speed (m s^{-1}) and acceleration (m s^{-2}) do not have special names, unlike most other units. The commonly used derived units in meteorology are:

newton for force, 1 N = 1 kg m s^{-2}

pascal for pressure, 1 Pa = 1 N m^{-2} = kg m^{-1} s^{-2}

joule for energy, 1 J = 1 kg m^2 s^{-2}

watt for power, 1 W = 1 kg m^2 s^{-3}

hertz for frequency, 1 Hz = 1 s^{-1}

It is difficult to change the widespread usage of other, nonmetric units. Some of these, still used in meteorology, are:

millibar for pressure, 1 mb = 100 Pa

millimeter mercury for pressure, 1 mmHg = 133 Pa

inch mercury for pressure, 1 inHg = 3378 Pa

knot for wind speed, 1 kt = 0.5144 m s^{-1}

degree Celsius for temperature(°C), C = K − 273.16

degree Fahrenheit for temperature (°F), F = 1.8 C + 32

For differences in temperature, 1°C = 1 K = 1.8°F

In an attempt to achieve worldwide standards, some of these units (millibar, degree Celsius) were promoted by the World Meteorological Organization long before the SI was introduced. However, after the SI was introduced (in 1974), even these are growing obsolete.

Among the most widely used nonmetric units are various units for time [such as minute (1 min = 60 s), hour (1 h = 3600 s), day (1 d = 86,400 s)] and for angles [degree (1° = (p/180) rad), minute (1′ = 1°/60) and second (1″ = 1′/60)]. The different abbreviations for minutes and seconds of time and angle should be noticed. The traditional units of °C, °F, and the angular units of degree, minute, and second are the only ones written behind the number, without a space in between.

Preferred time is the Universal Time Coordinated (UTC). This is the mean local time of the Greenwich meridian. "Coordinated" implies averaging the unequal apparent motion of the sun on the celestial sphere, the same as "mean" in the old designation. This time was known as Greenwich Mean Time (GMT) until 1987, when the designation UTC was introduced. Hours and minutes UTC are written as 1234 UTC, meaning 12 h and 34 min. "Midnight" UTC is preferably 0000, however 2400 should be used if a period ends at midnight. For example, a 24-h period may be described as: "0000 to 2400 on 22 March 1988." Synoptic observations in the world are made at fixed hours, preferably at 0000 and 1200 UTC. Occasionally, the time of various zones is used, especially to describe processes that have a strong diurnal variation.

It is advisable to spell out the names of months in dates since there are different ways of writing dates in different countries. The date 1-2-1993 indicates "first of February" in most of the world, but means "second of January" in the United States. The unambiguous date will be 1 February 1993. Similarly, the word *billion* should be avoided since in some countries it means 10^9 and in others 10^{12}.

Numbers sometimes include the computer style with E for the exponents of 10. In this way, for example, 1.2E–5 is equal to 1.2×10^{-5}.

Abbreviations for units are used without a period. The names of units are spelled with lowercase letters, although some abbreviations are capitalized. When the numbers become inconveniently large or small, new units are formed with decimal multipliers. Then the names of units obtain the following prefixes (abbreviations are in the last column):

10^{12}	tera	T
10^{9}	giga	G
10^{6}	mega	M
10^{3}	kilo	k
10^{2}	hecto	h
10	deka	da
10^{-1}	deci	d
10^{-2}	centi	c
10^{-3}	milli	m
10^{-6}	micro	μ
10^{-9}	nano	n
10^{-12}	pico	p
10^{-15}	femto	f
10^{-18}	atto	a

Some other non–SI units still widely used are:

gram, 1 g = 10^{-3} kg

geopotential meter, 1 gpm = 9.8 m^2 s^{-2}

geopotential dekameter, 1 gp dam = 10 gpm

liter, 1 l = 10^{-3} m^3

ton, 1 t = 10^3 kg

nautical mile, 1 n mi = 1852 m

statute mile (United States), 1 mi = 1609.344 m

foot, 1 ft = 0.3048 m

inch, 1 in. = 0.0254 m

The unit of geopotential (1 m^2 s^{-2}) has neither a shorter name nor an abbreviation. It is seldom used, since the practice is to use geopotential meters instead.

Concerning orthography, it is common to use abbreviations with numbers and spelled-out units in text without numbers. Thus it is proper to write "several newtons" and also "12.5 Pa." It is not proper to write: "several N" or "12.5 pascals."

Appendix C

Selected Equations of Dynamic Meteorology

This appendix lists several frequently used definitions and equations. Explanations of these equations are given in books on dynamic meteorology.

Many listed variables can be defined verbally. Such verbal definitions have two distinct disadvantages: They are difficult to memorize, and we cannot use the verbal definitions to calculate the needed values of the variables. Therefore, mathematical definitions are preferred. The variables that depend on water vapor are listed in Section 8-2.

Some of the following equations are definitions of computed variables. Definitions can be represented as mathematical identities, but often the "=" sign is used for "≡." There is not much difference whether we use a definition or an equation. A definition yields "true,"—that is, correct—values in the mathematical sense that both sides of the identity are always equal. On the other hand, if we use an equation to calculate one of the variables, there is no guarantee that the calculation will give a result equal to the measurement of that variable. For example, if we use the gas equation $p = R\rho T$ to calculate the pressure p from density ρ and temperature T, we may obtain a different value than from a barometer.

Part of a well-rounded education is to know, or even to memorize, several equations that are often used. For example, we make many conclusions on the basis of equations of motion. When we need these equations, there is usually not enough time to rotate the coordinate system and to figure out whether it is $+fu$ or $-fu$ in the first equation of motion, or is it fv? For this reason, several basic equations should be memorized. This group of equations also includes a short list of useful constants. The candidate equations for memorization follow.

Potential temperature θ, from temperature and pressure:

$$\theta = T(p_0/p)^{R/c_p} \qquad \text{(C-1)}$$

Here p_0 is 1000 mb, if p is used in millibars too. R is defined below.

The equation of state for ideal gasses, also called the "gas equation":

$$p = R\rho T \qquad \text{(C-2)}$$

Hydrostatic equation in three coordinate systems:

$$p\text{:} \quad \frac{\partial z}{\partial p} = -\frac{1}{g\rho} = -\frac{RT}{gp} \qquad \text{(C-3)}$$

$$z: \quad \frac{\partial p}{\partial z} = -g\rho = -\frac{gp}{RT} \qquad \text{(C-4)}$$

$$\theta: \quad \frac{\partial M}{\partial \theta} = c_p \frac{\partial T}{\partial \theta} \qquad \text{(C-5)}$$

Hypsometric equation (integrated hydrostatic):

$$z_2 - z_1 = -\frac{RT}{g} \ln \frac{p_2}{p_1} \qquad \text{(C-6)}$$

Geostrophic wind, from the geopotential height z, in the p-system:

$$\mathbf{V}_g = \frac{g}{f} \mathbf{k} \times \nabla z \qquad \text{(C-7)}$$

Components of geostrophic wind:

$$u_g = -\frac{g}{f} \frac{\partial z}{\partial y}$$

$$v_g = \frac{g}{f} \frac{\partial z}{\partial x} \qquad \text{(C-8)}$$

Geostrophic wind, from density and pressure distribution in z coordinates:

$$\mathbf{V}_g = \frac{1}{f\rho} \mathbf{k} \times \nabla p \qquad \text{(C-9)}$$

Vorticity (vertical component):

$$\zeta = \frac{\partial v}{\partial x} - \frac{\partial u}{\partial y} \qquad \text{(C-10)}$$

Absolute vorticity:

$$\eta = \zeta + f \qquad \text{(C-11)}$$

Potential vorticity follows in three forms for various applications. The barotropic, shallow-water form:

$$P = \frac{\zeta + f}{\delta h} \qquad \text{(C-12)}$$

z-system:

$$P = \frac{1}{\rho} \frac{\partial \theta}{\partial z} (\zeta + f) \qquad \text{(C-13)}$$

p-system:

$$P = -g \frac{\partial \theta}{\partial p} (\zeta + f) \qquad \text{(C-14)}$$

The choice of a minus sign in the last definition is such that P stays positive most of the time ($\partial\theta/\partial p$ is usually negative and $\zeta + f$ is usually positive in the Northern Hemisphere).

Thickness, from the geopotential of two specified pressure levels:

$$H = z_1 - z_2 \qquad \text{(C-15)}$$

Thermal wind, difference type, from $\mathbf{V}_g$ at two specified levels:

$$\mathbf{V}_T = \mathbf{V}_{g1} - \mathbf{V}_{g2} \qquad \text{(C-16)}$$

Thermal wind, derivative type:

$$\mathbf{V}_T = \frac{\partial \mathbf{V}}{\partial z} \qquad \text{(C-17)}$$

The following three equations of the thermal wind are often used. They hold in a hydrostatic and geostrophically balanced atmosphere. These equations relate the derivative type of thermal wind to the distribution of the temperature. In the p-system,

$$\frac{\partial \mathbf{V}}{\partial p} = -\frac{R}{fp} \mathbf{k} \times \nabla T \qquad \text{(C-18)}$$

in the z-system,

$$\frac{\partial \mathbf{V}}{\partial z} = \frac{g}{fT} \mathbf{k} \times \nabla T \qquad \text{(C-19)}$$

and in the θ-system,

$$\frac{\partial \mathbf{V}_g}{\partial \theta} = \frac{R}{f} \frac{p^{\kappa-1}}{p_0^{\kappa}} \mathbf{k} \times \nabla p \qquad \text{(C-20)}$$

It is possible to calculate the thermal wind directly, by evaluating $\mathbf{V}_g$ at two different levels, subtracting, and dividing by Δz. This procedure may give a different numerical result than the above equations.

Two-dimensional advection of the quantity a (this is an identity):

$$-\mathbf{V} \cdot \nabla a = -u \frac{\partial a}{\partial x} - v \frac{\partial a}{\partial y} \qquad \text{(C-21)}$$

Minus signs determine the common rule that a positive advection of a contributes to the local increase of a. Three-dimensional advection is similar, supplemented by the term with the third space derivative $-\omega\, \partial a/\partial p$.

Eulerian expansion of the total derivative with respect to time:

$$\frac{da}{dt} = \frac{\partial a}{\partial t} + u\frac{\partial a}{\partial x} + v\frac{\partial a}{\partial y} + \omega\frac{\partial a}{\partial p} \qquad \text{(C-22)}$$

Equation of motion in the p-system, in vector form, for the three-dimensional velocity $\mathbf{V}$:

$$\frac{\partial \mathbf{V}}{\partial t} + \mathbf{V}\cdot\nabla\mathbf{V} + \omega\frac{\partial V}{\partial p} = -g\,\nabla z - \mathbf{k}g - 2\Omega\times\mathbf{V} + \mathbf{F} \qquad \text{(C-23)}$$

The scalar form of the equations of motions follows, in three-coordinate systems. The form of the equations is similar in these systems, only vertical motion (ω, w, or $\dot{\theta}$), vertical derivatives, and the pressure force are different.

In the p-system,

$$\begin{aligned}\frac{\partial u}{\partial t} + u\frac{\partial u}{\partial x} + v\frac{\partial u}{\partial y} + \omega\frac{\partial u}{\partial p} &= -g\frac{\partial z}{\partial x} + fv\\ \frac{\partial v}{\partial t} + u\frac{\partial v}{\partial x} + v\frac{\partial v}{\partial y} + \omega\frac{\partial v}{\partial p} &= -g\frac{\partial z}{\partial y} - fu\end{aligned} \qquad \text{(C-24)}$$

In the z-system,

$$\begin{aligned}\frac{\partial u}{\partial t} + u\frac{\partial u}{\partial x} + v\frac{\partial u}{\partial y} + w\frac{\partial u}{\partial z} &= -\frac{1}{\rho}\frac{\partial p}{\partial x} + fv\\ \frac{\partial v}{\partial t} + u\frac{\partial v}{\partial x} + v\frac{\partial v}{\partial y} + w\frac{\partial v}{\partial z} &= -\frac{1}{\rho}\frac{\partial p}{\partial y} - fu\end{aligned} \qquad \text{(C-25)}$$

In the θ-system,

$$\begin{aligned}\frac{\partial u}{\partial t} + u\frac{\partial u}{\partial x} + v\frac{\partial u}{\partial y} + \dot{\theta}\frac{\partial u}{\partial \theta} &= -\frac{\partial M}{\partial x} + fv\\ \frac{\partial v}{\partial t} + u\frac{\partial v}{\partial x} + v\frac{\partial v}{\partial y} + \dot{\theta}\frac{\partial v}{\partial \theta} &= -\frac{\partial M}{\partial y} - fu\end{aligned} \qquad \text{(C-26)}$$

The pressure force in the preceding three sets of equations is expressed in terms of geopotential height z, pressure p, and Montgomery stream function M, respectively. Isopleths of each of these variables are the geostrophic streamlines on charts in corresponding coordinate systems. The definition of the Montgomery stream function is

$$M = c_p T + gz \qquad \text{(C-27)}$$

Numerous useful conclusions are made on the basis of the short form of the equations of motion:

$$\begin{aligned}\frac{du}{dt} &= -g\frac{\partial z}{\partial x} + fv\\ \frac{dv}{dt} &= -g\frac{\partial z}{\partial y} - fu\end{aligned} \qquad \text{(C-28)}$$

Another practical form of these equations is obtained when the definition of geostrophic wind (C-8) is used:

$$\begin{aligned}\frac{du}{dt} &= -fv_g + fv\\ \frac{dv}{dt} &= fu_g - fu\end{aligned} \qquad \text{(C-29)}$$

In these equations the pressure force is expressed with the product of f and geostrophic wind.

The equation of mass continuity is easiest in p-coordinates since the time derivative and density are not used:

$$\frac{\partial u}{\partial x} + \frac{\partial v}{\partial y} = -\frac{\partial \omega}{\partial p} \qquad \text{(C-30)}$$

Otherwise, in z coordinates the continuity equation is

$$\frac{1}{\rho}\frac{d\rho}{dt} = -\nabla\cdot\mathbf{V} - \frac{\partial w}{\partial z} \qquad \text{(C-31)}$$

or

$$\frac{\partial \rho}{\partial t} = -\nabla_3\cdot(\mathbf{V}_3\rho) \qquad \text{(C-32)}$$

First Law of thermodynamics, three forms:

$$\frac{dQ}{dt} = c_p\frac{dT}{dt} - \alpha\frac{dp}{dt} \qquad \text{(C-33)}$$

$$\frac{dQ}{dt} = c_v\frac{dT}{dt} + p\frac{d\alpha}{dt} \qquad \text{(C-34)}$$

$$\frac{dQ}{dt} = c_p\left(\frac{p}{p_0}\right)^{\kappa}\frac{d\theta}{dt} \qquad \text{(C-35)}$$

The first law is often used in the adiabatic form, with $dQ = 0$. The entropy form of this law is

$$\frac{dE}{dt} = \frac{c_p}{T}\left(\frac{p}{p_0}\right)^{\kappa}\frac{d\theta}{dt} \qquad \text{(C-36)}$$

or

$$\frac{dE}{dt} = c_p \frac{d}{dt} \ln \theta \qquad \text{(C-37)}$$

where dE is the differential of entropy

$$dE = \frac{dQ}{T} \qquad \text{(C-38)}$$

The form (C-37) is suitable for integration since dE is an exact differential. The similar differential dQ in (C-36) is not exact; that is, the expression on the right-hand side of (C-33) cannot be brought into the form $d(\,)$.

The simplest adiabatic form (subscript *ad*) of the first law follows from (C-37) as

$$\left(\frac{d\theta}{dt}\right)_{\text{ad}} = 0 \qquad \text{(C-39)}$$

From (C-37), there follows

$$E \propto \ln \theta \qquad \text{(C-40)}$$

This is reason that the isopleths of potential temperature θ are called isentropes.

In the case of condensation, the differential of heating dQ may be set equal to the release of latent heat:

$$dQ = -L dw_s \qquad \text{(C-41)}$$

in (C-30), first form. The equation for the moist adiabatic process becomes

$$\frac{dT}{dp} = \frac{\alpha}{c_p} - L \frac{dw_s}{dp} \qquad \text{(C-42)}$$

One of the forms of the vorticity equation is

$$\frac{\partial \zeta}{\partial t} + \mathbf{V} \cdot \nabla(\zeta + f) + \omega \frac{\partial \zeta}{\partial p} = -(\zeta + f)\nabla \cdot \mathbf{V} \qquad \text{(C-43)}$$

It should be noticed that (C-43) contains a time derivative of ζ and is different from the definition of vorticity (C-10).

Definition of the Coriolis parameter:

$$f = 2\Omega \sin \phi \qquad \text{(C-44)}$$

Here Ω is the angular speed of rotation of the earth and ϕ is the latitude.

Acceleration of gravity:

$$g = 9.806 \ldots \text{ m s}^{-2} \qquad \text{(C-45)}$$

This value of g varies a little with the geographical latitude and elevation, but this variation is usually ignored.

Geopotential (an approximate definition):

$$\phi = gz' \qquad \text{(C-46)}$$

where z' is the geometric height above the sea level. When a nondimensional constant

$$g_0 = 9.8 \qquad \text{(C-47)}$$

is introduced, the geopotential height can be used as

$$z = z' \frac{g}{g_0} \qquad \text{(C-48)}$$

The value of 9.8 is exact in the SI. The numerical values of geopotential z in $\text{m}^2\ \text{s}^{-2}$ are almost exactly equal to the corresponding values of height z'. With the preceding definitions it is useful to notice that

$$g_0 z = gz' \qquad \text{(C-49)}$$

is an exact equation. Different notation is used in various books for values of g, g_0, z, and z'. The prime with z is not standard. Also, g is often used when g_0 is meant. Sometimes, we cannot recognize whether gz means the product on the left- or on the right-hand side of (C-49). Fortunately, the result is equal, and it is not necessary to inquire about the exact meaning each time.

It is useful to memorize or have at hand the following constants, since they are often used in evaluation of computed variables:

Gas constant for dry air:

$$R = 287 \text{ m}^2\ \text{s}^{-2}\ \text{K}^{-1} = 287 \text{ J kg}^{-1}\ \text{K}^{-1}$$

Specific heat with constant pressure:

$$c_p = 1004 \text{ m}^2\ \text{s}^{-2}\ \text{K}^{-1}$$

Specific heat with constant volume:

$$c_v = 717 \text{ m}^2\ \text{s}^{-2}\ \text{K}^{-1}$$

Gas constant for water vapor:

$$R_v = 462 \text{ m}^2 \text{ s}^{-2} \text{ K}^{-1}$$

It should be noticed that $R = c_p - c_v$ and that the ratios between these values are given as ratios of rather small integers: $c_p : c_v : R = 7:5:2$. The listed gas constants are specific for dry air and water vapor. The specific gas constants are useful in the gas equations of the form (C-2) in which the molecular weight is not used. The *universal gas constant* $R' = 8.3143 \text{ J K}^{-1} \text{ mol}^{-1}$ should be used in the more general gas equation $p = m\rho R'T$, where m is the molecular weight. This more general form of the gas equation is seldom used in meterology, since the chemical composition of air is nearly constant.

Appendix D

Vectors

This Appendix presents general mathematical properties of vectors, as needed in meteorology. Particular attention is given to the gradient vector.

Vectors are quantities that consist of several mutually related components. The components are scalar, even if we sometimes draw them as vectors. Such drawings then contain the products of the components with corresponding unit vectors. A vector itself does not depend on the orientation of the coordinates. However, the components vary greatly as the orientation of the coordinates changes.

There is always an orientation of the coordinate system for which only one component of the vector is different from zero. This property of vectors is different from the "vectors" in computer science. Those "vectors" represent arrays for "vector processing."

The vectors used in this book are two-dimensional, unless specifically indicated otherwise. This means that their third component is equal to zero. In this way, the cross product is still defined, notwithstanding its original definition that calls for three-dimensional vectors.

There are several customary styles of vector notation. In this book, bold capital letters are used, as is most common in meteorological literature. The unit vectors, though, will be the lowercase bold **i**, **j**, and **k**. Elsewhere, other styles are often used, such as with an arrow over the symbol or with matrix notation with the indices.

The coordinate systems in which vector operations are defined are usually right-handed, although our most widely used system, the *p*-system, is left-handed. The *p*-coordinate points opposite of z, while x and y are about the same. However, we use the same algebraic definitions in the *p*-system, so our equations maintain the same form in both systems.

Addition of vectors is defined as addition of respective components. Thus when the vectors $\mathbf{V}_1$ and $\mathbf{V}_2$ are added to form $\mathbf{V}_3$, we have

$$\mathbf{V}_3 = \mathbf{V}_1 + \mathbf{V}_2$$

which means that the components are added as

$$u_3 = u_1 + u_2$$
$$v_3 = v_1 + v_2$$

Subtraction of vectors is defined as addition where one vector takes the opposite sign, that is,

$$\mathbf{V}_4 = \mathbf{V}_1 - \mathbf{V}_2 = \mathbf{V}_1 + \mathbf{V}_5$$

where the new vector $\mathbf{V}_5$ is defined in the direction opposite to $\mathbf{V}_2$:

$$\mathbf{V}_5 = -\mathbf{V}_2 = \begin{pmatrix} -u_2 \\ -v_2 \end{pmatrix}$$

Otherwise, there are no "negative vectors," even if we put a minus sign before some of them. This is analogous to the wind; it is never "negative."

Multiplication of vectors is defined in several ways.

(a) The scalar product S is obtained by the dot operator between two vectors:

$$S = \mathbf{V}_1 \cdot \mathbf{V}_2 = u_1 u_2 + v_1 v_2$$

As shown in this definition, the result of the scalar product is a scalar sum of the products of respective components.

(b) The cross-product $\mathbf{V}_6$ of the vectors $\mathbf{V}_1$ and $\mathbf{V}_2$ is defined with the determinant:

$$\mathbf{V}_6 = \mathbf{V}_1 \times \mathbf{V}_2 = \begin{vmatrix} \mathbf{i} & \mathbf{j} & \mathbf{k} \\ u_1 & u_2 & 0 \\ v_1 & v_2 & 0 \end{vmatrix} = \mathbf{k}(u_1 v_2 - u_2 v_1)$$

(c) The dyadic tensor $\nabla\mathbf{V}$, without a sign between the vectors ∇ and $\mathbf{V}$, is defined such that the expression $\mathbf{V} \cdot \nabla\mathbf{V}$ in the equation of motion is equal to the following other expressions:

$$\mathbf{V} \cdot \nabla\mathbf{V} = \mathbf{V} \cdot (\nabla\mathbf{V}) = (\mathbf{V} \cdot \nabla)\mathbf{V}$$

By comparison, the dyadic product is seldom used; for this reason its complete definition will be postponed until it becomes necessary (in the theory of turbulence).

All preceding products are defined in the same way when one of the vectors is the del operator (∇, or, to stress its vector nature, $\vec{\nabla}$). This vector operator is defined as

$$\nabla = \begin{pmatrix} \dfrac{\partial}{\partial x} \\ \dfrac{\partial}{\partial y} \\ \dfrac{\partial}{\partial p} \end{pmatrix}$$

$\nabla \cdot \mathbf{V}$ is the divergence (a scalar):

$$\nabla \cdot \mathbf{V} = \frac{\partial u}{\partial x} + \frac{\partial v}{\partial y}$$

$\nabla \times \mathbf{V}$ is the three-dimensional vorticity (or curl) of $\mathbf{V}$, but since $\mathbf{V}$ is a two-dimensional vector, only the vertical component differs from zero:

$$\nabla \times \mathbf{V} = \mathbf{k}\left(\frac{\partial v}{\partial x} - \frac{\partial u}{\partial y}\right)$$

This component (scalar) is the vorticity

$$\zeta = \frac{\partial v}{\partial x} - \frac{\partial u}{\partial y}$$

The most apparent feature on weather charts is the presence of isopleths. The gradient is a convenient measure for the direction and density of isopleths at any point. However, using the word *gradient* causes some ambiguity since different books have different definitions of the gradient. The following three definitions of the gradient of a scalar a are used:

$$\begin{aligned} &\text{(a)} \quad \nabla a \\ &\text{(b)} \quad -\nabla a \\ &\text{(c)} \quad |\nabla a| \end{aligned} \qquad \text{(D-1)}$$

The first definition is prevalent in mathematics and in parts of meteorology developed by mathematicians. The second definition was introduced in meteorology in the nineteenth century when meteorologists were not adequately educated in mathematics. They noticed that most forces acted "down the gradient." Therefore, they defined the gradient in the direction of decrease of various potentials (gravitational, electric, pressure).

Another traditional remnant of the gradient's old definition is found in the vertical component of the three-dimensional gradient. Often, this component is called the lapse rate and is defined as the opposite of the vertical derivative. For example, the vertical lapse rate of pressure in z-coordinates is defined as

$$-\frac{\partial p}{\partial z}$$

The advantage of this definition is that the lapse rate of pressure is a positive quantity. The definition of the vertical lapse rate of temperature is similar. However, its advantage of being presumably a positive quantity is often missing (in inversions). In agreement with the second definition (b) of the gradient, sometimes the word *ascendent* is used for ∇a. This distinguishes it from the gradient in the opposite direction.

Definition (c) of the gradient refers only to the magnitude of the gradient and can be used verbally. For example, we often refer to a "strong gradient" for a region where the isopleths are densely spaced on a weather chart. This third definition is used in this book whenever the word *gradient* is used without reference to the direction.

Because we need quantitative evaluations of direction and magnitude of the gradient, the word *gradient* should be avoided. The word *del* should be used for ∇, since there is never any ambiguity about the direction of ∇a (a being a scalar). Ambiguity exists only with the word *gradient*.

Two important ways to express ∇a are in the form of components or with magnitude and direction. The direction of ∇a is toward most increasing values of a, that is, normal to the isopleths of a. The magnitude of ∇a can be computed as a square root of the sum of the squares of both components:

$$|\nabla a| = \left[\left(\frac{\partial a}{\partial x}\right)^2 + \left(\frac{\partial a}{\partial y}\right)^2\right]^{1/2}$$

It can also be found by the evaluation of one derivative,

$$|\nabla a| = \frac{\partial a}{\partial n}$$

where n is the normal coordinate. Since the magnitude of ∇a is always positive, it is assumed here that n is taken in the sense of increasing a. It is possible to choose the other sense of n. In that case, the last definition should be taken with the absolute value on the right-hand side.

The standard way to represent the wind vector is by the *wind barb* (Fig. D-1). The barb consists of the *wind shaft*, *feathers*, and *pennants*. The shaft is attached to the station circle on the chart (or to a point that represents the observation point). In each case, the shaft stretches in the direction from which the wind is coming. The wind speed is indicated by feathers and pennants on the shaft. A rudimentary decimal system is used: a short feather for 5 kt, a long feather for each 10 kt (up to 40), and a pennant for each 50 kt.

The orientation of feathers or pennants on the shaft is chosen on the side where the lower pressure is expected, assuming geostrophic balance. Therefore, the wind barbs appear different in the Northern and Southern Hemispheres. The feathers and pennants may be thought of as little vectors that approximately show the direction of the pressure force. The observation of *calm* is indicated by a circle around the station circle.

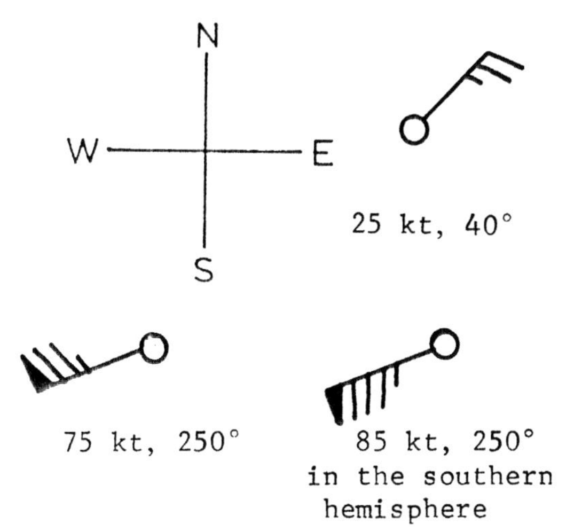

FIGURE D-1. Examples of wind barbs.

The word *velocity* usually indicates the vector of air motion. Therefore, the word *speed* should be used if the direction is not implied.

It is a property of English language (also of most other languages!) to refer to "stronger wind" when faster wind is meant. However, the wind force also implies the density of air: Wind in denser air is stronger than wind in less dense air when both have the same speed. At the earth's surface and not too high above sea level, the density of air does not vary much. People without instruments cannot determine variations in density of air. Under such conditions it seems equivalent to refer to wind speed or to wind force. However, very often we use wind observations from the upper troposphere or stratosphere where the density may be orders of magnitude smaller than at sea level. Then there is a great difference between wind speed and wind force. Therefore, when reference to wind speed is made, the proper expressions are faster wind or slower wind.

When comparing two del vectors [gradients, definition (c) in (D-1)], it is common to use the expression *stronger gradient* for the one that is absolutely greater, even if no forces are implied. The smaller vector will then be the weaker gradient.

Appendix E

Weather Reports

Transmission of weather data started in the middle of nineteenth century, when the telegraph became available. Since that time a worldwide communication network has been established. A few of the most common procedures for transmission of weather data and plotting data on weather charts are reviewed in this appendix. The information here is not sufficient to explain all the procedures; this is only a short review that gives an idea how work on weather charts is organized. This appendix gives only samples of procedures to introduce readers to the basic principles of handling weather data.

Organization of Observations

All countries maintain public weather services whose mission is to provide standardized weather observations and to broadcast the observed values ("observations") to users in all countries.

Observations and data exchange in the United States are done by the National Weather Service (NWS), with significant support from the Federal Aviation Administration and a few other organizations. The headquarters of the NWS are in Washington, D.C., and branch offices are located in all states.

Transmission of complete weather charts is also widespread. It is effected by a variety of computer and facsimile techniques for creation, storage, and transmission of computer graphics. Details of data transmission will not be described in this book since they vary very much between computer systems.

International Cooperation

International norms for observation and transmission of weather information are regulated by the World Meteorological Organization (WMO), a specialized agency of the United Nations. The WMO is the only organization in the world where all nations collaborate, including those nations that are not members the United Nations or WMO itself for various political reasons. Weather data keep flowing freely across borders. Meteorologists are rightfully proud that the WMO is in the forefront among organizations that promote peaceful cooperation between nations.

Form of Reports

Standard codes have been introduced in meteorological practices so that all users can construct weather charts from reports. The complete set of instructions makes a small library. Studying the weather code is needed when the codes are used. Besides the coded form of observations, there are many complete weather charts in digital form.

Traditionally, the code for transmission of observations contains groups of five digits. In times before widespread telephone use, telegraph companies charged for messages by the number of words, and five digits counted as one word. There is no necessity for such grouping in today's transmission between computers.

The next section contains an example of a weather report, representative of all codes. Students are not expected to memorize it. Only an interpretation of the text should be studied, without all the details.

SYNOP Code

The international code for weather reports from surface observing stations (known as the SYNOP code) is shown next as a sequence of letters. Actual reporting is in numbers in the places of code letters. The meaning of the letters is explained in the following list. The interpretation may seem cumbersome, but readers should be aware that this explanation is not nearly complete. Numerous specialized tables are needed for a complete interpretation of the code. The words *table needed* are mentioned below in such cases. Those tables are not shown here, but can be found in specialized texts. One such text is the manual by Alcorn (1990).

The formal form of the SYNOP code is

IIiii $i_R i_x$hVV Nddff $1s_n$TTT $2s_nT_dT_dT_d$ $3P_0P_0P_0P_0$ 4PPPP 5appp $6RRRt_R$ $7wwW_1W_2$ $8N_hC_LC_MC_H$

The meaning of letters is as follows:

II — Block number that shows the region of the reporting station. The blocks and station numbers are shown in Fig. G-1 in Appendix G, together with the illustration of the Mercator projection.

iii — Station number. Lists of stations and their numbers are available in the Weather Service. Table needed.

i_R — Indicator about availability and meaning of the later group $6RRRt_R$. For example, if $i_R = 1$, it means that the group $6RRRt_R$ is transmitted and precipitation was measured. Table needed.

i_x — Type of station. The code $i_x = 1$ is used on manned stations, $i_x = 4$ on automatic stations. Other values of this code show if the group $7wwW_1W_2$ and a few other relevant pieces of information are included. Table needed.

h — Cloud base above the station. Ten possibilities can be reported, from a specialized table. The code h = 9 indicates very high cloud base or no clouds. Table needed.

VV — Visibility. Table needed.

N — Sky cover. The codes used are N = 0 for clear sky, N = 8 for overcast and the intermediate values show partly cloudy sky. The code N = 9 is used for sky not visible (as in fog). Table is on p. 253.

1 — Here and later some digits are standard in the first place in the group. These are indicators of groups. They are introduced to assist interpreting reports with some data scrambled.

s_n — Sign of temperature. Here, $s_n=0$ for positive temperature, and $s_n=1$ for negative values in degrees Celsius.

TTT — Air temperature, with one decimal. For example, 12.3°C is reported as 123; 1.2°C is reported as 012, with a leading zero. The decimal point is omitted.

2 Group indicator.

s_n — Sign of the dew point, as above for temperature.

$T_dT_dT_d$ — Dew point, encoded like temperature above.

3 Group indicator.

$P_0P_0P_0P_0$ — Station pressure. The last four digits are used, down to 0.1 hPa. The decimal point is omitted.

4 Group indicator.

PPPP — Sea-level pressure. Encoded like the station pressure above.

5 Group indicator.

a — Form of the pressure trace on the barograph. A specialized table (p. 253) gives the 10 possibilities.

ppp — Pressure tendency, down to 0.1 hPa $(3\ h)^{-1}$, with omitted decimal point.

6 Group indicator.

RRR — Precipitation since the last report. Table needed.

t_R — Time interval in which the precipitation RRR accumulated. Table needed.

7 Group indicator.

ww — Present weather. Encoded with choices from a table that has listings such as "haze," "fog depositing rime, sky visible," "shower(s) of rain and snow mixed, slight," and 97 others. Table is on pp. 250–51.

W_1 Most significant weather since the last report. Table is on p. 253.

W_2 Second most significant weather since the last report. Table needed.

8 Group indicator.

N_h Amount of C_L or C_M clouds. Table is on p. 253.

C_L Prevalent genus and species of stratus, stratocumulus, cumulus, or cumulonimbus. Table is on p. 252.

C_M Prevalent genus and species of altostratus, altocumulus, or nimbostratus. Table is on p. 252.

C_H Prevalent genus and species of cirrus, cirrostratus, or cirrocumulus. Table is on p. 252.

The preceding listing does not exhaust all possibilities for surface reports. There are more elements to report, variations for types of stations (sea, mountain, automatic, and so forth) and other factors involved. Instructions for the reports of surface observing stations are contained in a 150-page book.

An Example of a SYNOP Report

A weather report in the international SYNOP code may read as

74451 11620 22316 10137. . .

An interpretation of this report follows:

74451: Station No. 451 (Dodge City, Kansas), in region 74.

11620: The leading "1" shows that there was rain at the station and the groups numbered 6 and 7 will be included in this report. However, for the sake of brevity, these groups are omitted in this example. The other "1" shows that this is a station with persons–observers. Cloud base is between 1000 and 1500 m (code 6). Visibility is 2 km (code 20).

22316: Sky cover $\frac{2}{10}$ or $\frac{3}{10}$ (code 2); wind direction 230° (code 23); wind speed 16 kt (code 16)

10137: Group indicator "1"; temperature is +13.7°C (zero is for "+").

More groups usually follow. Their interpretation follows a pattern similar to the example so far.

There is some flexibility in the code. Some groups may be omitted, as, for example, is pointed out in the meaning of i_x. In such cases, the group indicators are very useful. More groups can be added, according to a more complete description that is not shown here.

TEMP

The name *TEMP* indicates the results of soundings. This information also proceeds in codes of five digits. There are codes for pressure, temperature, and humidity at standard levels and separately for significant levels where the observers noticed a change of regime in the vertical distribution of weather elements.

Airways Code

The *airways code* is used in the United States for hourly weather reports. This code consists of a sequence of weather data, where the items are separated by a space or by a solidus. The style of this code is not as formal as the SYNOP code. It is illustrated in the following example:

CLL SA 0756 M5 OVC 4F 112/67/54/1624/001 15//
20082

The following explanation of these symbols can give an idea about the transmission of data:

CLL	abbreviated station name (College Station, Texas)
SA	type of observation (this one is hourly)
0756	hour and minute of the observations
M5	method of ceiling determination (M = measured) and ceiling (600 ft)
OVC	cloud cover (overcast)
4F	visibility (4 mi) and type of obstruction (fog)
112	Sea-level pressure (1011.2 mb)
67	air temperature (67°F)
54	dew point (54°F)
1624	wind direction and speed (160°, 24 kt)
001	altimeter setting (30.01 in Hg)
15	cloud form (code CL No. 5)

METAR

Similar to the airways code is the international code METAR. An example of this code follows, with the same data as given above in the airways code. It should be noticed that some information in one code (airways or METAR) does not appear in the other code. The spaces separate groups of digits or letters. A solidus separates some related elements. An example is:

KCLL 0756 16024KT/36 6000 42 8ST005 19/12 Q1011

The interpretation of this report is as follows:

KCLL	All U.S. stations have the prefix K; otherwise CLL is College Station as above in the airways code.
0756	time of observation, UTC
160	wind from 160°
24KT	wind speed in knots. Also m s^{-1} may be used, then this reads 12MPS. If KT or MPS is omitted, m s^{-1} is assumed. Three digits may be used, if the speed is higher than 99 m s^{-1} or 99 kt.
/36	maximum wind gust (may be omitted)
6000	visibility, m
42	fog, sky visible, from the ww SYNOP code
8	overcast, as in the SYNOP code
ST	cloud genus (here it is stratus)
005	ceiling, in 100 ft (here it is 500 ft)
19/12	temperature and dew point, both in °C
Q1011	sea-level pressure, in millibars

Representation of Data on a Weather Chart

The international representation of weather on a surface chart is illustrated in Fig. E-1. The station position is shown by a circle about 3 mm wide. The cloudiness is shown by filling the circle in a prescribed pattern. An empty circle shows a clear sky; a filled circle shows overcast. There is room for three cloud forms, selected from 30 species. These are entered directly under and above the station circle, in the places of CL, CM, and CH. The available symbols are shown at the end of this appendix. Their descriptions must be looked up in specialized instructions. The cloud in Fig. E-1a in the place CL (under the station circle) is a cumulonimbus. The present weather is a thunderstorm, shown by the kinked arrow on the left-hand side.

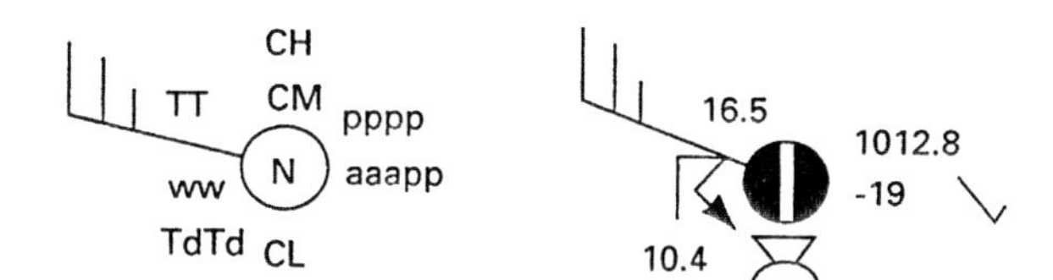

FIGURE E-1. The standard plotting of surface weather reports on the weather chart. Placement of items is on the left and an example of data on the right. This example shows a thunderstorm in progress.

The data are plotted similarly on the upper-level charts, although there are fewer elements on these charts.

The form of the station "circle" is round for the radiosonde reports, an asterisk for the wind evaluated from satellite observations, and square for the wind from aircraft reports (AIREP or ASDAR). The wind report is plotted everywhere with the same conventional barb as on the surface chart. Satellite and AIREP wind reports contain the level (LL) and the nearest hour (HH) when the observation was made.

Radiosonde reports are routinely plotted on constant-pressure charts. A plot for each station contains the temperature and dew point, both plotted as on the surface chart. The illustration for the constant-pressure charts is shown in Fig. E-2. A filled circle is used for a dew-point depression of 5°C or less, simulating a likely overcast sky.

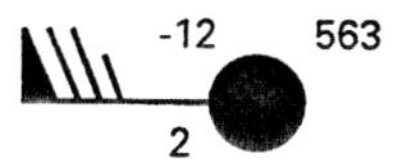

FIGURE E-2. An example of plotted weather data on a 500-mb chart. This example shows west wind, 75 kt, temperature −12°C, dew-point depression 2°C, and geopotential height of 5630 gpm.

Satellite wind observations are plotted with an asterisk, as illustrated in Fig. E-3a. Observations from aircraft are plotted with a square for station location (Fig. E-3b).

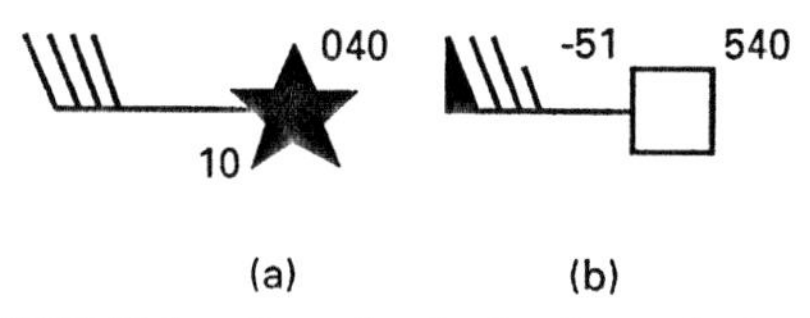

FIGURE E-3. Standard plotting of data on constant-pressure charts, distribution of elements, and an example. (a) Satellite wind; (b) AIREP.

The satellite wind illustrated in Fig. E-3a is approximately at an elevation of 4000 ft (coded as 040). This observation was made at about 1000 UTC, as noted by the number 10 placed southwest of the station. This item is plotted on an 850 mb chart of 1200 UTC, since this level (850-mb) is closest to the elevation of 4000 ft and 1200 UTC is the closest hour.

AIREP in Fig. E-3b contains the temperature (−51°C) and elevation of the aircraft (5400 m). This

TABLE E-1

Symbols used on weather charts

	0	1	2	3	4
00	Cloud development NOT observed or NOT observable during past hour.	Clouds generally dissolving or becoming less developed in past hour.	State of sky on the whole unchanged during past hour.	Clouds generally forming or developing during past hour.	Visibility reduced by smoke.
10	Light fog.	Patches of shallow fog at station, NOT deeper than 6 feet on land.	More or less continuous shallow fog at station, NOT deeper than 6 feet on land.	Lightning visible, no thunder heard.	Precipitation within sight, but NOT reaching the ground.
20	Drizzle (NOT freezing and NOT falling as showers) during past hour, NOT at time of observation.	Rain (NOT freezing and NOT falling as showers) during past hour, but NOT at time of observation.	Snow (NOT falling as showers) during past hour, but NOT at time of observation.	Rain and snow (NOT falling as showers) during past hour, but NOT at time of observation.	Freezing drizzle or freezing rain (NOT falling as showers) during past hour, NOT at time of observation.
30	Slight or moderate duststorm or sandstorm, has decreased during past hour.	Slight or moderate duststorm or sandstorm, no appreciable change during past hour.	Slight or moderate duststorm or sandstorm, has increased during past hour.	Severe duststorm or sandstorm, has decreased during past hour.	Severe duststorm or sandstorm, no appreciable change during past hour.
40	Fog at distance at time of observation, but NOT at station during past hour.	Fog in patches.	Fog, sky discernible, has become thinner during past hour.	Fog, sky NOT discernible, has become thinner during past hour.	Fog, sky discernible, no appreciable change during past hour.
50	Intermittent drizzle, NOT freezing, slight at time of observation.	Continuous drizzle, NOT freezing, slight at time of observation.	Intermittent drizzle, NOT freezing, moderate at time of observation.	Continuous drizzle, NOT freezing, moderate at time of observation.	Intermittent drizzle, NOT freezing, thick at time of observation.
60	Intermittent rain, NOT freezing, slight at time of observation.	Continous rain, NOT freezing, slight at time of observation.	Intermittent rain, NOT freezing, moderate at time of observation.	Continuous rain, NOT freezing, moderate at time of observation.	Intermittent rain, NOT freezing, thick at time of observation.
70	Intermittent fall of snowflakes, slight at time of observation.	Continuous fall of snowflakes, slight at time of observation.	Intermittent fall of snowflakes, moderate at time of observation.	Continuous fall of snowflakes, moderate at time of observation.	Intermittent fall of snowflakes, heavy at time of observation.
80	Slight rain shower(s).	Moderate or heavy rain shower(s).	Violent rain shower(s).	Slight shower(s) of rain and snow mixed.	Moderate or heavy shower(s) of rain and snow mixed.
90	Moderate or heavy shower(s) of hail, with or without rain or rain and snow mixed, not associated with thunder.	Slight rain at time of observation; thunderstorm during past hour, but NOT at time of observation.	Moderate or heavy rain at time of observation: thunderstorm during past hour, but NOT at time of observation.	or / Slight snow or rain and snow mixed or hail at time of ob.: thunderstorm during past hour, but NOT at time of observation.	or / Moderate or heavy snow, or rain and snow mixed, or hail at time to ob.: thunderstorm during past hour, but NOT at time of ob.

TABLE E-1 Cont.

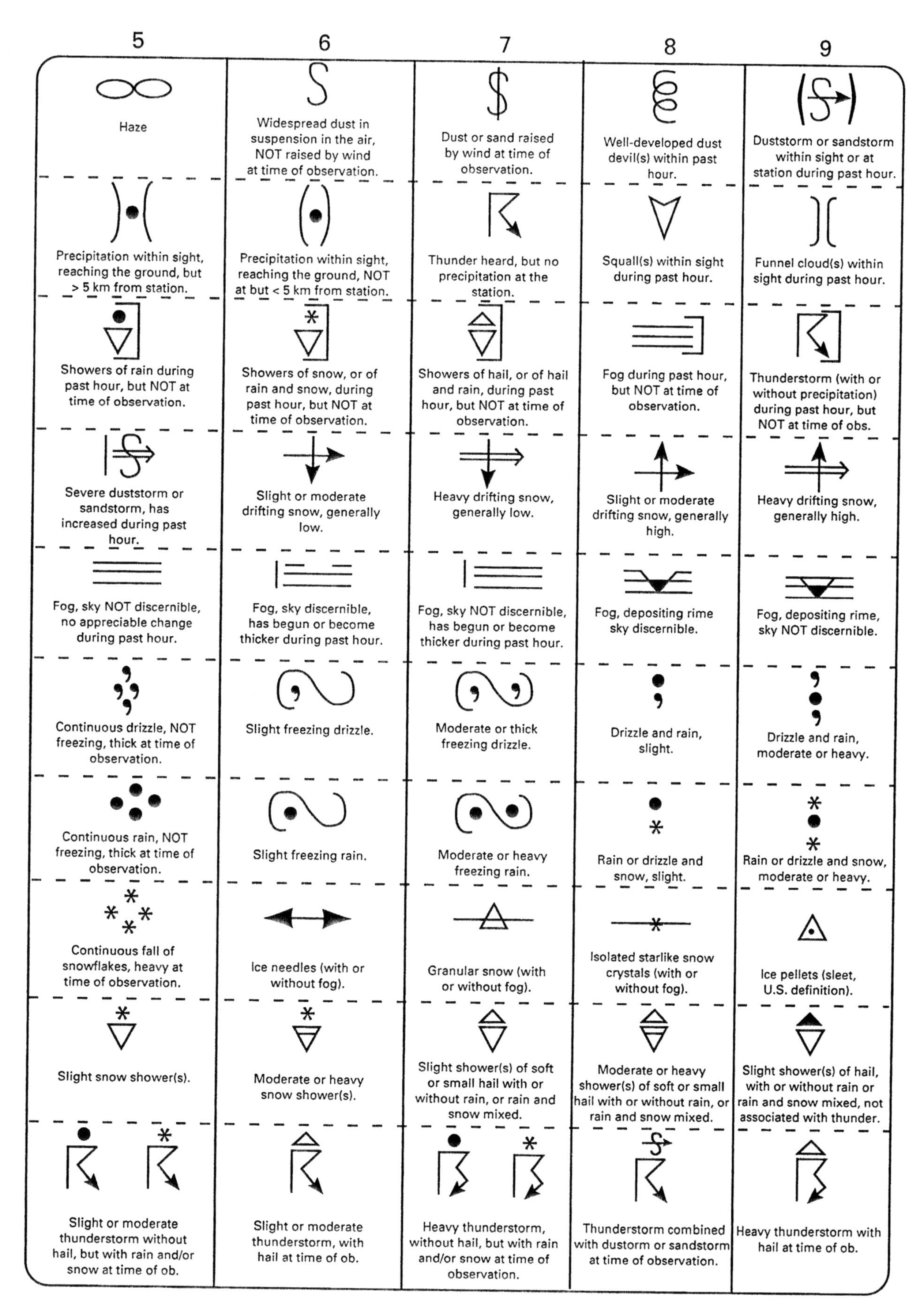

5	6	7	8	9
Haze	Widespread dust in suspension in the air, NOT raised by wind at time of observation.	Dust or sand raised by wind at time of observation.	Well-developed dust devil(s) within past hour.	Duststorm or sandstorm within sight or at station during past hour.
Precipitation within sight, reaching the ground, but > 5 km from station.	Precipitation within sight, reaching the ground, NOT at but < 5 km from station.	Thunder heard, but no precipitation at the station.	Squall(s) within sight during past hour.	Funnel cloud(s) within sight during past hour.
Showers of rain during past hour, but NOT at time of observation.	Showers of snow, or of rain and snow, during past hour, but NOT at time of observation.	Showers of hail, or of hail and rain, during past hour, but NOT at time of observation.	Fog during past hour, but NOT at time of observation.	Thunderstorm (with or without precipitation) during past hour, but NOT at time of obs.
Severe duststorm or sandstorm, has increased during past hour.	Slight or moderate drifting snow, generally low.	Heavy drifting snow, generally low.	Slight or moderate drifting snow, generally high.	Heavy drifting snow, generally high.
Fog, sky NOT discernible, no appreciable change during past hour.	Fog, sky discernible, has begun or become thicker during past hour.	Fog, sky NOT discernible, has begun or become thicker during past hour.	Fog, depositing rime sky discernible.	Fog, depositing rime, sky NOT discernible.
Continuous drizzle, NOT freezing, thick at time of observation.	Slight freezing drizzle.	Moderate or thick freezing drizzle.	Drizzle and rain, slight.	Drizzle and rain, moderate or heavy.
Continuous rain, NOT freezing, thick at time of observation.	Slight freezing rain.	Moderate or heavy freezing rain.	Rain or drizzle and snow, slight.	Rain or drizzle and snow, moderate or heavy.
Continuous fall of snowflakes, heavy at time of observation.	Ice needles (with or without fog).	Granular snow (with or without fog).	Isolated starlike snow crystals (with or without fog).	Ice pellets (sleet, U.S. definition).
Slight snow shower(s).	Moderate or heavy snow shower(s).	Slight shower(s) of soft or small hail with or without rain, or rain and snow mixed.	Moderate or heavy shower(s) of soft or small hail with or without rain, or rain and snow mixed.	Slight shower(s) of hail, with or without rain or rain and snow mixed, not associated with thunder.
Slight or moderate thunderstorm without hail, but with rain and/or snow at time of ob.	Slight or moderate thunderstorm, with hail at time of ob.	Heavy thunderstorm, without hail, but with rain and/or snow at time of observation.	Thunderstorm combined with dustorm or sandstorm at time of observation.	Heavy thunderstorm with hail at time of ob.

TABLE E-1 **Cont.**

	Clouds C_L	Clouds C_M	Clouds C_H
0	No Sc, St, Cu, or Cb clouds.	No Ac, As, or Ns clouds.	No Ci, Cc, or Cs clouds.
1	Ragged Cu, other than bad weather, or Cu with little vertical development and seemingly flattened, or both.	As, the greatest part of which is semitransparent through which the sun or moon may be faintly seen as through ground glass.	Filaments, strands or hooks of Ci, not increasing.
2	Cu of considerable development, generally towering with or without other Cu or Sc; bases at same level.	As, the greatest part of which is sufficiently dense to hide the sun or moon, or Ns.	Dense Ci in patches or twisted sheaves, usually not increasing or Ci with towers resembling cumuliform tufts.
3	Cb with tops lacking clear-cut outlines, but distinctly not cirriform or anvil-shaped: Cu, Sc, or St may be present.	Thin Ac, mostly semitransparent: other than crenelated or in cumuliform tufts; cloud elements change but slowly with bases at same level.	Ci, often anvil-shaped, derived from or associated with Cb.
4	Sc formed by spreading out of Cu; Cu may also be present.	Patches of semitransparent Ac that are at one or more levels; cloud elements are continuously changing.	Ci, hook-shaped and/or filaments, spreading over the sky and generally becoming dense as a whole.
5	Sc not formed by spreading out of Cu.	Semitransparent Ac in bands, or Ac in one more or less continuous layer gradually spreading over sky and usually thickening as a whole; the layer may be opaque or a double sheet.	Ci, often in converging bands, and Cs or Cs alone but increasing and growing denser as a whole; the continuous veil not exceeding 45° above the horizon.
6	St in a more or less continuous layer and/or ragged shreds, but no Fs of bad weather.	Ac formed by the spreading out of Cu.	Ci, often in converging bands, and Cs, or Cs alone, but increasing and growing denser as a whole; the continuous veil exceeds 45° above the horizon but sky not totally covered.
7	Fs and/or Fc of bad weather (scud) usually under As and Ns.	Double-layered Ac or an opaque layer of Ac, not increasing over the sky; or Ac coexisting with As or Ns or with both.	Veil of Cs completely covering the sky.
8	Cu and Sc (not formed by spreading out of Cu); base of Cu at a different level than base of Sc.	Ac with sprouts in the form of small towers or battlements, or Ac having the appearance of cumuliform tufts.	Cs not increasing and not completely covering the sky.
9	Cb having a clearly fibrous (cirriform) top, often anvil-shaped, with or or without Cu, Sc, St, or scud.	Ac generally at several layers in a chaotic sky, dense Ci is usually present.	Cc alone, or Cc accompanied by Ci and/or Cs, but Cc is the predominant cirriform cloud.

TABLE E-1 **Cont.**

	W	N	a
0	Cloud covering 1/2 or less of sky throughout the period.	No clouds.	Rising, then falling, greater than, or the same as, 3 hours ago.
1	Cloud covering more than 1/2 of sky during part of period and covering 1/2 or less during part of period.	One-tenth or less, but not zero. One-eighth.	Rising, then steady; or rising, then rising more slowly. Now higher than 3 hours ago.
2	Cloud covering more than 1/2 of sky throughout the period.	Two-or three-tenths. Two-eighths.	Rising steadily, or unsteadily. Now higher than 3 hours ago.
3	Sandstorm, or duststorm, or drifting or blowing snow.	Four-tenths. Three-eighths.	Falling or steady, then rising; or rising, then rising more quickly. Now higher than 3 hours ago.
4	Fog, or thick haze.	Five-tenths. Four-eighths.	Steady, same as 3 hours ago.
5	Drizzle.	Six-tenths. Five-eighths.	Falling, then rising. Now lower than, or the same as, 3 hours ago.
6	Rain.	Seven- or eight-tenths Six-eighths.	Falling, then steady, or falling, then falling more slowly. Now lower than 3 hours ago.
7	Snow, or rain and snow mixed, or ice pellets (sleet).	Nine-tenths, or more, but not ten-tenths. Seven-eighths.	Falling steadily, or unsteadily. Now lower than 3 hours ago.
8	Shower(s).	Ten-tenths. Eight-eighths.	Steady, or rising, then falling; or falling, then falling more quickly. Now lower than 3 hours ago.
9	Thunderstorm, with or without precipitation.	Sky obscured or cloud amount cannot be determined.	Not used for pressure tendency.

observation is plotted on a 500 mb chart at the time nearest to the observation time. Unfortunately, uniformity is not complete: Some reports are in feet, some are in meters.

Plotting Symbols for a Surface Chart

Table E-1 shows symbols for use on surface synoptic charts. These pages are from the manual by Alcorn (1990). Another good source of information about symbols on weather charts is by Decker (1982).

Some more frequently used symbols are for rain [dot(s)], snow [asterisk(s)], and showers (∇, "del operator"). Kinked arrows are used for lightning or thunder. Multiple symbols generally indicate stronger precipitation. Readers may see the complete description of these symbols in official manuals of the Weather Service.

Appendix F

Enhancement of Infrared Satellite Images

General Information

Each pixel (''point'') in satellite images contains information about the temperature of the radiating surface. This information is conventionally represented on images using various brightness of points on the images.

A usual rule used in all curves is that warmer radiating surfaces are represented by darker tones and colder surfaces are represented by brighter tones. Exceptions are made in some temperature domains to emphasize the contours of some clouds.

If the brightness is exactly proportional to the temperature, the image is *unenhanced*. While such images correctly represent the temperature, the details are hard to discern. Therefore, unenhanced images are seldom used. Instead, certain temperature domains are allotted more contrast, to make cloud patterns more easily recognized. These are the *enhanced IR images* (IR for *infrared*). Various accepted ways of image enhancement are usually represented by enhancement curves. Several of these curves are shown in this appendix and many more have been developed by the National Environmental Satellite Service (NESS).

Image Header

Disseminated images have a header that contains pertinent information. This interpretation is for the header of Fig. F-1, with comments that also generally apply:

0701 27NO89	Time (UTC) and date
29E	Satellite type and sector of the image (in a special code)
2	2-km resolution (between observing pixels)
ZA	Enhancement curve used. Other choices are MB, CC, and so forth.

Other numbers give detailed information about the coordinates of the satellite and the station that received and processed the image.

A strip of gray shades under the header shows the shades used on the particular image. The temperature scale goes along with this strip; however, this scale is not linear and therefore should be considered only qualitatively. A better scale of temperature (al-

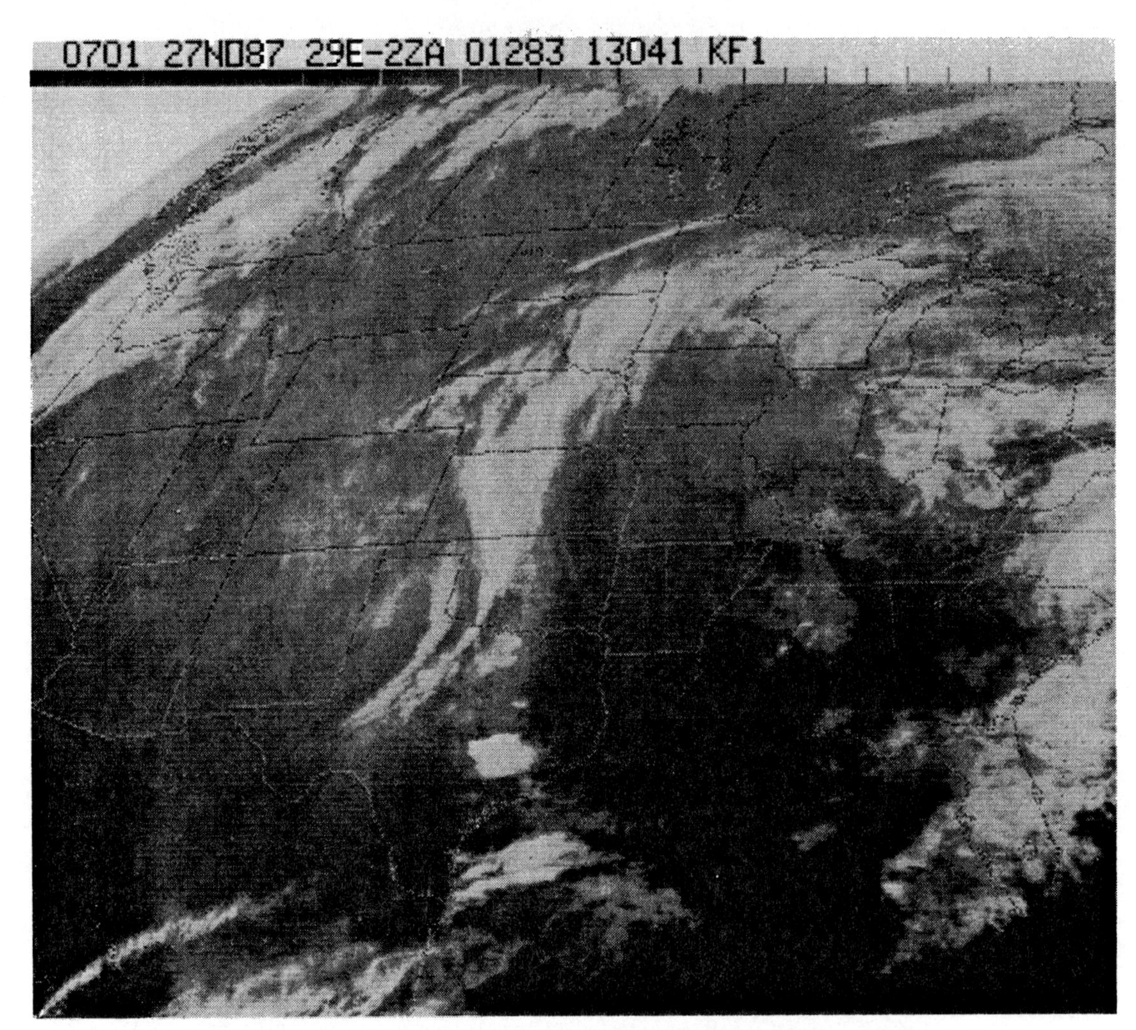

FIGURE F-1. IR satellite image of 0701 UTC 27 November 1987, in ZA enhancement. The homogeneous black surface is the warm ocean surface. The highest cloud tops are the brightest surfaces.

though rounded off) is given in Fig. F-2 for three common enhancement curves.

ZA Curve

The enhancement by this curve is weak, as shown by an almost monotonous slope of the curve. This kind of image is practical for general meteorological considerations.

Temperature ranges above 29°C and below −76°C are deemed of so little interest that they are ignored on the scale of brightness. These values are represented with black (above 29°C) and white (below −76°C). The enhancement is reserved for the range between −76 and 29°C, the range in which most clouds appear. Stronger enhancement is used in the ranges between −76 and −56°C and between 7 and 29°C, as can be seen from the steeper slope of the curve. Most clouds appear in these extreme ranges. Comparatively few clouds end development in the middle troposphere.

The lower part (7 to 29°C) gives a better depiction of low-level clouds. In addition, continent–sea differences can be seen better with this enhancement than in unenhanced images. The cold end of the temperature range gives enhancement of high cloud tops. This shows thunderstorms and the large cloud systems of extratropical cyclones well.

An example of a satellite image with ZA enhancement is shown in Fig. F-1. Most noticeable are the cloud bands that stretch from Texas through several states to Minnesota and Wisconsin. The widest and most prominent part, over Oklahoma and Kansas, is the *baroclinic leaf cloud*. The name derives from its similarity to a plant leaf. This cloud precedes cyclogenesis in the middle latitudes.

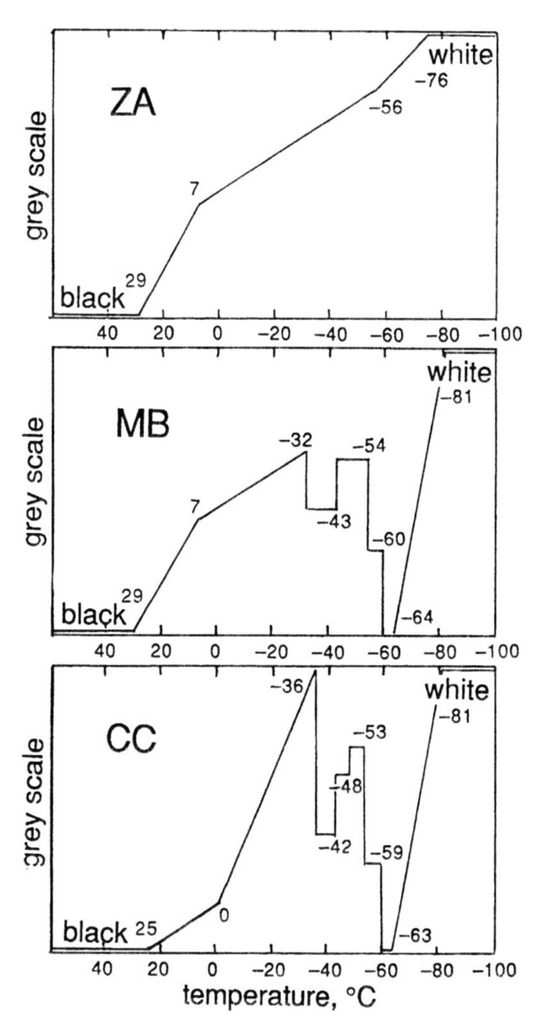

FIGURE F-2. The three curves used most often for enhancement of IR satellite imagery. The gray scale along the ordinate goes from black on the bottom to white on the top. The temperature (°C) is along the abscissa. The slanted parts of the curves are in the temperature ranges of gradual whitening of colder radiating surfaces. The jumps of the curves are at selected typical values of temperature that can be found on cumulonimbus tops. The value of temperature (rounded off) is entered at each kink or jump of the curves.

The bright, compact cloud over southeastern Texas is a tall cumulonimbus top, formed by joined anvils of several convective cells. The brightness of this cloud indicates that it is higher than the leaf cloud over Kansas.

MB Curve

The warmer radiation down to −32°C is represented exactly as on the ZA curve. Strong, discontinuous enhancement is applied in the colder radiation that comes from thunderstorm tops. The abrupt transitions at some temperature values give very clear contours that encompass selected temperature ranges. Something is lost in this enhancement: For example, we cannot tell if the cloud top temperature is −34 or −41°C since both values appear with the same degree of gray representation. However, the silhouettes of several cloud levels can be distinctly recognized. This is usually of more benefit than the distinction of temperature within one range on the enhancement curve.

An example of MB enhancement is given in Fig. F-3. This image is of a time within half an hour of the image in Fig. F-1. About the same clouds can be seen in both images. The enhancement in Fig. F-3 shows the first contour from white to gray at −32°C.

The second contour in Fig. F-3 appears as a transition to brighter gray at −43°C. The clouds in the central part of the image were not too cold at this hour, which is a sign that no severe storms were in progress. Several clouds in the Northwest did reach to the −60°C level, as shown by dark grey spots. The cumulonimbus over southeastern Texas has an area of −43 to −54°C, wider than the cloud over Kansas.

CC Curve

Stronger enhancement (steeper curve) in the range between 0 and −36°C is used in the CC curve (Fig. F-2, example in Fig. F-4). Also, jumps are provided at smaller temperature intervals for colder cloud tops. This will show more detail in cloud structure than the MB curve.

The example of CC enhancement in Fig. F-4 is for the same time as Fig. F-3, so the two can be compared. More contours can be noticed at the cloud tops than on the MB example, due to smaller intervals between kinks on the curve. The central brighter gray region on the top of the cloud in southeastern Texas (between −48 and −53°C) is noticeably wider than such an area over Kansas. The somewhat lower cloud (almost white in the image) that engulfs the baroclinic leaf over Kansas has a wider range of gray shades than the previous two examples. This is due to the steeper curve in the domain of 0 to −36°C.

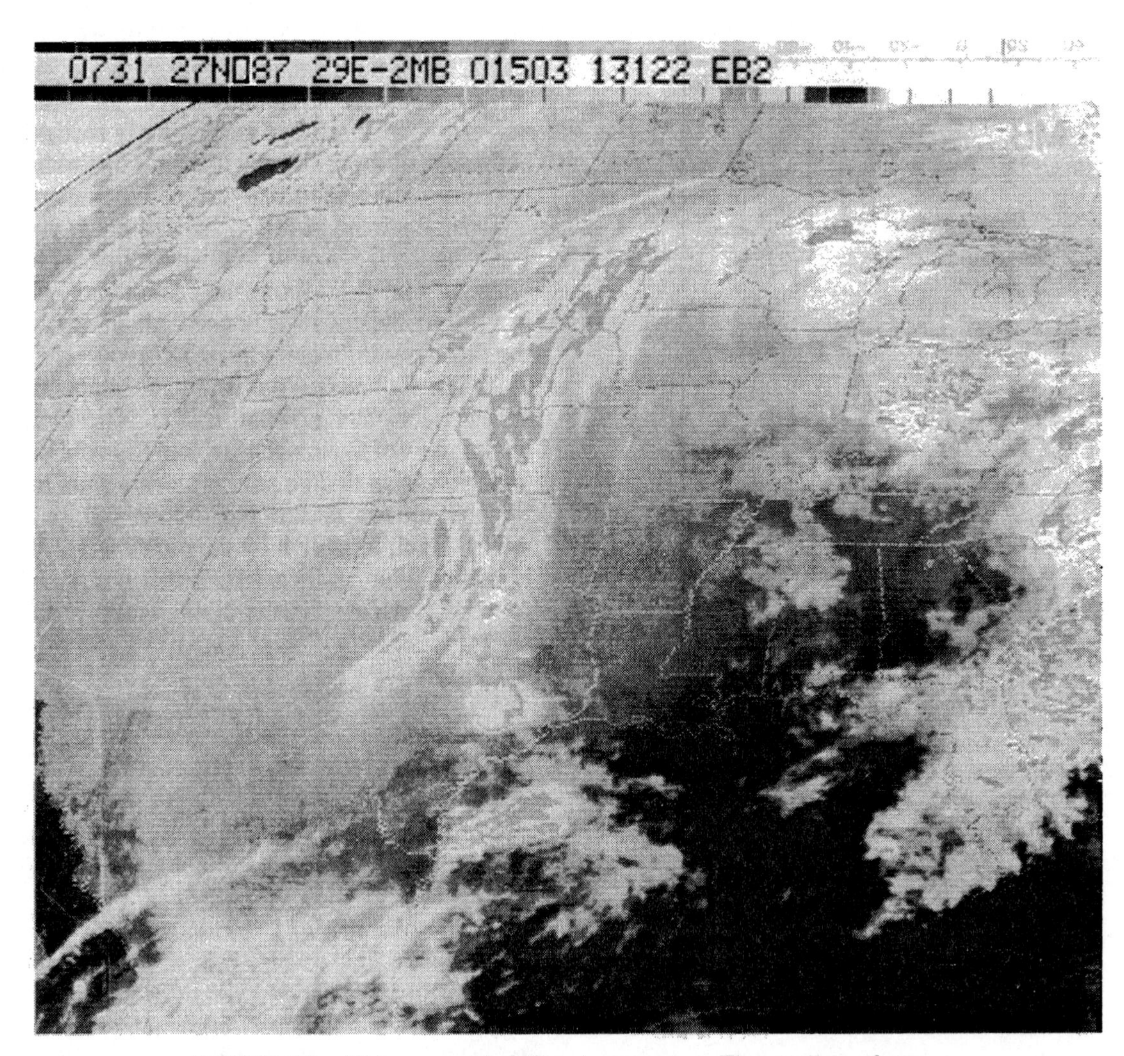

FIGURE F-3. IR image with MB enhancement. The outline of several levels (detected by temperature) is made clear by the abrupt transition between shades of gray. The temperature values at the transitions are given in Fig. F-2.

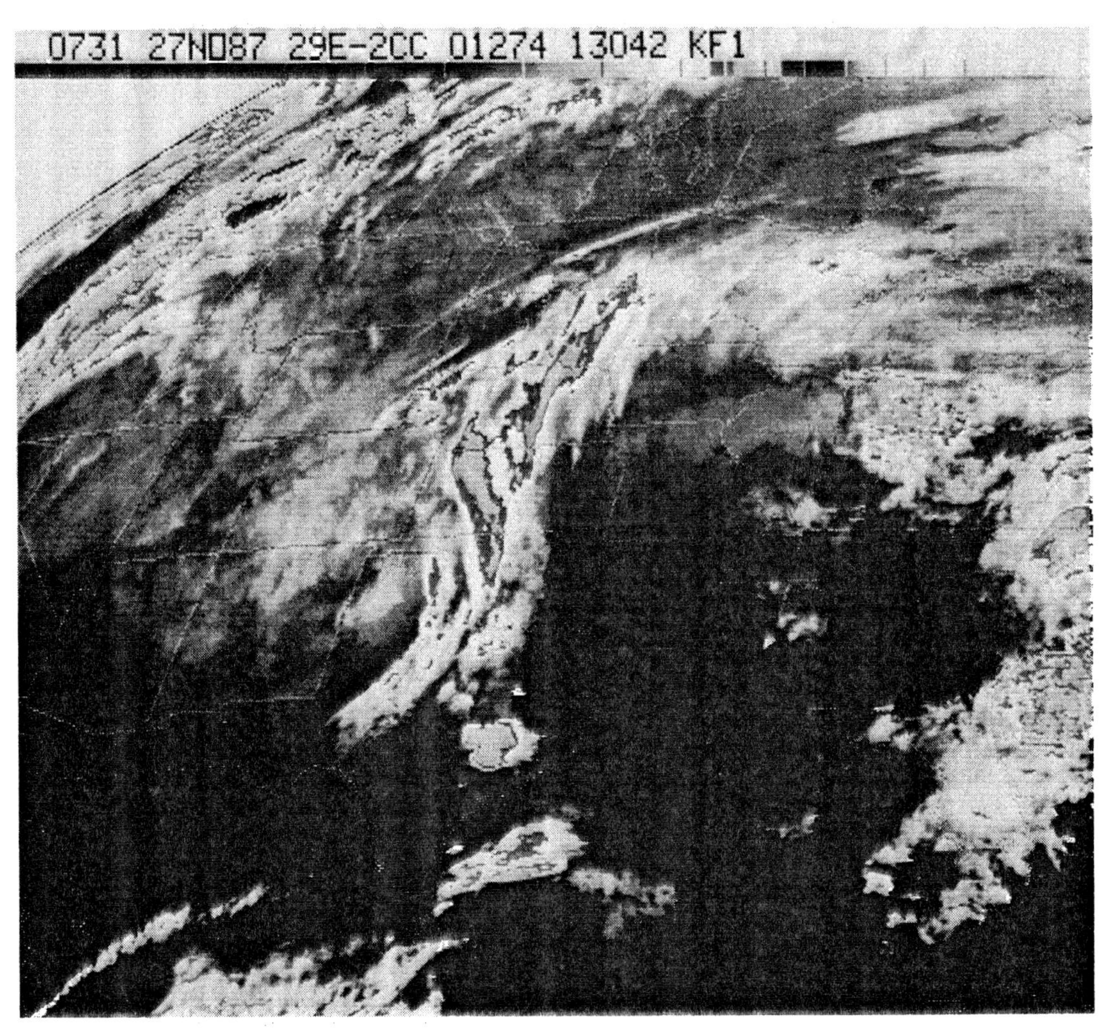

FIGURE F-4. IR image with CC enhancement at the same time as the image in Fig. F-2. More contours at selected temperature levels can be noticed than in Fig. F-2.

Appendix G

Geographic Maps

Because maps are widely used for representation of weather and for evaluation of various quantities, several elements of geographical cartography are shown in this appendix. The quantitative description is limited to conformal maps.

General Information

The (approximately) spherical surface of the earth cannot be correctly represented on flat paper. Since we greatly prefer flat paper to globes, various methods are needed to reduce the sphere to a flat projection. If we only want to show weather patterns, any popular map projection can be used successfully. However, we need various quantities with some accuracy; therefore, we have to be careful with measurements on maps.

Conformal (also called *orthomorphic*) *projections* are the most practical projections in meteorology. In these, the angles and shapes of small areas are correctly represented: The angles on the earth's surface are equal to the angles on the map. Only one number (*image scale*, S) is needed in every point to show the ratio of the distance on the map to the corresponding distance on the earth.

Nonconformal projections have different scales in different directions. If distances are needed with a higher accuracy than 5–10%, recourse to spherical trigonometry must be made. Some nonconformal projections preserve the ratios between areas on the map and areas on the earth; these are *equal area projections*. The angles on the earth are usually poorly shown on nonconformal projections.

Description of Conformal Maps

Conformal maps most often have the image scale that varies only with the latitude. The *map scale* (s) is equal to the image scale of the map (S) at one or two selected latitudes; these are the *standard latitudes*. The map scale is shown with a ratio such as 1 : 10,000,000 or so. The image scale of the map is formally represented by the product of the map scale s and the *map magnification factor* (*MMF*) m:

$$S = sm$$

The factor m is usually close to unity. Multiplication by m may be ignored in many approximate calculations. The MMF is also called *distortion*; even if the

TABLE G-1

Map magnification factor for main conformal projections

Projection:	*Mercator*		*Lambert*		*Polar stereographic*	
Standard latitudes:	0°	22.5°	30°, 60°	10°, 40°	90°	60°
MMF at 0°	1.00	0.92	1.28	1.06	2.00	1.87
5°	1.00	0.93	1.21	1.03	1.84	1.72
10°	1.02	0.94	1.15	1.00	1.70	1.59
15°	1.04	0.96	1.10	0.98	1.59	1.48
20°	1.06	0.98	1.06	0.97	1.49	1.39
25°	1.10	1.02	1.03	0.97	1.41	1.31
30°	1.15	1.07	1.00	0.97	1.33	1.24
35°	1.22	1.13	0.98	0.98	1.27	1.19
40°	1.31	1.21	0.97	1.00	1.22	1.14
45°	1.41	1.31	0.97	1.03	1.17	1.09
50°	1.56	1.44	0.97	1.07	1.13	1.06
55°	1.74	1.61	0.98	1.13	1.10	1.03
60°	2.00	1.85	1.00	1.21	1.07	1.00
65°	2.37	2.19	1.03	1.32	1.05	0.98
70°	2.92	2.70	1.08	1.48	1.03	0.96
75°	3.86	3.57	1.16	1.72	1.02	0.95
80°	5.76	5.32	1.29	2.16	1.01	0.94
85°	11.47	10.60	1.57	3.19	1.00	0.93
90°	∞	∞	∞	∞	1.00	0.93

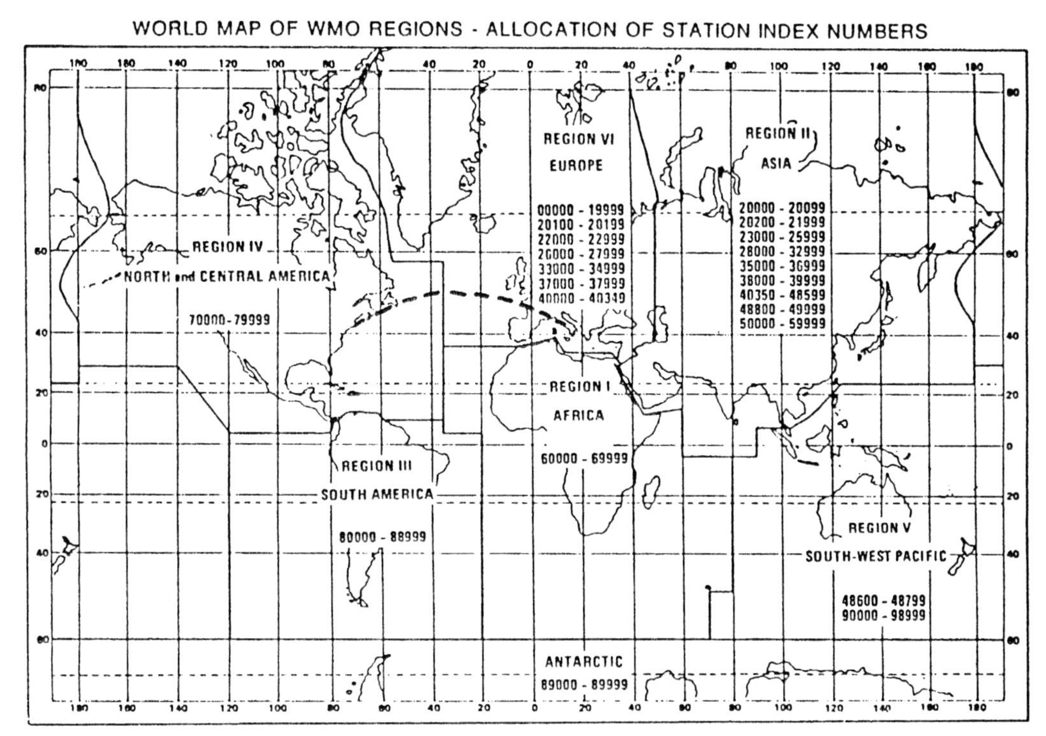

FIGURE G-1. An example of Mercator projection, representing almost the whole earth's surface (the regions near the poles cannot be shown). The station numbers shown are those used in the international exchange of weather data. The dashed line across the North Atlantic is a part of the great circle that connects New York and Rome.

shapes of small areas are not distorted, they are only reduced or enlarged. Evaluation of the MMF is shown in the next section. The results are summarized in Table G-1, which shows the variation of the MMF for the three most widely used conformal projections: *Mercator*, *Lambert conical*, and *polar stereographic*.

As is recommended by the World Meteorological Organization, the most commonly used standard latitudes are 22.5° for Mercator, 30° and 60° for Lambert conical, and 60° for polar stereographic. Standard latitudes of 10° and 40° are recommended for Lambert conical projection in the Southern Hemisphere. Table G-1 also contains the MMF for Mercator and polar stereographic projections for which the standard latitudes are 0° and 90°, respectively.

A desirable property of geographic maps is that the MMF is close to 1.0. This value indicates no magnification beyond the map scale. As can be seen from the table, values close to 1.0 can be found in different latitudes for different projections. If we take that $0.9 < m < 1.1$ is a satisfactory magnification, we can see that Mercator projection is very good for the tropical regions, Lambert conical projection for the middle latitudes, and polar stereographic projection for the charts that include the pole. The MMF is exactly 1.0 at standard latitudes. Mercator and Lambert projections cannot include the poles; there the magnification is infinite.

Several projections are illustrated in Figs. G-1 through G-3. The MMF in each varies only with latitude.

Mercator projection is shown in Fig. G-1. All parallels and meridians are straight. A peculiarity of Mercator projection is that Greenland appears much larger than Australia, when we know that the surface of Australia is more than 3.5 times larger than Greenland. The information on station numbers in Fig. G-1 is not related to the map projection; it is given here only as a potentially useful item. A reference to the station numbers is made in Appendix E.

Since Mercator projection is very good for a wide region in the middle part of the map, this projection is sometimes used to form other conformal projections. This is achieved by using a *slanted Mercator projection*, where some other great circle is used as the equator. One such projection is described at the end of this appendix.

An important property of Lambert conical projection is that the angles between meridians are smaller than their difference in longitude. The ratio between the angle on the map versus the difference in longitude is the *constant of the cone* (n). This constant is illustrated in Fig. G-2.

The MMF on a polar stereographic map (Fig. G-3) increases by a factor of two between the pole and the equator. Therefore, this projection is suitable for representation of one hemisphere, northern or southern. The MMF increases drastically in the opposite hemisphere. It tends to infinity when the map approaches the opposite pole.

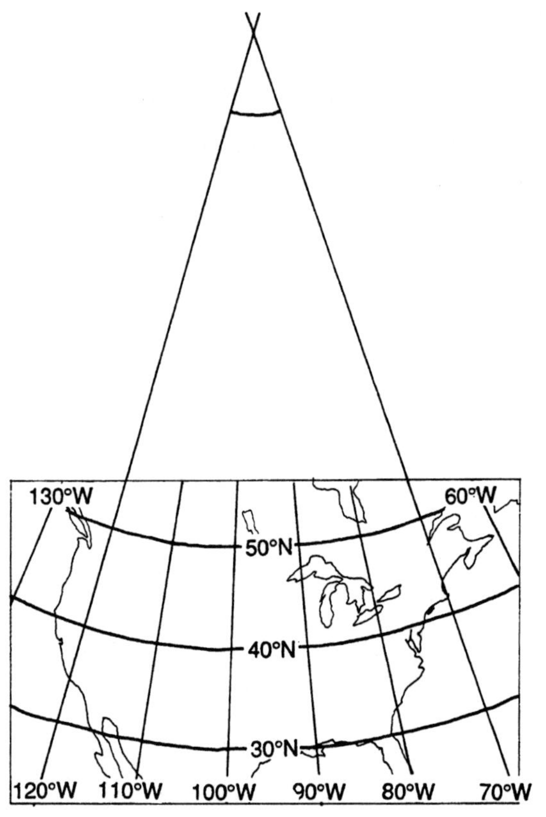

FIGURE G-2. An example of a map in Lambert conical projection. The angle between meridians of 70°W and 120°W amounts to 35.8°. The constant of the cone is $n = 35.8/50 = 0.7156$.

A *great circle* on a sphere is a circle whose center is in the center of the earth. A great circle through two points is the shortest distance between them, if the distance is measured on the surface of a sphere. In this way, a great circle plays the role of a straight line on a plane. All meridians are great circles, and so is the equator. Other geographical parallels are not great circles. Figure G-1 contains a part of a great circle (dashed) that shows the shortest distance (on the sphere) between New York and Rome. It illustrates that a straight line on a map is not the shortest distance between two points. It also illustrates that large features, such as the line New York–Rome, are distorted on conformal maps. Only "small" features are not distorted.

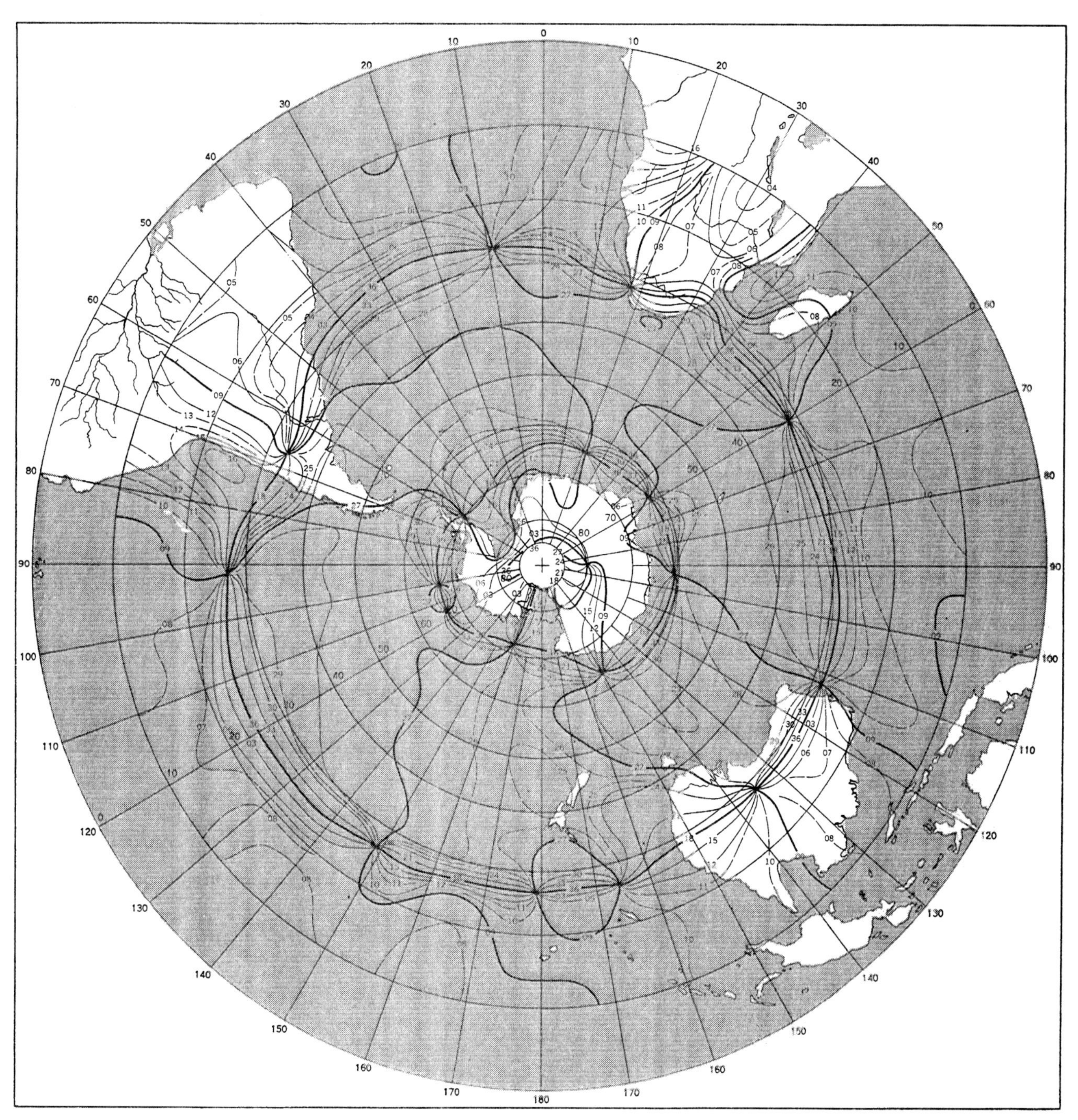

FIGURE G-3. Polar stereographic map of the Southern Hemisphere. Isogones of the surface geostrophic wind (average for July, degrees azimuth) are also shown, as an item of climatological interest. From Jenne and others (1971).

Construction of Main Conformal Maps

An exact formulation of map projections is important for proper usage of charts in numerical weather prediction and other applications. Except when the equations are in spherical coordinates, conformal projections are universally used for the representation of variables. Therefore, conformal projections are described next in some detail.

Lambert conical projection is rather general for conformal projections at which the MMF varies only with latitude. The other two (Mercator and polar ste-

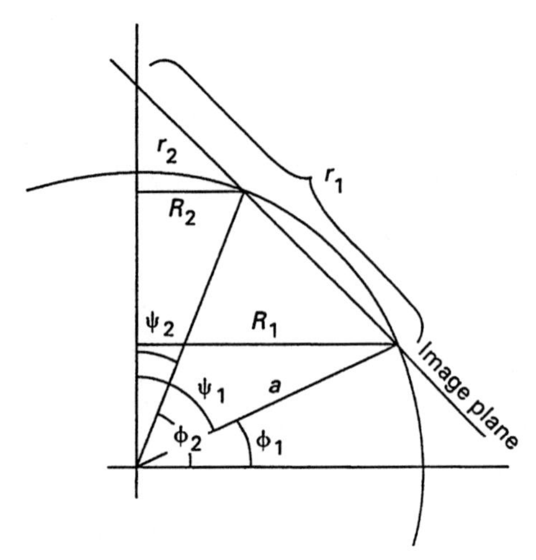

FIGURE G-4. The geometry of Lambert conical projection. The image plane is a cone that intersects the earth at standard latitudes and spreads flat on the map.

reographic) are special cases of Lambert projection when the two standard latitudes are set equal to each other and taken as 0° or 90° latitude. Lambert projection is illustrated in Fig. G-4. The circle is the section through the earth along a meridian. The secant "image plane" is part of the cone on which the projection is made. A developed cone is the conformal map.

The quantities needed are illustrated in Fig. G-4. The radius of the earth is a. It is assumed that the earth is an exact sphere. More sophisticated approximations to the shape of the earth are considered in geographical literature. However, in meteorology the assumption of a spherical earth is good for all practical considerations.

The colatitude $\psi = 90° - \phi$ is used several times below. The subscripted variables (with 1 and 2) refer to two standard latitudes. The radius of each geographical parallel on the map is r. The distance from the axis of the earth rotation is R. The map scale s will not be considered here, assuming that all distances are measured as on the earth. In this way, a distance along the image plane will be typically "several hundred kilometers," even if this distance takes only a few centimeters on the paper.

The increment along the cone is

$$dr = m_\lambda a \, d\psi \tag{G-1}$$

where m_λ is the MMF along the meridian, at a constant longitude λ, and a is the radius of the earth.

The developed cone does not cover the whole circle; instead, the nth part of it is as illustrated in Fig. G-2 for the angle between 70°W and 120°W. Thus the length L of the 360° arc on the map occupies n times the angle between longitudes. The ratio between the zonal arc on the map and the arc on the sphere (L_E, this latter one is a whole circle) is the MMF in the zonal direction (at constant latitude ϕ):

$$m_\phi = \frac{L}{L_E} = \frac{2nr\pi}{2R\pi} = \frac{nr}{a \sin \psi} \tag{G-2}$$

At standard latitudes $m_\phi = 1$ and

$$1 = \frac{nr_1}{a \sin \psi_1} = \frac{nr_2}{a \sin \psi_2}$$

The constant of the cone can be obtained from the last equations:

$$n = \frac{a \sin \psi_1}{r_1}$$

in terms of r_1. The same value is obtained with ψ_2 and r_2.

Conformity is achieved by the requirement $m_\phi = m_\lambda = m$ in (G-2). This gives a differential equation for the radius r of the geographical parallels on the map:

$$dr = ma \, d\psi = \frac{nr}{\sin \psi} d\psi \tag{G-3}$$

A useful solution of this equation is

$$r = r_1 \left(\frac{\tan \frac{\psi}{2}}{\tan \frac{\psi_1}{2}} \right)^n \tag{G-4}$$

where the indexed values (r_1, ψ_1) are at the standard latitude ϕ_1. A similar expression follows when the latitude ϕ_2 is used:

$$r = r_2 \left(\frac{\tan \frac{\psi}{2}}{\tan \frac{\psi_2}{2}} \right)^n \tag{G-5}$$

The constant of the cone n can be found from the two last expressions when $\tan(\psi/2)$ is eliminated:

$$n = \frac{\log \dfrac{\sin \psi_1}{\sin \psi_2}}{\log \dfrac{\tan \dfrac{\psi_1}{2}}{\tan \dfrac{\psi_2}{2}}} \tag{G-6}$$

This constant of the cone amounts to

$$n = 0.7155668$$

for $\phi_1 = 30°$ and $\phi_2 = 60°$, and

$$n = 0.4275969$$

for $\phi_1 = 10°$ and $\phi_2 = 40°$.

The MMF can be computed from the second equality in (G-3):

$$m = \frac{nr}{a \sin \psi}$$

Using the obtained expressions for r,

$$m = \frac{\sin \psi_1}{\sin \psi}\left(\frac{\tan \dfrac{\psi}{2}}{\tan \dfrac{\psi_1}{2}}\right)^n = \frac{\sin \psi_2}{\sin \psi}\left(\frac{\tan \dfrac{\psi}{2}}{\tan \dfrac{\psi_2}{2}}\right)^n \tag{G-7}$$

The values from this formula are listed in Table G-1, under Lambert conical projections.

Tangential conical projection has the property $\psi_1 = \psi_2$. In this projection, the constant of the cone reduces to

$$n = \cos \psi_1$$

where ψ_1 is the only standard colatitude. Mercator and polar stereographic projections also have one standard latitude ($\psi_2 = \psi_1$), even if these projections may be secantial instead of tangential. The constant of the cone n for these two projections can also be found from (G-6). In these cases, the ratio can be evaluated as $n = 1$ for polar stereographic projection, and $n = 0$ for Mercator projection. Mathematical work involves several trigonometric identities and L'Hospital's rule with $\psi_2 = \psi_1 + \varepsilon$ and $\varepsilon \to 0$. With these values of n, the MMF for polar stereographic projection follows from (G-7) as

$$m = C\frac{1}{1 + \cos \psi} \qquad C = \frac{\sin \psi_1}{\tan \dfrac{\psi_1}{2}}$$

The last expression reduces to $C = 1$ for the tangential polar stereographic projection in which $\phi_1 = 90°$ and $\psi_1 = 0°$. In terms of geographical latitude ϕ, the MMF for polar stereographic projection is

$$m = \frac{C}{1 + \sin \phi}$$

The MMF for Mercator projection also follows from (G-7) as

$$m = \frac{\sin \psi_1}{\sin \psi}$$

Transformation of Coordinates

Besides the information about the MMF, the above equations can be used for transformation of coordinates between a sphere and a map plane. This is illustrated in Fig. G-5. The spherical coordinates are longitude λ and latitude ϕ. The west longitude and south latitude are negative. The case of Lambert conformal

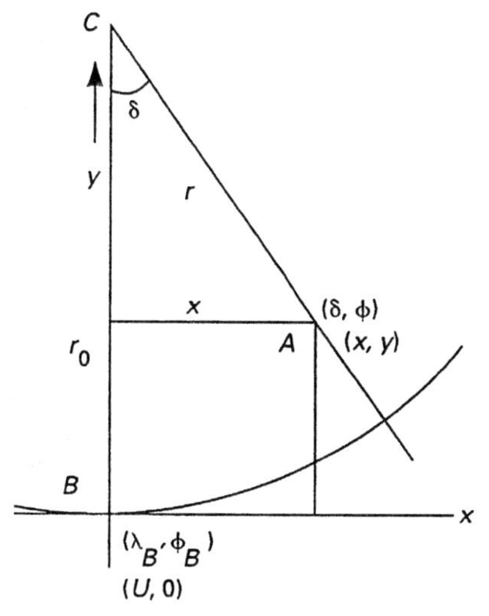

FIGURE G-5. Cartesian coordinates on Lambert conical projection. The polar coordinates on the map plane represent the spherical coordinates, with a provision that the angle δ is related to longitude λ as $\delta = n\lambda$.

projection is considered since this is the generalization of Mercator and polar stereographic projections as well.

There are two sets of coordinates on the flat map in Fig. G-5:

1. Cartesian coordinates chosen so that the y-axis coincides with one meridian, at longitude λ_B. The x-axis is chosen to intersect the y-axis and the meridian λ_B at the latitude ϕ_B.
2. Polar coordinates (δ, r) on the map plane.

Attention should be paid to the *polar angle* δ that looks like longitude on the chart, but is smaller than the corresponding longitude by the factor n. So, for example, if the longitude λ of point A is 50°E, measuring from the longitude of the base meridian through point B, then the corresponding polar angle δ is

$$\delta = n \times 50°$$

Assuming the standard latitudes at 30° and 60°, this gives

$$\delta = 0.7155668 \times 50° = 35.78°$$

The radius from the center C is given by (G-4) or (G-5). Then the position in the polar coordinate system on the flat chart is given by

$$\delta = n(\lambda + \lambda_B)$$

$$r = r_1 \left(\frac{\tan \frac{\psi}{2}}{\tan \frac{\psi_1}{2}} \right)^n$$

where λ_B is the longitude of the meridian that coincides with the y-axis. The radius r shows the distance from the "pole" C where the meridians meet. However, the pole must not be included on Lambert conical projection since the MMF is infinite in that point.

The Cartesian coordinates on a flat chart are

$$x = r \sin \delta$$
$$y = r_0 - r \cos \delta$$

where r_0 is the distance from the pole on the flat chart to the coordinate origin.

The radius r on Mercator projection is infinite. Therefore, measurement along meridians should be taken as the distance q from the equator, with

$$\Delta q = -\Delta r$$

This can be evaluated from (G-1) as

$$dq = -ma\, d\psi = -\frac{\sin \psi_1}{\sin \psi} a\, d\psi$$
$$q = a \sin \psi_1 \log(\csc \psi - \cot \psi)$$

Other Conformal Maps

Mercator projection is the most common basis for other conformal projections which use another great circle as the equator in Mercator projection. This is frequently used for representation of smaller regions. Geographers often use a meridian as the "equator" for constructing a new projection. This is *transversal Mercator projection.*

Slanted Mercator projection is the basis of the current Eta Model of the National Meteorological Center. It is based on spherical coordinates that are slanted by the angle α, as shown in Fig. G-6.

The coordinates of slanted Mercator projection can be evaluated when a new spherical coordinate system is introduced such that its "equator" coincides with the geographical parallel in the central point (C, Fig. G-6) of the region of interest. These new spherical

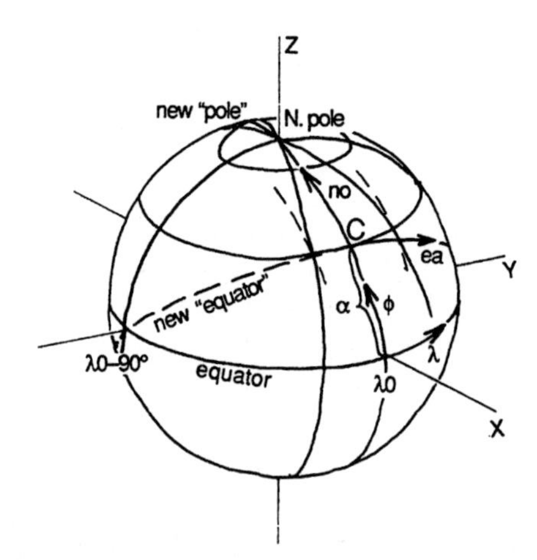

FIGURE G-6. Slanted spherical coordinates *ea* and *no* (dashed), compared with the usual geographical coordinates λ and ϕ (solid). The new coordinates *ea* and *no* have least "distortion" on a Mercator projection that is formed with the new "equator" through point C. The meridian through C is the same in both coordinate systems. All other meridians and all parallels do not coincide.

coordinates are the angles *ea* and *no*, resembling the east longitude λ and north latitude ϕ, respectively. The new coordinates are related to latitude ϕ and longitude λ as

$$ea = \tan^{-1} \frac{\cos \phi \sin \lambda}{\cos \alpha \cos \phi \cos \lambda + \sin \alpha \sin \phi}$$
$$no = \sin^{-1}(-\sin \alpha \cos \phi \cos \lambda + \cos \alpha \sin \phi) \tag{G-8}$$

where α is the angle by which the equator was "elevated" to the latitude of the central point C. When the regional geographic map is centered at C, least "distortion" ($m \approx 1$) occurs in the central part of the map. All previous considerations about Mercator projection now apply in the new coordinates *ea* and *no*. The radius r (height) does not need transformation.

Transformation to the slanted coordinates (G-8) can be developed when a Cartesian coordinate system xyz is placed in the center of the earth. The z-coordinate coincides with the axis of the earth rotation. Such a system is shown in Fig. G-6. The geographical coordinates $\lambda\phi r$ have the same origin in the center of the earth. Transformation between the two systems is given by

$$r = \sqrt{x^2 + y^2 + z^2}$$
$$\lambda = \tan^{-1} \frac{y}{x}$$
$$\phi = \sin^{-1} \frac{z}{r} \tag{G-9}$$

or the inverse

$$x = r \cos \phi \cos \lambda$$
$$y = r \cos \phi \sin \lambda$$
$$z = r \sin \phi$$

A new Cartesian system can be selected to keep the same y-axis and to rotate the two other axes by an angle α:

$$x_{new} = x \cos \alpha + z \sin \alpha$$
$$y_{new} = y$$
$$z_{new} = -x \sin \alpha + z \cos \alpha$$

Using these new coordinates in the place of x, y, and z in (G-9) gives the slanted coordinates (G-8).

Appendix H

Geometrical Properties of Vorticity and Divergence

The formulas for vorticity and divergence in natural coordinates are explained below, starting from the expressions in Cartesian coordinates, to show equivalency of the terms in Cartesian and natural coordinates. Several manual methods for determining divergence are shown to enhance the understanding of kinematics. Manual methods like this are not routinely used if computer methods are available. However, the manual methods are rather instructive and therefore are presented here.

Vorticity in Natural Coordinates

Vorticity can be conveniently studied using the example of a circular streamline (Fig. H-1) where the normal shear and curvature are significant. Stretching and diffluence can be ignored since they do not contribute to vorticity. The Cartesian coordinates are turned so that the x-axis coincides with the streamline in point A. The normal direction is in the positive y-direction.

The variation of the normal wind component v from A to B can be represented by the initial terms of a Taylor series (subscripts A and B show values in points A and B):

$$v_{\mathrm{B}} = v_{\mathrm{A}} + \frac{\partial v}{\partial l}\,\Delta l \approx v_{\mathrm{A}} + \frac{\partial v}{\partial x}\,\Delta x$$

Due to the orientation of the coordinate system, v_{A} vanishes, and

$$\frac{\partial v}{\partial x} \approx \frac{v_{\mathrm{B}}}{\Delta x} \approx \frac{v_{\mathrm{B}}}{\Delta l}$$

The similarity of triangles in Fig. H-1 gives

$$\frac{v_{\mathrm{B}}}{\Delta l} = \frac{V}{r}$$

where V is the wind speed and r is the radius of curvature. With these, the first term in the expression for vorticity becomes the curvature term

$$\frac{\partial v}{\partial x} \approx \frac{V}{r} \qquad \text{(H-1)}$$

All the above approximate equalities are exact when $\Delta x \rightarrow 0$.

The orientation of the coordinate system shows that the wind speed V in point A is exactly equal to the x-component u. Therefore, the other term in the definition of vorticity becomes equal to shearing in natural coordinates:

$$-\frac{\partial u}{\partial y} = -\frac{\partial V}{\partial n} \qquad \text{(H-2)}$$

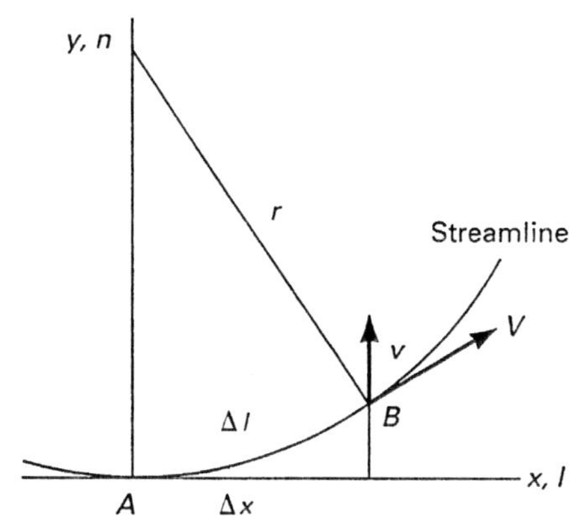

FIGURE H-1. Illustration of vorticity in natural coordinates. The Cartesian coordinate x coincides with the natural coordinate l at point A. The variation of direction between A and B is significant for vorticity.

Divergence in Natural Coordinates

Diffluence and stretching must be emphasized when considering divergence (Fig. H-2). Curvature and normal shear may be ignored in this case since they do not contribute to divergence.

Due to the coincidence of the x-axis and streamline at point A, the first term from the expression for divergence gives the stretching:

$$\frac{\partial u}{\partial x} = \frac{\partial V}{\partial l} \tag{H-3}$$

The variation of the y-component of wind and direction b between points A and B can be expressed by the first two terms of the Taylor series:

$$v_B = v_A + \frac{\partial v}{\partial y}\Delta y \qquad \beta_B = \beta_A + \frac{\partial \beta}{\partial y}\Delta y$$

β is the angle between the streamline at point B and the x-direction at point A. The orientation of the coordinates is selected such that

$$v_A = 0 \quad \beta_A = 0 \quad y = n \quad \Delta y = \Delta n$$

$$\frac{\partial v}{\partial y} = \frac{v_B}{\Delta n} \qquad \frac{\beta_B}{\Delta y} = \frac{\partial \beta}{\partial n}$$

From the triangle in Fig. H-2,

$$\beta_B = \frac{v_B}{V} \qquad \frac{\beta_B}{\Delta y} = \frac{v_B}{V\Delta n}$$

From these, the normal curvature term follows as

$$\frac{\partial v}{\partial y} = V\frac{\partial \beta}{\partial n}$$

It is practical to use azimuth for the wind direction. This is explained in Fig. H-2 by the arbitrary choice of the north direction. The wind direction is α. A comparison with the angle β shows that

$$\frac{\partial \beta}{\partial n} = -\frac{\partial \alpha}{\partial n}$$

With this, the last term in the expression for divergence becomes the normal curvature (or diffluence) term:

$$\frac{\partial v}{\partial y} = -V\frac{\partial \alpha}{\partial n} \tag{H-4}$$

Equations (H-1) through (H-4) illustrate the equivalency of the terms in Cartesian coordinates with the corresponding terms in natural coordinates, for both vorticity and divergence.

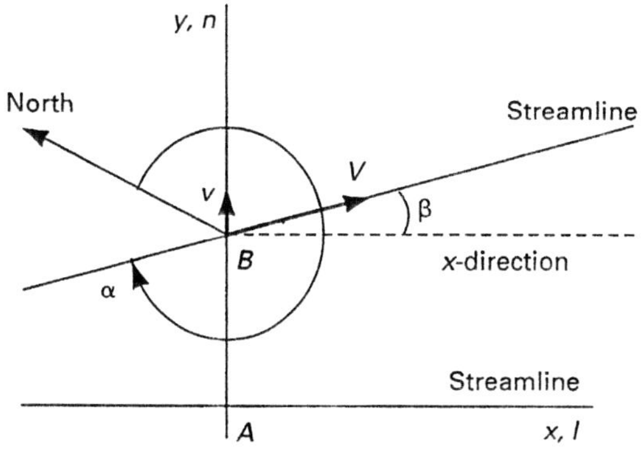

FIGURE H-2. An illustration of divergence in natural coordinates. The Cartesian coordinates are aligned with the streamline (and with the natural coordinates) at point A. Small variation in wind direction between points A and B is significant for divergence.

Divergence and Expansion of Material Surfaces

Divergence in two-dimensional flow is geometrically equivalent to the relative rate of increase of small material areas. This is expressed mathematically as

$$\text{div } \mathbf{V} = \frac{1}{A}\frac{dA}{dt} \tag{H-5}$$

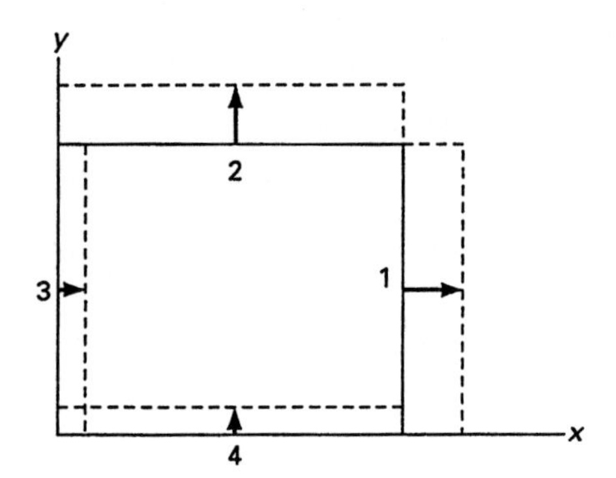

FIGURE H-3. Increase of area per unit time of a material rectangle. The dashed lines show increments of area on each side.

Area A is illustrated as a rectangle in Fig. H-3. The sides of the rectangle are equal to δx and δy, respectively, and they are numbered for convenience. The sides consist of fluid parcels that move with the velocity **V**. Components of flow **V** are drawn (with unit vectors) in the middle of each side. It may be assumed that the speed in the middle of a side is a fair representation of the average speed along the side.

Normal speed on all sides may be formally written as

$$u_3 = u \qquad v_4 = v$$

$$u_1 = u + \frac{\partial u}{\partial x}\,\delta x \qquad v_2 = v + \frac{\partial v}{\partial y}\,\delta y$$

The width of each increment is equal to the speed, since the increments are taken during a unit time and the length is related to speed and time as

$$\text{length} = \text{speed} \times \text{time}$$

Therefore, the increment is $u_1\,\delta y$ on side 1, $v_2\,\delta x$ on side 2, and so forth. The corners are ignored since they are smaller quantities of higher order. Consequently, area A increases (or decreases) per unit time in the amount

$$\frac{dA}{dt} = \left(u + \frac{\partial u}{\partial x}\,\delta x\right)\delta y + \left(v + \frac{\partial v}{\partial y}\,\delta y\right)\delta y - u\,\delta y - v\,\delta x$$

$$= \left(\frac{\partial u}{\partial x} + \frac{\partial v}{\partial y}\right)\delta x\,\delta y$$

Dividing both sides by area $A = \delta x\,\delta y$, equation (H-5) is proven. The equation (H-5) is rather general. Therefore it applies also for triangular regions, as shown next.

Divergence in Triangular Areas

The most direct evaluation of divergence on the basis of three-wind observation is illustrated in Fig. H-4. The wind is drawn as vector in the three observation points E, F, and G. After unit time, the vertices reach points E′, F′, and G′. The area of the triangle E′F′G′ is larger (smaller) than the area EGF if there is divergence (convergence) in the flow.

It may be of use to remember the formula for area P of a triangle, in terms of its sides a, b, and c:

$$P = \sqrt{s(s-a)(s-b)(s-c)}$$

where

$$s = \tfrac{1}{2}(a + b + c)$$

It is of interest to notice that the lengths and areas can be evaluated using measures on the chart, without respect to the map scale. This is justified, since the ratio between dA/dt and A is needed, and the scale cancels.

The length of the displacement vectors at vertices should be selected about three to five times smaller than the sides of the triangle. A typical triangle side is equal to the typical distance between neighboring radiosonde stations. This distance is of the order of 400 km in the United States. Therefore, a displacement of about 100 km will be convenient. If the wind speed is of the order of 10 m s^{-1} and a displacement of 100 km is wanted, the time that the trajectories of the vertices cover should be

$$\Delta t = \frac{\text{trajectory}}{\text{speed}} \approx \frac{10^5\ \text{m}}{10\ m\ s^{-1}} = 10^4\ \text{s}$$

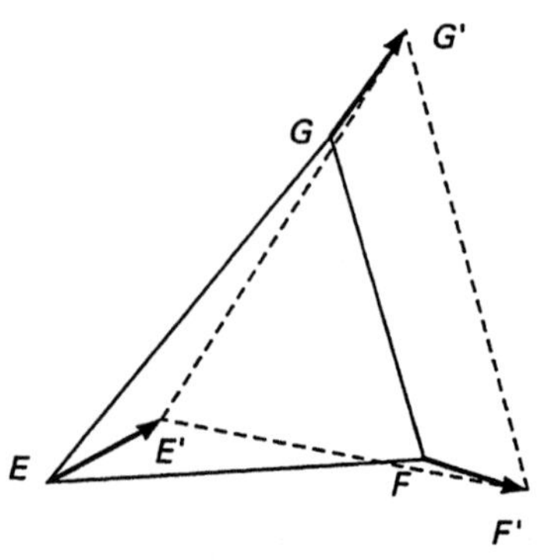

FIGURE H-4. Variation in the area of a triangle in unit time, assuming that the triangle moves with the air flow.

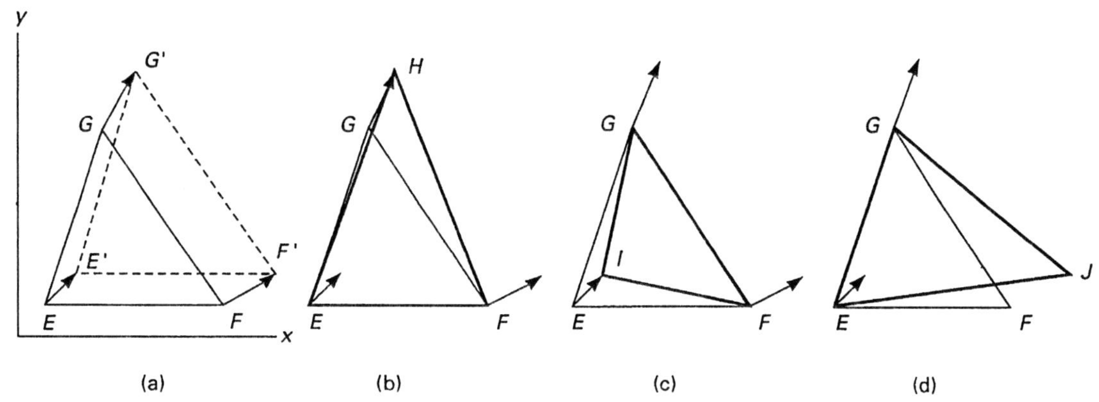

FIGURE H-5. Bellamy's method for evaluation of divergence. The variation of the area of the triangle EFG is calculated three times, once for each vertex [parts (b), (c), and (d). The total of the three area increments gives the increase of area A needed for a finite-difference evaluation of dA/dt in (H-5).

This choice of displacement of the vertices will yield the divergence in units of 10^{-4} s^{-1}.

Similar methods have been developed for evaluation of vorticity. These use the component of wind along the circumference of the triangle. However, vorticity is usually calculated much easier by grid-point methods. The use of triangles for vorticity is obsolete today and will not be described here.

Bellamy's Method

The preceding method is not overly practical, since it takes time to calculate all the steps. A variation of this method was developed by Bellamy (1949) that was widely used before computers were available. This method is still useful for a quick determination of divergence at selected points (small areas) on the charts. Instead of evaluating the "new" triangle E′F′G′ in Fig. H-5a, the expansion of the triangle EFG is first considered as if only one vertex moved (Fig. H-5b). The variation of the area is then evaluated by overlaying a nomogram on a transparent acetate. The nomogram gives the old and new area P of the triangle after the formula

$$P = \frac{ah}{2}$$

where a is the base and h is the height of the triangle. The ratio of the old area to the new area is exactly proportional to the ratio of the heights over the common base. Therefore, the evaluation of the area is very fast and suitable for manual work.

The increment of area is next calculated for the second and third vertex, assuming that the base is the opposite side (Fig. H-5c and d). The total increment of area is the sum of the three increments from three vertices.

It can be seen that Bellamy's method (Fig. H-5) does not yield the same result as direct evaluation of the area increment as in Fig. H-4. Other methods can also be devised; some examples are discussed by Schaefer and Doswell (1971). All the methods, of course, suffer from truncation error since they use finite increments to evaluate the continuous derivative dA/dt.

Appendix I

Thermal Wind and Thermal Advection

Thermal wind is the vertical shear of geostrophic wind. It is defined in two ways, like the general wind shear: as a derivative $\partial \mathbf{V}/\partial z$ or as the difference $\mathbf{V}_1 - \mathbf{V}_2$ (Section 3-2). In these expressions, $\mathbf{V}$ is geostrophic wind and z is the vertical coordinate. Index 1 indicates the upper level; index 2 indicates the lower level. In this way, points 1 and 2 are above each other. The relevant equations are in this appendix.

Derivative Type of Thermal Wind

The vertical derivative of wind can be reduced to a function of temperature. This follows from the hydrostatic equation in the form

$$\frac{\partial z}{\partial p} = -\frac{RT}{gp}$$

When the operator $(g/f)\mathbf{k} \times \nabla$ is applied, the result in p-coordinates is

$$\frac{\partial}{\partial p}\left(\frac{g}{f}\mathbf{k} \times \nabla z\right) = -\frac{R}{pf}\mathbf{k} \times \nabla T$$

The definition of geostrophic wind can be recognized in the parentheses on the left-hand side. Thus

$$\frac{\partial V_g}{\partial p} = -\frac{R}{fp}\mathbf{k} \times \nabla T \qquad \text{(I-1)}$$

If potential temperature is introduced, using $T = \theta\,(p/p_0)^\kappa$, (I-1) becomes

$$\frac{\partial \mathbf{V}_g}{\partial p} = -\frac{R}{fp}\left(\frac{p}{p_0}\right)^\kappa \mathbf{k} \times \nabla\theta \qquad \text{(I-2)}$$

With a transformation of coordinates in (I-1)

$$\frac{\partial}{\partial p} = -\frac{RT}{gp}\frac{\partial}{\partial z}$$

thermal wind in z coordinates becomes

$$\frac{\partial \mathbf{V}_g}{\partial z} = \frac{g}{fT}\mathbf{k} \times \nabla T \qquad \text{(I-3)}$$

$$\frac{\partial \mathbf{V}_g}{\partial z} = \frac{g}{f\theta}\mathbf{k} \times \nabla\theta \qquad \text{(I-4)}$$

Similarly, thermal wind in isentropic coordinates is

$$\frac{\partial \mathbf{V}_g}{\partial \theta} = \frac{R}{f} \frac{p^{\kappa-1}}{p_0^{\kappa}} \mathbf{k} \times \nabla p \qquad \text{(I-5)}$$

An important property of the preceding four equations is that they relate vertical derivatives of one variable (geostrophic wind) to the horizontal gradient of another variable (temperature). This property makes these equations very useful in several applications.

Each of the equations (I-1) through (I-5) is referred to as the *thermal wind equation*. This designation is also used for the thermal wind of difference type [see the next subsection, Eqs. (I-6) through (I-9)].

The frequent expression $\mathbf{k} \times \nabla T$ should be recognized as a vector that points along the isotherms, with higher values of T on the right-hand side. This is analogous to vector $\mathbf{k} \times \nabla z$ in the expression for geostrophic wind.

Difference Type of Thermal Wind

All five forms of the thermal wind equation (I-1) through (I-5) can be vertically integrated between levels 1 (higher level) and 2 (lower level). Formal steps in this procedure are shown next using (I-1):

$$\int_{p_1}^{p_2} \frac{\partial \mathbf{V}}{\partial p} dp = - \int_{p_1}^{p_2} \frac{R}{p_f} (\mathbf{k} \times \nabla T)_p \, dp$$

A very good approximation, $(\partial \mathbf{V}/\partial p)\, dp \approx d\mathbf{V}$, can be used here. Then the left-hand side can be readily integrated. The constants and temperature T may be taken out of the integral on the right-hand side. This is correct if the temperature is averaged over the pressure interval. Then it follows that

$$\int_{p_1}^{p_2} d\mathbf{V} = -\frac{R}{f} \mathbf{k} \times \nabla \overline{T} \int_{p_1}^{p_2} \frac{dp}{p}$$

$$\mathbf{V}_1 - \mathbf{V}_2 = -\frac{R}{f} \mathbf{k} \times \nabla \overline{T} \ln \frac{p_1}{p_2}$$

$$\mathbf{V}_1 - \mathbf{V}_2 = \frac{R}{f} \mathbf{k} \times \nabla \overline{T} \ln \frac{p_2}{p_1} \qquad \text{(I-6)}$$

In a similar way, this equation in p- and z-systems follows as

$$\mathbf{V}_1 - \mathbf{V}_2 = \frac{gh}{f\overline{T}} \mathbf{k} \times \nabla \overline{T} \qquad \text{(I-7)}$$

$$\mathbf{V}_1 - \mathbf{V}_2 = -\frac{Rp^{\gamma} \Delta p}{f p_0^{\kappa}} \mathbf{k} \times \nabla \overline{\theta} \qquad \text{(I-8)}$$

The variables with the overbar on the right-hand side are the averages for the layer over which the wind difference is taken. The thickness of the corresponding layer is Δp (in the p-system) or h (in the z-system).

A more practical equation for the wind difference can be obtained when the average temperature of the layer is expressed in terms of thickness. This is obtained by integrating the hydrostatic equation and solving it for T:

$$T = \frac{g}{R} \left(\ln \frac{p_2}{p_1} \right)^{-1} h$$

When this is inserted in (I-6), the equation for thermal wind expressed in terms of thickness is

$$\mathbf{V}_{\mathrm{T}} \equiv \mathbf{V}_1 - \mathbf{V}_2 = \frac{g}{f} \mathbf{k} \times \nabla h \qquad \text{(I-9)}$$

or in scalar form

$$u_{\mathrm{T}} = -\frac{g}{f} \frac{\partial h}{\partial y}$$

$$v_{\mathrm{T}} = \frac{g}{f} \frac{\partial h}{\partial x} \qquad \text{(I-10)}$$

Equations (I-9) and (I-10) can be obtained also from the definitions of $\mathbf{V}_g$ from Appendix E [(E.5) and (E.6)] and from the definition of thickness $z_1 - z_2 = h$ (E.3). The sequence of equations here illustrates the equivalence of the derivative and difference types of thermal wind.

Thermal Advection in Terms of Thermal Wind

Another important property of thermal wind is its relation to thickness advection

$$-\mathbf{V} \cdot \nabla h \qquad \text{(I-11)}$$

Here $\mathbf{V}$ is the average wind in the layer over which the thickness h of the layer is evaluated. As a good approximation, $\mathbf{V}$ is the arithmetic average of the wind at the top and bottom of the considered layer:

$$\mathbf{V} = \tfrac{1}{2}(\mathbf{V}_1 + \mathbf{V}_2) \qquad \text{(I-12)}$$

By the thermal wind equation (I-9) or (I-10)

$$\nabla h = -\frac{f}{g} \mathbf{k} \times \nabla T \qquad \text{(I-13)}$$

or, in scalar form,

$$\frac{\partial h}{\partial x} = \frac{f}{g} v_T$$
$$\frac{\partial h}{\partial y} = -\frac{f}{g} u_T$$

When (I-11) and (I-12) are inserted in (I-10), the thermal advection becomes

$$-\mathbf{V} \cdot \nabla h = \frac{f}{g} \mathbf{V} \cdot (\mathbf{k} \times \mathbf{V}_T)$$
$$= \frac{f}{g}\begin{pmatrix} u \\ v \end{pmatrix} \cdot \begin{pmatrix} -uv_T \\ vu_T \end{pmatrix} \qquad \text{(I-14)}$$

In the last equation the following approximations can be used

$$u \equiv \tfrac{1}{2}(u_1 + u_2)$$
$$u_T \equiv u_1 - u_2$$
$$v \equiv \tfrac{1}{2}(v_1 + v_2)$$
$$v_T \equiv v_1 - v_2$$

When these expressions [or (I-9) and (I-12) for the vector form] are introduced in (I-14) and the multiplication is performed, the result is

$$-\mathbf{V} \cdot \nabla h = \frac{f}{g} \mathbf{k} \cdot \mathbf{V}_1 \times \mathbf{V}_2$$
$$= \frac{f}{g}(u_1 v_2 - u_2 v_1) \qquad \text{(I-15)}$$

A geometrical interpretation of this equation is that the turning of wind with height indicates thermal advection: Turning to the right shows warm advection, turning to the left shows cold advection. Various cases are discussed in Section 4-9.

Thermal Vorticity

The operator $\mathbf{k} \cdot \nabla \times$ can be applied to each of equations (I-1) through (I-9). This yields the thermal vorticity. For example, (I-1) becomes

$$\frac{\partial}{\partial p} \mathbf{k} \cdot \nabla \times \mathbf{V}_g = -\frac{R}{fp} \mathbf{k} \cdot \nabla \times (\mathbf{k} \times \nabla T)$$

or

$$\frac{\partial}{\partial p} \zeta = \frac{R}{fp} \nabla^2 T$$

This is the equation of thermal vorticity. It shows that vorticity increases with *height* over cold centers. Anticyclonic (negative in the Northern Hemisphere) vorticity increases with height over warm centers.

Helicity

For an atmospheric layer of thickness Δz, the wind variation with height can be written as

$$u_1 = u_2 + \Delta z \frac{\partial u}{\partial z}$$
$$v_1 = v_2 + \Delta z \frac{\partial v}{\partial z}$$

When these expressions are introduced in (I-14), the thickness advection becomes

$$-\mathbf{V} \cdot \nabla h = \frac{f}{g} \Delta z \left(\mathbf{v} \frac{\partial u}{\partial z} - u \frac{\partial v}{\partial z} \right)$$

If we consider the environment of thunderstorms, the horizontal derivatives $\partial u/\partial x$, $\partial u/\partial y$, $\partial v/\partial x$, and $\partial v/\partial y$ may be ignored in comparison with vertical derivatives $\partial u/\partial z$ and $\partial v/\partial z$. In this case, the last equation may be written as

$$-\mathbf{V} \cdot \nabla h = \frac{f}{g} \Delta z \, \mathbf{V} \cdot \nabla_3 \times \mathbf{V} \qquad \text{(I-16)}$$

where ∇_3 is the three-dimensional del operator. The expression $\mathbf{V} \cdot \nabla_3 \times \mathbf{V}$ is the helicity. It shows the component of wind along the three-dimensional vorticity. As the last equation shows, helicity can also be used as a measure of thermal advection.

As shown in Section 6-5, thermal advection is an indicator of lifting in quasi-geostrophic flow. Therefore, since it is proportional to advection, helicity is also an indicator of lifting for a flow that is in quasi-geostrophic balance.

Sometimes helicity is evaluated for the air flow relative to the moving system. The coordinate system that moves with the thunderstorm environment is often used for this purpose. In this case, the velocity of the moving system $\mathbf{c}$ may be subtracted from the wind. The helicity then becomes

$$(\mathbf{V} - \mathbf{c}) \cdot \nabla_3 \times (\mathbf{V} - \mathbf{c})$$

Further, if this expression is scaled down by its own magnitude, this is the *relative helicity* H that is sometimes used as a measure of likelihood of thunderstorms:

$$H = \frac{(\mathbf{V} - \mathbf{c}) \cdot \nabla_3 \times (\mathbf{V} - \mathbf{c})}{|(\mathbf{V} - \mathbf{c}) \cdot \nabla_3 \times (\mathbf{V} - \mathbf{c})|} \qquad \text{(I-17)}$$

Intersections of Contours and Isotherms

Geostrophic advection can be conveniently estimated by the density of intersections between contours and isotherms. The density of intersections is inversely proportional to the size of elementary quadrangles between isopleths. This is illustrated in Fig. I-1 for several cases when the gradient of temperature ($|\nabla T|$) is constant in the domain and when geostrophic wind (thus also $|\nabla z|$) also is constant throughout the domain. These conditions are illustrated by equal distances between isotherms (distance d) and between contours (distance b). The increments in functions in Fig. I-1 may be selected, for example, as $\Delta T = 4°C$ and $\Delta z = 60$ gpm. The area A of the quadrangle enclosed by two neighboring isotherms and two neighboring contours is

$$A = \frac{bd}{\cos \alpha}$$

The number N of such quadrangles per unit area is

$$N = \frac{C_1}{A} = C_1 \frac{\cos \alpha}{bd}$$

where C_1 is a constant of proportionality. C_1 depends on the gradients $|\nabla T|$ and $|\nabla z|$ as well as on the choice of function increments ΔT and Δz between isopleths. Further similar constants C_2, C_3, and C_4 are introduced below.

The magnitudes of the gradients of geopotential height z and temperature T can be evaluated as

$$|\nabla z| = \frac{\Delta z}{b} \qquad |\nabla T| = \frac{\Delta T}{d}$$

and, therefore, the distances between contours are

$$b = \frac{\Delta z}{|\nabla z|} \qquad d = \frac{\Delta T}{|\nabla T|}$$

With these expressions, the number of quadrangles per unit area can be evaluated as

$$N = C_1 \frac{\Delta z}{|\nabla z|} \frac{\Delta T}{|\nabla T|} \cos \alpha$$

$$= C_2 \, |\nabla z| \, |\nabla T| \cos \alpha$$

where the constant C_2 contains the function increments Δz and ΔT.

Now geostrophic wind may be introduced as

$$|\nabla z| = C_3 \, |\mathbf{V}_g|$$

which, with $\mathbf{V} \approx \mathbf{V}_g$ and with another constant C_4, yields

$$\begin{aligned} N &= C_4 \, |\mathbf{V}| \, |\nabla T| \cos \alpha \\ &= C_4 \, |\mathbf{V} \cdot \nabla T| \\ &= C_4 \, |-\mathbf{V} \cdot \nabla T| \end{aligned}$$

The last expression shows the important conclusion that the density of intersections N (that is, the number of intersections per unit area) is proportional to the magnitude of thermal advection. The same conclusions can be used for advection of other quantities (thickness, humidity, and so forth).

The sign of advection (positive or "warm," negative or "cold") has been ignored in the above procedure. The intention was to demonstrate the magnitude of advection only. The sign of advection can be determined by the rule explained with Fig. 4-14.

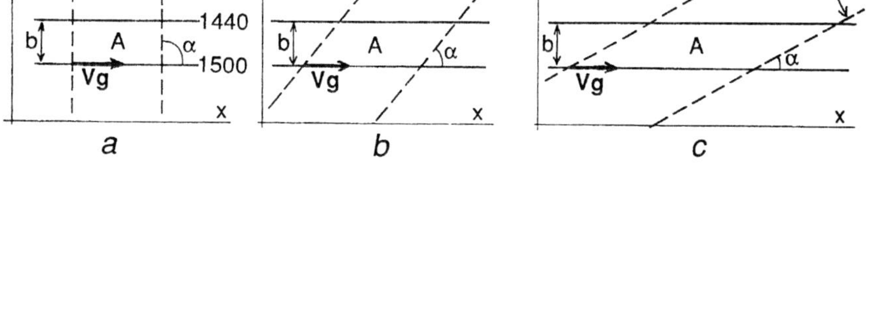

FIGURE I-1. Examples of thermal advection on an 850-mb chart. Contours (solid) are in gpm, isotherms (dashed) are in °C. The distances between isopleths of the same variable (b and d) are equal in all three cases. The angle α varies, and the area A increases as α decreases. The largest advection is in (a), where the contours and isotherms intersect at a right angle.

Appendix J

Physics of Fronts

Frontal Surface

The existence of a stationary sloping interface between air masses of different density can be described using hydrostatic and geostrophic balance. The sloping interface may stay stationary in the presence of the Coriolis force. The geometry of an ideally thin front is shown in Fig. J-1. A more realistic front of finite thickness will be described later.

The discontinuity in Fig. J-1 is along the line AB. There is a property a (for example, temperature, geopotential, or a gradient of these) that is not discontinuous across AB. The existence of a variable that is continuous across the front is the basis for the following formal arguments. As an example, the case of geopotential height will be considered. The geopotential of the isobaric surfaces is equal on both sides on the surface AB, since the space is filled only with air and there is no obstacle to prevent equalization of height and pressure. Temperature, though, may be different on both sides of the surface AB. The difference in temperature across the front is illustrated by isotherms. A simple case with horizontal isotherms is shown in Fig. J-1. There is an abrupt transition in temperature from one side of the front to the other, illustrated by kinks in the isotherms. Parts of isotherms lie in the frontal surface (line AB).

The discontinuity in temperature as shown in Fig. J-1 is a *zero-order* discontinuity, since the zero-order derivative (the function itself) is discontinuous. A *first-order* discontinuity occurs when the temperature is continuous across the front but its gradient is discontinuous. The gradient is equivalent to the first-order space derivative.

Figure J-1 shows a cold air mass under the front and a warm air mass above the front. The comparison between warm and cold is made horizontally (isobarically). Although it is usually cold aloft, this does not make a cold air mass. Only horizontal comparison is used for designating warm and cold air masses.

The variation of a from A to B can be represented mathematically as

$$\Delta a = \left(\frac{\partial a}{\partial x}\right)_c \Delta x + \left(\frac{\partial a}{\partial p}\right)_c \Delta p$$

$$= \left(\frac{\partial a}{\partial p}\right)_w \Delta p + \left(\frac{\partial a}{\partial x}\right)_w \Delta x$$

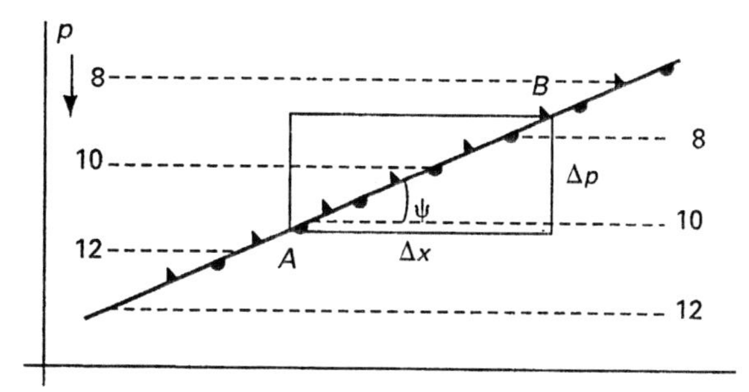

FIGURE J-1. The sloping interface on a vertical section. The front is shown by a line with conventional rounded and pointed tips. Isotherms (dashed, °C) illustrate a zero-order discontinuity at the front. Parts of isotherms are lying within the sloping front.

The increments Δa, Δx, and Δp are taken along the interface, x-direction, and p direction, respectively, all along the diagonal or the sides of the box in Fig. J-1. The quantities on the lower (cold) side are subscripted by c. The quantities on the upper (warm) side are subscripted by w.

The last of the above equalities can be solved for the ratio $\Delta p/\Delta x$ which is the slope s of the surface AB:

$$s = \frac{\Delta p}{\Delta x} = -\frac{\left(\frac{\partial a}{\partial x}\right)_{\mathrm{w}} - \left(\frac{\partial a}{\partial x}\right)_{\mathrm{c}}}{\left(\frac{\partial a}{\partial p}\right)_{\mathrm{w}} - \left(\frac{\partial a}{\partial p}\right)_{\mathrm{c}}} \tag{J-1}$$

The dimension of s is in Pa m^{-1}, or in other units of pressure per distance. If a similar derivation is made in z-coordinates, the slope is nondimensional (in radians).

Equation (J-1), and other forms of this equation that give the expressions for the slope s, is known as the *Margules equation,* after Austrian meteorologist Max Margules (1866–1919), who is also known for his pioneering work in atmospheric energetics and general theoretical meteorology.

The principal benefit of the Margules equation is to clarify our physical ideas about discontinuities in fluids. This equation should not be used to evaluate the slope of the front, which can be found more accurately by an analysis of weather reports. The significance of the Margules equation is similar to the significance of mathematical theorems of existence. These theorems do not give solutions, but give essential information about solutions.

Zero-Order Discontinuity

Zero-order discontinuity occurs when temperature is discontinuous on the front. Physical reasoning is maintained by finding a quantity that *is* continuous across the surface of discontinuity. The following case is for a continuous geopotential height of the isobaric surface z. Therefore, z is introduced for a in (J-1). The pressure derivatives in the denominator can be substituted from the hydrostatic equation [(C.19), Appendix C] as

$$\left(\frac{\partial z}{\partial p}\right)_{\mathrm{w}} = -\left(\frac{RT}{gp}\right)_{\mathrm{w}} \qquad \left(\frac{\partial z}{\partial p}\right)_{\mathrm{c}} = -\left(\frac{RT}{gp}\right)_{\mathrm{c}}$$

$$\left(\frac{\partial z}{\partial p}\right)_{\mathrm{w}} - \left(\frac{\partial z}{\partial p}\right)_{\mathrm{c}} = \frac{R}{gp}(T_{\mathrm{w}} - T_{\mathrm{c}})$$

With these, the slope can be written as

$$s = \frac{gp}{R}\,\frac{\left(\frac{\partial z}{\partial x}\right)_{\mathrm{w}} - \left(\frac{\partial z}{\partial x}\right)_{\mathrm{c}}}{T_{\mathrm{w}} - T_{\mathrm{c}}} \tag{J-2}$$

The values of g, p, R, and $T_{\mathrm{w}} - T_{\mathrm{c}}$ in (J-2) are positive. The algebraic difference

$$\left(\frac{\partial z}{\partial x}\right)_{\mathrm{w}} - \left(\frac{\partial z}{\partial x}\right)_{\mathrm{c}} \tag{J-3}$$

is positive for the slope s illustrated in Fig. J-1. If the sign is changed for $T_{\mathrm{w}} - T_{\mathrm{c}}$, the stratification is unstable (cold over warm). Such an atmosphere cannot persist in a stationary state.

The direction x is arbitrarily selected. In some books, it is chosen in a direction opposite of x in Fig. J-1. In such cases, the x-derivative changes sign. Then the slope is "negative" ($s < 0$) under the same stable physical circumstances.

A positive difference (J-3) is also significant, since it shows a trough on the front. Frequently, the geopotential height z increases *away* from the front on each side giving $(\partial z/\partial x)_{\mathrm{w}} > 0$ and $(\partial z/\partial x)_{\mathrm{c}} < 0$, and the difference (J-3) is positive. If one of the expressions $(\partial z/\partial x)_{\mathrm{w}}$ or $(\partial z/\partial x)_{\mathrm{c}}$ is zero or negligibly small, the other will prevail in the difference (J-3) and there will again be a trough on the front.

When the definition of potential temperature θ [(E-1), Appendix E] is used, the slope can be expressed as

$$s = \frac{gp}{R}\left(\frac{p_0}{p}\right)^{\kappa}\frac{\left(\frac{\partial z}{\partial x}\right)_{\mathrm{w}} - \left(\frac{\partial z}{\partial x}\right)_{\mathrm{c}}}{T_{\mathrm{w}} - T_{\mathrm{c}}} \tag{J-4}$$

Frontal slope is usually not considered in θ-coordinates, since the isentropes are normally concentrated within the frontal layer. In this way, the front is "horizontal" in θ-coordinates. The isentropic coordinates, however, are successfully used to gain insight into the structure of frontal layers.

The slope s in (J-2) and (J-4) is expressed in units of pressure per distance. A similar expression for slope can be derived in z-coordinates. Then, the pressure is the quantity that is equal on both sides of the front ($a = p$). The slope in z-coordinates is a nondimensional angle or its tangent; it is usually shown in books on dynamic meteorology.

For the sake of completeness, the expression for the frontal slope in z-coordinates is listed here, without derivation:

$$\tan \psi = \frac{f\theta}{g} \frac{V_{gw} - V_{gc}}{\theta_w - \theta_c} \quad \text{(J-5)}$$

using the designation ψ for the slope corresponding to the slope s in Fig. J-1.

The frontal slope is highly variable. Its usual magnitude (within a factor of 10) may be $s \approx 10^{-1}$ Pa m^{-1} or about 10^{-2} in z-coordinates.

Another important form of (J-2) is obtained from geostrophic wind [(E-3) from Appendix E] for the slope of the isobaric surfaces:

$$\frac{\partial z}{\partial x} = \frac{f}{g} v_g$$

The result is

$$s = \frac{pf}{R} \frac{v_{gw} - v_{gc}}{T_w - T_c} \quad \text{(J-6)}$$

and, since $T_w - T_c > 0$,

$$s = \text{const.} \times (v_{gw} - v_{gc})$$

This shows a discontinuous wind shear across a front. An inspection of signs shows that this wind shear is cyclonic.

Assuming that the wind [or geostrophic wind, as in (J-6)] is parallel with the front, only the y-component of wind (v_g) is different from zero. Even under these simple conditions, several cases of wind distribution may occur. In each case, the wind on the warm side v_{gw} is algebraically larger than the wind on the cold side v_{gc}. The circumstances are illustrated in Fig. J-2. The wind component v_g in all five cases stays the same. The other component (u_g) is zero. The situation, however, is more realistic when u_g is added. The two examples in Fig. J-3 illustrate the case e from Fig. J-2e with the added constant x-component u_g. The case of $u_g > 0$ is shown in Fig. J-3a, and the case with $u_g < 0$ is shown in Fig. J-3b. The y-component v_g is the same in both cases, as well as in Fig. J-2e. Contours of geopotential are added as thin lines, showing a trough on the front.

Tropopause

In addition to geopotential height, other quantities can be introduced for a in (J-1). For instance, if temperature is assumed equal on both sides of the discontinuity, the Margules equation (J-1) describes the form of isotherms at the tropopause. A simplified case of isotherms at a sloping tropopause is shown in Fig. J-4.

This situation is typical for a troposphere with more or less horizontal isotherms and a stratosphere

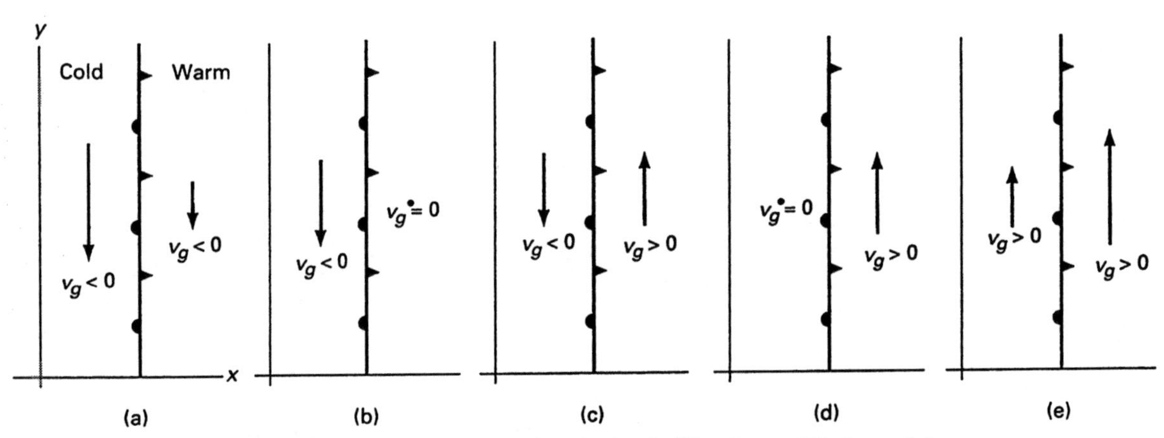

FIGURE J-2. Several cases of an isobaric ("horizontal") view of the front, with the y-component of geostrophic wind (drawn as $\mathbf{j}\ v_g$). Longer vectors show faster wind.

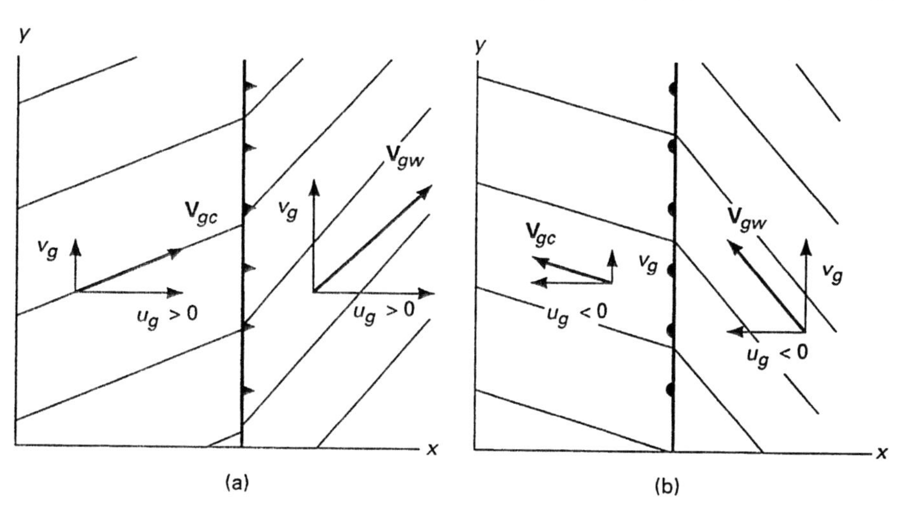

FIGURE J-3. Geostrophic wind distribution at the front. Both examples are based on the y-component (v_g) of wind from Fig. J-2e. The x-component (u_g) is added on both sides of the front: (a) $u_g > 0$, (b) $u_g < 0$. Contours that correspond to geostrophic wind are drawn as thin lines. Kinked contours show a sharp trough on the front in each case.

where the isotherms are often vertical. The "warm" side here is the upper side, the stratosphere. The "cold" side is the lower region, the troposphere. Appropriate substitutions in (J-1) are

$$a = T \qquad \left(\frac{\partial T}{\partial x}\right)_c = 0 \qquad \left(\frac{\partial T}{\partial p}\right)_w = 0$$

With these, (J-1) becomes

$$s = \frac{\Delta p}{\Delta x} = \frac{\left(\frac{\partial T}{\partial x}\right)_w}{\left(\frac{\partial T}{\partial p}\right)_c} \tag{J-7}$$

In a stable atmosphere, $(\partial T/\partial p)_c$ is negative. For a positive slope s, as shown in Fig. J-4, $(\partial T/\partial x)_w$ has to be negative. Thus, the higher segments of the tropopause are colder in a stationary state.

The wind distribution near a sloping tropopause can be inferred from thermal wind [(I-1) from Appendix I]. The second component of thermal wind shows that the y-component of wind v varies with pressure as

$$\frac{\partial v}{\partial p} = \frac{R}{fp}\frac{\partial T}{\partial x} \tag{J-8}$$

Vertical isotherms in the stratosphere (Fig. J-4) imply a significant derivative $\partial T/\partial x$ and, therefore, a significant vertical wind shear. Horizontal isotherms in the troposphere imply no vertical wind shear and vertical isotachs. Horizontal wind shear is strong near the jet stream, and there the tropopause is slanted. This situation is illustrated in Fig. J-5. The horizontal wind shear in the troposphere provides an outlet for the stratospheric isotachs that are kinked on the tropopause and enter the troposphere vertically. Such a bend on the sloping tropopause can be recognized on the equatorial side of the jet stream in practically all vertical sections that contain the jet stream and the tropopause.

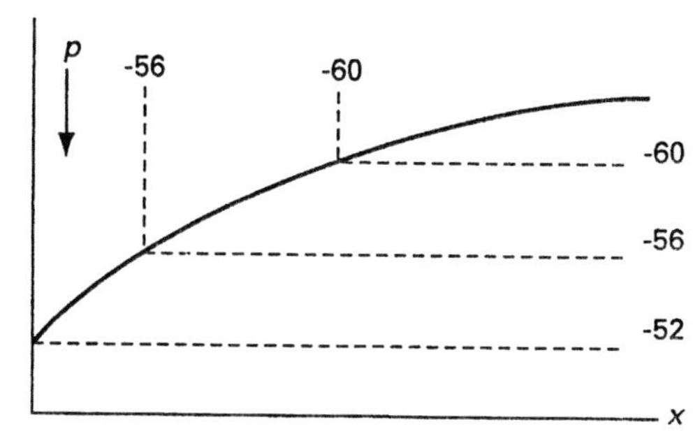

FIGURE J-4. Schematic isotherms (dashed, °C) at a sloping tropopause with $\partial T/\partial x = 0$ in the troposphere and $\partial T/\partial z = 0$ in the stratosphere.

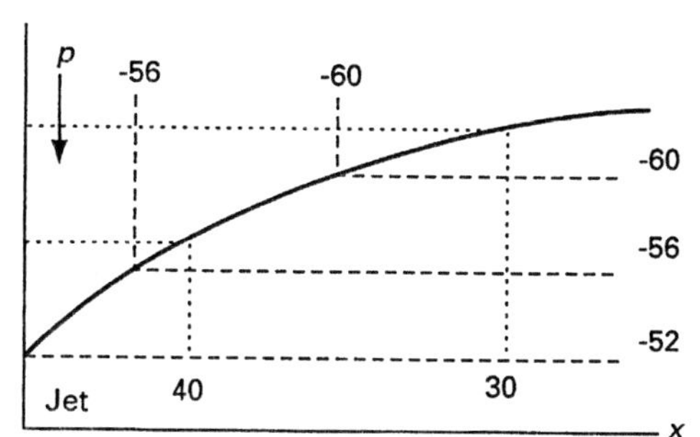

FIGURE J-5. Vertical section with wind speed (dotted, m s^{-1}) at the sloping tropopause from Fig. J-4.

Frontal Layer

A model of isotherms analogous to the tropopause can be found at the bottom of a frontal layer that has an appreciable thickness and isothermal stratification (Fig. J-6). Horizontal isotherms are shown under the frontal layer and vertical isotherms are shown within the layer. The orientation of the x-axis is taken opposite of Fig. J-1, so the shape of this front can be matched with the shape of the tropopause in Figs. J-4 and J-5. Usually, when the tropopause rises "to the right," the front under it slopes opposite (down) in that direction, "to the right."

The top of the layer shows an arrangement of isotherms with another kink. The equations that apply to the top of the frontal layer can be derived from (J-1) with the following substitutions:

$$a = T \qquad \left(\frac{\partial T}{\partial x}\right)_w = 0 \qquad \left(\frac{\partial T}{\partial p}\right)_c = 0$$

Here "warm" is above the front and "cold" is within the layer.

An inversion is commonly within the frontal layer. Then $(\partial T/\partial p)_c < 0$, and this term should be retained in (J-1). The isothermal case is selected here for the sake of brevity, since the inversion case does not alter the conclusions.

With the above assumptions, the slope s follows from (J-1) as

$$s = \frac{\left(\frac{\partial T}{\partial x}\right)_c}{\left(\frac{\partial T}{\partial p}\right)_w} \tag{J-9}$$

As in the previous case, temperature derivatives in (J-9) give information on wind shear as well. The thermal wind relation in (J-8) shows that vertical wind shear is strong in the frontal layer where the horizontal temperature gradient is strong. In Fig. J-6, $\partial T/\partial x < 0$ within frontal layer and, due to (J-8), $\partial v/\partial p > 0$. This increase of v with pressure implies that v decreases (algebraically, paying attention to the sign) with *height*. If this component v is negative on the bottom of the frontal layer, it is larger negative on the top of the layer.

The variation of wind around the front is illustrated in Fig. J-7, where the wind has only one component (v) due to the orientation of the coordinate system. The negative v ($v < 0$) indicates that the wind is going *into* the plane of the drawing. The direction x is selected to the left, such that the diagrams in Figs. J-4 through J-7 match each other. In Figs. J-4 to J-7, the

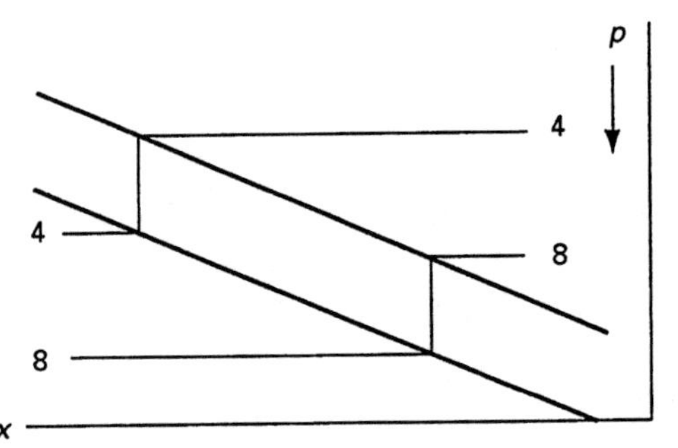

FIGURE J-6. Schematic isotherms in an isothermal frontal layer of finite thickness (vertical section). Isotherms (°C) are solid.

wind is faster on the left-hand side between the front and the tropopause.

An inspection of observed examples of sections through the front and jet stream shows how the jet stream fits between the front and the tropopause. The circled X in Fig. J-7 shows the arrow of the wind vector seen from the rear, giving a degree of three-dimensionality to the figure. The wind is faster on the top of the frontal layer due to the horizontal temperature gradient. This arrangement of isotherms and isotachs shows the very important conclusion that *the wind shear is cyclonic within the frontal layer.* This detail (cyclonic wind shear) is used as a powerful tool in analysis of weather charts.

Figures J-4 through J-7 also explain how wind shear is anticyclonic within the air mass above the front. That is the anticyclonic flank of the jet stream.

Isotachs within the frontal layer in Fig. J-7 are not parallel with the boundaries of the frontal layer. In this way, it is possible to accommodate the horizontal wind shear in warm air. Vertical isotachs in warm air descend from the tropopause and join the frontal layer. Their vertical structure is explained with the schematic tropopause in Fig. J-5.

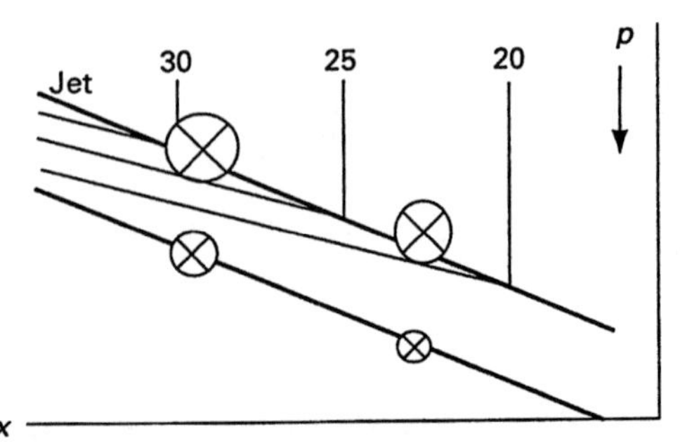

FIGURE J-7. Schematic isotachs (thin, m s^{-1}) at the frontal layer of Fig. J-6.

As Figs. J-5 and J-7 show, wind speed increases with height in the frontal layer and with proximity to the jet stream (drawn on the left-hand side in these figures). Maximum wind speed (core of the jet stream) is attained at the level where the sloping front reaches the tropopause. At that level,

$$\frac{\partial v}{\partial p} = 0$$

which is the mathematical expression for maximum wind in the vertical direction.

Thermal wind (J-7) implies that the temperature contrast vanishes at the jet level: $\partial T/\partial x = 0$. This derivative, if set in finite differences, implies that the thermal difference across the front vanishes as well at the jet level: $T_w = T_c$. From (J-2) it follows that the front is vertical when the temperature difference vanishes. This is commonly observed with jet streams and fronts.

Thickness of Finite Layers

If details are not needed, the frontal layer of finite width (Figs. J-6 and J-7) can be assumed to be very thin. Such a thin surface may occur in the atmosphere for a short time, until mixing of air produces a finite layer. However, if there are no detailed observations and if we want to simplify the analysis, the concept of a thin front is often used.

Mixing at the front is greatly enhanced by the outbreak of shearing instability in layers with high vertical wind shear $|\partial V/\partial p|$. This type of instability is *Kelvin–Helmholtz instability*. From dynamic meteorology it is known that this instability occurs when the Richardson number (*Ri*) decreases under the critical value of 0.25 (or 0.21, as indicated by some measurements). A short discussion of this subject is given in Chapter 3.

A convenient form of the Richardson number can be written in *z*-coordinates as

$$Ri = \frac{g}{\theta}\frac{\frac{\Delta\theta}{\delta}}{\left(\frac{\Delta V}{\delta}\right)^2} = \frac{g}{\theta}\frac{\delta\,\Delta\,\theta}{(\Delta V)^2}$$

where g is the acceleration of gravity, θ is the potential temperature, δ is the thickness of the frontal layer, V is the wind speed, and Δ is the difference operator across a layer where *Ri* is evaluated. Here this is a slanted frontal layer.

The Richardson number must stay greater than 0.25 in an atmosphere that is stable for shearing instability:

$$\frac{g\,\delta\,\Delta\theta}{\theta(\Delta V)^2} > 0.25$$

From this it follows that the thickness δ of the layer is limited to larger values

$$\delta > \frac{0.25\theta(\Delta V)^2}{g\,\Delta\theta}$$

Here $\Delta\theta$ can be eliminated using the Margules equation (J-4) in the form

$$\tan\psi = \frac{f\theta\Delta V}{g\,\Delta\theta}$$

where $\tan\psi$ is the observed frontal slope. The thickness of the frontal layer is then restricted to values larger than the right-hand side of:

$$\delta > \frac{0.25\,\Delta V\tan\psi}{f}$$

For typical values of $\Delta V = 20$ m s^{-1}, $\tan\psi = 10^{-2}$, $f = 10^{-4}$ s^{-1}, this expression yields $\delta > 500$ m.

A physical explanation for the described circumstances is as follows: First the strong shear causes turbulence and mixing. Next, this mixing partly equalizes the momentum through the mixed layer and thereby reduces shear. When a sufficiently thick mixed layer is established, the shear in it is reduced below the critical value of *Ri* and turbulence ceases. The mixed layer has achieved a stable finite width.

Exact values of *Ri* cannot be found on the basis of routine radiosonde observations. The relative sparsity of observations in the vertical direction often shows that turbulence breaks out when *Ri* values of 1 or lower are evaluated. This is due to the truncation error in computation with sparse data.

Appendix K

Passive Fronts

Model

A passive cold front moves under its weight. It is described in hydrodynamics as a *gravity current* or *density current*. Pressure differences around such a front are greatly determined by the distribution of density and wind. Dynamic influences from the jet-stream level are negligible.

The following example will consider the case where the density ρ is constant within each air mass: ρ_c in cold air and ρ_w in warm air. The vertical distribution of air masses is shown in Fig. K-1. The vertical decrease of density need not be considered since it does not contribute to the argument. The case of no vertical variation of density applies correctly for water, but the conclusions are the same for the atmosphere. The shape of a front does not necessarily look like that of Fig. K-1. A front frequently takes the shape shown in Fig. 7-28.

Hydrostatic Pressure in Cold Air

The example in Fig. K-1 assumes a very steep lower part of a front. The top of cold air rises to about 2 km and spreads horizontally in the distance. There is a horizontal variation of hydrostatic pressure within cold air that can be determined from the difference in density and from the depth of the air mass. This is illustrated by the 950-mb isobar that slopes in cold air. The excess surface hydrostatic pressure p in cold air can be evaluated by an integration of the hydrostatic equation [(C-19), Appendix C] from a level where pressure does not vary horizontally to the surface. In this example, integration is from 750 mb to the surface. The integration is performed twice: once in warm air and the other time through cold air. The integration through cold air goes in two segments, above and under the front, since there is a discontinuity at the front.

$$p_c = p_{750} + \rho_w g\,(H_{700} - H) + \rho_c g H$$
$$p_w = p_{750} + \rho_w g H_{700}$$

The indices c and w indicate cold and warm air, respectively, and H is the height of the front. A subtraction of the last two equations yields the pressure in cold air as

$$p_c = p_w + gH(\rho_c - \rho_w)$$

This distribution of hydrostatic pressure along the x-coordinate can be written as

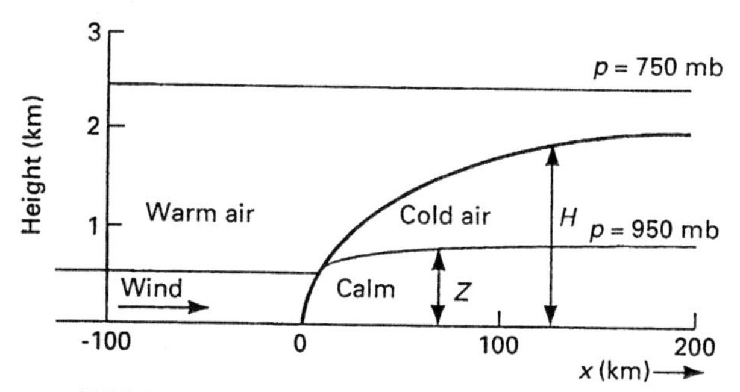

FIGURE K-1. A passive front. The geopotential gradient at 750 mb is assumed to be zero. The hydrostatic part of pressure varies with the depth of cold air, as shown by the isobar of 950 mb.

$$p_c(x) = p_w + gH(x)(\rho_c - \rho_w) \qquad \text{(K-1)}$$

where it is stressed that p_c and H vary in the x-direction.

It is useful to express the pressure variation in terms of the height of a relevant isobaric surface. Using the standard hydrostatic transformation from the z- to p-system in the form

$$\left(\frac{\partial p}{\partial x}\right)_z = g\rho\left(\frac{\partial z}{\partial x}\right)_p$$

and the integrated form

$$p(x) = g\rho z(x) + \text{const.}$$

there follows

$$z(x) = \frac{1}{g\rho}\, p(x) + \text{const.}$$

When $p(x)$ from (K-1) is inserted,

$$z(x) = H(x)\,\frac{\rho_c - \rho_w}{\rho_c} + \text{const.} \qquad \text{(K-2)}$$

The constant of integration (const.) determines the selection of the pressure level. The height z in the last equation can be used to compute or estimate the hydrostatic part of the pressure force $\sim g(\partial z/\partial x)$ when the height of the front H is known.

Dynamic Pressure in Warm Air

The assumption so far was that cold air was stationary. The vertical part of a front near the surface implies that the hydrostatic pressure abruptly increases on the cold side of the front. This situation can remain stationary if a corresponding buildup of pressure occurs on the warm side as well. Since the warm air mass is homogeneous in this model, there cannot be a hydrostatic horizontal pressure variation within warm air. However, a dynamic increase of pressure may occur on the warm side. At a point far from the front, the wind component toward the front may be u (illustrated by the wind vector in Fig. K-1). At the front itself, this wind component is reduced to zero. The air is deflected to stream above or around the front. This situation can be described by the first equation of motion [(E-22), Appendix E] in the form

$$\frac{du}{dt} = -g\,\frac{\partial z}{\partial x} \qquad \text{(K-3)}$$

This form, without the y-component v, is justified since the y-component of wind v may be equal to zero. The variable z on the right-hand side is the height of the isobaric surface (not the height of the front).

The left-hand side can be rewritten as

$$\frac{du}{dt} = \frac{du}{dx}\frac{dx}{dt} = u\,\frac{du}{dx} = \frac{d}{dx}\frac{u^2}{2} \qquad \text{(K-4)}$$

An alternative to this transformation is to use the Eulerian development of the total derivative

$$\frac{du}{dt} = \frac{\partial u}{\partial t} + u\,\frac{\partial u}{\partial x} + v\,\frac{\partial u}{\partial y} + w\,\frac{\partial u}{\partial z}$$

and to apply the assumption of stationarity ($\partial u/\partial t = 0$), to use $v = 0$ and to ignore the vertical advection of momentum $w(\partial u/\partial z)$. This reduces the right-hand side of the last equation to

$$= \frac{\partial}{\partial x}\left(\frac{u^2}{2}\right)$$

and since x is the only remaining coordinate, the total derivative with respect to x may be used:

$$= \frac{d}{dx}\left(\frac{u^2}{2}\right)$$

which is the same as the above result (K-4).

With this, the equation of motion (K-3) becomes

$$\frac{d}{dx}\frac{u^2}{2} = -\frac{d}{dx}(gz)$$

Integration of this equation from $-\infty$ to the front ($x = 0$) yields

$$\int_{-\infty}^{0} \frac{d}{dx}\frac{u^2}{2}\,dx = \int_{-\infty}^{0} \frac{d}{dx}(gz)\,dx$$

$$0 - \left(\frac{u^2}{2}\right)_{-\infty} = -g(z_0 - z_{-\infty})$$

Here it was assumed that the normal component of wind in warm air (u) is equal to zero at the front. The last equation can be solved for z_0 which shows the height increase (like the pressure increase in the z-system) in warm air immediately at the front:

$$z_0 = z_i + \frac{u_i^2}{2g} \qquad \text{(K-5)}$$

The notation introduced is u_i for "wind at infinity" and z_i for "geopotential height at infinity." The last equation shows that the height of an isobaric surface near the earth's surface increases to a higher value z_0 at the front, on the warm side. This is the *dynamic pressure* at an obstacle, also called *wind pressure*. It applies to other scales of flow as well (for example, on the side of a grass blade or on the side of a mountain). In the case of a stationary front, this pressure keeps the balance against the hydrostatic pressure on the cold side of the stationary passive front.

Limitations of the Theory

The above simplified theory does not give an explanation about how the speed u and the height of an isobaric surface z_0 vary between "infinity" and the stagnation point at a front. This variation depends on the shape and lateral extent of the front. Under usual circumstances, a wind speed 10–20 km from a front may play the role of wind at infinity. Wind at a greater distance has too small effect for such a front. Most variation of wind takes place only in the nearest kilometer from the front. The accompanying increase of pressure in warm air occurs so close to the front that the common notion is that the pressure increases only in cold air. Also, the accompanying convergence zone in warm air is practically reduced to a convergence line. The narrow convergence zone causes a narrow lifting line, and this lifting is often visible by the formation of a narrow band of cumulus (*narrow frontal cloud band*).

Moving Passive Front

Another important case occurs in nature: a passive front that advances against calm warm air. Warm air escapes mostly vertically out of the way of a front. This case may be considered in a coordinate system that moves with wind speed u_i, such that warm air at "infinity" does not move in this new coordinate system. The situation is equal to the case in Fig. K-1, except that the front moves with speed u_i against stagnant warm air. The physics are the same as in the previous case. The dynamic increase of geopotential height on the warm side of the front (K-5) is the same, only now it can be explained as a compression before the moving body of cold air. Therefore, (K-5) can be solved for u_i, which is now the speed of the front:

$$u_i = \sqrt{2g(z_0 - z_i)}$$

The difference $z_0 - z_i$ in warm air describes the dynamic pressure increase within warm air. It is balanced by the hydrostatic height increase in cold air. Therefore, $z_0 - z_i$ can be taken as equal to the hydrostatic pressure difference within the cold air:

$$z_0 - z_i = z_1 - z_f$$

This pressure difference in the cold air $z_1 - z_f$ is expressed in terms of the height of the isobaric surface z_1 at the "top" of the cold air mass and at the front z_f. Further, the hydrostatic height at the front z_f may conveniently be taken as zero. The height z_1 at high values of H is, as in (K-2),

$$z_1 = z_1 - z_f = H\frac{\rho_c - \rho_w}{\rho_c}$$

Therefore the height difference in warm air becomes

$$z_0 - z_i = H\frac{\rho_c - \rho_w}{\rho_c}$$

and the speed of the front becomes

$$u_i = \sqrt{2gH\frac{\rho_c - \rho_w}{\rho_c}} \qquad \text{(K-6)}$$

which gives the speed of a passive cold front against stagnant warm air.

While the above simplified theory gives us an understanding about front movement, this formula cannot be generally used to evaluate the speed of a front or even to predict it. The observations about height H and density are normally not sufficiently accurate to evaluate speed. In this sense, the meaning of the equation (K-6) should be understood similarly to the meaning of the Margules formula for the frontal slope: The purpose of (K-6) is to generalize our concepts about processes in the atmosphere. This equation cannot be used in routine weather analysis.

Other Cases of Gravity Current

The computed speed of front propagation applies also to other similar phenomena in fluids. Cold air behind the outflow boundary near thunderstorms spreads under such dynamics. A remarkable phenomenon of this type is a flash flood, when a flood wave rushes along a valley. In this case, the density ratio $(\rho_c - \rho_w)/\rho_c$ is about 1, since ρ_c must be interpreted as the density of water and ρ_w as the density of air. The other quantities are similarly interpreted: H is the depth of water and g is the acceleration of gravity. For a water depth of, say, 5 m, the above formula yields a flood propagation speed of 10 m s^{-1}.

The two considered cases (stationary front and stagnant warm air) are not the only possibilities; other speeds of the coordinate system may be considered as well. Many such cases have been observed in nature. Each time there may be a different partition of u_i between the wind in warm air and the movement of the front.

The above theory applies to the spreading of dense fluid over a horizontal surface. If the underlying surface is inclined, additional accelerations may become important. For example, a flash flood may rush faster in a steep valley than over a horizontal surface. Such a flood will assume a steady speed as well (albeit greater than on flat terrain), since the friction at the bottom increases with the square of water speed.

Appendix L

Calculation of Buoyant Energy

Numerical Evaluation of Definite Integrals

The vertical integral of temperature in (5-25) gives the energy (per unit mass) accumulated by buoyant air particles:

$$E = R \int_{p_1}^{p_2} [T_l(p) - T_a(p)]\, d \ln p \qquad \text{(L-1)}$$

This may serve as an example for estimation of a large class of integrals. The suggested method consists of estimating (or calculating) the average value of the argument $[T_l(p) - T_a(p)]$. This is fairly precise if the argument $[T_l(p) - T_a(p)]$ is evaluated at several equally spaced points and the arithmetic average is calculated. The average, let it be $\overline{\Delta T}$, is a constant. Therefore, it can be taken out of the integral:

$$E \approx R\, \overline{\Delta T} \int_{p_1}^{p_2} d \ln p$$

$$= R\, \overline{\Delta T} \ln \frac{p_2}{p_1} \qquad \text{(L-2)}$$

As an example, this integral will be evaluated for the example of positive energy between 750 and 255 mb in Fig. 5-5. These values are the limits of integration: $p_1 = 255$ mb and $p_2 = 750$ mb. Work with millibars is acceptable here, since only the ratio of pressure values is needed. Other quantities should be in the SI. The average $\overline{\Delta T} = \overline{T_l - T_a}$ can be found, for example, from the values at every 50 mb. In Fig. 5-5 we can read off the values of $T_l - T_a$ at indicated pressure levels; as shown in Table L-1

Addition and division by 9 gives the average $\overline{\Delta T} = 5.4$ K. Then the value of the integral (2) can be estimated as

$$E \approx 287 \text{ m}^2 \text{ s}^{-2} \text{ K}^{-1} \times 5.4 \text{ K} \times \ln \frac{750}{255}$$

$$\approx 1672 \text{ J kg}^{-1}$$

Methods of numerical integration that are much more sophisticated can be found in mathematical books. The above example is only a quick method for an estimate that is better than visual inspection.

TABLE L-1

An example of evaluation of ΔT from the sounding in Fig. 5-5

Pressure (mb):	300	350	400	450	500	550	600	650	700
$T_l - T_a$ (K):	2.7	5.1	3.7	7.2	7.7	8.5	6.2	5.7	1.8

Estimation of Overshooting Cumulonimbus Tops

Buoyancy considerations around the tropopause are considered in Fig. L-1. This is a thermodynamic diagram with coordinates z and T. The ambient temperature (T_a, solid uneven line) is approximately constant in the stratosphere. The lifted temperature is drawn for a parcel that arrived to the equilibrium level EL with significant vertical motion w. The parcel lifts along the dry adiabat (T_l, θ = const., dashed) and continues lifting through the stratosphere until the peak level $z = h$ is reached. At this level the parcel has consumed all the buoyant energy it accumulated earlier in the troposphere.

Negative buoyant energy above EL can be evaluated by (5-25) or (L-1). It is practical to adapt these equations as shown in (5-35):

$$E = g \int_0^h \frac{\Delta T}{T} dz$$

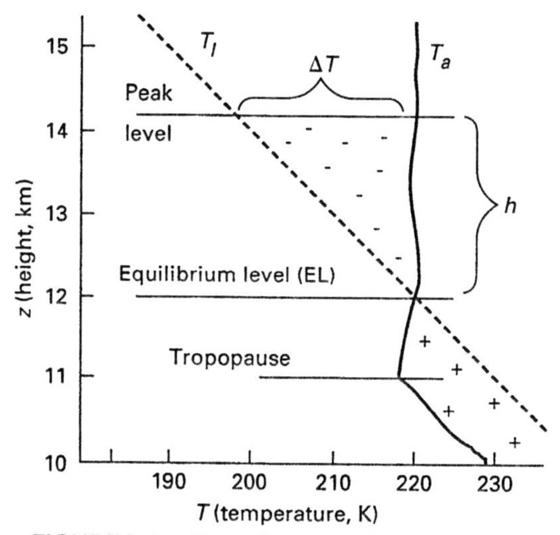

FIGURE L-1. Part of a sounding near the tropopause on a z, T thermodynamic diagram. The ambient temperature T_a (heavy uneven line) is approximately constant in the stratosphere. The lifted temperature T_l (dashed) is for a parcel from the lower troposphere. The parcel reaches the peak level when it expends all kinetic energy.

where the height is measured from EL ($z = 0$) to the peak level h.

For estimation purposes it is acceptable to use a constant value of T in the denominator. The other part of the argument (ΔT) can be computed as the difference between the lifted temperature and ambient temperature:

$$\Delta T = T_l - T_a$$

Lifted temperature follows from the first law of thermodynamics (adiabatic form):

$$0 = c_p\, dT - \alpha\, dp$$

Using the hydrostatic equation, this yields

$$c_p\, dT = -g\, dz$$

Integration between levels of EL (where $T = T_{EL}$ and $z = 0$) and any level z gives the lifted temperature as

$$T_l = T_{EL} - \frac{g}{cp} z$$

Now the temperature difference becomes

$$\Delta T = T_{EL} - \frac{g}{c_p} z - T_a$$

Since the ambient temperature T_a is constant and equal to T_{EL},

$$\Delta T = -\frac{g}{c_p} z$$

With this, the buoyancy energy amounts to

$$E = \frac{g}{T} \int_0^h \left(-\frac{g}{c_p} z\right) dz$$
$$= \frac{g}{T}\left(-\frac{g}{c_p}\right) \int_0^h d\frac{z^2}{2}$$
$$= -\frac{g^2 h^2}{2Tc_p}$$

The negative sign shows the negative buoyancy energy.

The parcel may rise through the equilibrium level with speed w. Its kinetic energy (per unit mass) is $\frac{1}{2}w^2$. Equating this energy with the absolute value of buoyant energy gives

$$\frac{g^2h^2}{2Tc_p} = \frac{1}{2}w^2$$

$$h = w\sqrt{\frac{Tc_p}{g^2}}$$

Using approximate numbers, the value of the square root is

$$\sqrt{\frac{Tc_p}{g^2}} \approx \sqrt{\frac{220\text{ K } 1004\text{ m}^2\text{s}^{-2}\text{K}^{-1}}{(9.8\text{ m s}^{-2})^2}} \approx 50\text{ s}$$

Therefore, the peak height achieved by a parcel is

$$h = w \times 50\text{ s}$$

A parcel that lifts at 50 m s^{-1} at the EL will reach the peak at 2500 m above EL.

Appendix M

Omega Equation in Terms of the Q Vector

The classical omega equation (6-20) was first derived by J. Charney and by B. Bolin in mid-1950s. It took more than 20 years until Hoskins, Draghici, and Davies (1978) rederived the old omega equation in the practical form shown below. They stress the distinction between terms that can be evaluated using geostrophic approximation and terms that are essentially ageostrophic. It is assumed that geostrophic approximation can be used only if the usual inadequacies in the data do not obscure the interesting processes. Thus vertical advection is ignored, since it cannot be evaluated with sufficient accuracy if geostrophic approximation is used. The other terms in the Eulerian expansion of the total derivative can be evaluated with the usual data from weather charts. Since the retained total derivative terms express the derivative of momentum, the following equations of motion are said to be in *geostrophic momentum* form.

The partial derivatives are shown below by indices that are coordinates (x, y, p, t). For the time being, an f-plane is considered, with $f =$ const. This approximation is justified for synoptic scale waves that are shorter than Rossby waves. The variation of the Coriolis parameter f will be mentioned below, with Trenberth's form of the omega equation.

$$\frac{\partial u}{\partial t} + \mathbf{V} \cdot \nabla u - f\,v_{ag} = 0 \;\Big|\; \frac{\partial}{\partial p}$$

$$\frac{\partial}{\partial t}\frac{\partial u}{\partial p} + u\frac{\partial}{\partial x}\frac{\partial u}{\partial p} + \frac{\partial u}{\partial p}\frac{\partial u}{\partial x} + v\frac{\partial}{\partial y}\frac{\partial u}{\partial p} + \frac{\partial v}{\partial p}\frac{\partial u}{\partial y} - f\frac{\partial}{\partial p}v_{ag} = 0$$

The scalar form of the thermal wind equation is:

$$\frac{\partial u}{\partial p} = \frac{h}{f}\theta_y \qquad \frac{\partial v}{\partial p} = -\frac{h}{f}\theta_x \qquad h \equiv \frac{R}{p}\left(\frac{p}{p_0}\right)^{\kappa}$$

Inserting the thermal wind equation into the preceding equation of motion yields

$$\frac{\partial}{\partial t}\theta_y + u\theta_{xy} + u_x\theta_y + v\theta_{yy} - \theta_x u_y - \frac{f^2}{h}\frac{\partial}{\partial p}v_{ag} = 0$$

We may use geostrophic continuity equation $u_x + v_y = 0$ in the third term with the result

$$\frac{\partial}{\partial t}\theta_y + u\theta_{xy} - v_y\theta_y + v\theta_{yy} - \theta_x u_y - \frac{f^2}{h}\frac{\partial}{\partial p}v_{ag} = 0 \qquad \text{(M-1)}$$

The adiabatic form of the first law of thermodynamics (the adiabatic equation) can be used with a constant stability θ_p:

$$\frac{\partial}{\partial t}\theta + u\theta_x + v\theta_y + \omega\theta_p = 0 \left| \frac{\partial}{\partial y} \right.$$

$$\frac{\partial}{\partial t}\theta_y + u_y\theta_x + u\theta_{xy} + v_y\theta_y + v\theta_{yy} + \omega_y\theta_p = 0 \qquad \text{(M-2)}$$

The terms $v_y\theta_y$ and $u_y\theta_x$ have opposite signs between (M-1) and (M-2), while several other terms have the same signs in the two equations. Assuming for a moment that the ageostrophic terms $(f^2/h)(\partial/\partial p)v_{ag}$ and $\omega_y\theta_p$ from (M-1) and (M-2) are canceled, the terms $v_y\theta_y$ and $u_y\theta_x$ are the only ones that are left when (M-1) and (M-2) are subtracted from each other. We may subtract the complete equations (M-1) and (M-2), canceling the time derivatives. The resulting equation is of diagnostic character (without time derivatives):

$$2u_y\theta_x + 2v_y\theta_y + \omega_y\theta_p + \frac{f^2}{h}\frac{\partial}{\partial p}v_{ag} = 0$$

For convenience, we define the second component of the **Q** vector as

$$Q_2 = -u_y\theta_x - v_y\theta_y = -\frac{\partial \mathbf{V}}{\partial y}\cdot\nabla\theta \qquad \text{(M-3)}$$

so that the preceding equation can be written as

$$-2Q_2 + \omega_y\theta_p + \frac{f^2}{h}\frac{\partial}{\partial p}v_{ag} = 0 \qquad \text{(M-4)}$$

Similarly, we have the following sequence of the second equation of motion and the adiabatic equation, the latter now derived after x:

$$\frac{\partial v}{\partial t} + \mathbf{V}\cdot\nabla v + fu_{ag} = 0 \left| \frac{\partial}{\partial p} \right.$$

$$\frac{\partial}{\partial t}\frac{\partial v}{\partial p} + \frac{\partial u}{\partial p}\frac{\partial v}{\partial x} + \frac{\partial v}{\partial p}\frac{\partial v}{\partial y} + u\frac{\partial}{\partial x}\frac{\partial v}{\partial p} + v\frac{\partial}{\partial y}\frac{\partial v}{\partial p} + f\frac{\partial}{\partial p}u_{ag} = 0$$

As above, with thermal wind and geostrophic continuity:

$$\frac{\partial}{\partial t}\theta_x + u\theta_{xx} - v_x\theta_y + v\theta_{xy} - \theta_x u_x - \frac{f^2}{h}\frac{\partial}{\partial p}u_{ag} = 0 \qquad \text{(M-5)}$$

When the adiabatic equation is derived after x, then

$$\frac{\partial}{\partial t}\theta_x + u_x\theta_x + u\theta_{xx} + v_x\theta_y + v\theta_{xy} + \omega_x\theta_p = 0 \qquad \text{(M-6)}$$

Subtraction of the last two equations yields

$$2u_x\theta_x + 2v_x\theta_y + \omega_x\theta_p + \frac{f^2}{h}\frac{\partial}{\partial p}u_{ag} = 0$$

Again, for the sake of convenience we define another component of the **Q** vector (this one is the first, along x) as

$$Q_1 = -u_x\theta_x - v_x\theta_y = -\frac{\partial \mathbf{V}}{\partial x}\cdot\nabla\theta \qquad \text{(M-7)}$$

and we have another diagnostic equation:

$$-2Q_1 + \omega_x\theta_p + \frac{f^2}{h}\frac{\partial}{\partial p}u_{ag} = 0 \qquad \text{(M-8)}$$

The next step is analogous to the derivation of the vorticity equation. The second equation of motion is derived after x, the first after y, and the two equations are subtracted. Taking $\partial/\partial x$ of (M-8) and $\partial/\partial y$ of (M-4) and subtracting them [we actually add them; the change of sign in one of the equations occurred in (M-5)], yields

$$-2\frac{\partial Q_1}{\partial x} - 2\frac{\partial Q_2}{\partial y} + \theta_p(\omega_{xx} + \omega_{yy}) + \frac{f^2}{h}\frac{\partial}{\partial p}\left(\frac{\partial}{\partial x}u_{ag} + \frac{\partial}{\partial y}v_{ag}\right) = 0$$

The continuity equation for the ageostrophic flow is

$$\frac{\partial}{\partial x}u_{ag} + \frac{\partial}{\partial y}v_{ag} = -\frac{\partial\omega}{\partial p}$$

This shortens the last term in the preceding equation. Using vector notation, we continue as follows:

$$-2\,\nabla\cdot\mathbf{Q} + \theta_p\,\nabla^2\omega - \frac{f^2}{h}\frac{\partial^2\omega}{\partial p^2} = 0$$

The vector $\mathbf{Q}$ is defined as

$$\mathbf{Q} = \begin{pmatrix} Q_1 \\ Q_2 \end{pmatrix}$$

and the components Q_1, Q_2 are defined by (M-7) and (M-3). Since θ_p is normally negative, the stability may be defined as a usually positive quantity:

$$\sigma = -h\theta_p$$

With this, the omega equation is conveniently written as

$$\sigma \nabla^2\omega + f^2 \omega_{pp} = -2\, h\nabla \cdot \mathbf{Q} \qquad \text{(M-9)}$$

This equation can be converted to z-coordinates using

$$\omega \approx -g\rho w \qquad \nabla^2\omega \approx -g\rho\, \nabla^2 w$$

$$\frac{\partial}{\partial p} = -\frac{1}{g\rho}\frac{\partial}{\partial z} \qquad \frac{\partial^2}{\partial p^2} = \frac{1}{g^2\rho^2}\frac{\partial^2}{\partial z^2}$$

Returning to the omega equation,

$$-g\rho\sigma\, \nabla^2 w - \frac{f^2}{g\rho} w_{zz} = -2h\, \nabla \cdot \mathbf{Q}$$

$$\sigma\rho^2 g^2\, \nabla^2 w + f^2 w_{zz} = -2hg\rho\, \nabla \cdot \mathbf{Q}$$

The definition of the Brunt–Väisälä frequency N,

$$N^2 = \frac{g}{\theta}\frac{d\theta}{dz} \approx -\frac{\rho g^2}{\theta}\frac{d\theta}{dp} = \sigma\rho^2 g^2$$

can be used to obtain

$$N^2\, \nabla^2 w + f^2 w_{zz} = 2\,\frac{g}{\theta}\, \nabla \cdot \mathbf{Q} \qquad \text{(M-10)}$$

$$\mathbf{Q} \equiv -\begin{pmatrix} u_x\theta_x + v_x\theta_y \\ u_y\theta_x + v_y\theta_y \end{pmatrix}$$

Mathematically, it is important that the coefficients N^2 and f^2 with the second derivatives of w both stay positive in (M-10). Only in this way is (M-10) elliptic and w real. There is no problem with f^2; this is always positive. The problem arises in an unstable atmosphere. In that case, the Brunt–Väisälä frequency N is imaginary and its square is negative. The above geostrophic and hydrostatic theory applies only in cyclone-scale and large-scale processes where hydrostatic and geostrophic considerations give good results.

We may get a good physical insight into the solution of elliptic equations such as (M-9) and (M-10) if we use the approximate spectral proportionality

$$w \propto -\nabla^2 w$$

Then the omega equation (M-10) assumes a particularly short form:

$$w \propto -\nabla \cdot \mathbf{Q} \qquad \text{(M-11)}$$

From this we see that lifting is proportional to convergence of $\mathbf{Q}$ (div $\mathbf{Q} < 0$). Conversely, the divergence of $\mathbf{Q}$ (div $\mathbf{Q} > 0$) signifies large-scale subsidence.

The name of the omega equation comes from (M-9), where ω is the variable that we need to evaluate.

Several instructive examples of application of the $\mathbf{Q}$ vector are given by Hoskins and Pedder (1980).

Trenberth's Form

The classical omega equation (Holton, 1992, p. 167) has two terms on the right-hand side:

$$\sigma\nabla^2\omega + f^2\omega_{pp} = f\frac{\partial}{\partial p}[\mathbf{V}\cdot\nabla(\zeta + f)] + h\,\nabla^2(\mathbf{V}\cdot\nabla\theta) \qquad \text{(M-12)}$$

giving the conclusion that lifting occurs in the regions of positive advections of absolute vorticity and temperature. Criticism of this equation points out that significant parts of the two terms on the right-hand side cancel. Therefore, the equation is too sensitive for estimates of vertical motion. This is acceptable for computer programs that accurately carry several digits for all quantities; however, it is not suitable for manual work when visual estimates are needed. Cancellation of parts of the terms in (M-12) has been achieved in the places where the time derivatives were eliminated after (M-1) and (M-2) and after (M-5) and (M-6). Similar cancellation is achieved when (M-12) is rewritten as follows:

$$\sigma\nabla^2\omega + f^2\omega_{pp} =$$

$$f\left[\frac{\partial\mathbf{V}}{\partial p}\cdot\nabla(\zeta + f) + \mathbf{V}\cdot\nabla\frac{\partial\zeta}{\partial p} + \frac{h}{f}\mathbf{V}\cdot\nabla(\nabla^2\theta) + \frac{h}{f}\nabla\theta\cdot\nabla^2\mathbf{V} + 2\frac{h}{f}(\nabla u\cdot\nabla\theta_x) + \nabla v\cdot\nabla\theta_y\right] \qquad \text{(M-13)}$$

Further formalities include the thermal wind, thermal vorticity, and geostrophic continuity equations:

$$\theta_x = -\frac{f}{h} v_p \qquad \theta_y = \frac{f}{h} u_p \qquad \nabla\theta = \frac{f}{h} \mathbf{k} \times \mathbf{V}_p$$

$$\nabla^2\theta = -\frac{f}{h}\frac{\partial\zeta}{\partial p} \qquad u_x + v_y = 0$$

The term with $\nabla(\nabla^2\theta)$ cancels the term with $\partial\zeta/\partial p$ in view of the thermal vorticity. This significantly shortens equation (M-13). The term with $\nabla^2\mathbf{V}$ can be transformed as follows:

$$\nabla\theta \cdot \nabla^2\mathbf{V} = \theta_x \nabla^2 u + \theta_y \nabla^2 v = \nabla^2 u \frac{f}{h}(-v_p) + \nabla^2 v \frac{f}{h} u_p$$

$$= \frac{f}{h}(-v_p u_{xx} - v_p u_{yy} + u_p v_{xx} + u_p v_{yy})$$

$$= \frac{f}{h}\left\{-v_p\left[\frac{\partial}{\partial x}(-v_y) - \frac{\partial}{\partial y} u_y\right] + u_p\left[\frac{\partial}{\partial x} v_x + \frac{\partial}{\partial y}(-u_x)\right]\right\}$$

$$= \frac{f}{h}\left[v_p \frac{\partial}{\partial y}(v_x - u_y) + u_p \frac{\partial}{\partial x}(v_x - u_y)\right]$$

$$= \frac{f}{h}\left(v_p \frac{\partial\zeta}{\partial y} + u_p \frac{\partial\zeta}{\partial x}\right) = \frac{f}{h}\frac{\partial\mathbf{V}}{\partial p} \cdot \nabla\zeta$$

This term adds with the first term on the right-hand side of (M-13). Based on observations, Trenberth (1978) showed that the terms in the last parentheses of (M-13) are smaller than other terms in usual weather situations. These terms depend on the second derivatives of potential temperature. Therefore, when those terms are ignored we find:

$$\sigma\nabla^2 \omega + f^2\omega_{pp} = 2f\frac{\partial\mathbf{V}}{\partial p} \cdot \nabla\zeta + f\frac{\partial\mathbf{V}}{\partial p} \cdot \nabla f \quad \text{(M-14)}$$

If there was a multiplier 2 in the last term, or if the number 2 was missing in the first term on the right-hand side, the forcing function could be interpreted as being proportional to the advection of absolute vorticity $\zeta + f$ with the thermal wind $\partial\mathbf{V}/\partial p$:

$$\sigma\nabla^2\omega + f^2\omega_{pp} \propto \frac{\partial\mathbf{V}}{\partial p} \cdot \nabla(\zeta + f)$$

However, even if this is not the case, the form (M-14) is very useful. It shows that the advection of relative vorticity plays a greater role than the advection of f, since the advection of z is multiplied by 2 and the advection of f is not. The term $\nabla\zeta$ is normally larger than ∇f, even if f is larger than ζ. This gives us further justification for neglecting the beta term, following the above approach of Hoskins, Draghici, and Davies. Therefore, we are justified in keeping only the advection of ζ in visual estimates of vertical motion.

Using the same simplifications as before, as well as switching from ω to w yields

$$\nabla^2\omega \propto -\omega \qquad w \propto -\omega \qquad \frac{\partial\mathbf{V}}{\partial p} \propto -\frac{\partial\mathbf{V}}{\partial z}$$

This yields the useful proportionality

$$w \propto -\frac{\partial\mathbf{V}}{\partial z} \cdot \nabla\zeta \quad \text{(M-15)}$$

which means that lifting occurs in the regions of cyclonic vorticity advection by thermal wind and sinking occurs in the corresponding anticyclonic advection. The major advantage of (M-15) over (M-1) is that there are not two separate advections of vorticity and temperature.

Appendix N

Frontogenesis

General Expression

The definition of frontogenesis can be written in the following simple form if the coordinate system is oriented so that θ increases along the y-axis:

$$\frac{d}{dt}\frac{\partial\theta}{\partial y} = \frac{\partial}{\partial t}\frac{\partial\theta}{\partial y} + u\frac{\partial}{\partial x}\frac{\partial\theta}{\partial y} + v\frac{\partial}{\partial y}\frac{\partial\theta}{\partial y} + \omega\frac{\partial}{\partial p}\frac{\partial\theta}{\partial y}$$

$$= \frac{\partial}{\partial y}\frac{\partial\theta}{\partial t} + u\frac{\partial}{\partial y}\frac{\partial\theta}{\partial x} + v\frac{\partial}{\partial y}\frac{\partial\theta}{\partial y} + \omega\frac{\partial}{\partial y}\frac{\partial\theta}{\partial p}$$

$$= \frac{\partial}{\partial y}\frac{\partial\theta}{\partial t} + \frac{\partial}{\partial y}\left(u\frac{\partial\theta}{\partial x}\right) + \frac{\partial}{\partial y}\left(v\frac{\partial\theta}{\partial y}\right) + \frac{\partial}{\partial y}\left(\omega\frac{\partial\theta}{\partial p}\right)$$

$$- \frac{\partial u}{\partial y}\frac{\partial\theta}{\partial x} - \frac{\partial v}{\partial y}\frac{\partial\theta}{\partial y} - \frac{\partial\omega}{\partial y}\frac{\partial\theta}{\partial p}$$

Here the term with $\partial\theta/\partial x$ disappears due to the orientation of the coordinate x along the isentrope. The other terms yield the following:

$$\frac{d}{dt}\frac{\partial\theta}{\partial y} = \frac{\partial}{\partial y}\left(\frac{\partial\theta}{\partial t} + u\frac{\partial\theta}{\partial x} + v\frac{\partial\theta}{\partial y} + \omega\frac{\partial\theta}{\partial p}\right) - \frac{\partial v}{\partial y}\frac{\partial\theta}{\partial y}$$

$$- \frac{\partial\omega}{\partial y}\frac{\partial\theta}{\partial p}$$

$$\frac{d}{dt}\frac{\partial\theta}{\partial y} = \frac{\partial}{\partial y}\frac{d\theta}{dt} - \frac{\partial v}{\partial y}\frac{\partial\theta}{\partial y} - \frac{\partial\omega}{\partial y}\frac{\partial\theta}{\partial p} \tag{N-1}$$

which is the frontogenesis equation (6-1).

Geostrophic Frontogenesis

Frontogenesis can be mathematically defined also in a different form:

$$Fg = \frac{d}{dt}|\nabla\theta|^2 \tag{N-2}$$

where, to stress its magnitude and to avoid first-order discontinuity at the zero value, the gradient is squared. If the total derivative in (N-2) is evaluated using geostrophic wind in the advection term, the vertical advection may be neglected:

$$\frac{d}{dt} \equiv \frac{\partial}{\partial t} + \mathbf{V}\cdot\nabla$$

With these expressions, frontogenesis is related to the familiar **Q** vector, as the steps below show. It is convenient to start with $\nabla\theta$ not squared.

$$\frac{d}{dt}\nabla\theta = \frac{\partial}{\partial t}\nabla\theta + u\frac{\partial}{\partial x}\nabla\theta + v\frac{\partial}{\partial y}\nabla\theta$$

$$= \nabla\frac{\partial\theta}{\partial t} + u\nabla\frac{\partial\theta}{\partial x} + v\nabla\frac{\partial\theta}{\partial y}$$

$$= \nabla\frac{\partial\theta}{\partial t} + \nabla\left(u\frac{\partial\theta}{\partial x} + v\frac{\partial\theta}{\partial y}\right) - \frac{\partial\theta}{\partial x}\nabla u - \frac{\partial\theta}{\partial y}\nabla v$$

$$= \nabla\left(\frac{\partial\theta}{\partial t} + u\frac{\partial\theta}{\partial x} + v\frac{\partial\theta}{\partial y}\right) - \begin{pmatrix}\theta_x u_x + \theta_y v_x \\ \theta_x u_y + \theta_y v_y\end{pmatrix}$$

The expression in the first parentheses is equal to zero with the adiabatic and geostrophic approximations. The second parenthesis contains the familiar **Q**:

$$\frac{d}{dt}\nabla\theta = \mathbf{Q}$$

Next, both sides are multiplied by $\nabla\theta$:

$$\nabla\theta \cdot \frac{d}{dt}\nabla\theta = \nabla\theta \cdot \mathbf{Q}$$

$$\frac{d}{dt}\frac{|\nabla\theta|^2}{2} = \mathbf{Q} \cdot \nabla\theta$$

The conclusion can be drawn that positive frontogenesis occurs in the regions where **Q** at least partly points along $\nabla\theta$. Frontolysis is characterized by opposite pointing **Q** and $\nabla\theta$.

A Shortcoming of Continuous Frontogenesis

If we assume that $\partial v/\partial y$ is constant (say that $\partial v/\partial y = -c$, negative for confluence) and if we consider the second term on the right-hand side of (N-1) only, (N-1) reduces to an equation for the intensity of the thermal gradient:

$$\frac{d}{dt}a = ca$$

Here a is written for the intensity of the gradient ($a = |\partial/\partial y|$). The general solution of this equation is

$$a = a_0\, e^{ct}$$

From this it follows that the doubling of the intensity ($a/a_0 = 2$) occurs after time

$$t = \frac{\ln 2}{c}$$

With a typical value of shrinking $c = \partial v/\partial y \approx 10^{-5}\ \mathrm{s}^{-1}$, this time is about 0.7×10^5 s, or nearly 1 day. This is too short a time to produce a discontinuous front where $\nabla\theta$ is larger by at least an order of magnitude from the usual, nonfrontal values. For this reason, other processes, such as separation of flow, are considered in Chapter 6.

References

ALCORN, M., 1990: Classroom and Laboratory Manual for Meteorological Instruments and Observations. Department of Meteorology, Texas A&M University.

BARNES, S. L., 1968: An empirical shortcut to the calculation of temperature and pressure at the lifted condensation level. *J. Appl. Meteor.*, **7**, 511.

BELLAMY, J. C., 1949: Objective calculations of divergence, vertical velocity and vorticity. *Bull. Amer. Meteor. Soc.*, **30**, 45–49.

BLACK, P. G., R. C. GENTRY, V. J. CARDONE, AND J. D. HAWKINS, 1985: Seasat microwave wind and rain observations in severe tropical and midlatitude marine storms. *Satellite Oceanic Remote Sensing,* Ed.: B. Saltzman, *Advances Geophys.*, **27**, 197-277.

BLUESTEIN, H. B., 1992: *Synoptic-dynamic meteorology in Midlatitudes. Volume I. Principles of kinematics and dynamics.* Oxford Univ. Press, 431 pp.

BLUESTEIN, H. B., 1993: *Synoptic-dynamic meteorology in Midlatitudes. Volume II. Observations and Theory of Weather Systems.* Oxford Univ. Press, 594 pp.

BLUESTEIN, H. B., AND M. H. JAIN, 1985: Formation of mesoscale lines of precipitation: Severe squall lines in Oklahoma during the spring. *J. Atmos. Sci.*, **42**, 1711–32.

BROWNING, K. A., AND G. A. MONK, 1982: A simple model for the synoptic analysis of cold fronts. *Quart. J. Roy. Meteor. Soc.*, **108**, 435–52.

BUSINGER, S., AND R. J. REED, 1989: Cyclogenesis in cold air masses. *Wea. Forecasting.*, **4**, 133–56.

BYERS, H. R., AND R. R. BRAHAM, JR., 1948: Thunderstorm structure and circulation. *J. Meteor.*, **5**, 71–86.

CARLSON, T. N., 1980: Airflow through midlatitude cyclones and the comma cloud pattern. *Mon. Wea. Rev.*, **108,** 1498–509.

CARLSON, T. N., 1991: *Mid-Latitude Weather Systems.* HarperCollinsAcademic, 507 pp.

CARR, F. H., AND J. P. MILLARD, 1985: A composite study of comma clouds and their association with severe weather over the Great Plains. *Mon. Wea. Rev.*, **113**, 370–87.

DECKER, F. W., 1981: *The Weather Workbook.* The Weather Workbook Co., 827 N. W. 31st St., Corvallis, Oregon 97330.

DEFANT, F., AND H. T. MÖRTH, 1978: *Synoptic Meteorology.* WMO Compendium of Meteorology, vol. I, part 3.

DJURIĆ, D., AND M. S. DAMIANI, 1980: On the formation of the low-level jet over Texas. *Mon. Wea. Rev.*, **108**, 1854–65.

DJURIĆ, D., AND D. S. LADWIG, 1983: Southerly low-level jet in the winter cyclones of the southwestern Great Plains. *Mon. Wea. Rev.*, **111**, 2275–81.

DOSWELL, C. A., III, 1982: *The Operational Meteorology of Convective Weather. Volume I: Operational mesoanalysis.* NOAA Techn. Mem. NWS NSSFC-5.

———, 1985: *The Operational Meteorology of Convective Weather. Volume II: Storm scale analysis.* NOAA Techn. Mem. ERL ESG-15.

DOVIAK, R., AND D. ZRNIĆ, 1993: *Doppler Radar and Weather Observations. Second Edition.* Acad. Press.

ELSBERRY, R. L. (ed.), 1987: *A Global View of Tropical Cyclones*. Naval Postgrad. School, Monterey, Ca.

FUJITA, T. T., AND H. R. BYERS, 1977: Spearhead echo and downburst in the crash of an airliner. *Mon. Wea. Rev.*, **105**, 129–46.

GODSKE, C. L., T. BERGERON, J. BJERKNES, AND R. C. BUNDGAARD, 1957: *Dynamic Meteorology and Weather Forecasting*. Amer. Meteor. Soc.

HOBBS, P. V., 1981: Mesoscale structure in midlatitude frontal systems. *Proc. IAMAP Symp. Nowcasting: Mesoscale Observations and Short-Range prediction, 1981*. Eur. Space Agency **SP–165**, 29–36.

HOLTON, J. R., 1992: *An Introduction to Dynamic Meteorology. Third Edition*. Acad. Press.

HOUZE, R. A., AND P. V. HOBBS, 1982: Organization and structure of precipitating cloud systems. *Advances Geophys.*, **24**, 225–315.

HOSKINS, B. J., I. DRAGHICI, AND H. C. DAVIES, 1978: A new look at the *w*–equation. *Quart. J. Roy. Meteor. Soc.*, **104**, 31–38.

HOSKINS, B. J., AND M. A. PEDDER, 1980: The diagnosis of middle latitude synoptic development. *Quart. J. Roy. Meteor. Soc.*, **106**, 707–19.

JENNE, R. L., H. L. CRUTCHER, H. VAN LOON, AND J. J. TALJAARD, 1971: *Climate of the Upper Air: Southern Hemisphere. Volume III, Isogon and Isotach Analyses*. NCAR TN/STR-58.

KEYSER, D., 1986: Atmospheric fronts: An observational perspective. *Mesoscale Meteorology and Forecasting*. Amer. Meteor. Soc., 216–258.

KEYSER, D., AND M. A. SHAPIRO, 1986: A review of the structure and dynamics of upper-level frontal zones. *Mon. Wea. Rev.*, **114**, 452–99.

MADDOX, R. A., K. W. HOWARD, D. L. BARTELS, AND D. M. RODGERS, 1986: Mesoscale convective complexes in the middle latitudes. *Mesoscale Meteorology and Forecasting*, Amer. Meteor. Soc., 390–413.

MANABE, S., AND R. F. STRICKLER, 1964: Thermal equilibrium of the atmosphere with a convective adjustment. *J. Atmos. Sci*, **21**, 361–85.

MASON, B. J., 1985: Progress in cloud physics and dynamics. *Recent Advances in Meteorology and Physical Oceanography*. Roy. Meteor. Soc., 1–14.

MCDONNELL, P. W., JR., 1979: *Introduction to Map Projections*. Dekker, 174 pp.

MCGUIRK, J. P., AND D. A. DOUGLAS, 1988: Sudden stratospheric warming and anomalous U. S. weather. *Mon. Wea. Rev.*, **116**, 162–74.

MENARD, R. D., AND J. M. FRITSCH, 1989: A mesoscale convective complex-generated inertially stable warm core vortex. *Mon. Wea. Rev.*, **117**, 1237–61.

MILLER, R. C., 1972: Notes on analysis and severe-storm forecasting procedures of the military weather warning center. *Tech. Rep.* **200**, Air Weather Service (MAC), Scott Air Force Base, Ill.

MOGIL, H. M., AND R. L. HOLLE, 1972: Anomalous gradient winds: Existence and implications. *Mon. Wea. Rev.*, **100**, 709–16.

MOORE, J. T., 1988: *Isentropic Analysis and Interpretation: Operational Applications to Synoptic and Mesoscale Forecast Problems*. Saint Louis University, Department of Earth and Atmospheric Sciences.

NASA, 1987: *LASA, Lidar Atmospheric Sounder and Altimeter*. Earth Observing System, Volume IId.

NEWTON, C. W., 1956: Mechanism of circulation change during a lee cyclogenesis. *J. Meteor.*, **13**, 528–39.

NEWTON, C. W., 1967: Severe convective storms. *Adv. Geophys.*, **12**, 257–308.

OGURA, Y., AND M.-T. LIOU, 1980: The structure of a midlatitude squall line: A case study. *J. Atmos. Sci.*, **37**, 553–67.

ORVILLE, R. E., R. W. HENDERSON, AND L. F. BOSART, 1983: An East Coast lightning detection network. *Bull. Amer. Meteor. Soc.*, **64**, 1029–37.

PAGNOTTI, V., AND L. F. BOSART, 1984: Comparative diagnostic case study of East Coast secondary cyclogenesis under weak versus strong synoptic-scale forcing. *Mon. Wea. Rev.*, **112**, 5–30.

PALMÉN, E., AND K. M. NAGLER, 1948: The formation and structure of a large-scale disturbance in the westerlies. *J. Meteor.*, **6**, 227–42.

PALMÉN, E., AND C. W. NEWTON, 1969: *Atmospheric Circulation Systems*. Acad. Press.

PARKE, P. S. (ED.), 1986: *Satellite Imagery Interpretation for Forecasters*. National Weather Association, vols. 1, 2, and 3.

PEPPLER, R. A., 1988: A review of static stability indices and related thermodynamic parameters. Illinois State Water Survey Division, Climate and Meteorology Division, SWS Misc. Publ., **104.**

PETTERSSEN, S., 1956: *Weather Analysis and Forecasting. Second Edition*. McGraw Hill. vols. 1 and 2.

PETTERSSEN, S., AND S. J. SMEBYE, 1971: On the development of extratropical cyclones. *Quart. J. Roy. Meteor. Soc.*, **97**, 457–82.

POWELL, M. D., 1982: The transition of the Hurricane Frederic boundary-layer wind field from the open Gulf of Mexico to landfall. *Mon. Wea. Rev.*, **110**, 1912–32.

RADINOVIĆ, D., 1965: On forecasting of cyclogenesis in the West Mediterranean and other areas bounded by mountain ranges by baroclinic models. *Arch. Met. Geoph. Biokl.*, **A 14**, 279–299.

—, 1986: On the development of orographic cyclones. *Quart. J. Roy. Meteor. Soc.*, **112**, 927–51.

RAY, P. S. (ed.), 1986: *Mesoscale Meteorology and Forecasting*. Amer. Meteor. Soc.

READ, W. L., AND R. A. MADDOX, 1983: Apparent modification of synoptic-scale features by widespread convection. *Mon. Wea. Rev.*, **111**, 2123–28.

REED, R. J., 1988: Polar lows. *Seminar Proceedings, the Nature and Prediction of Extra Tropical Weather Systems 7–11 September* 1987. Volume I. European Centre for Medium-Range Weather Forecasts, 213–36.

RILEY, G. T., AND L. F. BOSART, 1987: The Windsor Locks, Connecticut tornado of 3 October 1979: An analysis of an intermittent severe weather event. *Mon. Wea. Rev.*, **115**, 1655–77.

RINEHART, R. E., 1991: *Radar for Meteorologists*. Dept. of Atmos. Sciences, Univ. of North Dakota, Grand Forks, ND 58202-8216.

ROTUNNO, R, J. B. KLEMP, AND M. L. WEISMAN, 1988: A theory for strong, long-lived squall lines. *J. Atmos. Sci.*, **45**, 463–85.

SAUCIER, W. J., 1955: *Principles of Meteorological Analysis*. Univ. of Chicago, 438 pp.

SCHAEFER, J. T., 1986: The dryline. *Mesoscale Meteorology and Forecasting,* Amer. Meteor. Soc., 549–572.

SCHAEFER, J. T., AND C. A. DOSWELL III, 1979: On the interpolation of a vector field. *Mon. Wea. Rev.*, **107**, 458–476.

SHAPIRO, M. A., 1978: Further evidence of the mesoscale and turbulent structure of upper level jet stream-frontal zone systems. *Mon. Wea. Rev.*, **106**, 1100–1111.

—, T. HAMPEL AND A. J. KRUEGER, 1987: The arctic tropopause fold. *Mon. Wea. Rev.*, **115**, 444–454.

—, AND J. T. HASTINGS, 1973: Objective cross-section analyses by Hermite polynomial interpolation on isentropic surfaces. *J. Appl. Meteor.*, **12**, 753–762.

TRENBERTH, K. E., 1978: On the interpretation of the diagnostic quasi-geostrophic omega equation. *Mon. Wea. Rev.*, **106**, 131–37.

UCCELLINI, L. W., 1976: Operational diagnostic applications of isentropic analysis. *Nat. Wea. Digest*, **1**, 4–12.

U.S. AIR FORCE, 1978: *Weather, Use of the skew T, log p Diagram in Analysis and Forecasting*. AWSM 105–24.

VELASCO, I., AND J. M. FRITSCH, 1987: Mesoscale convective complexes in Americas. *J. Geophys. Res.*, **92**, 9591–613.

WEISMAN, M. L., AND J. B. KLEMP, 1982: The dependence of numerically simulated convective storms on vertical wind shear and buoyancy. *Mon. Wea. Rev.*, **110**, 504–520.

WHITTAKER, T. M., 1977: Automated streamline analysis. *Mon. Wea. Rev.*, **105**, 786–88.

WILHELMSON, R. B., AND J. B. KLEMP, 1981: A three-dimensional numerical simulation of splitting severe storms on 3 April 1964. *J. Atmos. Sci.*, **38**, 1581–600.

WILSON, J., AND H. P. ROESLI, 1985: Use of Doppler radar and radar networks in mesoscale analysis and forecasting. *ESA J.*, **9**, 125–46.

WOOD, V. T., AND R. A. BROWN, 1986: Single Doppler velocity signature interpretation of nondivergent environmental winds. *J. Atmos. Oceanic Technol.*, **3**, 114– 28.

YOUNG, M. V., G. A. MONK, AND K. A. BROWNING, 1987: Interpretation of satellite imagery of a rapidly developing cyclone. *Quart. J. Roy. Meteor. Soc.*, **113**, 1089–1115.

YOUNG, G. S., AND J. M. FRITSCH, 1989: A proposal for general conventions in analyses of mesoscale boundaries. *Bull. Amer. Meteor. Soc.*, **70**, 1412–1421.

Index

C

D

E

F

G

H

I

J

K

L

M

N

O

P

Q

R

S

T

U

V

W

Z